AUTISM
A Reappraisal
of Concepts and Treatment

AUTISM
A Reappraisal
of Concepts and Treatment

Edited by
Michael Rutter
University of London

and
Eric Schopler
University of North Carolina at Chapel Hill

PLENUM PRESS·NEW YORK AND LONDON

Library of Congress Cataloging in Publication Data

Main entry under title:

Autism, a reappraisal of concepts and treatment.

  "Based on papers presented at the International Symposium on Autism held in
St. Gallen, Switzerland, July 12-15, 1976."
  Includes bibliographies and index.
  1. Autism—Congresses. I. Rutter, Michael. II. Schopler, Eric. III. International Sym-
posium on Autism, St. Gallen, Switzerland, 1976.
RJ506.A9A89   618.9'28'982   77-26910
ISBN 0-306-31096-1

First Printing – April 1978
Second Printing – May 1979
Third Printing – November 1979

Based on papers presented at the International Symposium on Autism
held in St. Gallen, Switzerland, July 12–15, 1976

© 1978 Plenum Press, New York
A Division of Plenum Publishing Corporation
227 West 17th Street, New York, N.Y. 10011

Printed in the United States of America

Preface

This volume aims to provide the reader with an up-to-date account of knowledge, research, education, and clinical practice in the field of autism, from an international perspective. The emphasis throughout is on the growing points of knowledge and on the new developments in practice. We have tried to keep a balance between the need for rigorous research and systematic evaluation and the importance of expressing new ideas and concepts so that they may influence thinking at a stage when questions are being formulated and fresh approaches to treatment are being developed.

The book had its origins in the 1976 International Symposium on Autism held in St. Gallen, Switzerland but it is not in any sense a proceedings of that meeting. Most papers have been extensively rewritten to provide a fuller coverage of the topic and also to take account of the issues raised at the meeting. Discussion dialogues have been revised and restructured to stand as self-contained chapters. Many significant contributions to the conference have not been included in order to maintain the balance of a definitive review; however a few extra chapters have been added to fill crucial gaps.

We hope the result is a vivid picture of the current state of the art. As editors we have been most impressed by the advances since the 1970 international conference held in London. Clearly the topic of autism continues to attract some of the liveliest minds and the result has been not only significant advances in knowledge but also the much needed development of better services for autistic children and their families.

Our purpose in preparing this volume has been to bring together these advances and developments in one volume which we hope will be of interest to all who are concerned with autistic children in either a research or caring capacity.

Michael Rutter
London

Eric Schopler
Chapel Hill

Acknowledgments

It is a pleasure to acknowledge the help given by Dr. Erwin Witkin, president of Physicians Associated for Continuing Education, Inc., and his wife Helen Witkin. The original idea of the international conference which gave rise to this book was theirs and they were responsible for the conference arrangements. We are also grateful to Dr. Hermann Städeli, whose hospitality and care in local planning did so much to make St. Gallen a good venue for the meeting. Extensive help throughout the planning of the meeting and the preparation of this volume has been ably given by Christine Lemberg, Joy Maxwell, and Nancy Park. We would also like to thank Peter Clark for producing the index.

M.R.
E.S.

Contents

1. Diagnosis and Definition   1
   Michael Rutter

Social Characteristics

2. Social, Behavioral, and Cognitive Characteristics: An
   Epidemiological Approach   27
   Lorna Wing

3. The Partial Noncommunication of Culture to Autistic
   Children—An Application of Human Ethology   47
   John Richer

4. The Assessment of Social Behavior   63
   Patricia Howlin

Psychological and Physiological Studies

5. Language: The Problem Beyond Conditioning   71
   Don W. Churchill

6. Language Disorder and Infantile Autism   85
   Michael Rutter

7. Language: What's Wrong and Why   105
   Paula Menyuk

8. Neurophysiologic Studies   117
   Edward M. Ornitz

9. Images and Language   141
   Beate Hermelin

10. Research Methodology: What Are the "Correct Controls"?   155
    William Yule

Biological Investigations

11. Biochemical and Hematologic Studies: A Critical Review   163
    Edward R. Ritvo, Karen Rabin, Arthur Yuwiler, B. J. Freeman, and Edward Geller

12. A Report on the Autistic Syndromes   185
    Mary Coleman

13. Biochemical Strategies and Concepts   201
    R. Rodnight

14. A Neuropsychologic Interpretation of Infantile Autism   207
    G. Robert DeLong

15. A Twin Study of Individuals with Infantile Autism   219
    Susan Folstein and Michael Rutter

16. Biological Homogeneity or Heterogeneity?   243
    Edward M. Ornitz

Family Characteristics

17. Personality Characteristics of Parents   251
    Wm. George McAdoo and Marian K. DeMyer

18. Family Factors   269
    Dennis P. Cantwell, Lorian Baker, and Michael Rutter

19. Limits of Methodological Differences in Family Studies   297
    Eric Schopler

Psychotherapy

20. Psychotherapeutic Work with Parents of Psychotic Children   303
    Irving N. Berlin

21. Play, Symbols, and the Development of Language   313
    Austin M. DesLauriers

22. Etiology and Treatment: Cause and Cure 327
Michael Rutter

Biological Treatments

23. Pharmacotherapy 337
Magda Campbell

24. Therapy with Autistic Children 357
Theodore Shapiro

Behavioral Treatments

25. Parents as Therapists 369
O. Ivar Lovaas

26. Treating Autistic Children in a Family Context 379
R. Hemsley, P. Howlin, M. Berger, L. Hersov, D. Holbrook, M. Rutter, and W. Yule

27. Changing Parental Involvement in Behavioral Treatment 413
Eric Schopler

Education

28. Educational Approaches 423
Lawrence Bartak

29. Individualized Education: A Public School Model 439
Margaret D. Lansing and Eric Schopler

30. Educational Aims and Methods 453
Maria Callias

Follow-Up and Outcome

31. Long-Term Follow-Up of 100 "Atypical" Children of Normal Intelligence 463
Janet L. Brown

32. Follow-Up Studies 475
V. Lotter

33. Developmental Issues and Prognosis   497
     Michael Rutter

Conclusion

34. Subgroups Vary with Selection Purpose   507
     Eric Schopler and Michael Rutter

Index

Contributors

Lorian Baker, The Neuropsychiatric Institute, Center for the Health Sciences, University of California, Los Angeles, California

Lawrence Bartak, Faculty of Education, Monash University, Clayton, Victoria, Australia

M. Berger, Institute of Education, London, England

Irving N. Berlin, Department of Psychiatry, School of Medicine. University of California, Davis, California

Janet L. Brown, James Jackson Putnam Children's Center, Boston, Massachusetts

Maria Callias, Department of Psychology, Institute of Psychiatry, London, England

Magda Campbell, New York University Medical Center, New York, New York

Dennis P. Cantwell, The Neuropsychiatric Institute, Center for the Health Sciences, University of California, Los Angeles, California

Don W. Churchill, Indiana University School of Medicine, Indianapolis, Indiana

Mary Coleman, Washington, D. C.

G. Robert DeLong, Pediatric Neurology Unit, Massachusetts General Hospital, Boston, Massachusetts

Marian K. DeMyer, Indiana University School of Medicine, Indianapolis, Indiana

Austin M. DesLauriers, Center for Autistic and Schizophrenic Children, The Devereux Foundation, Devon, Pennsylvania

B. J. Freeman, Mental Retardation and Child Psychiatry Program, Department of Psychiatry, School of Medicine, University of California, Los Angeles, California

Susan Folstein, Division of Child Psychiatry, Johns Hopkins Hospital, Baltimore, Maryland

Edward Geller, Mental Retardation and Child Psychiatry Program, Department of Psychiatry, School of Medicine, University of California, Los Angeles, California

R. Hemsley, Department of Psychology, Institute of Psychiatry, London, England

Beate Hermelin, MRC Developmental Psychology Unit, London, England

L. Hersov, The Maudsley Hospital, London, England

D. Holbrook, The Maudsley Hospital, London, England

Patricia Howlin, Departments of Psychology and Child and Adolescent Psychiatry, Institute of Psychiatry, London, England

Margaret D. Lansing, Division TEACCH, University of North Carolina, Chapel Hill, North Carolina

O. Ivar Lovaas, University of California, Los Angeles, California

V. Lotter, University of Guelph, Guelph, Ontario, Canada

Wm. George McAdoo, Indiana University School of Medicine, Indianapolis, Indiana

Paula Menyuk, Applied Psycholinguistics Program, Boston University, Boston, Massachusetts

xi

Edward M. Ornitz, Mental Retardation and Child Psychiatry Division, Department of Psychiatry, and Brain Research Institute, School of Medicine. University of California, Los Angeles, California

Karen Rabin, Mental Retardation and Child Psychiatry Program, Department of Psychiatry, School of Medicine, University of California, Los Angeles, California

John Richer, Smith Hospital, Henley on Thames, Oxon, England. Present Address: Psychology Department, Park and Warneford Hospitals, Oxford, England

Edward R. Ritvo, Mental Retardation and Child Psychiatry Program, Department of Psychiatry, School of Medicine, University of California, Los Angeles, California

R. Rodnight, Department of Biochemistry, Institute of Psychiatry, London, England

Michael Rutter, Department of Child and Adolescent Psychiatry, Institute of Psychiatry, London, England

Eric Schopler, Department of Psychiatry, Division of Health Affairs, University of North Carolina School of Medicine, Chapel Hill, North Carolina

Theodore Shapiro, Department of Psychiatry, Cornell University Medical College; Department of Child and Adolescent Psychiatry, Payne Whitney Clinic, New York, New York

Lorna Wing, MRC Social Psychiatry Unit, Institute of Psychiatry, London, England

William Yule, Departments of Psychology and Child and Adolescent Psychiatry, Institute of Psychiatry, London, England

Arthur Yuwiler, Mental Retardation and Child Psychiatry Program, Department of Psychiatry, School of Medicine, University of California, Los Angeles, California

1

Diagnosis and Definition

MICHAEL RUTTER

Most of the chapters in this book take for granted the definition of infantile autism and the criteria to be used in its diagnosis. That is right and proper, but the questions of definition and diagnosis have given rise to such controversy over the years that it is necessary to set the scene for what follows by some discussion of the issues involved.

IDENTIFICATION OF A SYNDROME

Any account of the definition of autism must start with Kanner's (1943) careful and systematic observations on eleven children with a previously unrecognized syndrome. He noted in clear careful prose a variety of behavioral features which seemed both to be characteristic of all eleven children and also to differentiate them from children with other psychiatric disorders. These features included an inability to develop relationships with people, a delay in speech acquisition, the noncommunicative use of speech after it develops, delayed echolalia, pronominal reversal, repetitive and stereotyped play activities, an obsessive insistence on the maintenance of sameness, a lack of imagination, a good rote memory, and a normal physical appearance. Furthermore, abnormalities were already evident in infancy which, as Kanner noted, made the disorder different from all previously described varieties of schizophrenia or child psychosis. During the next decade workers in the United States and Europe reported observations on children with similar features (Despert,

MICHAEL RUTTER · Department of Child and Adolescent Psychiatry, Institute of Psychiatry, De Crespigny Park, Denmark Hill, London, SE5 8AF, England.

1951; Van Krevelen, 1952; Bosch, 1953; Bakwin, 1954). In addition, several earlier accounts of autistic children were discovered, although the syndrome had not been recognized as such at the time (Darr & Worden, 1951; Vaillant, 1962; see also Wing, 1976). There was no doubt that autistic children existed but equally there was considerable confusion over the boundaries of the syndrome as well as on its nature and causation (see Rutter, 1974). With the knowledge now available it is possible to see how this confusion arose.

To begin with, there was the unfortunate choice of name (that is "autism") which immediately led to confusion with Bleuler's use of the same term to refer to the active withdrawal into fantasy shown by schizophrenic patients (Bleuler, 1911). This was confusing (see Bosch, 1970; Wing, 1976) because, firstly, it suggested a *withdrawal* from relationships whereas Kanner had actually described a failure to *develop* relationships; secondly it implied a rich fantasy life whereas Kanner's observations suggested a lack of imagination; and thirdly, it postulated a link with schizophrenia as evident in adults. This last confusion was further compounded by a tendency among child psychiatrists to use childhood schizophrenia, autism, and child psychosis as interchangeable diagnoses (Laufer & Gair, 1969).

A further problem arose from Eisenberg and Kanner's later observation (1956) that the same syndrome can arise after apparently normal development in the first one to two years of life. This observation has been confirmed by a number of other investigators (Rutter *et al.*, 1967; Lotter, 1966) and it is now generally accepted that the limits for age of onset extend up to 30 months (Rutter, 1971a, 1972). However, an unfortunate consequence of this finding was an increasing tendency to ignore age of onset as a necessary criterion for the diagnosis (e.g., Creak, 1961), with the result that disorders beginning in early infancy were grouped together with psychoses not emerging until later childhood or adolescence (e.g., Bender, 1947; Vrono, 1974).

Lastly, many difficulties have followed from Eisenberg and Kanner's (1956) reduction of the essential symptoms to just two, "extreme aloneness" and "preoccupation with the preservation of sameness" (together with an onset in the first two years). Curiously, the simplified criteria entirely omitted the peculiar abnormalities of language to which Kanner had drawn attention in both his original description (Kanner, 1943) and in later writings (Kanner, 1946). Of course, the attempt to determine the essential features of autism was entirely right (as we shall see), but their efforts were sometimes taken as a licence to use these criteria without any reference to the carefully described clinical phenomena to which they referred and sometimes as a licence to change altogether the criteria.

Thus, Schain and Yannet (1960) omitted preservation of sameness from their diagnostic criteria. Tinbergen and Tinbergen (1972) went further in placing their main emphasis on avoidance of eye-to-eye gaze, which not only ignored all the other clinical features described by Kanner but also took just one social feature out of context and without cognizance of the developmental and behavioral characteristics of the autistic child's "aloneness" (as described by Kanner) which differentiated it from the many, many other types of social difficulty shown by children and adults (see also Wing & Ricks, 1976). Ornitz and Ritvo (1968) kept closer to Kanner's description but elevated "disturbances of perception" to the primary symptom and in so doing placed an emphasis on disturbed motility patterns which added a new element not included by Kanner. Rendle-Short (1969) produced a list of fourteen manifestations and maintained that seven should be present for the diagnosis to be considered. However these seven might consist of lack of fear about danger, strong resistance to learning, use of gesture, laughing and giggling, overactivity, difficulty in playing with other children, and acts as deaf—a list which includes none of Kanner's criteria.

As a result of these very varying sets of diagnostic criteria the literature is full of clinical accounts and pieces of research which deal with rather different kinds of problems, all placed under the same label—"autism." That seems to lead to these questions. "What is the true definition of autism?" or "How should autism really be defined?" However, to ask these questions is to completely mistake the process of scientific inquiry—so elegantly and convincingly described by Popper (1959, 1972). As he puts it, definitions should be read from right to left, and not from left to right. In other words the question is not "What is autism?" but rather "To what set of phenomena shall we apply the term autism?" There is no point in starting with the word "autism" and then defining it. It is merely a word and like any other word it means just what we want it to mean—no more and no less. In short, the word "autism" is merely a convenient substitute or short-hand term for Kanner's long prose description, and no information is to be gained by analyzing it.

This may seem a mere semantic quibble but in fact the distinction has important practical consequences (Campbell, 1976). The word "autism" could be used to describe children who merely avert eye-to-eye gaze or for that matter to describe old wooden tables. There is nothing intrinsically wrong in doing that but it does cause considerable confusion if the impression is given that the word is being used in the same way as that employed by Kanner when evidently it is not.

But Kanner's use of the term autism was more than a simple label, and that is where the trouble really increases. It was also a hypothesis—a suggestion that behind the behavioral description lay a disease entity. In short, he suggested (quite properly) that the particular grouping of behaviors he chose to call autism had a validity in the sense that children with these behaviors differed from children with other psychiatric disorders in some important sense. Now, if that hypothesis is correct, it raises the further possibility that Kanner might not have picked the most appropriate symptoms by which to recognize the hypothesized disease-state. As a result there must be some means of testing the validity of the proposed behavioral grouping and of modifying the diagnostic criteria in the light of the research findings. That process and the research findings which result from it, constitute the basis for the rest of this chapter.

SPECIFICITY OF SYMPTOMATOLOGY

By suggesting that autism constituted a syndrome, Kanner meant two things: first, that there were certain behaviors which tended to group together and second, that these behaviors differed from those found in other psychiatric conditions. Accordingly, the first step was to determine by comparative studies how far this was true in order to clarify the diagnostic criteria. There was a need to find out which symptoms were both *universal* and *specific*—that is, those which were present in all or nearly all autistic children and also which were relatively *in*frequent in children who did not have the syndrome. A differentiation had to be made between behaviors which *could* occur in autism (but which also occurred in other conditions) and those behaviors which were specifically characteristic of autism. In doing this it was obviously important to control for age, sex, IQ, and presence of psychiatric disorder in order to ensure that any differences found were not merely a reflection of the fact that children's behavior varies according to these nonspecific features (Rutter *et al.*, 1970b; Rutter, 1971b).

When this was done, it was found that three broad groups of symptoms were found in all (or almost all) children diagnosed as suffering from infantile autism (or infantile psychosis) *and* were also much less frequent in children with other psychiatric disorders (Rutter, 1966; Rutter & Lockyer, 1967). These symptoms were a profound and general failure to develop social relationships; language retardation with impaired comprehension, echolalia and pronominal reversal; and ritualistic or compulsive phenomena (i.e., "an insistence on sameness"). In addition stereotyped repetitive movements (especially hand and finger mannerisms), a short attention span, self-injury, and delayed bowel control were also more common in autistic children, but these symptoms did not occur in all cases. In short, the findings broadly confirmed Kanner's original diagnostic criteria and provided a working definition of the syndrome in terms of universal and specific symptoms. However, before proceeding to test the validity of the syndrome, certain other issues needed clarifying. The most important of these concerned IQ, age of onset, neurological status, and details of symptomatology.

INTELLECTUAL FUNCTIONING

Originally, Kanner (1943) had thought that autistic children were really of normal intelligence and that their poor functioning was simply a secondary consequence of their autistic failure to make relationships. The children's often good rote memory, their serious facial expression, and lack of physical stigmata were in keeping with this view, which many other writers followed. It was important to test this hypothesis by further examination of the nature of intellectual performance in autistic children. This was done in several different ways.

First, it was necessary to determine if the IQ score had the same properties in autistic children as it did in other children. It was found that it did. Thus, several independent studies have all shown considerable stability of IQ scores throughout middle childhood and adolescence (Lockyer & Rutter, 1969; Gittelman & Birch, 1967; Mittler *et al.*, 1966; DeMyer *et al.*, 1974), a stability which is closely similar to that found in nonautistic children. Furthermore, as in normal or retarded children, the IQ score in autistic youngsters proved to be a reasonable predictor of later educational attainments (Lockyer & Rutter, 1969; Bartak & Rutter, 1971, 1973; Rutter & Bartak, 1973; Mittler *et al.*, 1966). This was so in autistic children who received very little schooling and also in those who received skilled special education.

The second approach was to determine how intellectual performance varied with psychiatric state. Cowan *et al.* (1965) produced evidence indicating that hospitalized autistic children of low IQ exhibited negativism. They suggested that the children's poor intellectual performance might be attributable to motivational factors. Of course, motivational factors will influence the performance of autistic children just as they do the performance of other children. However, the question is whether motivational factors are sufficient to account for the low IQ scores of many autistic children. The findings available so far suggest they are not. First, Clark and Rutter (1977), in a replication of the Cowan study, did *not* find negativism in a group of autistic children who had received special education. As most of the children did show low IQ scores, clearly negativism could not explain their poor intellectual performance. Second, Hingtgen and Churchill (1969, 1971) showed that even when motivation had been greatly increased using operant techniques, the cognitive skills of many autistic children were still considerably below normal. Third, the long-term follow-up study of autistic children attending The Maudsley Hospital showed that the IQ level remained much the same even when there was a major improvement in the autism (Lockyer & Rutter, 1969).

The third approach to intelligence concerned the proportion of children who were untestable on the usual set of IQ tests. Were these children not performing because they could not, or was their failure due to unwillingness to attempt the task? In the short-term, Alpern found that if simple enough test items were used the children would and could do them reliably, but they continued to fail the age-appropriate ones (Alpern, 1967; Alpern & Kimberlin, 1970). Similarly, in the long-term, Lockyer and Rutter (1969) found that the children who were untestable initially later behaved in the same fashion as did the severely retarded.

In summary, all evaluations agree in showing that the IQ in autistic children functions in much the same way as in any other group of individuals. Autistic children with low IQs are just as retarded as anyone else with a low IQ and the score means much the same thing. In short autism and mental retardation frequently coexist. This finding has very important implications both for diagnosis and research. In the first place, it means that mental age must be taken into account in assessing behavior. It is not enough to base the diagnosis of autism on lack of social responsiveness and impaired language. If a four-year-old child has a mental age of six months, obviously he cannot be expected to show friendship patterns and communicative speech. To regard these as "autistic" is to totally ignore developmental considerations. Lack of social responsiveness and impaired language can *only* be taken as indicators of autism if the impairment is out of keeping with the child's mental age *and* if it shows the special features characteristic of autism rather than normal development.

In the second place, studies which set out to compare autistic and nonautistic children must control for mental age as well as for chronological age. A failure to do this means that the investigator is quite unable to determine whether any differences found relate to autism or to mental retardation. Unfortunately, a large number of otherwise sound pieces of research are virtually uninterpretable because they fail to take mental age into account either in defining symptoms or in comparing groups (Hingtgen & Bryson, 1972).

AGE OF ONSET

A further consideration in the definition of infantile autism is whether age of onset should be included in the diagnostic criteria and, if it should, what age should set the limits. This question can be approached in several different ways. First, it is useful to consider the age distribution for onset of psychoses in childhood. Studies in Britain (Kolvin, 1971), Japan (Makita, 1966), and the USSR (Vrono, 1974) all agree in showing a bipolar distribution. There is one large peak for children whose disorders begin before three years of age and a second large peak for those whose psychoses are first evident in early adolescence or just before that. Psychoses beginning in middle childhood are much rarer. The second approach is to determine whether cases with early onset and late onset differ in symptomatology, etiology, or outcome. Kolvin's careful studies (Kolvin, 1971) clearly indicate that the two groups differ in a host of important ways. Less is known about cases beginning after age three but before adolescence. However, although these disintegrative disorders have some similarities with the syndrome of autism, the behavioral manifestations also tend to differ in some crucial respects (Heller, 1930; Corbett *et al.*, 1977). Furthermore, a high proportion show clear-cut evidence of gross structural brain pathology, sometimes of a progressive kind (Malamud, 1959; Creak, 1963; Ross, 1959; Kanner, 1957; Corbett *et al.*, 1977), which is quite uncommon in autism. On both these grounds, it seems highly desirable to differentiate these disorders from autism.

There is a greater difficulty, however, in deciding whether to differentiate within the group of autism-like conditions which are evident before age three, between those where apparently there has been a prior period of normal development, and those where there has been none. One of the problems is the great difficulty in any individual case of deciding whether or not early development has been trouble-free in all respects. In practice, unless the child has clearly advanced to conversational speech and imaginative play the distinction is of dubious validity. Furthermore, the very limited available evidence suggests that below age three, the age of onset is of no predictive value. On the other hand, as some of the disintegrative psychoses begin about age three there is something to be said for having the onset age limit for autism slightly lower. The World Health Organization draft glossary takes 30 months as its cut-off point for the onset of autism and in the absence of further data that seems a reasonable solution.

A special problem exists with respect to the differentiation of autism and autistic psychopathy, a condition first described by Asperger (1944). Autistic psychopathy is said to resemble autism in many respects but it is thought to be a personality trait not evident until the third year of life or later and with a good social prognosis (Van Krevelen & Kuipers, 1962; Van Krevelen, 1971). Intelligence is unimpaired but coordination and visuo-spatial perception are poor, there are gross social impairments, and obsessive preoccupations or circumscribed interest patterns (Mnukhin & Isaev, 1975; Isaev & Kagan, 1974). Such cases undoubtedly exist but it remains uncertain whether they constitute a distinct syndrome different from mild childhood autism.

NEUROLOGICAL DISORDER

Another decision is required on whether to take the presence of neurological disorder into account when making the diagnosis. Eisenberg (1966, 1972) has argued convincingly on the importance of knowing whether or not autism is associated with some identifiable brain disease. Obviously, this distinction is of crucial importance when considering etiology. On the other hand, long-term follow-up studies have clearly indicated that a sizable proportion of autistic children who have *no* evidence of neurological disorder in early childhood nevertheless develop epileptic fits in adolescence (Rutter, 1970; Creak, 1963; Lotter, 1974). Thus out of the nine children in Kanner's original eleven cases who were followed into adult life, two developed fits (Kanner, 1971). A few also go on to develop signs of diseases such as tuberose sclerosis which must have been present when the autism began even though neurological examination was normal (Lotter, 1974). The net effect of these findings is that many autistic children who are diagnosed as free of organic brain disease in early childhood will later turn out to have a neurological disorder which may be presumed to have played an important role in the etiology of the autism.

Of course, it could be argued that these errors in diagnosis would not have occured if the neurological assessment had included "soft signs" or other more subtle indicators of minimal brain dysfunction. Unfortunately, the concepts of both "soft signs" and also "minimal brain dysfunction" have proved quite unsatisfactory (Rutter *et al.*, 1970a; Rutter, 1977). Moreover, such signs are associated with low intellectual level (Goldfarb, 1974) and yet have only a weak connection with social outcome at follow-up (Goldfarb, 1974).

The evidence suggests that it would be unwise to attempt to define autism in terms of the absence of organic brain dysfunction. Such a definition would inevitably have a very low validity. On the other hand, the presence of diagnosable brain disease must be noted in any satisfactory classification. This matter is discussed further later in the paper.

DETAILS OF SYMPTOMATOLOGY

The last major issue with respect to the preliminary definition of the syndrome of autism concerns refinement of the clinical criteria for symptomatology. One of the most important findings derives from the several follow-up studies which have all shown that autistic children's behavior often changes markedly as they grow older (see Rutter, 1970; DeMyer *et al.*, 1973; Kanner *et al.*, 1972). Although social, conceptual and obsessive difficulties frequently persist, they do so in forms which are rather different from those shown in the early years.

However, even in the first five years many investigations have added precision to the early behavioral descriptions (Kanner, 1946; Wolff & Chess, 1964, 1965; Hutt *et al.*, 1965; Hutt & Ounsted, 1966; Cunningham, 1968; Sorosky *et al.*, 1968; Hermelin & O'Connor, 1970; Ornitz *et al.*, 1970; Wing, 1971; Churchill & Bryson, 1972; DeMyer *et al.*, 1972).

Impaired Social Relationships

Several studies (e.g., Wolf & Chess, 1964; Hutt & Vaizey, 1966; Sorosky *et al.*, 1968; Wing, 1969; Churchill & Bryson, 1972; Bartak *et al.*, 1975) have shown that autistic children's social development has a number of rather distinctive features. First, there is a lack of attachment behavior and a relative failure of bonding which is most marked in the first five years. Unlike the normal toddler, autistic children tend not to follow their parents about the house and they do not run to greet them when the parents return after having been out. They tend not to go to their parents for comfort when they are hurt or upset, and almost always they do not develop the bedtime kiss and cuddle routine followed by so many normal children. However, they do *not* usually physically withdraw from people and may enjoy a tickle or a rough and tumble. In the first year, quite often they do not take up an anticipatory posture or put up their arms to be picked up in a way that normal children do. On the other hand, especially in the more intelligent autistic children, social abnormalities may not be obvious until well into the second year of life. Of course, a failure in bonding is also seen in conditions other than autism. But it should be noted that the style of social interaction is different in these conditions. Thus children reared in poor quality institutions with multiple caretakers tend to be indiscriminate in their relationships with people and often do not develop personal bonds. On the other hand, quite unlike autistic children, they do show marked attachment behavior and are often clinging and attention-seeking (Tizard & Rees, 1975).

Lack of eye-to-eye gaze is usually said to be particularly characteristic of autistic children. However, clinical observation suggests that it is not so much the amount of eye-to-eye gaze (which may be normal) but rather the way eye-to-eye gaze is used which is characteristic (this observation has yet to be confirmed by systematic studies). The normal child, like the normal adult, uses eye-to-eye gaze in a highly discriminating fashion, looking up at people's faces when he wants to gain their attention, when he wants to be picked up, when he is being aggressive toward them, or when he is being spoken to. What is striking about the autistic child is that he does not use eye-to-eye gaze in these ways. It should be noted that this pattern of eye-to-eye gaze is also different from that of the very shy or very anxious child whose gaze avoidance is characteristically highly specific to intense social interaction.

After age five or thereabouts, many of the social impairments may no longer be evident (at least, not to the same degree), but serious social difficulties continue. This is most evident in (i) a lack of cooperative group play with other children, (ii) a failure to make personal friendships, and (iii) a lack of empathy and a failure to perceive other people's feelings and responses. The last abnormality often results in the child saying or doing socially inappropriate things.

Language and Prelanguage Skills

Not only are autistic children usually markedly delayed in their acquisition of speech, but also their pattern of language development and their usage of language is strikingly different, both from normal children and from children with other language disorders (see Rutter, 1965, 1966; Ricks, 1975; Ricks & Wing, 1975; Bartak *et al.*, 1975). First of all there are serious impairments in a variety of skills which are often thought to underlie or precede language. For example, autistic children usually fail to show much social imitation. They do not wave "bye-bye," they do not participate in imitative games like pat-a-cake, and they are less likely than other children to copy or follow their parents' activity (in terms of cleaning, vacuuming, mowing the lawn, household repairs, etc.). They are delayed in their meaningful use of objects so that when they are very young they may spin the wheels of a toy car or put the car in their mouths rather than use it in the intended way. They are also much delayed in their appropriate use of miniature objects (such as a toy brush or tea set) and, most of all, they lack imaginative or make-believe play. Thus, they are unlikely to engage in pretend games, such as mothers and fathers, schools, tea parties, or cowboys and Indians.

Sometimes an autistic child does have some make-believe actions in which he engages, but, if so, these tend to be stereotyped and repetitive rather than imaginative, creative and ever-changing like the pretend play of normal children.

Frequently, but not always, patterns of babble are also impaired or abnormal (Ricks, 1975; Bartak *et al.*, 1975). This tends to be especially the case toward the end of the second year when normal children (if they are not speaking) are engaging in a rich, varied pattern of babble with the cadences of speech. While this occurs in some autistic children, it is decidedly unusual.

Almost always autistic children are impaired in their understanding of the spoken language. Whereas they may follow simple instructions if given in a familiar social context or with the aid of gesture they usually do not follow instructions which lack these cues or which involve the combination of two or more ideas (e.g., "Please fetch my book which is on the table by my bed"). Equally characteristically, autistic children lack gesture and mime. They tend to make their needs known by taking the adult by the wrist (not usually by the grasped hand). Often they do not point (and, if they do, it is usually with the hand rather than with the extended index finger) and rarely is it accompanied by mime, demonstration, or symbolic gesture.

About half of autistic children, especially those who are also mentally retarded, never gain useful speech. However, in those who do learn to speak there are a variety of characteristic abnormalities. First, immediate echolalia and the delayed repetition of stereotyped phrases is usual for quite a long period after speech first develops. Characteristically, this is accompanied by I-You pronominal reversal (e.g., "You want biscuit" meaning "I want a biscuit"). This phenomenon is closely associated with echolalia (Bartak & Rutter, 1974) and seems to be a function of the echoing tendency (Rutter, 1968). Secondly, speech tends not to be used in the usual way for social communication. Thus, the autistic child tends to talk much less than the normal child of a comparable level of language development. Although the normal toddler may well be silent in the presence of a stranger, he is likely to chatter nonstop as he follows his mother about the house—autistic children rarely do this. Indeed they strikingly lack the ordinary to and fro chatter with the reciprocal interaction which is so characteristic of normal conversation. What they say is less often related to what they have heard—they give the impression of talking *to* someone rather than *with* someone. Also they are usually very poor in talking about anything outside the immediate situation so that they do not, for example, tell their parents what they have done at school during the day. While speech is still developing autistic children show much the same kinds of immaturities of grammar that are found in any child with limited speech. There may also be some difficulties in articulation but in neither of these respects is the speech of the autistic child characteristic. However, often the autistic child's use of words is somewhat unusual, with curious metaphors and odd ways of putting things (Kanner, 1946; Rutter, 1965).

"Insistence on Sameness"

The term "insistence on sameness" is not a very satisfactory one, in that it involves inferences. However, since Kanner's original description it has been widely used to cover a variety of stereotyped behaviors and routines. Characteristically, in early childhood, there are rigid and limited play patterns which lack both variety and imagination. Thus, the children may endlessly line up toys, or make patterns of household implements, or collect curious objects such as tins, or stones of a special shape. Second, there may be intense attachments to these objects, so that the children *have* to carry around with them a piece of grit in the fold between the thumb and the index finger, or they *have* to have a particular belt at all times. Usually these attachments persist in spite of extreme distortions in the size or shape of the object, so that the function of the object is irrelevant to the attachment (Marchant *et al.*, 1974). The attachment is to a specific object and the children protest if it is removed. However, if the object is not eventually returned to the child, attachment to a new object frequently takes place. Third, especially in middle childhood and later, many autistic children have unusual preoccupations which they follow to the exclusion of other activities. Typically these involve things like bus routes, train timetables, colors, numbers, and patterns. Sometimes, the preoccupation takes the form of repeatedly asking stereotyped questions to which specific answers must be given. Fourth, ritualistic and compulsive phenomena are very common. In early and middle childhood these usually take the form of rigid routines but in adolescence it is not infrequent for them to develop into frankly obsessional symptoms, with touching compulsions and the like. Fifth, there is sometimes a marked resistance to changes in the environment so that the child becomes extremely distressed if furniture in the house is moved or if the ornaments are changed.

VALIDITY OF AUTISTIC SYNDROME

Let us summarize so far. The term "childhood autism" is used for disorders beginning before the age of 30 months, in which there may or may not be associated intellectual retardation or neurological dysfunction, but in which the key clinical features are a particular form of impaired social and language development (which must have a number of specified characteristics *and* which must be out of keeping with the child's general intellectual level), together with an "insistence on sameness" as shown by stereotyped play patterns, abnormal preoccupations, and resistance to change. The next question is "Does this syndrome have any validity?" As Davis and Cashdan (1963) argued in another connection, the differentiation of a behavioral syndrome can only be justified as having meaning if the differentiation also carries information about *other* differences in terms of etiology, prognosis, or treatment. Does the syndrome of autism have any meaning in that sense? That question is best considered by comparing autism with the various other conditions with which it might be grouped, namely mental retardation, schizophrenia, neurosis, and the developmental disorders of receptive language.

AUTISM AND MENTAL RETARDATION

Numerous studies have compared autistic children with mentally retarded children or with others of comparable age, sex, and level of general intelligence. For example, Rutter and Lockyer (Rutter *et al.*, 1967; Lockyer & Rutter, 1969, 1970) found that the diagnosis of autism was associated with a distinctive pattern of scores on IQ tests, with the more frequent development of epileptic fits in adolescence, with persisting language delay, and with poor employment prospects. Autism was also much more often found in children from middle-class or professional families. Similarly, Hermelin and O'Connor (1970) in a series of well-planned psychological studies demonstrated a host of ways in which autistic children differed from matched groups of retarded children. The autistic children made less use of meaning in their memory processes, were impaired in their use of concepts, and were generally limited in their powers of coding and categorizing. Other investigators, too, have found autism to be associated with specific cognitive deficits (e.g., Gillies, 1965; Tubbs, 1966; Schopler, 1966). The findings all indicate that, unlike mentally retarded children, autistic children have a particular cognitive deficit which involves language and central coding processes. The results of these clinical and experimental studies leave no doubt that the differentiation of autism from mental retardation is valid, meaningful, and of practical utility.

AUTISM AND SCHIZOPHRENIA

The next issue is whether autism differs from schizophrenia. In order to avoid this becoming a tautology, it is better to redefine the question in terms of whether psychoses beginning in the first three years of life (most of which fulfill the criteria for autism) differ from psychoses beginning during or after pubescence (most of which fulfill the criteria for schizophrenia). The key studies in this connection are those by Kolvin and his colleagues (Kolvin, 1971). They found that children with early and with late onset psychoses differed significantly in terms of social class, family history of schizophrenia, evidence of cerebral dysfunction, symptom patterns, and level of intelligence. Like Makita (1966) and Vrono (1974) they also found that the age of onset of psychoses followed a markedly bipolar distribution with one peak in infancy and another in adolescence. The course of autism and schizophrenia is different: an episodic course with remissions and relapses is much more characteristic of schizophrenia (Rutter, 1968), and follow-up studies have indicated that autistic individuals rarely develop delusions and hallucinations when they reach adulthood (Rutter, 1970). Again, the evidence leaves no alternative but to conclude that there are so many differences between autism and schizophrenia that they should be regarded as separate conditions (Rutter 1972, 1974).*

AUTISM AND NEUROSIS

In view of the fact that autism and neurosis are phenomenologically so unalike, it might be thought unnecessary to examine the differentiation of the two conditions. However, Tinbergen and Tinbergen (1972) have argued that anxiety is the key element in both, O'Gorman (1970) has suggested that autism is merely an exaggeration of a normal defense mechanism and, of course, elective mutism and autism do both share the common feature of lack of speech. The findings show that whereas autism predominately occurs in boys (Rutter, 1967), neurosis has a roughly equal sex ratio in childhood and a marked female preponderance in adulthood (Rutter *et al.*, 1970b; Rutter, 1970). The prognosis for emotional disorders in childhood is generally good (Robins, 1972) whereas for autism it is relatively poor (Rutter, 1970). Autism is commonly associated with both low IQ and organic cerebral dysfunction whereas both are much less frequent with neurosis. More specifically, children with elective mutism do not usually show any impairment in language comprehension and they speak normally in some social circumstances (see Rutter & Martin, 1972). Neither is true of autistic children. It is apparent that the behavioral difference between autism and neurosis carries with it a multitude of other differences.

AUTISM AND DEVELOPMENTAL LANGUAGE DISORDERS

Lastly, we must consider the differences between autism and the developmental language disorders, an issue discussed in greater detail in Chapter 6. Comparative studies clearly demonstrate the marked differences between autism and developmental receptive dysphasia.

*This has been disputed (Bender, 1971; Miller, 1974), but the contrary "evidence" relies on reports which fail to make systematic comparisons by age of onset (e.g., Bomberg *et al.*, 1973; Ornitz, 1971) and/or which use concepts of schizophrenia that are so wide as to be virtually meaningless (Bender, 1947, 1960).

It may be concluded that there is strong evidence for the validity of the syndrome of childhood autism as originally outlined by Kanner and confirmed by numerous other workers. However, that conclusion must be the beginning, not the end, of nosological research. Having demonstrated the validity of the syndrome it is necessary to go on to ask whether it represents a unitary disorder or a group of conditions, whether the findings allow a better definition of the conditions, whether any subclassification is possible, how autism should be defined and classified now in the light of the research over the last 30 years, and what directions future research should take.

TYPE OF DISORDER

It is sometimes (quite wrongly) assumed (Miller, 1974) that writers who have argued that autism is not synonymous with childhood schizophrenia (e.g., Kolvin, 1971; Rutter, 1972) also imply that autism is a homogeneous condition. Not only is this wrong in logic, but also the same writers have been at pains to point out that it is not yet known whether autism is a single disease entity, a syndrome of biological impairment, or a collection of symptoms which may be due to a heterogeneous group of influences both biological and psychosocial (Rutter, 1974).

Much medical research has been planned on the basis that autism will turn out to be a single condition such as Down's syndrome or phenylketonuria (Rimland, 1964). However, it has already been shown that autism can arise on the basis of conditions as pathologically diverse as congenital rubella (Chess *et al.*, 1971), "hypsarrhythmia" (Taft & Cohen, 1971), congenital syphilis (Rutter & Lockyer, 1967), and tuberose sclerosis (Lotter, 1974). These findings demonstrate that the behavioral syndrome of autism can develop in children with quite heterogeneous disease states. On the other hand, it could still be that the majority of autistic children (who do not have these varied diseases) have a different, still to be identified, medical condition which is common to them all. At the moment, there is no good evidence to support this view but in the absence of any sound knowledge on the biological basis of autism it is important that the possibility continue to be examined.

On the other hand, autism could turn out to be a behavioral syndrome without a single cause but nevertheless with a common biological causation—as is the case with cerebral palsy. If this is the case, of course, the syndrome may include several different but distinct conditions (as occurs with mental retardation). There is certainly very good reason to suppose that autism may often (perhaps usually or even always) arise on the basis of some type of organic brain disorder but very little is known about what type of brain disorder it might be (Rutter, 1974). It has been variously suggested that the locus of abnormality may be in the reticular system, the midbrain, or in the left hemisphere. So far there is insufficient evidence to justify placing the lesion in any one particular area of the brain. Hauser *et al.* (1975) have produced important pneumoencephalographic evidence of enlargement of the left temporal horn in cases of autism (see also Chapter 14). However, other workers (Aarkrog, 1968; Boesen & Aarkrog, 1967) have reported that AEG abnormalities are no more common in autism than in other child psychiatric disorders, and it remains uncertain whether or not left temporal horn dilatation is *specific* to autism. The plasticity of function in the infant's brain (Rutter *et al.*, 1970a) makes it almost certain that a bilateral lesion would be required to produce the deficits found in autism, but little else can be concluded. The existence of language impairment does little to help localize the lesion because the deficit is not just an impairment of speech and, furthermore, it is not yet known whether the disorder is of language as such or whether the language disability stems from a more widespread cognitive deficit affecting skills needed for language.

A third possibility is that autism may be the end-product of a wide and heterogeneous range of factors, both biological and psychosocial (just as language delay or reading difficulties may be due to brain damage, genetic influences, or lack of stimulation). The chief reason for doubting this suggestion is the failure so far to demonstrate any psychosocial variables which can cause autism in the absence of biological impairment. At present, the second alternative of a nonspecific syndrome of biological impairment seems the most likely but it is far too early to regard the matter as settled.

SUBCLASSIFICATION OR REDEFINITION OF SYNDROME

We must now turn to the questions of whether a useful and valid subclassification of childhood autism is possible and whether the research findings allow an improved redefinition of autism. In this connection several points must be borne in mind. First, not all psychoses arising during the first 30 months of life fulfill the criteria for autism (Rutter, 1972; Miller, 1974). Many are quite similar to autism in several crucial respects but some appear rather different. Second, uncertainty remains about the nosological status of disintegrative psychoses arising about the age of three to five after an initial period of apparently normal development. Third, there is good evidence that nonpsychotic psychiatric abnormalities in childhood are present in many individuals who develop schizophrenic psychoses in adolescence or adult life (Offord & Cross, 1969; Rutter,

1972). However, these abnormalities take rather varied forms and so far there are no good criteria for diagnosing these prepsychotic disorders. Fourth, there are psychotic conditions which arise in childhood which fit none of the identified syndromes of autism, disintegrative psychosis, or schizophrenia (Rutter, 1972; Miller, 1974). Many questions on definition and classification remain.

Kanner's Syndrome and Other Infantile Psychoses

Prior and her colleagues (Prior and Macmillan, 1973; Prior *et al.*, 1975a, 1975b) have used taxonomic techniques to investigate the question of whether it is justifiable to differentiate between Kanner's syndrome and other infantile psychoses. They found that the best differentiation (in numerical taxonomy terms) was between, on the one hand, children whose disorders began in the first two years and were characterized by impaired social development, language abnormalities, and stereotyped play, and on the other, children with disorders that usually began later and which showed a more varied symptomatology. This differentiation is similar to, but slightly broader than, the definition of autism used for the purpose of this paper. Within the early onset group, the children with Kanner's syndrome (as defined by scores on Rimland's E2 scale—see below) differed only in having more symptoms suggestive of "insistence on sameness" and in being more likely to have islets of high ability and good fine motor coordination. However, the results showed that the non-Kanner's syndrome early onset psychoses had more in common with Kanner's syndrome than with the later onset psychoses.

Rimland (1964, 1968, 1971) has produced a behavioral questionnaire which he claims differentiates between infantile autism and other disorders. Scores on the questionnaire have been found to differentiate between autistic and retarded children (Douglas & Sanders, 1968) and also, in Rimland's hands, to differentiate biochemical findings (Boullin *et al.*, 1971) and response to megavitamin therapy (Rimland, 1973). Unfortunately the biochemical findings have not been confirmed by others (Yuwiler *et al.*, 1975) and the megavitamin results have yet to be replicated. The usefulness of Rimland's E2 scale to differentiate *within* the infantile psychoses has yet to be demonstrated.

Presence/Absence of Mental Retardation

Because the syndrome of autism has been found to occur in children of all levels of intelligence ranging from the profoundly retarded to superior, it has usually been thought that the condition is basically similar regardless of IQ level. However, it is necessary to consider the possibility that mentally retarded and normally intelligent autistic children may have rather different conditions (Bartak & Rutter, 1976).

In the first place, the prognosis is both worse and different for retarded autistic children (Rutter, 1970). They are much less likely to gain speech and far fewer acquire skills in the "3 R's." Thus, Bartak and Rutter (1976) found that nearly three quarters of autistic children with an IQ of 70 or more acquired competence in all four basic arithmetical skills (multiplication, division, etc.) whereas less than a fifth of those with an IQ below 70 did so. Similarly, half those with normal intelligence go on to higher education or acquire regular paid employment whereas scarcely any of those with an IQ below 70 do so. Moreover, a third of the mentally retarded autistic youngsters developed epileptic fits whereas very few of those with normal intelligence did so. This is in keeping with the (rather less satisfactory) evidence that an overt neurological disorder is more often found in cases of autism with severe mental handicap (Goldfarb, 1961; Rutter, 1970; Chess *et al.*, 1971; Rutter *et al.*, 1971).

Second, several studies have shown that the pattern of cognitive deficit is somewhat different in very low IQ autistic children (Hermelin & O'Connor, 1970). In general, the mentally retarded ones show a wider cognitive deficit involving general difficulties in sequencing and feature extraction, whereas in the normally intelligent autistic children the defect mainly affects verbal skills.

Third, there are also some differences in symptomatology. Although low IQ and high IQ autistic children are closely similar in terms of the main phenomena specifically associated with autism, the mentally retarded autistic children show a more severe disorder of social development and are more likely to exhibit deviant social responses (such as smelling people), stereotypies, and self-injury.

The evidence is not decisive, in that it is not yet clear whether these differences represent different ends of the same continuum or rather a qualitative distinction. However, the differences are sufficiently great that it is certainly necessary for all investigators to make clear distinctions between autism in children of normal (nonverbal) intelligence and autism associated with mental retardation.

Neurological Abnormality

As already indicated, no sharp distinctions have been found in connection with the presence or absence of overt neurological disorder, other than the fact that when there is a neurological disorder there is more likely to be mental retardation as well as autism. Moreover, the diagnosis of neurological abnormality is much influenced by age, as epileptic fits usually do not develop until early adolescence. For these reasons it would be most unwise to limit the diagnosis of childhood autism to cases where there are no indications of neurological disorder. On the other hand, if the possibility of a single disease causing autism is to be investigated it is obviously necessary to differentiate between cases where the autism is already known to be secondary to some disease (such as rubella), those where there is nonspecific evidence of some brain abnormality (such as when there are fits), and those without any indication of organic brain pathology.

Redefinition of Autistic Syndrome

When it is known which symptoms in autism are primary and which secondary, or when the causal mechanisms are understood, it will be essential to redefine the syndrome. As things stand at the moment, there are good pointers for the elucidation of both primary symptoms and causal mechanisms (Rutter, 1974). However, even the best ideas remain at the hypothesis stage and it would still be premature to redefine autism in terms of any of the existing views.

CONCLUSIONS

In summary, the definition of childhood autism in terms of four essential criteria in relation to the child's behavior before age five still seems to be the best procedure. The four criteria are (1) an onset before the age of 30 months; (2) impaired social development which has a number of special characteristics and which is out of keeping with the child's intellectual level; (3) delayed and deviant language development which also has certain defined features and which is out of keeping with the child's intellectual level; and (4) "insistence on sameness" as shown by stereotyped play patterns, abnormal preoccupations, or resistance to change. The syndrome as defined in this way has been shown to be valid and meaningful in that it differs markedly from other clinical syndromes in a host of respects. It is strongly recommended that, in order to insure comparability, all investigators define their samples in this way.

However, it is not enough to use this definition. Cases must also be described in terms of IQ level and neurological or medical status. Studies of classification of child psychiatric disorders have repeatedly shown the advantages of using a multiaxial approach in which the behavioral syndrome, the intellectual level, the medical conditions, and the psychosocial situation are coded (or described) on separate independent axes (Rutter *et al.*, 1969; Tarjan *et al.*, 1972; Rutter *et al.*, 1975). The same applies to childhood autism (Rutter, 1972) and much confusion could be avoided if these different dimensions of diagnosis were clearly differentiated.

For similar reasons it is essential that studies comparing autistic and nonautistic samples should always take intellectual level into account. Investigations which fail to equate groups for mental age are almost uninterpretable because any differences found may relate to autism or mental retàrdation or both. Such studies should no longer be undertaken.

While the methodological considerations in relation to diagnosis and classification in the light of present knowledge are reasonably straightforward, it is also clear that many major questions have yet to be answered. It is not known whether autism is a single disease entity or a behavioral syndrome with several causes. It remains uncertain whether autism should be subclassified and, if it should, how it should be divided. Knowledge is lacking on whether autism in children of normal intelligence is the same thing as autism in those who also have mental retardation. Disagreement continues on how best to classify children who show some features of autism (as defined), but not all. It is unclear which symptoms are primary and which secondary. Lastly, of course, the cause or causes of autism and the psychological mechanisms which underlie its development have still to be established. There are good leads on all these questions and the knowledge accumulated from research during the last decade has considerably sharpened the issues to be studied during the next ten years. While we are still far from a complete understanding of childhood autism, we are very much nearer that elusive goal, as is shown by other chapters in this book.

REFERENCES

Aarkrog, T. Organic factors in infantile and borderline psychoses: a retrospective study of 46 cases subject to pneumoencephalography. *Danish Medical Bulletin*, 1968, *15*, 283–288.

Alpern, G. D. Measurement of "untestable" autistic children. *Journal of Abnormal Psychology*, 1967, *72*, 478–496.

Alpern, G. D. & Kimberlin, C. C. Short intelligence test ranging from infancy levels through childhood levels for use with the retarded. *American Journal of Mental Deficiency*, 1970, *75*, 65–71.

Asperger, H. Die "Autistischen Psychopathen" im Kindesalter. *Archiv Fur Psychiatrie und Nervenkrankheiten*, 1944, *117*, 76–136.

Bakwin, H. Early infantile autism. *Journal of Pediatrics*, 1954, *45*, 492–497.

Bartak, L. & Rutter, M. Educational treatment of autistic children. In M. Rutter (Ed.), *Infantile autism: Concepts, characteristics and treatment*. London: Churchill-Livingstone, 1971. Pp. 258–280.

Bartak, L. & Rutter, M. Special educational treatment of autistic children: A comparative study. I. Design of study and characteristics of units. *Journal of Child Psychology and Psychiatry*, 1973, *14*, 161–179.

Bartak, L., & Rutter, M. Use of personal pronouns by autistic children. *Journal of Autism and Childhood Schizophrenia*, 1974, *4*, 217–222.

Bartak, L., & Rutter, M. Differences between mentally retarded and normally intelligent autistic children. *Journal of Autism and Childhood Schizophrenia*, 1976, *6*, 109–120.

Bartak, L., Rutter, M., and Cox, A. A comparative study of infantile autism and specific developmental receptive language disorder. I. The children. *British Journal of Psychiatry,* 1975, *126,* 127–145.

Bender, L. Childhood schizophrenia. Clinical study of one hundred schizophrenic children. *American Journal of Orthopsychiatry,* 1947, *17,* 40–56.

Bender, L. Treatment in early schizophrenia. *Progressive Psychotherapy,* 1960, *5,* 177–184.

Bender, L. Alpha and omega of childhood schizophrenia. *Journal of Autism and Childhood Schizophrenia,* 1971, *1,* 115–118.

Bleuler, E. *Dementia praecox oder gruppe der schizophrenien.* Deutiche, 1911. (Translated by J. Zinkin. New York: International University Press, 1950.)

Boesen, U., & Aarkrog, T. Pneumo-encephalography of patients in a child psychiatric department. *Danish Medical Bulletin,* 1967, *14,* 210–218.

Bomberg, D., Szurek, S. A., & Etemad, J. G. A statistical study of a group of psychotic children. In S.A. Szurek & I. M. Berlin (Eds.), *Clinical studies in childhood psychoses.* London: Butterworth, 1973.

Bosch, G. Über primären autismus in kindesalter 1953. Cited by G. Bosch in *Infantile autism.* New York: Springer-Verlag, 1970.

Bosch, G. *Infantile autism.* New York: Springer-Verlag, 1970.

Boullin, D. J., Coleman, M., O'Brien, R. A., & Rimland, B. Laboratory prediction of infantile autism based on 5-hydroxytryptamine efflux from blood platelets and their correlation with the Rimland E2 score. *Journal of Autism and Childhood Schizophrenia,* 1971, *1,* 63–71.

Campbell, E. J. M. Basic science, science and medical education. *Lancet,* 1976, *1,* 134–136.

Chess, S., Korn, S. J., & Fernandez, P. B. *Psychiatric disorders of children with congenital rubella.* New York: Brunner/Mazel, 1971.

Churchill, D. W., & Bryson, C. Q. Looking and approach behavior of psychotic and normal children as a function of adult attention and preoccupation. *Comparative Psychiatry,* 1972, *13,* 171–177.

Clark, P., and Rutter, M. Compliance and resistance in autistic children. *Journal of Autism and Childhood Schizophrenia,* 1977, *7,* 33–48.

Corbett, J., Harris, R., Taylor, E., & Trimble, M. Psychiatric aspects of the neurodegenerative disorders. *Journal of Child Psychology and Psychiatry,* 1977, *18,* 211–220.

Cowan, P. A., Hoddinott, B. A., & Wright, B. A. Compliance and resistance in the conditioning of autistic children: An explanatory study. *Child Development,* 1965, *36,* 913–923.

Creak, M. Schizophrenia syndrome in childhood: Progress report of a working party. *Cerebral Palsy Bulletin,* 1961, *3,* 501–504.

Creak, E. M. Childhood psychosis: A review of 100 cases. *British Journal of Psychiatry,* 1963, *109,* 84–89.

Cunningham, M. A. A comparison of the language of psychotic and nonpsychotic children who are mentally retarded. *Journal of Child Psychology and Psychiatry,* 1968, *9,* 229–244.

Darr, G. C., & Worden, F. G. Case report twenty-eight years after an infantile autistic disorder. *American Journal of Orthopsychiatry,* 1951, *21,* 559–570.

Davis, D. R., & Cashdan, A. Specific dyslexia. *British Journal of Educational Psychology,* 1963, *33,* 80–82.

DeMyer, M. K., Alpern, G. D., Barton, S., DeMyer, W., Churchill, D. W., Hingtgen, J. M., Bryson, C. W., Pontius, W., & Kimberlin, C. Imitation in autistic, early schizophrenic and non-psychotic subnormal children. *Journal of Autism and Childhood Schizophrenia,* 1972, *2,* 264–287.

DeMyer, M. K., Barton, S., Alpern, G.D., Kimberlin, C., Allen, J., Yang, E., & Steele, R. The measured intellingence of autistic children. *Journal of Autism and Childhood Schizophrenia,* 1974, *4,* 42–60.

DeMyer, M. K., Barton, S., DeMyer, W. E., Norton, J. A., Allen, J., & Steel, R. Prognosis in autism: A follow-up study. *Journal of Autism and Childhood Schizophrenia,* 1973, *3,* 199–245.

Despert, J. L. Some considerations relating to the genesis of autistic behaviour in children. *American Journal of Orthopsychiatry,* 1951, *21,* 335–350.

Douglas, V. I., & Sanders, F. A. A pilot study of Rimland's diagnostic check list with autistic and mentally retarded children. *Journal of Child Psychology and Psychiatry,* 1968, *9,* 105–109.

Eisenberg, L. The classification of the psychotic disorders in childhood. In L. D. Eron (Ed.), *Classification of behaviour disorders.* Chicago: Aldine, 1966.

Eisenberg, L. The classification of childhood psychosis reconsidered. *Journal of Autism and Childhood Schizophrenia,* 1972, *2,* 338–342.

Eisenberg, L., & Kanner, L. Early infantile autism 1943–55. *American Journal of Orthopsychiatry,* 1956, *26,* 556–566.

Gillies, S. M. Some abilities of psychotic children and subnormal controls. *Journal of Mental Deficiency Research,* 1965, *9,* 89–101.

Gittelman, M., & Birch, H. G. Childhood schizophrenia: Intellect, neurologic status, perinatal risk, prognosis and family pathology. *Archives of General Psychiatry,* 1967, *17,* 16–25.

Goldfarb, W. *Childhood schizophrenia.* Cambridge, Mass: Harvard University Press, 1961.

Goldfarb, W. *Growth and change of schizophrenic children—A longitudinal study.* Washington: Winston, 1974.

Hauser, S. L., DeLong, G. R., & Rosman, N. P. Pneumographic findings in the infantile autism syndrome: A correlation with temporal lobe disease. *Brain,* 1975, *98,* 667–688.

Heller, T. About dementia infantilis. Reprinted in J. G. Howells (Ed.), *Modern perspectives in international child psychiatry.* Edinburgh: Oliver & Boyd, 1930.

Hermelin, B., & O'Connor, N. *Psychological experiments with autistic children.* London: Pergamon, 1970.

Hingtgen, J. N., & Bryson, C. Q. Recent developments in the study of early childhood psychoses: Infantile autism, childhood schizophrenia and related disorders. *Schizophrenia Bulletin,* 1972, *5,* 8–53.

Hingtgen, J. N., & Churchill, D. W. Identification of perceptual limitations in mute autistic children. *Archives of General Psychiatry,* 1969, *21,* 68–71.

Hingtgen, J. N., & Churchill, D. W. Differential effects of behaviour modification in four mute autistic boys. In D. W. Churchill, C. D. Alpern, & M. DeMyer (Eds.), *Infantile autism.* Springfield, Ill: Charles C Thomas, 1971.

Hutt, C., & Ounsted, C. The biological significance of gaze aversion with particular reference to the syndrome of infantile autism. *Behavioural Science,* 1966, *11,* 346–356.

Hutt, C., & Vaizey, M. J. Differential effects of group density on social behaviour. *Nature, London,* 1966, *209,* 1371–1372.

Hutt, S. J., Hutt, C., Lee, D., & Ounsted, C. A behavioural and electroencephalographic study of autistic children. *Journal of Psychiatric Research,* 1965, *3,* 181–197.

Isaev, D. N., & Kagan, V. E. Autistic syndromes in children of adolescents. *Acta Paedopsychiatrica,* 1974, *40,* 182–190.

Kanner, L. Autistic disturbances of affective contact. *Nervous Child,* 1943, *2,* 217–250.

Kanner, L. Irrelevant and metaphorical language in early infantile autism. *American Journal of Psychiatry,* 1946, *103,* 242–245.

Kanner, L. *Child psychiatry* (3rd ed.). Oxford: Blackwell Scientific Publications, 1957.

Kanner, L. Follow-up study of eleven autistic children originally reported in 1943. *Journal Autism and Childhood Schizophrenia.* 1971, *1,* 119–145.

Kanner, L., Rodriquez, A., & Ashenden, B. How far can autistic children go in matters of social adaptation? *Journal of Autism and Childhood Schizophrenia,* 1972, *2,* 9–33.

Kolvin, I. Psychoses in childhood—a comparative study. In M. Rutter (Ed.), *Infantile autism: Concepts, characteristics and treatment.* London: Churchill-Livingstone, 1971. Pp. 7–26.

Laufer, M. W., & Gair, D. S. Childhood schizophrenia. In L. Bellak & L. Loeb (Eds.), *The schizophrenic syndrome.* New York: Grune & Stratton, 1969.

Lockyer, L., & Rutter, M. A five to fifteen year follow-up study of infantile psychosis. III. Psychological aspects. *British Journal of Psychiatry, 1969, 115,* 865–882.

Lockyer, L., & Rutter, M. A five to fifteen year follow-up study of infantile psychosis. IV. Patterns of cognitive ability. *British Journal of Social and Clinical Psychology, 1970, 9,* 152–163.

Lotter, V. Epidemiology of autistic conditions in young children. I. Prevalence. *Social Psychiatry, 1966, 1,* 124–137.

Lotter, V. Factors related to outcome in autistic children. *Journal of Autism and Childhood Schizophrenia, 1974, 4,* 263–277.

Makita, K. The age of onset of childhood schizophrenia. *Folia Psychiatrica et Neurologica Japonica, 1966, 20,* 111–121.

Malamud, N. Heller's disease and childhood schizophrenia. *American Journal of Psychiatry, 1959, 116,* 215–218.

Marchant, R., Howlin, P., Yule, W., & Rutter, M. Graded change in the treatment of the behaviour of autistic children. *Journal of Child Psychology and Psychiatry, 1974, 15,* 221–227.

Miller, R. T. Childhood schizophrenia: A review of selected literature. *International Journal of Mental Health, 1974, 3,* 3–46.

Mittler, P., Gillies, S., & Jukes, E. Prognosis in psychotic children. Report of follow-up study. *Journal of Mental Deficiency Research, 1966, 10,* 73–83.

Mnukhin, S. A., & Isaev, D. N. On the organic nature of some forms of schizoid or autistic psychopathy. *Journal of Autism and Childhood Schizophrenia, 1975, 5,* 99–108.

Offord, D. Q., & Cross, L. A. Behavioral antecedents of adult schizophrenia. *Archives of General Psychiatry, 1969, 21,* 267–283.

O'Gorman, G. *The nature of childhood autism.* London: Butterworth, 1970.

Ornitz, E. M. Childhood autism: A disorder of sensorimotor integration. In M. Rutter (Ed.), *Infantile autism: Concepts, characteristics and treatment.* London: Churchill-Livingstone, 1971. Pp. 50–68.

Ornitz, E. M., Brown, M. B., Sorosky, A. D., Ritvo, E. R., & Dietrich, L. Environmental modifications of autistic behavior. *Archives of General Psychiatry, 1970, 22,* 560–565.

Ornitz, E. M., & Ritvo, E. R. Perceptual inconstancy in early infantile autism. *Archives of General Psychiatry, 1968, 18,* 76–98.

Popper, K. *The logic of scientific discovery* (2nd ed.). London: Hutchinson, 1959.

Popper, K. *Conjectures and regulations: The growth of scientific knowledge* (4th ed.). London: Routledge and Kegan Paul, 1972.

Prior, M., & Macmillan, M. B. Maintenance of sameness in children with Kanner's syndrome. *Journal of Autism and Childhood Schizophrenia, 1973, 3,* 154–167.

Prior, M., Boulton, D., Gajzago, C., & Perry, D. The classification of childhood psychoses by numerical taxonomy. *Journal of Child Psychology and Psychiatry, 1975, 16,* 321–330.(a)

Prior, M., Perry, D., & Gajzago, C. Kanner's syndrome or early-onset psychosis: A taxonomic analysis of 142 cases. *Journal of Autism and Childhood Schizophrenia, 1975, 5,* 71–80.(b)

Rendle-Short, J. Infantile autism in Australia. *Medical Journal of Australia, 1969, 2,* 245–249.

Ricks, D. M. Vocal communication in pre-verbal normal and autistic children. In N. O'Connor (Ed.), *Language, cognitive deficits and retardation.* London: Butterworth, 1975.

Ricks, D. N., & Wing, L. Language, communication, and the use of symbols in normal and autistic children. *Journal of Autism and Childhood Schizophrenia,* 1975, *5,* 191–222.

Rimland, B. *Infantile autism.* New York: Appleton-Century-Crofts, 1964.

Rimland, B. On the objective diagnosis of infantile autism. *Acta Paedopsychiatrica,* 1968, *35,* 146–161.

Rimland, B. The differentiation of childhood psychoses: An analysis of checklists for 2,218 psychotic children. *Journal of Autism and Childhood Schizophrenia,* 1971, *1,* 161–174.

Rimland, B. High dosage levels of certain vitamins in the treatment of children with severe mental disorders. In D. R. Hawkins and L. Pauling (Eds.), *Orthomolecular psychiatry.* San Francisco: W. H. Freeman, 1973.

Robins, L. N. Follow-up studies of behavior disorders in children. In H. C. Quay and J. S. Werry (Eds.), *Psychopathological disorders of childhood.* New York: Wiley, 1972.

Ross, I. S. Presentation of a clinical case: An autistic child. Pediatric Conference, The Babies Hospital Unit, United Hospital, Newark, New Jersey, 1959, *2,* No. 2.

Rutter, M. Speech disorders in a series of autistic children. In A. W. Franklin (Ed.), *Children with communication problems.* London: Pitman, 1965.

Rutter, M. Behavioural and cognitive characteristics of a series of psychotic children. In J. Wing (Ed.), *Early childhood autism.* London: Pergamon, 1966. Pp. 51–81.

Rutter, M. Psychotic disorders in early childhood. In A. Coppen & A. Walk (Eds.), *Recent developments in schizophrenia.* British Journal of Psychiatry Special Publication. Ashford, Kent: Headley Bros., 1967. Pp. 133–158.

Rutter, M. Concepts of autism: A review of research. *Journal of Child Psychology and Psychiatry,* 1968, *9,* 1–25.

Rutter, M. Autistic children: Infancy to adulthood. *Seminars in Psychiatry, 1970, 2,* 435–450.

Rutter, M. The description and classification of infantile autism. In D. W. Churchill, G. D. Alpern, & M. K. DeMyer (Eds.), *Infantile autism.* Springfield, Ill: Charles C Thomas, 1971.(a)

Rutter, M. Psychiatry. In J. Wortis (Ed.), *Mental retardation: An annual review* (Vol. III). New York: Grune & Stratton, 1971. Pp. 186–221.(b)

Rutter, M. Childhood schizophrenia reconsidered. *Journal of Autism and Childhood Schizophrenia,* 1972, *2,* 315–337.

Rutter, M. The development of infantile autism. *Psychological Medicine,* 1974, *4,* 147–163.

Rutter, M. Brain damage syndromes in childhood: Concepts and findings. *Journal of Child Psychology and Psychiatry,* 1977, *18,* 1–21.

Rutter, M., & Bartak, L. Special educational treatment of autistic children: A comparative study. II. Follow-up findings and implications for services. *Journal of Child Psychology and Psychiatry,* 1973, *14,* 241–270.

Rutter, M., Bartak, L., & Newman, S. Autism—a central disorder of cognition and language? In M. Rutter (Ed.), *Infantile autism: Concepts, characteristics and treatment.* London: Churchill-Livingstone, 1971. Pp. 148–171.

Rutter, M., Graham, P., & Yule, W. *A neuropsychiatric study in childhood.* Clinics in Developmental Medicine Nos. 35/36. London: Heinemann/SIMP, 1970.(a)

Rutter, M., Greenfeld, D., & Lockyer, L. A five to fifteen year follow-up study of infantile psychosis. II. Social and behavioural outcome. *British Journal of Psychiatry,* 1967, *113,* 1183–1199.

Rutter, M., Lebovici, L., Eisenberg, L., Sneznevsky, A. V., Sadoun, R., Brooke, E., & Lin, T-Y. A tri-axial classification of mental disorder in childhood. *Journal of Child Psychology and Psychiatry,* 1969, *10,* 41–61.

Rutter, M., & Lockyer, L. A five to fifteen year follow-up study of infantile psychosis. I. Description of sample. *British Journal of Psychiatry,* 1967, *113,* 1169–1182.

Rutter, M., & Martin, J. A. M. (Eds.), *The child with delayed speech.* Clinics in Developmental Medicine No. 43. London: Heinemann/SIMP, 1972.

Rutter, M., Shaffer, D., & Shepherd, M. *A multi-axial classification of child psychiatric disorders.* Geneva: WHO, 1975.

Rutter, M., Tizard, J., & Whitmore, K. (Eds.), *Education, health and behaviour.* London: Longmans, 1970.(b)

Schain, R. J., & Yannet, H. Infantile autism: An analysis of 50 cases and a consideration of certain relevant neurophysiologic concepts. *Journal of Pediatrics,* 1960, *57,* 560–567.

Schopler, E. Visual versus tactual receptor preference in normal and schizophrenic children. *Journal of Abnormal Psychology,* 1966, *71,* 108–114.

Sorosky, A. D., Ornitz, E. M., Brown, N. B., & Ritvo, E. R. Systematic observations of autistic behavior. *Archives of General Psychiatry,* 1968, *18,* 439–449.

Taft, L. T., & Cohen, H. J. Hypsarrhythmia and infantile autism: A clinical report. *Journal of Autism and Childhood Schizophrenia,* 1971, *1,* 327–336.

Tarjan, G., Tizard, J., Rutter, M., Begab, M., Brooke, E., De La Cruz, F., Lin, T-Y, Montenegro, H., Strotzka, H., & Sartorius, N. Classification of mental retardation: Issues arising in the 5th WHO Seminar on Psychiatric Diagnosis, Classification and Statistics. *American Journal of Psychiatry,* 1972, *128,* 11, 34–45 (Supplement).

Tinbergen, E. A., & Tinbergen, N. Early childhood autism: An etiological approach. In Advances in ethology, 10, Supplement to *Journal of Comparative Ethology.* Berlin and Hamburg: Verlag Paul Parry, 1972.

Tizard, B., & Rees, J. The effect of early institutional rearing on the behaviour problems and affectional relationships of four year old children. *Journal of Child Psychology and Psychiatry,* 1975, *16,* 61–74.

Tubbs, V. K. Types of linguistic disability in psychotic children. *Journal of Mental Deficiency Research,* 1966, *10,* 230–240.

Vaillant, G. E. John Haslam on early infantile autism. *American Journal of Psychiatry,* 1962, *119,* 376.

Van Krevelen, D. A. Early infantile autism. *Acta Paedopsychiatrica,* 1952, *91,* 81–97.

Van Krevelen, D. A. Early infantile autism and autistic psychopathy. *Journal of Autism and Childhood Schizophrenia,* 1971, *1,* 82–86.

Van Krevelen, D. A., & Kuipers, C. The psychopathology of autistic psychopathy. *Acta Paedopsychiatrica,* 1962, *29,* 22–31.

Vrono, M. Schizophrenia in childhood and adolescence. *International Journal of Mental Health,* 1974, *2,* 7–116.

Wing, J. K. Kanner's syndrome: A historical introduction. In L. Wing (Ed.), *Early childhood autism: Clinical, educational and social aspects* (2nd ed.). Oxford: Pergamon Press, 1976.

Wing, L. The handicaps of autistic children—a comparative study. *Journal of Child Psychology and Psychiatry,* 1969, *10,* 1–40.

Wing, L. Perceptual and language development in autistic children: A comparative study. In M. Rutter (Ed.), *Infantile autism: Concepts, characteristics and treatment.* London: Churchill-Livingstone, 1971. Pp. 173–197.

Wing, L., & Ricks, D. M. The aetiology of childhood autism: A criticism of the Tinbergen's ethological theory. *Psychological Medicine,* 1976, *6,* 533–544.

Wolff, S., & Chess, S. A behavioural study of schizophrenic children. *Acta Psychiatrica Scandinavica,* 1964, *40,* 438–466.

Wolff, S., & Chess, S. An analysis of the language of fourteen schizophrenic children. *Journal of Child Psychology and Psychiatry,* 1965, *6,* 29–41.

Yuwiler, A., Ritvo, E., Geller, E., Glousman, R., Schneiderman, G., & Matsuno, D. Uptake and efflux of serotonin from platelets of autistic and nonautistic children. *Journal of Autism and Childhood Schizophrenia,* 1975, *5,* 83–98.

2

Social, Behavioral, and Cognitive Characteristics: An Epidemiological Approach

LORNA WING

Kanner, who observed and then named the syndrome of early childhood autism (Kanner, 1943; Kanner & Eisenberg, 1955) believed that the two principal diagnostic criteria (in his view the source of the other clinical manifestations) are extreme self-isolation and an obsessive insistence on sameness. The latter is manifested by resistance to change in aspects of the environment and the daily routine, and by the child's own stereotyped activities, varying from simple repetitive movements to very elaborate but equally repetitive rituals.

Kanner (1943, 1973) described in graphic detail the autistic child's air of detachment and indifference to others, especially his peers, made even more marked by poor eye contact. Kanner also noted how, when someone interferes with the child's solitary and repetitive but skillful manipulation of objects, he reacts on the instant with a storm of rage and distress that, once the interference ceases, is turned off as rapidly as it began, leaving behind no trace of any disturbance.

This picture is seen most clearly in the younger children, especially before the age of five (Wing, 1971). Follow-up studies (DeMyer, *et al.*, 1973; Eisenberg, 1957; Kanner, 1973; Kanner & Eisenberg, 1955, 1956; Rutter, 1970) have shown that changes may occur with increasing age. Although some of the children make no progress and a few deteriorate,

LORNA WING · MRC Social Psychiatry Unit, Institute of Psychiatry, De Crespigny Park, London SE5 8AF, England.

about 20 to 25% become less resistant to change, less involved with stereotyped activities, lose their indifference to people, and want to join in social life, but are handicapped by marked social ineptness when trying to relate to others.

The social behavior of Kanner's autistic children in both early and later childhood is very different from that of the shy normal child, who relates normally to adults and children with whom he feels secure, or the child with elective mutism (Reed, 1963). It is also different from the pattern seen in the young child suffering from the effects of removal from his family (Ainsworth, 1962; Bowlby, 1951; Rutter, 1972) or the child deprived of the opportunity of forming social bonds with his adult caretakers (Tizard & Rees, 1975; Wolkind, 1974).

Although the abnormalities of social interaction and the insistence on sameness are among the most striking characteristics of autistic children, the syndrome described by Kanner (1943, 1946, 1951, 1954, 1973) includes a number of other abnormalities, namely unusual responses to various kinds of sensory inputs; cognitive problems affecting communication, symbolic language and imaginative play; and impairments of motor organization and imitation. Kanner also emphasized the presence of positive skills, that is, dexterity in the manipulation of objects requiring good ability with visuo-spatial tasks and an excellent rote memory. His choice of the primary problems is based on an unproven hypothesis.

Different workers have different views as to which aspects of the syndrome are fundamental, but virtually all consider the social aloofness to be of great importance, regardless of how they try to explain it. The insistence on sameness has tended to receive rather less attention in the development of causal hypotheses (see the review of etiological theories by Wing, 1976). Some authors regard the poor social contact as one of a number of abnormalities all of which are produced by one underlying cause, physical or psychological, for example, faulty conditioning (Ferster, 1961), an abnormality of the vestibular system (Ornitz, 1973), or a lesion in and around the head of the nucleus of the tractus solitarius in the posterior brain stem (McCulloch & Williams, 1971).

Other writers, while acknowledging the importance of the abnormality of social behavior as a diagnostic feature, suggest that it is a *consequence* of the impairment of other psychological functions. Churchill (1972), Bartak *et al.* (1975), Cox, Rutter, Newman & Bartak (1975), and Wing (1969) have postulated that a severe global language problem affecting comprehension and use of all forms of language may prevent the development of normal social relationships. Hermelin & O'Connor (1970) suggest that there is a central deficit, affecting the encoding of stimuli and concept formation, that severely impairs all complex behavior, including social interaction. DeMyer (1976) considers that problems of motor imitation affect social development and make the autistic child socially withdrawn or inept.

Finally, there are workers who, like Kanner, take the view that self-isolation is a primary problem causing the development of other aspects of the autistic syndrome. Hutt, Hutt, and Richer and their colleagues (Hutt & Hutt, 1970; Hutt *et al.*, 1965; Hutt & Ounsted, 1966; Hutt *et al.*, 1975; Richer, 1974) have worked out a theory of this kind in some detail, based on an ethological approach, in which they attempt to explain both the poor social contact and the children's stereotyped activities. They concentrate upon the simple repetitive activities, rather than elaborate, complex rituals, or resistance to change in the environment and routines.

In the view of these workers, autistic children react to social approaches by becoming overaroused. Direct eye contact is especially likely to produce even further overarousal. Autistic children are therefore motivated to avoid social approaches and eye-to-eye gaze. The resulting poverty of social interaction (and, in the view of Tinbergen & Tinbergen, 1972, the prevention of social bonding), beginning very early in life, affects the parent-child relationship and seriously impairs the learning of language and other skills. Stereotypies are prominent because they are a mechanism for reducing the level of arousal.

Hutt & Hutt (1970) also state that the nonverbal components of an autistic child's social behavior (lifting arms to be picked up, leading an adult by the hand, and so on) are normal in form when they occur, but this is markedly at variance with the reports of other workers (DeMyer, 1976; Kanner, 1943; Ricks, 1975; Ricks & Wing, 1975; Wing, 1969).

One reason why workers disagree in their opinions on the relative importance of the different aspects of the autistic syndrome is the variation in the children they select for study. Early childhood autism has, at present, to be defined in terms of an abnormal behavior pattern, and there are no objective tests on which to base a diagnosis. Differences in the way the label is applied are particularly likely if brief summaries of the clinical features are used for identification of cases, instead of careful comparison with Kanner's lengthy clinical histories. It is not surprising if people prefer to use brief summaries, since Kanner's case histories, though fascinating to read, do not present the details in any logical framework, and it is a laborious process to sort out the pattern well enough to match individual children with it. Add to this the fact that the children's behavior changes with increasing age and it can be seen that the problems are formidable indeed.

There seems to be, so to speak, a pool of children all of whom have at least some aspects of the autistic syndrome and a few of whom show the full classic picture. Different workers have dipped into the pool and have come up with different samples of children with different age ranges.

These samples overlap with each other, but there are large variations, which affect the theories put forward. The many ways in which the terms childhood autism, psychosis, and schizophrenia are used do not allow readers to be sure which children are being studied unless detailed clinical descriptions are given.

One way of attempting to bring some order into the scene is to adopt the epidemiological approach and investigate the contents of the whole pool. My colleague, Judith Gould, and I have attempted to identify and study *all* the children in one geographical area, of all levels of intelligence, who might possibly show some or all of the elements of Kanner's syndrome, so that we could look at the problem in context across a wide age range, and consider the boundaries of the condition, as well as the classic, nuclear cases.

This chapter describes the associations, within this population of children, between, on the one hand, social aloofness, and certain aspects of insistence on sameness (the two abnormalities considered by Kanner to be of primary importance) and, on the other hand, the level of development of certain relevant cognitive skills, namely visuo-manual abilities, comprehension of language, and symbolic, representational play. The relationship between degree of social contact and the types of organic conditions that were diagnosed in the children studied is also discussed. These preliminary data provide the necessary basis for the further investigation of the etiological significance of these aspects of the autistic syndrome.

SUBJECTS

The children studied had home addresses in the former borough of Camberwell, a predominantly working-class inner suburb of southeast London. The population was approximately 155,000 in 1971, of whom nearly 35,000 were under the age of 15.

The area has a comprehensive range of clinics, hospitals, schools, and day and residential services for mentally handicapped or disturbed children, provided by the local authority and supplemented by voluntary bodies. The education authority has a statutory duty to provide schooling for all children, however severely retarded or disturbed in behavior. Children below school age (five) who are suspected of being handicapped or abnormal in development are reported, by pediatricians or health visitors, to the Observation and Handicap Register kept by the local health authority. The seven local day nurseries, run by the social services department, have special units for preschool handicapped or disturbed children. The services are used by all social classes.

Since 1964, the Medical Research Council's Social Psychiatry Unit has maintained a cumulative psychiatric and mental retardation case register in Camberwell, which collects information from all the above sources (Wing & Hailey, 1972). The epidemiological study of Camberwell children aged under 15 on December 31, 1970, was carried out using the facilities offered by the register and by the extensive services.

Preliminary observations showed that items of autistic behavior, especially stereotyped movements, are very common, although by no means universal, among children known to the services as severely mentally retarded, whereas they are comparatively rare among the mildly retarded. It was therefore decided to include in the study all severely retarded children, whether or not they had autistic behavior, so that the nonautistic severely retarded children could be used as a comparison group. It would have been of interest to investigate also those with mild retardation but without autistic behavior, but the numbers involved were too large for the available resources.

From the Camberwell register, and with the kind and willing cooperation of the various local authorities and voluntary bodies, 145 children under 15 years of age, known to the services as severely mentally retarded, were identified, though four died before they could be seen for the study. All but 18 of these children had intelligence quotients below 50 on formal tests. One other child who had Down's syndrome, but who was in a school for mildly retarded children, was included, so as to have a complete population of children with this condition.

In order to find children with at least some aspects of autistic behavior, but who were not administratively classified as severely retarded, the following methods were used. First the Camberwell register was searched for any child with a diagnosis of autism, psychosis or schizophrenia, or with descriptive labels such as "severe language impairment," "echolalia," or even "a very strange child." Second, all children (769 altogether) were screened, who were attending special schools, classes, or clinics for any kind of physical or mental handicap (other than severe retardation) or because of slow learning, behavior disturbances, or speech impairments. This involved a brief structured interview with the teacher, inspection of all medical, educational, and social case notes, and observation in the classroom (Wing *et al.*, 1976). As a result a further 25 children were found, who showed at least one autistic feature. Of these, 11 were in schools for children with "mild educational subnormality," six were in schools for autistic children, three were attending schools for the deaf or partially hearing, two were in units for partially sighted/partially hearing children, one was in a school for the partially sighted, one was in a school for delicate children, and one was in a day nursery. This made a grand total of 167 who were intensively investigated as will be described below. Children with physical handicaps or impairments of hearing or vision were included as long as they fitted the criteria for acceptance.

Ideally, the total population of 35,000 children under the age of 15 should have been screened, but this was a practical impossibility. Lotter (1966) in his study of 78,000 children in the former county of Middlesex, England, found only one autistic child attending a school for normal children. In the present study, all the children with relevant diagnoses who were found through the Camberwell register because they had attended psychiatric clinics were also in special schools or were known to the preschool services as needing special education. One autistic child was later transferred to a school for normal children, but was in a special school on the census day.

To insure that no children were overlooked, the search for subjects continued during the entire course of the study, which lasted from 1972 to 1975. It was possible that one or two children with autistic behavior were missed because they had always attended schools for normal children and had never been diagnosed as autistic by any service making returns to the register, but it was very unlikely that greater numbers than this were undiscovered.

Two children who had temporary toxic confusional states and two, both mildly retarded, who developed psychotic behavior in their early teens were excluded. At the time of the detailed interview (see below) ages ranged from 2 years 2 months to 18 years. The investigators have remained in touch with the children since the interviews and tests were completed.

METHOD

Information Collected

In the intensive study, each child's teacher or nurse and, wherever possible, his mother were interviewed using a structured schedule (Wing, 1975; Wing & Gould, in press) concerning his current level of development on a wide range of functions such as self-care, language and practical skills, and current abnormalities of behavior such as stereotypies, echolalia, and aggressiveness. The informants were encouraged to describe the children's behavior in concrete detail, and the final rating was made by the interviewer (that is, Judith Gould or the present author). The children were also observed by the interviewers at school and, where possible, at home.

Tests were given of language comprehension, mainly the Reynell language scales (Reynell, 1969), and of nonverbal skills, that is, fitting and assembly tasks not requiring understanding or use of symbolic language, mainly from the Merrill-Palmer scale (Stutsman, 1931). Full details of testing are given by Gould (1976).

The children also had neurological, biochemical, chromosomal, and EEG investigations carried out by Corbett and colleagues (Corbett *et al.*, 1975). Birth records and medical, psychiatric, and educational case notes were examined, and details of developmental history were collected.

Ratings of behavior

In the present chapter, ratings are based mainly on the child's behavior in school or nursery. Social contact is measured by combining ratings on response to physical contact, engaging in social play, recognition of and differentiation between familiar people and strangers, and pointing things out for adults to look at. The combined score is collapsed into a 5-point scale ranging from 1 (no interest in social contact) to 5 (enjoys and initiates social contact).

Problems of eye contact are rated as present if the teacher feels she has a definite and continuing problem in obtaining eye-to-eye gaze with the child in appropriate circumstances.

Only one aspect of insistence on sameness will be discussed here, namely, repetitive stereotyped behavior, which was described more often and in greater detail by the informants than resistance to change in the environment or the daily routine. Simple stereotypies (such as flicking of fingers or pieces of string, tapping or scratching on surfaces, arm flapping, jumping or self-spinning) are rated as present if they are seen most of the time the child is unoccupied. Repetitive routines are described as elaborate if they involve manual dexterity and organization of materials and the environment. Even within this category, it is possible to define some elaborate routines as more complex than others (Wing *et al.*, 1976) but, in the present paper, they will all be considered together.

Symbolic play is classified into three groups (Wing *et al.*, 1977). "No symbolic play" means that the child does not use toys or materials to represent real objects or situations. "Stereotyped symbolic play" means that the child uses toys or other materials to represent real objects but repeats the same narrow range of activities over and over again without regard to suggestions from other children. "True symbolic play" is representational play (Lowe, 1975; Sheridan, 1969) that can change in theme and develops in complexity and range as the child matures.

RESULTS

Classification of the Children

From the information collected, it was evident that Kanner's two main criteria for the diagnosis of autism, namely lack of affective contact, and resistance to change shown by repetitive, stereotyped activities, can both occur in varying degrees of severity. A child may show these problems in different forms at different stages of his life.

Lack of affective contact can take the form of extreme social aloofness and active withdrawal from others. This shades into a milder degree of social withdrawal in which the children concerned rarely or never initiate contact with people, but can be pushed or pulled into other children's games. They take a completely passive part, perhaps as the baby in a game of mothers and fathers, but are quite amiable. The mildest form in which the problem of social contact appears is in those children who actually initiate social approaches, but have so little know-how about how to get on with other people that they are soon left out by other children, and never make any friendships.

Repetitive, stereotyped behavior also varies in its severity. It takes a variety of forms, including simple stereotypies, elaborate rituals, stereotyped representational play, or stereotyped speech. This last involves repetitive delayed echolalia, or repetitive questioning, or concentration upon one topic regardless of the demands of the situation.

Excluding eight children who were too profoundly physically and mentally handicapped to show any of the behavior described above, and one other child, who at the time he was assessed, had an exacerbation of a neurological condition that produced a marked, though temporary, effect on his behavior, the remaining 158 children were divided into two groups. The first contains 84 children who have both of Kanner's criteria to some degree, and will be called here the "psychotic" group, for want of a better term. The remaining 74 children are not socially aloof, taking mental age into account. There are 17 children in this nonpsychotic group with very low language comprehension ages on formal tests. Six of these have simple stereotypies and three have a low sociability score on the crude measure used. Nevertheless, in the light of their mental ages, they show appropriate pleasure in nonverbal ways when approached by other people, in marked contrast to those psychotic children who have the same low mental age.

The above classification was made regardless of level of IQ, the presence of additional physical handicaps, or known organic etiology. This is because techniques of testing cognitive skills and methods of detecting organic conditions have improved considerably since Kanner first described his group in the early 1940s. If Kanner's view that autistic children have normal cognitive potential and no organic pathology is adhered to, the prevalence rate of the autistic syndrome may appear to dwindle steadily as methods of diagnosis and testing continue to improve in accuracy.

It should be emphasized that the assignment of children to the two groups was based on clinical judgment, utilizing detailed information collected in a systematic fashion.

Seventeen of the "psychotic" group (4.8 per 10,000 of all children aged 0 to 14) had the syndrome of early childhood autism using the system of selection, based on Kanner's descriptions, devised by Lotter (1966). Social aloofness, especially to peers, *and* elaborate rituals (defined above) were the main criteria. Both of these problems had to be present currently, or in marked degree from before age two and one-half and continuing up to at least age seven. Simple stereotypies (defined above) did not, by themselves, qualify a child for this group.

The children were also divided, independently of whether or not they were considered to be "psychotic," into three categories, depending on mental age on nonverbal, that is nonsymbolic, visuo-spatial skills, and on tests of language comprehension. The first category contains 43 children whose nonverbal age and language comprehension age are below 20 months. The second contains 27 children with nonverbal age of 20 months or above and language comprehension below that level. The third contains 88 children with both nonverbal and language comprehension ages of 20 months or above. There were no children with a language age of 20 months or above but with a nonverbal mental age of 0 to 19 months. The cutoff point of 20 months is used because normal children, by this age, have the beginnings of symbolic inner language, shown in representational, imaginative play (Lowe, 1975; Sheridan, 1969).

Associations Between Variables

Table 1 gives the association between nonverbal and language comprehension age categories and sociability score. In the psychotic children, sociability score is associated very significantly with mental age. Language comprehension age and nonverbal age make an equal contribution to the association. In the nonpsychotic retarded children, nonverbal age (tested on fitting and assembly tasks) appears to be more important, but the numbers of children with a low social score, or with a discrepancy between language comprehension and nonverbal age are too small to draw any firm conclusions.

Table 1. Association Among Sociability Score, Language Comprehension Age,
and Nonverbal Age

Categories of children based on cognitive skills		Psychotic group sociability score			Nonpsychotic group sociability score		
Language comprehension age	Mental age on nonverbal tests	Low	High	Total	Low	High	Total
1. 0–19 months	0–19 months	27	4	31	3	9	12
2. 0–19 months	20+ months	11	11	22	—	5	5
3. 20+ months	20+ months	4	27	31	—	57	57
	Total	42	42	84	3	71	74
Significance of association		$X^2 = 34.13, df = 2, p < .001$	—		—	—	—

Note: The figures in the body of this and subsequent tables refer to absolute numbers (not percentages).

Table 2. Association Among Problems in Obtaining Eye Contact, Language
Comprehension Age, and Nonverbal Age

Categories of children based on cognitive skills		Psychotic group eye contact			Nonpsychotic group eye contact		
Language comprehension age	Mental age on nonverbal tests	Problem	No problem	Total	Problem	No problem	Total
1. 0–19 months	0–19 months	18[a]	4	22	4[b]	7	11
2. 0–19 months	20+ months	16[c]	3	19	2	3	5
3. 20+ months	20+ months	10	21	31	9[d]	47	56
	Total	44	28	72	15	57	72
Significance of association		$X^2 = 19.09, df = 2, p < .001$			$X^2 = 3.50, df = 2, NS$		

[a] Nine with severe visual impairments excluded.
[b] One with severe visual impairment excluded.
[c] Three with severe visual impairments excluded.
[d] One with severe visual impairment excluded.

Poor eye contact is associated with poor cognitive skills. Table 2 shows that, in the psychotic children, the association is very significant and that language comprehension age is the more important factor. In the nonpsychotic children the association with mental age is not significant. The children in this group with poor eye contact with the teacher and language comprehension age above 20 months are very shy and anxious, or else are stubborn and rebellious with adults. They have normal eye contact with their parents and their peers.

Simple stereotypies occur most often in the more retarded children. Table 3 shows that the association is very significant in both the psychotic and nonpsychotic children. In the former, language comprehension age is the more important factor.

Elaborate routines, however, occur in the psychotic children only and, as would be expected from the skill needed to carry them out, are almost confined to those with nonverbal ages above 19 months. The association with higher nonverbal mental age is very significant (Table 4).

Problems affecting symbolic play are marked in the psychotic children. Only one, a girl of 15 with partial deafness and developmental receptive speech disorder, has true symbolic play. Table 5 shows that stereotyped symbolic play in psychotic children is very significantly associated with higher mental age, especially language comprehension. In

Table 3. Association Among Simple Stereotypies, Language Comprehension Age, and Nonverbal Age

Categories of children based on cognitive skills		Psychotic group simple stereotypies			Nonpsychotic group simple stereotypies		
Language comprehension age	Mental age on nonverbal tests	Marked	Minor or absent	Total	Marked	Minor or absent	Total
1. 0–19 months	0–19 months	25	6	31	6	6	12
2. 0–19 months	20+ months	19	3	22	0	5	5
3. 20+ months	20+ months	4	27	31	3	54	57
	Total	48	36	84	9	65	74
Significance of association		$X^2 = 39.43, df = 2, p < .001$			$X^2 = 19.31, df = 2, p < .001$		

Table 4. Associations Among Elaborate Routines, Language Comprehension Age, and Nonverbal Age

Categories of children based on cognitive skills		Psychotic group elaborate routines			Nonpsychotic group elaborate routines		
Language comprehension age	Mental Age on nonverbal tests	Present	Absent	Total	Present	Absent	Total
1. 0–19 months	0–19 months	1	30	31	0	12	12
2. 0–19 months	20+ months	12	10	22	0	5	5
3. 20+ months	20+ months	11	20	31	0	57	57
	Total	24	60	84	0	74	74
Significance of association		$X^2 = 17.75, df = 2, p < .001$			—	—	—

the nonpsychotic children, true symbolic play occurs in all those with language comprehension age of 20 months or above.

The types of organic conditions associated with the different groups are shown in the footnote to Table 6. The conditions classified under the heading "Type A" are all those found, in this study, to be particularly associated with psychotic behavior. For many of these conditions this association is recorded elsewhere in the literature and references are given in the table.

Table 5. Association Among Symbolic Play, Language Comprehension Age, and Nonverbal Age

Categories of children based on cognitive skills		Psychotic group play				Nonpsychotic group play			
Language comprehension age	Mental age on nonverbal tests	No symbolic play	Stereo. symbolic play	True symbolic play	Total	No symbolic play	Stereo. symbolic play	True symbolic play	Total
1. 0–19 months	0–19 months	28	3	0	31	12	0	0	12
2. 0–19 months	20+ months	17	5	0	22	5	0	0	5
3. 20+ months	20+ months	3	27	1	31	0	0	57	57
	Total	48	35	1	84	17	0	57	74
Significance of association		$X^2 = 46.09, df = 2, p < .001$				$X^2 = 73.70, df = 2, p < .001$			

Table 6. Association Between Sociability Score and Organic Condition

	Psychotic group organic conditions			Nonpsychotic group organic conditions		
		Type B[b]			Type B[b]	
Sociability score	Type A[a]	+ not known	Total	Type A[a]	+ not known	Total
Low	25	17	42	2	1	3
High	6	36	42	12	59	71
Total	31	53	84	14	60	74
Significance of association	$X^2 = 16.56, df = 1, p < .001$			—	—	—

[a]Type A = Maternal rubella (Chess, 1971), phenylketonuria (Jervis, 1963), tuberose sclerosis (Critchley & Earl, 1932; Earl 1943), kernicterus, retrolental fibroplasia (Keeler, 1958), or a history of infantile spasms (Taft & Cohen, 1971), encephalitis, meningitis, severe perinatal problems without cerebral palsy. (See also Creak & Pampiglione, 1969; Kolvin *et al.,* 1971.)

[b]Type B = Down's syndrome, cerebral palsy not associated with a Type A condition, chromosomal and metabolic abnormalities not in Type A, multiple congenital abnormalities, and others not in Type A.

Table 6 shows that the association between "Type A" organic conditions and low sociability score is highly significant in the psychotic children, though not in the nonpsychotic group.

The associations described above can be seen within the Kanner's syndrome children as well as the psychotic group as a whole, though the numbers are too small for statistical analysis.

DISCUSSION

The results show major differences between the "psychotic" and "nonpsychotic" children, both on formal testing and on the information obtained at interview.

Studies of this kind are an essential preliminary to more precise classification and to structured and controlled laboratory investigations (Jones, 1972). The grouping together of all the children from one geographical area who have behavior that can be called "psychotic" is not intended to imply that there is only one syndrome of childhood psychosis. It is likely that many different conditions have been put together in this study. Kanner's syndrome, for example, may cover one or perhaps two specific disorders. Nevertheless, a consideration of the distribution of certain types of behavior throughout the group as a whole underlines the fact that working with different samples of children is likely to influence the construction of hypotheses. For example, social avoidance is most obvious in the more severely handicapped groups, whereas the language abnormalities and poor development of symbolic function stand out in the more able children.

It is clear that, among the psychotic children, low sociability, poor eye contact, and simple stereotypies are associated with a low level of cognitive development, especially language comprehension. These findings, on their own, do not resolve the problem of which came first—the social indifference or the impairment of language.

The idea that a low language comprehension level causes social indifference is too simple a hypothesis. Seventeen children in the nonpsychotic group also have language comprehension below 20 months. As observed in the study, although some of them have simple stereotypies, and all lack symbolic play, they differ from the psychotic children with the same language comprehension age in that they respond to social approaches with appropriate smiling and, if not too physically handicapped, they also use gestures such as pointing, facial expression, and vocalizations on a simple level to convey their requests and point out interesting things to others. They have some nonverbal communication even if their language comprehension age on formal tests is low, whereas the withdrawn psychotic child's language problem appears to affect all aspects of communication, nonverbal as well as verbal.

Some clue as to the nature of the problem in the psychotic children can be obtained from the behavior of those who are more sociable. The differences between the less able and the more able children observed in the cross section in this study are also seen in the form of changes in those Kanner's syndrome children who improve with increasing chronological age.

These children become less withdrawn, but they do not become normal. They are friendly, but naive and totally at sea with the rules of social conduct (Dewey & Everard, 1974; Ricks & Wing, 1975; Rutter, 1970). They do not avoid eye contact, but, if impressions gained in the present study are correct, they gaze longer than normal children at adults during social interaction. When given a new task by a teacher they will fix their eyes on her face and not on the work in hand, as if seeking for information that eludes them. The simple stereotypies of the less able child have their counterparts in the elaborate repetitive routines and the stereotyped speech of the more able children. When spoken language develops, it is characterized by very special abnormalities, suggesting that it is learned by rote without understanding of the rich and subtle associations of words (Eisenberg, 1956; Kanner, 1973; Rutter, 1970; Ricks & Wing, 1975). If symbolic play occurs at all, it is stereotyped, repetitive, and does not develop in range and complexity as symbolic play does in the normal child, or the nonpsychotic retarded child with language age above 19 months.

A nonpsychotic retarded child with symbolic play, though very limited compared to a normal child even of the same language comprehension age, appears lively and imaginative when contrasted with an autistic child.

These observations suggest that, although sociability does improve in some children, the cognitive aspect of social communication on both a verbal and a nonverbal level remains impaired. If the language problem is due to social avoidance, why does not development proceed normally once the child begins to accept or even initiate social approaches?

Perhaps the psychotic child's language and cognitive skills develop abnormally because his early social avoidance causes him to start learning much later than the normal child. But there are some accounts of children deprived in their early years of the normal opportunity to learn language by close contact with adults, and these children, though they may sometimes be retarded, do not behave like psychotic children (Ainsworth, 1962; Clarke, 1972; Koluchova, 1972; Rutter, 1972).

The idea that the language problems are caused by social avoidance appears as oversimplified as its converse. One hypothesis that does seem to fit the facts is that the development of symbolic language, together with a mental store of ideas and associations, *and* the development of social interaction are both dependent upon some other underlying factor. This factor may be the ability, possibly inbuilt in the normal human brain, to seek out and actively endeavor to make sense of experience. Normal babies show an eagerness to explore the world from a very early age and so do nonpsychotic retarded children, though later and less intensely. This idea is discussed in more detail by Ricks and Wing (1975).

The classically autistic child lacks this ability, but he can do fitting and assembly tasks and has a good rote memory. Even if a child of this kind makes good progress, he never really manages to make sense of the world, but tries to cope by rigidly following rules the reasons for which he does not comprehend. He cannot cope with novelty, so obtains his enjoyment from constant repetition of the same activities and experiences, and his sense of security from the maintenance of the familiar environment and routine, in obsessive detail. The mute, withdrawn, severely retarded child may have this problem as one of a wide range of cognitive handicaps, sometimes including impairments of hearing and vision. The psychotic child with stereotyped speech and play may have an impairment or distortion of the ability to make sense of the world, but not a total absence.

The social avoidance and poor eye contact of the most withdrawn psychotic child can be seen as arising from the absence of any mechanism for understanding the environment. A child of this kind can do very little except to engage in simple stereotypies. Trying to involve him in situations he cannot comprehend may well produce distress and consequent overarousal, followed by attempts at avoidance. Withdrawal is much less in evidence when a task is within the child's capabilities (DeMyer, 1976).

If this hypothesis holds up in other studies, this still leaves unanswered the question as to the cause of the central problem.

It has been suggested that the parents of autistic children are often intellectuals, emotionally detached from their children (Eisenberg, 1957; Kanner, 1954), but the evidence for this detachment is unconvincing (Cox *et al.*, 1975). Analyses of the data from the present study that are in progress concerning the children's families indicate that the mothers of the psychotic children are just as likely as the mothers of the nonpsychotic children to wish to keep their handicapped child living at home with them, or to keep in close and regular contact if residential care is necessary.

The other relevant finding is that, among the psychotic children, a low sociability score is very significantly associated with certain types of organic conditions, for instance a history of encephalitis, meningitis, maternal rubella, infantile spasms, or tuberose sclerosis—but *not* with certain other conditions, notably Down's syndrome. The data from the present study therefore tend to support the view that the problem is organic rather than psychological in origin.

In summary, the hypothesis suggested here is that the social aloofness, the repetitive stereotyped behavior and the abnormalities affecting comprehension and use of all forms of communication and the development of symbolic thought are facets of the same underlying impairment of cognitive development. Arguments as to which of these aspects of the autistic syndrome are primary can therefore be regarded as purely semantic.

The very close relationship among these three abnormalities is not self-evident, but the epidemiological study I have described provides evidence of this association and also suggests that an organic lesion is the basic etiology. It is a necessary first step in clearing the ground for future studies. In order to pursue these questions further, improved techniques for classifying subgroups of psychotic children, and for describing and measuring eye contact, social interaction, symbolic play, and development of inner language are needed, and are currently being studied by Judith Gould. Investigation of the precise pathological effects of the organic conditions that tend to be associated with psychotic behavior in children might provide some clues. The most interesting and most difficult problem of all is to identify the neurobiological basis of social and symbolic development in the normal infant, and studying "psychotic" children may help to elucidate the origins of these most fundamental aspects of human behavior.

REFERENCES

Ainsworth, M. D. The effects of maternal deprivation: A review of findings and controversy in the context of research strategy. In *Deprivation of maternal care: A reassessment of its effects*. Geneva: World Health Organization, 1962.

Bartak, L., Rutter, M., & Cox, A. A comparative study of infantile autism and specific developmental receptive language disorder. I. The children. *British Journal of Psychiatry*, 1975, *126*, 127—145.

Bowlby, J. *Maternal care and mental health*. Geneva: World Health Organization, 1951.

Chess, S. Autism in children with congenital rubella. *Journal of Autism and Childhood Schizophrenia*, 1971, *1*, 33–47.

Churchill, D. W. The relationship of infantile autism and early childhood schizophrenia to developmental language disorders of childhood. *Journal of Autism and Childhood Schizophrenia*, 1972, *2*, 182–197.

Clarke, A. D. B. Commentary on Koluchova's "Severe deprivation in twins: A case study." *Journal of Child Psychology and Psychiatry*, 1972, *13*, 115–120.

Corbett, J. A., Harris, R., & Robinson, R. G. Epilepsy. In J. Wortis (Ed.), *Mental retardation and developmental disabilities. Vol. VII*. New York: Brunner Mazel, 1975.

Cox, A., Rutter, M., Newman, S., & Bartak, L. A comparative study of infantile autism and specific developmental receptive language disorder. II. Parental characteristics. *British Journal of Psychiatry*, 1975, *126*, 146–159.

Creak, M., & Pampiglione, G. Clinical and E.E.G. studies on a group of 35 psychotic children. *Developmental Medicine and Child Neurology*, 1969, *11*, 218–227.

Critchley, M., & Earl, C. J. C. Tuberose sclerosis and allied conditions. *Brain*, 1932, *55*, 311–346.

DeMyer, M. K. Motor, perceptual-motor and intellectual disabilities of autistic children. In L. Wing (Ed.), *Early childhood autism: Second edition*. Oxford: Pergamon, 1976.

DeMyer, M. K., Barton, S., DeMyer, W. E., Norton, J. A., Allen, J., & Steele, R. Prognosis in autism: A follow-up study. *Journal of Autism and Childhood Schizophrenia*, 1973, *3*, 199–246.

Dewey, M. A., & Everard, M. P. The near-normal autistic adolescent. *Journal of Autism and Childhood Schizophrenia*, 1974, *4*, 348–356.

Earl, C. J. C. The primitive catatonic psychosis of idiocy. *British Journal of Medical Psychology*, 1943, *14*, 230–253.

Eisenberg, L. The autistic child in adolescence. *American Journal of Psychiatry*, 1956, *112*, 607–612.

Eisenberg, L. The fathers of autistic children. *American Journal of Orthopsychiatry*, 1957, *27*, 715–724.

Ferster, C. B. Positive reinforcement and behavioural deficits of autistic children. *Child Development*, 1961, *32*, 437–456.

Gould, J. Language impairments in severely retarded children: An epidemiological study. *Journal of Mental Deficiency Research*, 1976, *20*, 129–145.

Hermelin, B., & O'Connor, N. *Psychological experiments with autistic children*. Oxford: Pergamon, 1970.

Hutt, C., Forrest, S., & Richer, J. M. Heart rate and stereotypies. *Acta Psychiatrica Scandinavica*, 1975, *51*, 361–372.

Hutt, S. J., & Hutt, C. Social behaviour. In *Direct observation and measurement of behaviour*. Springfield: Charles C Thomas, 1970.

Hutt, S. J., Hutt, C., Lee, D., & Ounsted, C. A behavioural and electroencephalographic study of autistic children. *Journal of Psychiatric Research*, 1965, *3*, 181–197.

Hutt, C., & Ounsted, C. The biological significance of gaze aversion with particular reference to the syndrome of infantile autism. *Behavioural Science*, 1966, *11*, 346–356.

Jervis, G. A. The clinical picture. In F. L. Lyman (Ed.), *Phenylketonuria*. Springfield: Charles C Thomas, 1963.

Jones, N. B. Characteristics of ethological studies of human behaviour. In N. Blurton Jones (Ed.), *Ethological studies of child behaviour*. London: Cambridge University Press, 1972.

Kanner, L. Autistic disturbances of affective contact. *Nervous Child*, 1943, *2*, 217–250.

Kanner, L. Irrelevant and metaphorical language in early childhood autism. *American Journal of Psychiatry*, 1946, *103*, 242–246.

Kanner, L. The conception of wholes and parts in early infantile autism. *American Journal of Psychiatry*, 1951, *108*, 23–26.

Kanner, L. To what extent is early childhood autism determined by constitutional inadequacies? *Proceedings of the Association for Research in Nervous and Mental Diseases*, 1954, *33*, 378–385.

Kanner, L. *Childhood psychosis: Initial studies and new insights*. Washington: Winston, 1973.

Kanner, L., & Eisenberg, L. Notes on the follow-up studies of autistic children. In P. H. Hoch & J. Zubin (Eds.), *Psychopathology of childhood*. New York: Grune & Stratton, 1955.

Kanner, L., & Eisenberg, L. Early infantile autism, 1943–1955. *American Journal of Orthopsychiatry,* 1956, *26,* 55–65.

Keeler, W. R. Autistic patterns and defective communication in blind children with retrolental fibroplasia. In P. H. Hoch & J. Zubin (Eds.), *Psychopathology of communication.* New York: Grune & Stratton, 1958.

Koluchova, J. Severe deprivation in twins: A case study. *Journal of Child Psychology and Psychiatry,* 1972, *13,* 107–114.

Kolvin, I., Ounsted, C., & Roth, M. Cerebral dysfunction and childhood psychosis. *British Journal of Psychiatry,* 1971, *118,* 407–414.

Lotter, V. Epidemiology of autistic conditions in young children—I. Prevalence. *Social Psychiatry,* 1966, *1,* 124–137.

Lowe, M. Trends in the development of representational play infants from one to three years: An observational study. *Journal of Child Psychology and Psychiatry,* 1975, *16,* 33–48.

McCulloch, M. J., & Williams, L. On the nature of infantile autism. *Acta Psyciatrica Scandinavica,* 1971, *47,* 295–314.

Ornitz, E. M. Childhood autism: A review of the clinical and experimental literature. *California Medicine,* 1973, *118,* 21–47.

Reed, G. F. Elective mutism in children: A re-appraisal. *Journal of Child Psychology and Psychiatry,* 1963, *4,* 99–107.

Reynell, J. *Reynell Developmental Language Scales.* Slough, Bucks: N.F.E.R., 1969.

Richer, J. M. *The social and stereotyped behaviour of autistic children.* Ph.D. Thesis, Reading, 1974.

Ricks, D. M. Verbal communication in pre-verbal normal and autistic children. In N. O'Connor (Ed.), *Language, cognitive deficits and retardation.* London: Butterworths, 1975.

Ricks, D. M., & Wing, L. Language, communication and the use of symbols in normal and autistic children. *Journal of Autism and Childhood Schizophrenia,* 1975, *5,* 191–221.

Rutter, M. Autistic children: Infancy to adulthood. *Seminars in Psychiatry,* 1970, *2,* 435–450.

Rutter, M. *Maternal deprivation reassessed.* Harmondsworth: Penguin, 1972.

Sheridan, M. D. Playthings in the development of language. *Health Trends,* 1969, *1,* 7–10.

Stutsman, R. *Merrill-Palmer Scale of Mental Tests.* New York: Harcourt, Brace & World, 1931.

Taft, L. T., & Cohen, H. J. Hypsarrhythmia and childhood autism: A clinical report. *Journal of Autism and Childhood Schizophrenia,* 1971, *1,* 327–336.

Tinbergen, E. A., & Tinbergen, N. Early childhood autism: An ethological approach. *Beihefte zur Zeitschrift fur Tierpsychologie,* No. 10, 1972.

Tizard, B., & Rees, J. The effect of early institutional rearing on the behaviour problems and affectional relationships of four-year-old children. *Journal of Child Psychology and Psychiatry,* 1975, *16,* 61–73.

Wing, J. K., & Hailey, A. (Eds.), *Evaluating a community psychiatric service: The Camberwell Register 1964–71.* London: Oxford University Press, 1972.

Wing, L. The handicaps of autistic children: A comparative study. *Journal of Child Psychology and Psychiatry,* 1969, *10,* 1–40.

Wing, L. Perceptual and language development in autistic children: A comparative study. In M. Rutter (Ed.), *Infantile autism: Concepts, characteristics and treatment.* London: Churchill-Livingstone, 1971.

Wing, L. A study of language impairments in severely retarded children. In N. O'Connor (Ed.), *Language, cognitive deficits and retardation.* London: Butterworth, 1975.

Wing, L. Epidemiology and theories of etiology. In L. Wing (Ed.), *Early childhood autism.* Second edition. Oxford: Pergamon, 1976.

Wing, L., & Gould, J. The handicaps, behaviour and skills (HBS) structured interview schedule for use with mentally retarded and psychotic children. I. Level of agreement between parents and professional informants. *Journal of Autism and Childhood Schizophrenia,* in press.

Wing, L., Gould, J., Yeates, S. R., & Brierley, L. M. Symbolic play in mentally retarded and in autistic children. *Journal of Child Psychology and Psychiatry,* 1977, *18,* 167–178.

Wing, L., Yeates, S. R., Brierley, L. M., & Gould, J. The prevalence of early childhood autism: A comparison of administrative and epidemiological studies. *Psychological Medicine,* 1976, *6,* 89–100.

Wolkind, S. N. The components of "affectionless psychopathy" in institutionalized children. *Journal of Child Psychology and Psychiatry,* 1974, *15,* 215–220.

3

The Partial Noncommunication of Culture to Autistic Children—An Application of Human Ethology

JOHN RICHER

Autistic children's incompetence in the use of language and symbols has attracted much attention in recent years. Fifty percent of autistic children are mute, and others show characteristic peculiarities of language such as reversal of the pronouns "I" and "you," echolalia, repetitive and out of context speech, few questions or spontaneous utterances, and the idiosyncratic use of words. As Kanner (1943) remarked, they rarely use language to communicate. Rutter (1974) has found that nearly all autistic children show language peculiarities and retardation and in fact he suggests this as a diagnostic criterion.

Rutter (1974), among others, gives these language-related problems central importance in autism, classifying them as a "cognitive language handicap." Ricks and Wing (1975), too, describe the central problem of autism as an impairment of complex symbolic functioning affecting all forms of communication. Not all human communication is symbolic, and in babies very little is. Even so, these authors are pointing to vital aspects of autistic children's behavior which must be explained, not least because the effect of these incompetencies on an autistic child's development is so disastrous. Both Kanner (1971) and Rutter (1974) found that of the autistic children who became self-supporting in adulthood all spoke before they were five years old.

JOHN RICHER · Smith Hospital, Henley on Thames, Oxon, England. Present address: Psychology Department, Park and Warneford Hospitals, Oxford, England.

It should be made clear that the phrase "cognitive-language handicap" refers to a classification of behavior in which language skills are impaired. It leaves open the question of how the language handicap arose.

Language is a social communication skill and for this reason social deficits could well impede the development of language and communication competence (of course the reverse also applies). Accordingly, it might be fruitful to study autistic children's patterns of social interaction in order to see what implications they might have for both socialization and communication. That is the purpose of this chapter. It is not a study of how autism is first caused but rather a study of social processes in children already autistic for whatever reason. Nevertheless, an understanding of these processes may have implications for the development of autism and also for its treatment.

In normal children language develops slowly in the first years of life. A child gradually becomes competent in the use of language and the other communication skills of his culture through observation of, and interaction with, other people. As Newson & Newson (1975) concluded, "the origins of symbolic functioning should be sought . . . in those idiosyncratic shared understandings which he [the child] first evolves during his earliest social encounters with familiar human beings who are themselves steeped in human culture." So it would seem promising to move from accounts of the causes or cause of autism to detailed analyses of autistic children's ontogeny, tracing some of the paths they took to become incompetent in the use of language and other communication skills. This is often forgotten when "primary" handicaps are sought.

Let me widen the problem by suggesting that an autistic child's retardation is most apparent in his entry to his *culture*. Of course, there are many conceptions of culture. We might follow Geertz (1965) and say that culture is a system of shared symbols or meanings, or we might follow Harris (1968) and say that it is a system of socially transmitted behavior patterns that serve to relate human communities to their ecological setting. However, it is pretty clear that if a child does not learn the meanings of his culture's symbols, including language, then he is not a fully paid-up member of that culture, and if he further fails to acquire many of the skills for communication and cooperation he is further behind with his subscription. Autistic children, perhaps like the "feral" children with whom they have often been compared, are, I am suggesting, such "dyscultural" children.

How is it then that autistic children especially fail to acquire those skills which would make them full members of their culture? Many factors are candidates for consideration, but I shall concentrate on describing how autistic children's social behavior, in particular their predominant avoidance of social interactions, works against their learning these skills.

Of course, some autistic children do learn to speak and communicate, and to cooperate in other ways, and a few become highly proficient in the use of some symbol systems such as numbers (Wing, 1971) which, incidentally, are less used in face-to-face interactions. My concern is with some of the mechanisms that work against normal developmental processes to a greater or lesser degree to produce autistic children with varying degrees of "culture incompetence." I stress that this is not a search for the "causes of autism." It is an account, at times a little speculative and lacking direct quantitative support, of some effects of avoidance on the children's unfolding ontogeny, *once they are autistic,* and how they are progressively handicapped in acquiring the skills of their culture. In concentrating on some of the effects of social avoidance behavior, I am not implying that these are the only factors retarding language development. As with any behavior pattern we must look for many causes and many effects, and we must attempt to plot its complex and unfolding ontogeny.

AUTISTIC CHILDREN'S SOCIAL BEHAVIOR

Autistic children's social behavior was partly studied by prolonged direct observation of autistic children in their every-day environments— particularly hospitals and schools. Such observation is a necessary first step in any investigation of behavior, not least to discover and objectively describe the phenomena to be explained (still a problem in autism).

Human ethologists and others have considered that the most useful description of autistic children's social behavior is that they predominantly avoid social interactions. They do this, for instance, by moving away, turning away, gaze-averting, hanging the head down, being on the edge of areas, relinquishing objects, and so on. This was given quantitative support in a comparison of their behavior with that of age- and sex-matched 7–11-year -old disturbed nonautistic children in a school playground (Richer, 1976). This predominant avoidance is *not* a consequence of autistic children being aggressed against more often than other children (Hutt & Vaizey, 1966; Richer, 1976). Instead it suggests that autistic children, compared to nonautistic children, have a *lowered threshold* for avoidance. Unlike other children, autistic children frequently show social avoidance after nonthreatening approaches by other people (Fig. 1, see also Figs. 3 and 8 below).

On the comparatively rare occasions when autistic children make social approaches there is usually great ambivalence. Approach may swiftly alternate with moving away. The child may approach but with his head hung down or he may approach backwards or from behind. He may stand side-on to the adult, often performing a stereotypy. He may sit on

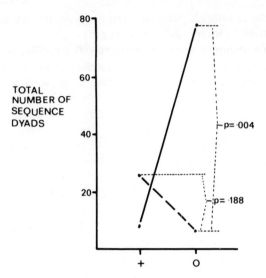

Fig. 1. Behavior of other people which preceded avoidance behavior in autistic and nonautistic groups (data from Richer, 1976). Avoidance behaviors = move away, defensive postures, and immediately relinquish object. + = "violent" and "threatening" approaches by other people preceding avoidance. 0 = "nonthreatening" approaches or just coming near preceding avoidance. (The difference between + and 0 is operationally defined by the data from the *non*autistic children, viz., + precedes avoidance more often than 0.) This shows that avoidance was preceded by "nonthreatening" approaches or just coming near more often in autistic group; · ——— · autistic children, · — — · nonautistic children.

an adult's lap, hold tight, but do little else. This is often called "empty clinging," a misleading description. Much of autistic children's social behavior may be viewed as the net result of conflicting approach and avoidance motivations, where avoidance motivation is much stronger than it is in normal children.

IMMEDIATE CAUSAL FACTORS

Next I want to consider some *immediate causal factors* which influence whether or not an autistic child will approach or avoid. Two are particularly important.

One is the difficulty of, or uncertainty in, the mutual activity. The more uncertain the activity, the more likely are autistic children to avoid. As Churchill (1971) found, failure at a task leads to avoidance, while success is not so often followed by avoidance (see also Richer, 1977).

The second factor is the behavior of other people. Not surprisingly an autistic child is more likely to avoid someone who threatens him compared with someone who does not. But in addition, an autistic child is more likely to avoid someone who stares at him than someone who does not (Richer & Coss, 1976); and he is more likely to avoid if someone smiles, or, more generally, "reacts," after eye contact than if there is no smile or "reaction" (Richer & Richards, 1975).

The child's responsiveness to these two factors fluctuates. In other words, his threshold for avoidance changes. I shall not discuss the factors associated with this save to note that the threshold is lowered if the child has just been avoiding.

IMMEDIATE EFFECTS

A study of the *immediate effects* of avoidance shows that in some situations the child's avoidance is often followed by an adult's approach. Matched pairs of 4–9-year-old autistic and nonautistic disturbed and retarded children were observed with their own class teachers (Richer, 1977). Although IQ tests had not been performed on the autistic children, it is likely that most were retarded to some degree. With both groups, the teacher's approaches sometimes followed avoidance by the child, but this

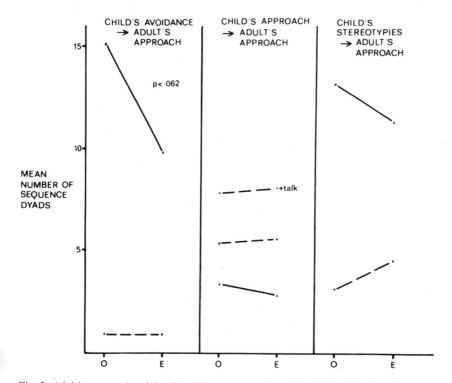

Fig. 2. Adult's approaches following different categories of the child's behavior. Comparison of observed with expected mean number of sequence dyads. Differences were not significant in all but one case, viz., avoidance in the autistic group is followed more often than expected by approaches by the adults; · ——— · autistic children, · — — · nonautistic children, 0 = observed score, E = expected score.

tendency was stronger with the autistic children (Fig. 2). Although the autistic children avoided much more frequently and approached less than the nonautistic children (Table 1) teachers approached them about as frequently and in fact tried to look into their faces more, than with the nonautistic children (Table 1). The autistic children tended to react to these approaches with further avoidance, whereas the nonautistic children tended to react with approaches (Fig. 3).

Table 1. *Frequencies of Social Approach and Avoidance in Autistic and Nonautistic Groups and Frequencies of Adults' Behaviors*

	Mean % time sample scores		
Behavior element	Autistic group	Nonautistic group	P = (sign test)
CHILDREN			
Social approach behavior			
Orient to teacher	6	23	.062
Talk	0	41	.031
Point	0	7	.031
Hold for help	9	2	ns
Hold	1	1	ns
Touch	1	.3	ns
Embrace	13	3	ns
Means of summed approach behavior	4	11	.008
Social avoidance behavior			
Orient away	13	2	.062
Look up	13	2	ns
Hang head	12	1	ns
Leave seat	.5	4	ns
Means of summed avoidance behavior	11	2	.008
ADULTS			
Call attention to a specific aspect of task	30	35	ns
Call attention to whole task	37	19	ns
Congratulate	8	19	.062
Tell the child he is wrong	3	7	ns
Touch	6	1	.031
Hold	6	1	ns
Embrace	4	1	ns
Laugh	1	6	ns
Guide hand	4	0	ns
Look	30	5	.062
Means of summed behavior scores	13	9	ns

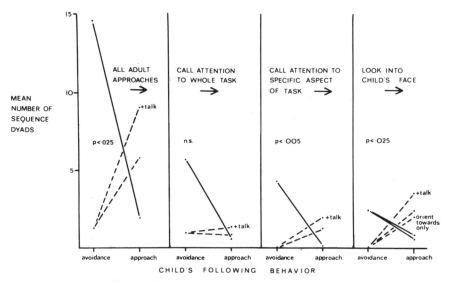

Fig. 3. Child's behavior following the adult's approaches. Comparison of the two groups' avoidance and approach (*p* values give the significance of the "groups' following behavior" interaction term in an analysis of variance); · ——— · autistic children, · — — · nonautistic children.

We may deduce from this that the teacher's approaches are aversive to the autistic children—they avoided.* Thus, up to the point when they eventually succeed in escaping, autistic children are continually frustrated in this social encounter. This is likely to reinforce their tendency to avoid such encounters.

Yet the child would not have been frustrated had he not been avoidance-motivated. It seems that an autistic child's strong tendency to avoid social encounters such as these *reinforces and maintains itself* because the adult makes approaches when the child is avoiding. This may be called a *reflexive mechanism* because the avoidance reinforces itself—albeit via the reactions of adults. It is a positive feedback reaction.

So whether the child is autistic or not, avoidance sometimes provokes approaches from adults, and furthermore the persistent avoidance of autistic children is followed slightly more often by adult approaches than the avoidance of nonautistic children (Fig. 2). Perhaps this is because the nonautistic children often themselves make approaches which the adults can wait for, whereas the autistic children make very few approaches, so the adult takes the initiative.

Of course, it should be emphasized that there is nothing unusual about the adult's reactions involved in this reflexive mechanism. The

---

*However, because the children's responses were only studied after approach by the teacher, it is not possible to know whether approach was more or less aversive than other behavior by the teacher.

teachers whose reactions are part of the hypothesized mechanism, helping to maintain autistic children's avoidance, are the same teachers who participate in successful social interactions with nonautistic children.

ANALOGOUS REACTIONS IN NORMAL
INFANT-MOTHER PAIRS

Analogous reactions have been found by Kenneth Kaye (quoted in Bruner, 1975) observing mothers with their normal children under the age of one year. He asked each mother to teach her child to take an object from behind a transparent screen. Mothers actively drew their child's attention to the object, but, more importantly for my argument, these attention-attracting maneuvers greatly increased above chance level when the child looked *away*.

Neither is the child's reaction pattern confined to autistic children. Although the avoidance seems less intense, the sequence of child's nonattention or avoidance leading to mother's approach leading to increased nonattention or avoidance by the child has been observed during interactions between normal mothers and their normal infants. Brazelton *et al.* (1974) and Stern (1971, 1974), for example, have reported a cycle of infant social attention and nonattention. The infant alternates between being sociable and attending to his mother, then not attending, then being sociable again, and so on. Some mothers cue into the baby's behavior and make approaches only when the baby is in the sociable phase. But other mothers also make a significant number of approaches when the baby is in the nonattentive phase. The result of this is usually an increase in the baby's nonattention and even clear avoidance of the mother's approaches. Here then is the mechanism operating in the interactions between some mothers and their normal children analogous to the reflexive mechanism helping to maintain avoidance in autistic children, although, of course, the avoidance is less intense and persistent in the normal children. Just how intense this avoidance must be to trigger the reflexive mechanism, and under what conditions, is not yet known, but at least it is a straightforward empirical question.

EARLY SOCIAL BEHAVIOR OF AUTISTIC CHILDREN

If we accept the argument so far, we can ask how long this mechanism has been in operation in autistic children? There is considerable evidence that autistic children were strongly avoidance-motivated from an early age. Mothers report that their child was uncuddly, did not adopt the anticipatory posture before being picked up, rarely smiled or cried (a distress call), and seemed happiest when left alone (Kanner, 1943; Clancy & McBride, 1969; Ruttenberg, 1971). My own observations of an 18-month-old autistic boy support this. The child intensely avoided his mother. Also, his mother made repeated approaches to him when he was avoiding, and this further intensified his avoidance. However, on the rare occasions when he made social approaches, her behavior changed, and then tended not to provoke avoidance. In addition, with the boy's twin sister who was normal, her behavior did not provoke avoidance. It seems that the child's avoidance elicited behavior from his mother which only reinforced his avoidance.

EFFECTS OF SOCIAL AVOIDANCE ON "CULTURAL COMPETENCE"

What are the implications of this for a child's acquisition of language and the other skills for communication and cooperation of his culture?

Firstly, it should be noted that autistic children are not entirely without social skills: Their avoidance and curtailment of social interactions is sometimes subtle and sophisticated.* But it is avoidance.

To try and explain why autistic children rarely communicate we might simply argue that they are inhibited from doing so by excessive avoidance motivation. Certainly, supposedly mute autistic children have been heard to say words when their other behavior is sociable too, and the speech of speaking autistic children becomes less characteristically autistic when they are less avoiding and more sociable. But although autistic children sometimes display new skills which surprise their caretakers, this inhibition is only a small part of the story, and we must consider those interactions in which communication takes place and in which communicative competence is acquired. This will, in passing, throw some light on why autistic children are called "noncommunicating."

These "communication interactions" are often characterized by turn-taking and by pausing and looking at the other person at the end of a turn (Kendon, 1967; Newson & Newson, 1975). One person (A) does something directed by the other person (B) and then pauses and looks at B, waiting for the reply. Then B replies, often in the full glare of A's gaze, and himself then pauses and looks at A, and so on (Fig. 4).

This interaction needs to be examined more closely to suggest why autistic children should especially avoid it, such that, perhaps, they are

*It is not suggested that autistic children necessarily have the skills to engage in social interaction but fail to do so because of avoidance. Of course, they lack many social skills. Nevertheless, the two cannot be considered independent. What is suggested is that some aspects of autistic children's social behavior are likely to retard their acquisition of further social skills.

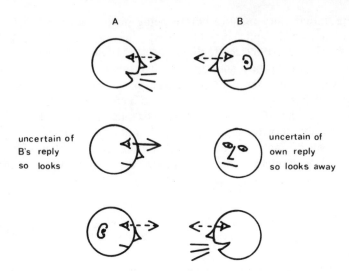

Fig. 4. Schematic representation of the turn-taking and gaze pattern in a communication interaction. (The communicative behavior need not be speech.)

called noncommunicating. The changeover point between A's and B's turns is crucial. For each participant it is one of maximum uncertainty relative to the rest of the interaction. A is uncertain of B's reply so he looks for it, and B is uncertain of what to reply.

Suppose now an autistic child were the first person (A) in the interaction (Fig. 5)—an uncommon event. After some sociable behavior he ought to look at B to get feedback and to signal "it's your turn now." However, this is not only a time of uncertainty, which is an immediate

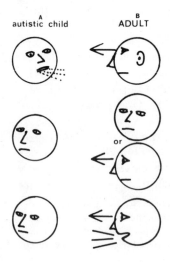

Fig. 5. Schematic representation of communication interaction with an autistic child as A.

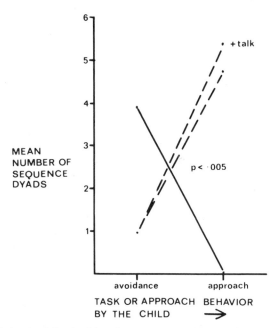

Fig. 6. Child's behavior following his or her own approach or task behavior. Comparison of the two groups' approach and avoidance (meaning of *p* value as in Fig. 3); · ——— · autistic children, · — — · nonautistic children.

causal factor promoting avoidance, but also the other person is being reactive—he is replying—which further promotes avoidance. Autistic children usually avoid at this point. This was found in a study of autistic and nonautistic children videotaped with their class teachers (Richer, 1977). Autistic children tended to avoid after behaving sociably or doing the required task, whereas nonautistic children tended to look at the teacher or make some other approach (Fig. 6).

Suppose now an autistic child is the second person (B) in the communication interaction and that he has been approached but now it is his turn to reply (Fig. 7)—a more usual state of affairs. At this changeover point there is not only uncertainty as before, but also the child is being looked at which further promotes avoidance.

In the same study of autistic and nonautistic children with their teachers, it was found that the teacher's approaches to a child and her looks at him tended to be followed by the autistic child turning away but by the nonautistic child looking at the teacher (Fig. 8).

The main point is that the turn-taking, uncertainty, looking, and reactiveness, characteristic of communication interactions make them a type of social encounter in which autistic children are very likely to avoid. It is argued that the avoidance of these interactions retards autistic children's

A
ADULT

B
autistic child

Fig. 7. Schematic representation of communication interaction with an autistic child as B.

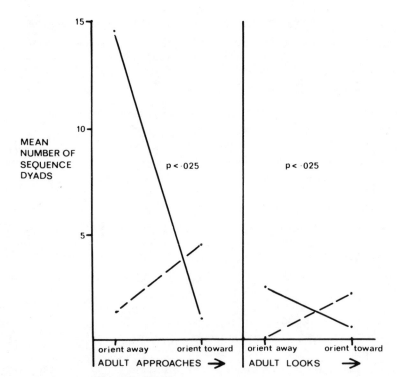

Fig. 8. Child's behavior following adult's approaches or looks only. Comparison of the two groups' approach and avoidance (meaning of *p* values as in Fig. 3); · ——— · autistic children, · — — · nonautistic children.

acquisition of language and other communication and cooperation skills. Of course, the child's relative communicative incompetence also makes social interactions difficult, so that avoidance is not diminished by interactions becoming easier—a vicious circle.

Bruner (1975) has described many processes by which a child proceeds from prespeech communication to language. Many of these processes will be impaired by an autistic child's predominant avoidance. By way of example only, the "reciprocal mode" of social behavior seen toward the end of the first year may be considered. In this reciprocal mode there is continual eye-checking, vocalization, and turn-taking by infant and mother; their *joint* tasks themselves become the objects of *joint* attention. It is partly through interactions such as these that normal children develop communicative competence, negotiate relationships with others, and eventually acquire the meanings, i.e., the uses, of words and symbols. Yet, as I have described, autistic children are highly avoiding if placed in such encounters, and they avoid at the point normal children would have been getting feedback, or at the point they should be replying and thereby "trying out" a social communicative behavior. In addition they rarely initiate these encounters. The negotiations of "shared understandings" (Newson & Newson, 1975), the shared meanings of words, for instance, rarely take place. Autistic children are particularly retarded in their acquisition of the *meanings* of words (e.g., Rutter *et al.*, 1971).

But we must add yet another nail to an autistic child's "communication coffin." For the reflexive mechanism maintaining avoidance seems, from direct observation, especially likely to operate in interactions such as these, not only because autistic children strongly avoid in them, but also because adults make many approaches and in particular they try to look into a child's face.

No claim is made for the primacy of avoidance behavior. Rather, attention has been paid to the effects of social avoidance on language, and cultural competence (together with a reverse effect, whereby language incompetence exacerbates avoidance). These ideas would predict that language incompetence and avoidance should be positively correlated. This is precisely what Wing (see Chapter 2) finds when she notes that language competence and sociability (the converse of avoidance) are correlated.

In summary, the hypothesis put forward is that autistic children's incompetence is greatest in the area of language, symbol use, and the other communication and cooperation skills of their culture just because a number of factors conspire together to produce their greatest effect here. In particular, not only do autistic children strongly avoid communication interactions so that acquisition of communication competence is retarded, but also if they are in such interactions their avoidance is very likely to be reinforced by the reflexive mechanism described. Catch 22.

A POSTSCRIPT FOR ANTHROPOLOGISTS

There was a time when anthropologists were fond of speculating on the nature of "man without culture," "the savage in a state of nature," but as Fox (1973) says, man is biologically a cultural animal and attempts to discover purely biological man are a waste of time. To the same end, Geertz (1965) wrote that, "undirected by cultural patterns—organized systems of significant symbols—a man's behavior would be virtually ungovernable, and mere chaos of pointless acts and exploding emotions, his experience virtually shapeless."

Although this is *not* accurate as an *objective* description of autistic children, Geertz unwittingly comes close to how autistic children are seen and described by many people, people who are themselves, of course, members of a culture. Being "dyscultural," autistic children throw into disarray our everyday ways of describing each other, since they communicate much less in the categories of our culture, and their behavior is less constrained by those categories. In consequence we find their behavior confusing, a confusion which some theorists project onto autistic children asserting that they (the children) cannot "make sense of sensory input." This is a theory which is as incoherent as it is untestable.

REFERENCES

Brazelton, T. B., Kaslowski, B., & Main, M. The origins of reciprocity: The early mother infant interaction. In M. Lewis & L. A. Rosenblum (Eds.), *The effect of the infant on its caregiver.* New York: Wiley Interscience, 1974.

Bruner, J. From communication to language—a psychological perspective. *Cognition,* 1975, *3*(3), 255–287.

Churchill, D. W. The effects of success and failure in psychotic children. *Archives of General Psychiatry,* 1971, *25,* 208–214.

Clancy, H., & McBride, G. The autistic process and its treatment. *Journal of Child Psychology and Psychiatry,* 1969, *10,* 233–244.

Fox, R. *Encounter with anthropology.* New York: Harcourt Brace Jovanovitch, 1973.

Geertz, C. The impact of the concept of culture on the concept of man. In J. R. Platt (Ed.), *New views on the nature of man.* Chicago: University of Chicago Press, 1965.

Harris, M. *The rise of anthropological theory—a history of theories of culture.* New York: Cowell, 1968.

Hutt, C., & Vaizey, M. J. Differential effects of group density on social behaviour. *Nature,* 1966, *209,* 1371–1372.

Kanner, L. Autistic disturbances of affective contact. *Nervous Child,* 1943, *2,* 217–250.

Kanner, L. Follow up study of eleven autistic children originally reported in 1943. *Journal of Autism and Childhood Schizophrenia,* 1971, *1,* 119–145.

Kendon, A. Some functions of gaze-direction in social interaction. *Acta Psychologica,* 1967, *26,* 22–63.

Newson, J., & Newson, E. Intersubjectivity and the transmission of culture: On the social origins of symbolic functioning. *Bulletin of the British Psychological Society,* 1975, *28,* 437–446.

Richer, J. M. The social avoidance behaviour of autistic children. *Animal Behaviour,* 1976, *24,* 898–906.

Richer, J. M. The self perpetuation of autism in a normal social environment. Submitted for publication, 1977.

Richer, J. M., & Coss, R. G. Gaze aversion in autistic and normal children. *Acta Psychiatrica Scandinavica,* 1976, *53,* 193–210.

Richer, J. M., & Richards, B. Reacting to autistic children: The danger of trying too hard. *British Journal of Psychiatry,* 1975, *127,* 526–529.

Ricks, D., & Wing, L. Language, communication, and the use of symbols in normal and autistic children. *Journal of Autism and Childhood Schizophrenia,* 1975, *5,* 191–221.

Ruttenberg, B. A. A psychoanalytic understanding of infantile autism and its treatment. In D. W. Churchill, G. D. Alpern, & M. K. DeMeyer (Eds.), *Infantile autism.* Springfield, Ill.: Charles C Thomas, 1971.

Rutter, M. The development of infantile autism. *Psychological Medicine,* 1974, *4,* 147–163.

Rutter, M., Bartak, L., & Newman, S. Autism—A central disorder of cognition and language. In M. Rutter (Ed.), *Infantile autism: Concepts, characteristics and treatment.* London: Churchill-Livingstone, 1971.

Stern, D. N. A micro-analysis of mother-infant interaction: Behavior regulating social contact between a mother and her 3½ month old twins. *Journal of the American Academy of Child Psychiatry,* 1971, *10,* 501–517.

Stern, D. N. Mother and infant at play: The dyadic interaction involving facial, vocal, and gaze behaviours. In M. Lewis & L. A. Rosenblum (Eds.), *The effect of the infant on its caregivers.* New York: Wiley Interscience, 1974.

Wing, L. *What is an autistic child?* London: National Society for Autistic Children, 1971.

4

The Assessment of Social Behavior

PATRICIA HOWLIN

In his earliest description of autistic children Kanner (1943) stated that "the outstanding 'pathognomonic' fundamental disorder is the children's *inability* to *relate themselves* in the ordinary way to people and situations from the beginning of life."

Recently attempts have been made to delineate more precisely the social handicaps shown by these children (see Chapters 2 and 3). One important outcome of such studies is the recognition that autistic children should not be treated as a homogeneous group, with regard either to outcome, or even to current symptomatology. For, while all children show, at least to some extent, impairment of social and communication skills, the severity of these symptoms varies considerably from child to child and has important implications for prognosis.

Goldfarb (1974) in his extremely detailed three-year longitudinal study of autistic children has shown that communication skills and responses to sensory stimuli, although, on the whole, deviant in comparison with normal children, differ very much according to the age, sex, neurological status, and intelligence of the child. Wing's study (Chapter 2) provides additional valuable information on both retarded and psychotic children. She reports that nonverbal mental age and language comprehension are highly related to sociability ratings and eye contact, as well as to other behaviors such as imaginative play. Work of this nature is an important reminder that several factors, such as age, nonverbal intelligence, and language impairment must be considered before making generalized

PATRICIA HOWLIN · Departments of Psychology and Child and Adolescent Psychiatry, Institute of Psychiatry, Denmark Hill, London SE5 8AF, England.

statements about social skills or, for that matter, any other behaviors in autistic children.

Although detailed studies of this nature are necessary in identifying the particular handicaps of different groups of autistic children, and in making predictions for the future, they are, as Wing herself makes clear, only a preliminary to more precise classification, and to controlled experimental investigations.

Richer has discussed the outcome of his attempts to analyze, within an ethological framework, the behavior of autistic children in relatively structured conditions (Richer & Coss, 1976; Richer & Richards, 1975; Richer, 1976a; see also Chapter 3). Results of experimental investigations both in the laboratory and the classroom showed that autistic children avoided eye contact and social contact more than normal children, and it is suggested that the way in which parents and teachers respond to such behaviors may actually have the effect of increasing withdrawal. However, it is important to note that the children did not avoid contact entirely and detailed analyses of their responses revealed that they spent up to 30% of their time in social-approach behavior, and over 50% on task behaviors.

Social responsiveness, even in autistic children, is a complex affair. For example, Hutt and Ounsted (1966) found that although autistic children rarely initiated social encounters, once contact with an adult was made they often tolerated a greater degree of proximity and much closer physical contact than other children. Hermelin and O'Connor (1970), too, showed that while autistic children tended to look less at human faces than nonautistic controls, they also looked less at everything and their responses to people or images of people were not significantly different from their responses to objects. Castell (1970) reported that autistic children made fewer and briefer visual contacts with a familiar adult than nonautistic children, but in many other ways their behavior toward the adult, in terms of proximity and approach, was very similar. A study by Churchill and Bryson (1972) also failed to find any significant differences in amount of visual fixation and physical avoidance shown by normal and autistic children, regardless of the amount of adult attention given to them.

Ethological approaches have provided us with valuable information on the behavior patterns of animals and, more recently, of human subjects (Hutt & Hutt, 1963, 1970; Hutt & Vaizey, 1966; Tinbergen and Tinbergen, 1972). As Hutt and Hutt point out (1970) it is important to observe subjects in their natural habitats, and to look at the *interactions* between many different behaviors, rather than counts of individual behaviors. Extrapolations from brief observations of small groups of children can lead to oversimplistic and even erroneous conclusions. In trying to learn more about the abnormal social responses of autistic children, it is not merely the quantity of their responses but the quality which needs to be assessed.

What, for example, is the exact nature of the "avoiding behavior" shown by autistic children? Is it avoidance in the active sense, with the child deliberately pulling away from the adult? This may be so with some younger children, but the work by Hutt and Ounsted (1966) and Castell (1970) shows that it is not the case with all autistic children. Negativistic behavior, that is, the child simply refusing to cooperate with the adult's attempts to make contact, is another possible explanation of the children's apparent aloofness. Clark's work (Clark & Rutter, 1977), however, fails to support this assumption. Also with many older children and adolescents, it is clear that they are highly motivated to make social contact; it is just the way they go about it which is all wrong (Dewey & Everard, 1974; Ricks & Wing, 1975; Rutter, 1970).

Abnormal eye-to-eye gaze is only one of the characteristics of the deviant social behaviors shown by autistic children, but even this has not been adequately investigated. That young autistic children tend to make less eye contact than their normal or mentally retarded peers is supported by considerable experimental evidence, although much research on this has been conducted in the laboratory, and not in the child's accustomed environment. Older children, however, may show no avoidance but still fail to make normal contact. Thus, they may stare long and hard at the adult but in a way which is inappropriate to the conversation or the social context in which they are involved. Again the quality as well as the quantity of the behavior needs closer analysis.

Richer (Chapter 3) himself stresses the complexity of the behaviors involved in a two-way interpersonal interaction, and it may be that autistic children are not deliberately avoiding social contact, but that they lack the skills which are necessary in any reciprocal relationship. Failure to understand and obey the multiplicity of roles involved in even the most superficial of social contacts is likely to make the child appear aloof and uninvolved, and even to precipitate his withdrawal from the situation. If the contact is purely physical, however, and makes no demands of the child, he may maintain it even longer than a nonautistic child.

Epidemiological studies, such as those by Wing, and follow-up studies by Lotter (1974), Lockyer and Rutter (1969), Bartak and Rutter (1976), and DeMyer *et al.* (1973) all show that the severity and duration of the social handicap is greatly influenced by the child's cognitive and linguistic skills. The fact that the more intelligent and verbal children have fewer social problems than those who are less intelligent or nonspeaking indicates that poor social interaction is attributable more to the child's *inability* to comprehend and become involved in social situations than to an unwillingness to do so. Investigations of the ways in which autistic children can be helped to cope more adequately in social situations are badly needed and closer examination of the factors influencing the child's abnormal responses may well be of practical value in treatment.

Experimental work by DeMyer (1976), Churchill (1971), and Coss (1972) has shown that withdrawal and pathological behaviors increase when children are given tasks which are too difficult for them. If attempts are made to ensure that the social demands made of the child are not beyond his capabilities and that interaction is kept at a relatively simple level, then withdrawal should be reduced. Koegel and Covert (1972) showed that autistic children's responsiveness to their environment is less if they are allowed to indulge in ritualistic behaviors, and, again, reducing stereotypes may improve social awareness. Hutt and Vaizey (1966) also found that autistic children made more physical contact (although not more eye-gaze) with adults as group density in a room was increased.

More attention also needs to be given to the responses of both members of the dyad in assessing interaction. The study by Gardner (1977) showed that although there were no significant differences between the number of glances toward adults made by autistic children and normal children, the amount of time spent in mutual mother-child facings was actually greater in autistic child-mother pairs. Gardner also found differences in the behaviors of mothers toward autistic children. Mothers terminated mutual facing more frequently with autistic children than with normal children, but they spent longer time in such contact before terminating it.

Richer, in a number of different studies, has analyzed children's responses following social contact by adults. However, his "approach-avoidance—increased approach" model is rather difficult to accommodate within any learning theory paradigm and conflicts with the results of other ethological studies. Hutt and Ounsted (1966), for example, found that a failure to make eye-gaze has a very marked inhibitory effect on social interaction. Richer's hypothesis is also very much at odds with the work by Greenman (1963), Robson (1967), Vine (1973), or Jaffe *et al.* (1973) on normal mother-baby interaction, or with the findings of Freedman (1964) on mothers' contact with blind children.

Although, as already suggested, social interaction with autistic children is probably best kept at a simple level, Richer advises that parents and teachers of autistic children should not attempt to make eye contact with them as this will tend to maintain the autistic withdrawal. It is not easy to predict how children can ever begin to develop social skills if everyone avoids looking at them, and advice of this nature gives no positive practical help on how to achieve more normal relationships with the child. For years parents of autistic children have been accused of maintaining, or even causing, their child's handicap by not interacting enough; now they are told by Richer and Tinbergen and Tinbergen (1972) that they are at fault in interacting too much.

Great care should be taken in basing theories of treatment on limited observation of small groups of children, and one of the problems of ethological studies is that such caution is not always heeded. Experimental evidence certainly fails to support Richer's conclusions. For example, studies by McConnell (1967), Currie and Brannigan (1970), and Lovaas *et al*. (1965) have shown that deliberately bringing eye-gaze and social contact under operant control, far from having a deleterious effect on autistic children, greatly improves their social, motivational, and emotional behavior.

Finally, perhaps, we might seek to learn more about the precise abnormalities of the social behavior of autistic children by comparing their responses with those of normal children. Are social reactions, such as eye-gaze, related to their proximity to another individual, to the type of approach made by an adult, or to the amount of anxiety or security which they experience in a given situation? Although the findings of Hutt and Vaizey (1966), Gardner (1977), and Castell (1970) are important in this context, their analyses are limited to estimates of frequency or duration of behaviors such as eye contact. More complex analyses which take into account differences in the quality of interaction between autistic and normal children are badly needed, and studies similar to those carried out by Argyle (see Argyle & Cook, 1976) might be of value here. It is almost certain that the responses of autistic children are qualitatively not the same as those shown by normal children, but understanding more about *how* their behaviors differ, and how social responses are affected by environmental conditions, could have important implications for treatment. Again, the social behavior of autistic children might be compared with that shown by other groups of handicapped children. Work by Wing (1969), Tizard and Rees (1975), and Rutter (1968) suggests, for example, that the responses of autistic children to an approach by adults certainly differ from the responses of mentally handicapped, Down's syndrome, or socially deprived children, but the precise nature of these differences has so far not been explored.

By reducing the analysis of social behavior in autistic children and their caretakers to the simplified reflexive model postulated by Richer, more detailed information on their social responses, which he himself states can be quite subtle and sophisticated, will be lost. If autistic children are to learn how to acquire the skills of their culture, we need to study in far more detail the abnormalities which they show and how others respond to these, and how different aspects of the social interaction may be modified. It is also necessary, as the work by Wing and others shows, that many other factors, not directly related to social skills, be taken into account before reaching firm conclusions about the extent of the social handicaps in autism, and possible ways in which these might be overcome.

REFERENCES

Argyle, M., & Cook, M. *Gaze and mutual gaze*. Cambridge: Cambridge University Press, 1976.

Bartak, L., & Rutter, M. Differences between mentally retarded and normally intelligent autistic children. *Journal of Autism and Childhood Schizophrenia*, 1976, *6*, 109–120.

Castell, R. Physical distance and visual attention as measures of social interaction between child and adult. In S. J. Hutt & C. Hutt (Eds.), *Behaviour studies in psychiatry*. Oxford: Pergamon, 1970.

Churchill, D. Effects of success and failure in psychotic children. *Archives of General Psychiatry*, 1971, *25*, 208–214.

Churchill, D. W., & Bryson, C. Looking and approach behavior of psychotic and normal children as a function of adult attention or preoccupation. *Comprehensive Psychiatry*, 1972, *13*, 171–177.

Clark, P., & Rutter, M. Compliance and resistance in autistic children. *Journal of Autism and Childhood Schizophrenia*, 1977, *7*, 33–48.

Coss, R. G. Eye-like schemata: Their effect on behaviour. Ph.D. thesis. University of Reading, 1972.

Currie, K., & Brannigan, C. Behavioural analysis and modification with an autistic child. In S. J. Hutt and C. Hutt (Eds.) *Behaviour studies in psychiatry*. Oxford: Pergamon, 1970.

DeMyer, M. K. Motor, perceptual motor and intellectual disabilities of autistic children. In L. Wing (Ed.), *Early childhood autism*, (2nd ed.). Oxford: Pergamon, 1976.

DeMyer, M. K., Barton, S., DeMyer, W. E., Nort, J. A., Allan, J., & Steele, R. Prognosis in autism: A follow-up study. *Journal of Autism and Childhood Schizophrenia*, 1973, *3*, 199–246.

Dewey, M., & Everard, P. The near normal autistic adolescent. *Journal of Autism and Childhood Schizophrenia*, 1974, *4*, 348–356.

Freedman, D. G. Smiling in blind infants and the issue of innate vs acquired. *Journal of Child Psychology and Psychiatry*, 1964, *5*, 171–184.

Gardner, J. Three aspects of childhood autism. Ph.D. thesis. University of Leicester, 1977.

Goldfarb, W. *Growth and change of schizophrenic children: A longitudinal study*. New York: John Wiley & Sons, 1974.

Greenman, G. Visual behavior of newborn infants. In A. Solnit and S. Provence (Eds.): *Modern perspectives in child development*. New York: International Universities Press, 1963.

Hermelin, B., & O'Connor, N. *Psychological experiments with autistic children*. Oxford: Pergamon Press, 1970.

Hutt, C., & Hutt, S. J. Effects of environmental complexity on stereotyped behaviours of children. *Animal Behaviour*, 1963, *13*, 1–4.

Hutt, C., & Ounsted, C. The biological significance of gaze aversion with particular reference to the syndrome of infantile autism. *Behavioral Science*, 1966, *11*, 346–356.

Hutt, S. J., & Hutt, C. *Behaviour Studies in Psychiatry*. Oxford: Pergamon Press, 1970.

Hutt, S. J., & Vaizey, M. Differential effects of group density on social behaviour. *Nature*, 1966, *209*, 1371–1372.

Jaffe, J., Stern, D., & Peery, J. "Conversational" coupling of gaze behaviour in prelinguistic human development. *Journal of Psycholinguistic Research*, 1973, *2*, 321–329.

Kanner, L. Autistic disturbances of affective contact. *The Nervous Child*, 1943, *2*, 217–250.

Kogel, R. L., & Covert, A. The relationship of self-stimulation to learning in autistic children. *Journal of Applied Behavioral Analysis*, 1972, *5*, 381–387.

Lockyer, L., & Rutter, M. A five to fifteen year follow-up study of infantile psychosis. 3. Psychological aspects. *British Journal of Psychiatry*, 1969, *115*, 865–882.

Lotter, V. Factors related to outcome in autistic children. *Journal of Autism and Childhood Schizophrenia*, 1974, *4*, 263–277.

Lovaas, O. I., Schaeffer, B., & Simmons, J. Q. Experimental studies in childhood schizophrenia: Building social behavior in autistic children by the use of electric shock. *Journal of Experimental Research in Personality*, 1965, *1*, 99–109.

McConnell, O. L. Control of eye contact in an autistic child. *Journal of Child Psychology and Psychiatry*, 1967, *8*, 249–255.

Richer, J. The social avoidance behaviour of autistic children. *Animal Behaviour*, 1976, *2*, 898–906. (a)

Richer, J. The self perpetuation of autism in a normal social environment. Submitted for publication, 1976. (b)

Richer, J. & Coss, R. Gaze aversion in autistic and normal children. *Acta Psychiatric Scandinavica*, 1976, *53*, 193–210.

Richer, J., & Richards, M. Reacting to autistic children: The danger of trying too hard. *British Journal of Psychiatry*, 197, *27*, 526–529.

Ricks, D., & Wing, L. Language communication and the use of symbols. *Journal of Autism and Childhood Schizophrenia*, 1975, *5*, 191–221.

Robson, K. The role of eye-to-eye contact in maternal-infant attachment. *Journal of Child Psychiatry*, 1967, *8*, 13–25.

Rutter, M. Concepts of autism: A review of research. *Journal of Child Psychology and Psychiatry*, 1968, *9*, 1–25.

Rutter, M. Autistic children: Infancy to adulthood. *Seminars in Psychiatry*, 1970, *2*, 435–450.

Tinbergen E. A., & Tinbergen, N. Early childhood autism: An ethological approach. *Advances in Ethology*, 1972, *10*.

Tizard, B., & Rees, J. The effect of early institutional rearing on the behaviour problems and affectional relationships of four-year old children. *Journal of Child Psychology and Psychiatry*, 1975, *16*, 61–73.

Vine, I. The role of facial-visual signalling in early social development. In M. von Cronach & I. Vine (Eds.), *Social communication and movement*. New York: Academic Press, 1973.

Wing, L. The handicaps of autistic children: A comparative study. *Journal of Child Psychology and Psychiatry*, 1969, *10*, 1–40.

5

Language: The Problem Beyond Conditioning

DON W. CHURCHILL

It is now widely accepted that disturbance of language is a central feature of childhood autism, despite continuing dispute concerning the nature of the language disturbance and how relevant it is to the panoply of symptoms displayed (Rutter, 1965, 1968; Hermelin, 1968; Wing, 1969; Rutter *et al.*, 1971; Halpern, 1971; Hermelin & Frith, 1971; Cobrinik, 1974; Bartak *et al.*, 1975). A severe language disorder has been found in all autistic children studied by the author with an experimental nine-word language (9WL) and other procedures (Churchill, 1972, 1973). This language disorder, considered common to all autistic children, is proposed as a central explanatory concept to account for those behavioral features which autistic children have in common: the profound impairment of speech and gestures for communication, nonfunctional use of objects, and severe impairment of interpersonal relations. Other demonstrable deficits (of perception, memory, and motor integration) as well as, of course, special features of social history may better account for the ways in which autistic children differ among themselves. Be that as it may, this chapter is limited to a consideration of some special features of language disability which have become apparent in autistic children through detailed, systematic, hierarchical testing of language and those conditionable functions which subserve it. It will focus not on conditioning problems but on a higher level (syntactical?) language disorder with illustrations from just two autistic children in whom we could discern no evidence of difficulty in their basic conditionability using the 9WL.

DON W. CHURCHILL · Indiana University School of Medicine, Indianapolis, Indiana.

Over 15 years ago, experimental psychologists began working clinically with autistic children (Ferster, 1961). The rigorous conditioning procedures brought to the field in this way led to important clinical and
research advances. On the one hand, it was found possible to do much
more by way of training adaptive behavior in even the lowest-functioning
autistic children than was previously thought possible. On the other hand,
achieving a larger measure of behavioral control over the conditioned,
precision responding of autistic children permitted clearer delineation of
the capabilities—and, perhaps more importantly, of the limits, the
disabilities—of these children in responding to well-defined stimuli
(Hermelin & O'Connor, 1964; Ottinger *et al.*, 1965; Hermelin, 1966;
Schopler, 1966; Frith & Hermelin, 1969; Hingtgen & Churchill, 1969;
Bryson, 1970; DeMyer, 1971). It often has been found possible to train
previously mute or echolalic children to use spoken or written words to
designate objects, actions, and wants, to answer simple questions, and to
follow instructions. This is basically a process of conditioning, i.e., of
establishing and maintaining regular stimulus-response associations. The
clinical value of even such modest accomplishment has been attested by
many. But a child so trained (conditioned), even when echolalia has been
converted to so-called "functional speech," can scarcely be said to have
developed language (Risley & Wolf, 1967; Weiss & Born, 1967; De
Villiers & Naughton, 1974).

Psycholinguists have cogently argued that a child's development of
language competence, i.e., the ability to generate and to understand a
virtually limitless number of sentences, cannot be accounted for on the
basis of conditioning alone. There seems rather to be some inherent
capacity of the brain to apprehend without formal teaching the implicit
syntactical rules which govern whatever language a child is exposed to
(Bellugi & Brown, 1964; McNeill, 1966; Lenneberg, 1967). We heartily
agree with this, both on logical grounds and on the basis of our empirical
studies. Yet, conditionability, on both logical and empirical grounds, is an
essential substrate function prerequisite to language competence in any
child. For unless a child is able to make clear and consistent discriminations among auditory and visual stimuli, clear and consistent motor and
vocal differential responses, and consistent associations across sensory
and expressive modalities, all is chaos for child and observer alike. Therefore, our training-testing effort has attempted to span both (a) the basic
conditionability of autistic children as well as (b) some of the lower
reaches of their grammatical (syntactical) performance in order to delineate with each child an ability-disability profile with respect to the full
development of language.

PROCEDURE

The training-testing procedure has been described elsewhere (Churchill, 1972, 1973), and a full protocol is available. Presented here is a synoptic description. Basic conditionability is explored by means of an experimental nine-word language (9WL) which uses nine objects: three blocks, three wooden rings, and three sticks. One each of the blocks, rings, and sticks is painted either red, yellow, or blue. This basic nine-object stimulus array thus presents three nouns (block, ring, stick) and three adjectives (red, yellow, blue). We add to this three verbs ("give," "tap," "slide"), and arrive at a basic nine-word vocabulary containing three different parts of speech. A child learning to associate consistently to each of the nine words can then be trained and examined further with two-word phrases and, finally, a very rudimentary grammar, i.e., three-word sentences consisting of verb-adjective-noun. By associating particular hand-signs to each of the nine words, a child's visual as well as auditory reception can be examined, and a child who can be trained to make the hand-signs as well as speak the words can be examined with respect to two expressive modalities. This is the basic 9WL, and it allows systematic exploration of a child's conditionability with respect to receptive visual, receptive auditory, expressive vocal, and expressive motor functions.

Thereafter, we attempt to explore higher-order language functions. First, there is a series of generalization tests using novel stimuli and designed to see whether children can generalize their performance beyond the particular 9WL objects with which they have been trained. Second, there is a series of cross-referencing tasks—again using the original objects of the 9WL. In this, a child is required to use a discriminative stimulus to respond selectively to either only the color or only the shape of the 9WL objects. From there, we begin to introduce other parts of speech, first prepositions and then pronouns. Finally, we explore a child's ability to generate and to understand novel, simple sentences using familiar parts of speech. From the beginning of work with a particular child, and for as long as necessary to maintain adequate responding, we work within a strict operant conditioning paradigm. One last word about this most cursory of procedural descriptions: It is time consuming and must be highly individualized for each child. Time consuming, because even with two or three sessions per day it ordinarily takes several months to examine a child across the entire range just described. Highly individualized, because even when following the overall training-testing model the individual error patterns and learning/training difficulties encountered with each child are almost as unique as a set of fingerprints, and each child's training presents special problems. By the same token, it is primarily a descriptive, highly idiographic approach, one which is poorly suited to generating group data, although some general statements about the group tested do seem warranted.

SUBJECTS

Out of 16 children trained and tested with the 9WL, 13 were diagnosed autistic according to the DeMyer classification (DeMyer *et al.*, 1971). Two were mentally retarded but not autistic, and one had an expressive aphasia. All but two of the 16 were boys. The mean age at the midpoint of testing was 74 months (range 59–108 months). The mean "full scale" IQ was 43.5 (range 22–92); mean performance quotient 60.2 (range 23–115); mean verbal IQ 41.4 (range 17–76, n = 13). The mean number of training-testing sessions with the 9WL was 108 (range 33–256).

RESULTS

Understanding some of the difficulties encountered by lower-functioning children on the 9WL gives helpful perspective on the higher level disorders. Each of the low-functioning autistic children reached an impasse at some point on the 9WL, although the particular point of difficulty varied from child to child. Some seemed to have especial difficulty with particular parts of speech—for instance, one child who looked dyspraxic clinically, quickly mastered adjectives and nouns but had extraordinary difficulty in responding to verbs. Other children encountered especial difficulty when a particular sensory or expressive modality was required—for instance, a child might do well with receptive visual parts of the 9WL but have extreme difficulty on receptive auditory parts, or vice versa. Still others, while demonstrating no difficulty with any of the three parts of speech or receptive-expressive modalities, displayed great difficulty in dealing with more than one word (part of speech) at a time. In other words, these various low-functioning children display highly individualistic and sharply defined abilities and disabilities in their handling of the 9WL tasks. What they have in common is that they are stumbling at what must be considered a very low level of conditionability, and their difficulty is so specific that it is likely to be overlooked unless they are examined in this kind of systematic detail. Two general points are worth noting about such children. First, when they reach such an impasse on a basic conditioning task, it has been our experience that it is usually insurmountable by any degree of extra effort and inventiveness that we have been able to muster. The stuck point seems to represent a true impasse which does not yield to the combined efforts of child and adult. Second, trial-by-trial examination of records during which a child is working at such a point of impasse is interesting with respect to both the child's "error patterns" and his "strategies" as he tries to master the task (or perhaps to maximize his rate of positive reinforcement). For the errors often are systematic, and close scrutiny invites inferences concerning the child's strategy, so that we are often led to conclude that even the lowest-functioning child—unless he is precipitated into overwhelming frustration and despair by inept training methods—is carrying on a kind of primitive experimentation. These observations about low-functioning autistic children have been reported elsewhere (Churchill, 1973) and will shortly be reported in still greater detail.

Five higher-functioning autistic children plus the one who had been diagnosed as having expressive aphasia mastered all of the 9WL tasks with little or no hesitation. And yet, when they were trained on the cross-referencing tasks or introduced to prepositions and pronouns and expected to manipulate simple syntactical structures, they demonstrated their own special difficulty in dealing with language at these higher levels. The cases of Leon and Jonathan illustrate this.

Leon and Cross-Referencing: Any object can be viewed as having more than one stimulus dimension. For example, the objects of the 9WL have a color dimension and a shape dimension. Leon demonstrated flawless responses to both the color and the shape dimensions of all nine objects. We asked if he could respond selectively to *either* the shape dimension *or* the color dimension alone upon request. The adult would hold up an object and say either "color" or "object," and Leon would be reinforced only for responding accurately to the relevant stimulus dimension. Our expectation was that, having breezed through the 9WL, Leon would quickly master the cross-referencing task. This never happened in spite of extensive training. The various error patterns and strategies demonstrated by Leon are an interesting story in themselves and are to be reported elsewhere. What can be emphasized here is that in spite of 213 training sessions over a three-and-a-half-month period in which we used every training device we could think of Leon demonstrated absolutely no progress in his ability to use an auditory discriminative stimulus to select and respond to a particular stimulus dimension. His visual-vocal associations to the objects remained robust, i.e., he never misnamed the objects or colors which were presented to him (as some autistic children under similar stress come to do); but if a yellow stick were held up and he were asked to say the color, he would be just as likely to say "stick" as "yellow." Colloquially, it might be said of Leon's performance on the cross-referencing task that what he says is correct, but he is not answering the question. In one training modification we switched from the 9WL objects to more commonplace objects which Leon knew the names for: a red car, a yellow cup, a blue comb, etc. Then, rather by accident, we introduced a *visual* discriminative stimulus: If the *object* were to be named, the adult would make an appropriate action with the object such as drinking from the cup or combing hair; if the *color* were to be named, the object would simply be presented motionless. Almost at once Leon's success rate jumped from a chance level to over 90%! We found that even "inappropriate" action with an object (drinking from a comb, combing one's hair with a shoe) still served as an adequate discriminative stimulus for Leon: He would name either object only or color only with nearly 100% accuracy. We tried subtler visual discriminative stimuli such as horizontal motion with an object *versus* vertical motion with an object and Leon's performance again dropped to chance levels of correct responding across 19 training sessions. Reintroduction of appropriate action with an object when a noun response was wanted promptly led to 90% correct responding. Figure 1 shows Leon's performance on this cross-referencing task.

Throughout this training with visual discriminative stimuli our efforts to associate the usable visual discriminative stimuli (action with objects) with auditory discriminative stimuli ("object" or "color") was unsuccessful. Leon's failure to improve suggests that he had "no idea" that the auditory discriminative stimulus had any bearing on the "correctness" of the responses he made to the visual stimuli, i.e., he was not integrating auditory and visual information. With further auditory training employing more conspicuous auditory discriminative stimuli, e.g., buzzer and no buzzer, it is possible that Leon's responses might have come under control of an auditory stimulus. Regardless, it seemed clear that Leon had significant impairment of his ability to use verbal auditory discriminative stimuli beyond what one would expect in a normal child, even though his auditory-visual and auditory-vocal associations were demonstrably faultless on the 9WL. We did not dwell further on the cross-referencing task, electing rather to do more training with prepositions.

Leon and Prepositions (Double Classification): Knowing that Leon could make consistent, accurate auditory-visual and visual-vocal associations to objects, we explored whether he could relate objects to each other through the medium of language, i.e., whether he could grammatically relate substantives (objects) through the use of prepositions. For more than two years, Leon had had occasional training with the prepositions "on," "in," "under," and "beside." He had not made much progress. In response to a four-word sentence paradigm, "Put [object] on/in/under/

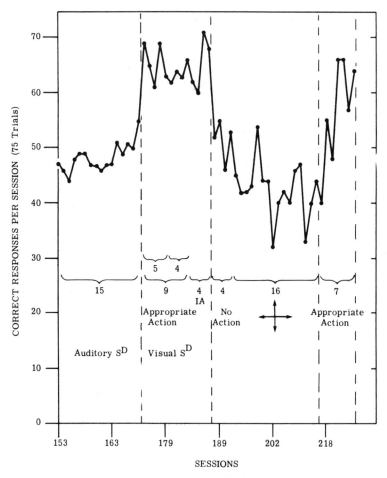

Fig. 1. Leon's performance on cross-referencing task using common objects. (IA = inappropriate action; $+$ = horizontal or vertical motion.)

beside [object]," five types of errors emerged. The two most common of these are relevant here: (1) incorrect responding to the preposition itself, e.g., asked to "Put fork *beside* shoe," Leon would put fork *in* shoe and (2) a reversal of the direct object and the object of the preposition, e.g., asked to "Put *shoe* on *cup*," Leon would put *cup* on *shoe*. We would eliminate one type of error, but always other types of errors then appeared. Structuring our training so as to eliminate the new errors would lead to reappearance of the old errors. Three identical preposition tests given during 11 months of training indicated that Leon was rearranging his errors but not proceeding toward task mastery. It was again as if Leon could master one dimension of the task if all other dimensions were held constant, but if two dimensions of the task were varied simultaneously he was lost.

In such situations, it is our wont to work regressively, i.e., to simplify the task until the child attains a high level of correct responding. To do this we turned again to the objects of the 9WL and concentrated on Leon's two commonest types of errors: preposition errors and reversal of objects. Using the 9WL materials, Leon had 80 training sessions over the next two months. The same four-word sentence paradigm and operant conditioning methods were used, but the stimulus array was initially limited to two objects at a time and to a single preposition, "beside." Within six sessions, all nine objects were reintroduced in the stimulus array, and Leon was making no reversal of object type errors. However, it seemed likely that by using only the one preposition no attention to the semantic import of "beside" was required of Leon; its significance had simply been absorbed into a stereotyped motor response. We then established the same consistent responding using only the preposition "on." The next step, of course, was to admix both the substantive (object) and preposition variables, i.e., "put [object] on/beside [object]." The result was that *both* preposition *and* reversal of object type errors reappeared! We were unable to eliminate these.*When returned to two blocks of different colors, one attached to the table, and asked to put the one block "on" or "beside" the other, i.e., when the independent variable was reduced to a single dimension, Leon immediately responded with 100% success and made only one error in the next three sessions!

From this point, by means of minutely programed steps of successive approximation, Leon was able in the next 17 sessions to make differential responses to simultaneously changing substantives and prepositions using all 9WL objects and the two prepositions while maintaining 88% correct responses. When, in the last phase of work with prepositions, Leon had eight more sessions using the two prepositions "on" and "beside" and small common objects (three at a time) his correct responses dropped to 74%. Seventy-six percent of his errors were reversals of object, and 16%were preposition errors.

A final administration of the same preposition test which had been given three times before showed essentially no progress over the performance demonstrated at the outset 15 months earlier.

Thus, we see that with very limited and carefully structured condi-

---

*At best, Leon developed a predictable strategy of systematically rotating through all possible errors—a method which led to a correct response on every fourth trial. While this very systematic trial-and-error approach was certainly one way of solving the task (and establishing an FR4 schedule of reinforcement), we could adduce no evidence that Leon was making auditory discriminations between words which referred *both* to a substantive *and* to a preposition.

tions Leon did show some capacity to respond correctly to discriminative auditory stimuli across two dimensions simultaneously. But his difficulty here is striking, nevertheless, and it must have profound effect on his developing functional language.

Jonathan and Pronouns: Jonathan had been diagnosed as high-autistic according to the DeMyer classification (DeMyer *et al.*, 1971). He was 7 years 9 months old at the midpoint of his work with the 9WL. He was a higher functioning child than Leon. Of special note is that he had extraordinarily good speech. It was loud and clear and syntactically correct. However, his speech was almost entirely echolalic, showed pronominal reversal, and we knew Jonathan to be an extremely concrete child. His unusually good rote memory was exemplified by his reciting "The Night Before Christmas."

Jonathan also breezed through the entire 9WL, never faltering for more than a moment on any part. He moved with equal ease through two cross-referencing tasks, using first an auditory then a visual discriminative stimulus. Finally, he did fairly well on all four generalization tests. In other words, Jonathan is another example of an autistic child who has extremely good speech, is eminently conditionable, and demonstrates full possession of those substrate functions which are explored with the 9WL. Nor did Jonathan demonstrate any difficulty with prepositions of the sort displayed by Leon.

Yet, as we attempted to move from echolalic speech and rote performance to a more functional language, we encountered difficulties with Jonathan similar to those encountered with Leon, albeit at a higher level of complexity. Some of our work employed a question and answer format. Thus, Jonathan learned to answer correctly a number of questions concerning his dress, the weather, body parts, etc. We found that at times, when asked questions about, say, his dress, Jonathan would respond with "weather" answers—correct descriptions of the weather at the moment, mind you. It was as if Jonathan was giving right answers, but to the wrong set of questions (a phenomenon not unknown among college students at examination time).

Part of a recorded, semistructured interview, again employing a question and answer format, exemplifies one of Jonathan's problems with language. Jonathan was initially asked a series of questions which he had been trained to answer very well: his name, address, telephone number, etc. Gradually, unfamiliar questions were introduced, whereupon the quick, clear, well-articulated responses changed. Hesitation, drop in voice volume, and poor articulation appeared. Again, identifiable "strategies" were very evident and will be reported elsewhere. Shifting back to familiar questions resulted in an immediate return to prompt, loud, clear responding. Pronouns were then introduced, specifically inquiring about the colors of "your shirt" and of "my shirt." Now here again, Jonathan, who in other contexts is absolutely clear about colors, demonstrated confusion and inaccuracy as the questions went back and forth. This continued until his hand was placed on the shirt about which he was to speak. Thereupon, he readily named the colors. He clearly knew what a shirt is, and he clearly could name colors, but he seemed not to know whose shirt he was talking about.

Jonathan and Double-Classification: Next in the same interview, still another variable was introduced. The situation was this: Jonathan was sitting down and given a yellow block, while I was standing up and held a blue cup. Now Jonathan previously had been taught to discriminate sitting down and standing up, and he could consistently and accurately describe the two positions (as well as many others), could follow an instruction to sit down or stand up, etc. He also clearly could discriminate and name a yellow block and a blue cup (as well as hundreds of other objects). At this point Jonathan was asked just four different questions: "What do you have?" [yellow block]; "What do I have?" [blue cup]; "What are you doing?" [sitting down]; "What am I doing?" [standing up].

In brief, through more than 40 repetitions of these various questions, Jonathan showed no evidence of giving correct answers at better than a chance level. The particular kind of difficulty to be noted is that he can *either* discriminate and name objects *or* he can discriminate and describe positions, but he is patently befuddled when expected to do *both*. Formally, the challenge might be described as a double-classification system, as portrayed in Figure 2.

Once a set is established for talking about what he is doing *or* what I am doing *or* what he has *or* what I have, Jonathan is well able to give correct answers endlessly, even when dozens of different objects or positions are introduced—so long as those objects or positions stay within the *same response set*. But to talk interchangeably about what he is doing *and* what I am doing *and* what he has *and* what I have leaves him demon-

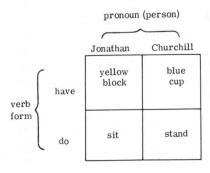

Fig. 2. Model of double classification task attempted with Jonathan.

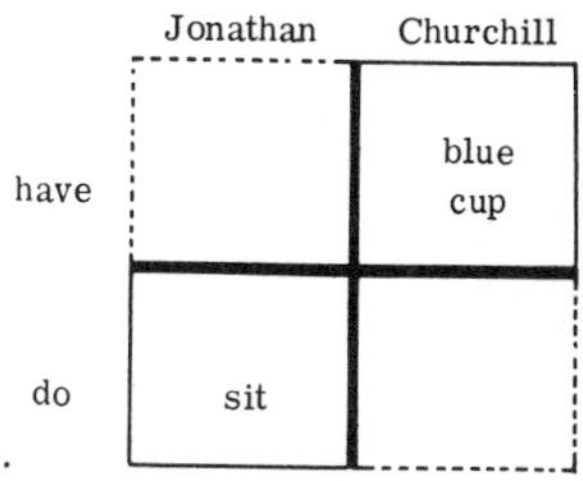

Fig. 3. Jonathan's performance on double classification task.

strably incompetent. Under the sustained pressure of repeatedly rejecting incorrect answers, one solution Jonathan falls upon is to establish a separate set for each of us, i.e., he appears to pay attention only to what he is "doing" and to what I "have." He seems unable to simultaneously recognize or express that he also *has* something or that I also am *doing* something. Figure 3 illustrates what happens to our "conversation."

It is as if only two of the four cells exist for Jonathan, while the other cells are nonexistent or irretrievable.

DISCUSSION

Conditionability is prerequisite for language competence, but language competence requires something more. The illustrative material chosen from work with Leon and Jonathan suggests that even among those autistic children who demonstrate no disorder in their basic conditionability there are, nevertheless, peculiar disabilities which must have profound effects on their developing language competence. For example, practical implications of such a simple task as cross-referencing are suggested if we reflect on our own moment-by-moment experience. We are constantly presented with stimuli, most more complex than those of the 9WL, to which various responses are possible. But whether the response which we selectively emit is likely to be positively reinforced depends on our ability to make use of various contextual cues, i.e., discriminative stimuli. We have here the distinction between adaptive or "appropriate" behavior and that which is nonadaptive, "inappropriate," or even "crazy." Many autistic children—some at much lower levels of functioning than Leon—demonstrate a fair repertoire of precise, conditional responses, but these responses seem not to be under control of discriminative stimuli—at least, not the ones which are familiar to us. Or again, given a stimulus sentence like, "Put yellow stick beside red block," Leon clearly discriminates the colors and shapes. He can also be taught to make distinctions between what we speak of grammatically as "direct object" and "object of the preposition." Furthermore, if no other demands are made, he can readily make auditory discriminations and appropriate differential responses to two or more prepositions. But when he is expected to make these distinctions simultaneously, the task is clearly too much for him except under extremely simple and artificial circumstances. And even extraordinary training leads not to "learning" the fundamental task, but, at best, to a systematic but inefficient (and, from the standpoint of functional language, virtually useless) strategy of trial and error. Leon's difficulty in the preposition task may point to the same kind of difficulty encountered with the cross-referencing task, only now several discriminative stimuli must be dealt with in close succession.

The problem is similar although still more complex for Jonathan struggling with pronouns. There is an underlying grammatical rule about the use of pronouns. "I" and "my" always refer to the person speaking, while "you" and "your" always refer to the person spoken to. In other words, the referent for a pronoun shifts according to who is speaking to whom. In normal language development, children gradually apprehend this rule by themselves. No one teaches them. The child who for some reason does not grasp this underlying rule is faced with a hopeless ambiguity concerning the referents of such words. At best, it must be for Jonathan like trying to guess the outcome of a tossed coin. Confusion about such matters may not represent so-called identity confusion; instead, as Bartak and Rutter (1974) note, it may reflect an inability to grasp the underlying grammatical rule—and a resort to echolalia may be seen as a compromise strategy when faced with an otherwise insoluble task.

Both Leon and Jonathan can display perfect competence in tasks which require responding to a single dimension. It is when, so to speak, a second dimension is introduced (colors *or* shapes; substantives *and* prepositions; substantives *and* pronouns; pronouns *and* verbs) that we see a *sudden break* from absolute mastery to utter incompetence. It represents a discontinuity—in task, and in manifest ability. Koegel and Wilhelm (1973) describe selective attention in some autistic children who are presented with multiple visual cues, but there is a difference between that and the syntactical or cross-referencing problems described here. It is the difference between multiple cues which are or are not syntactically related. While most autistic children initially attend selectively to the multiple stimuli of the 9WL, most can also be trained to attend to additional stimuli *seriatim*. It is when these multiple stimuli come to have syntactical relationship to each other that even the highest-functioning autistic child demonstrates incompetence.

Because of the lack of normal control data, the question raised by Menyuk (1974) remains as to whether this "disability" is unique to autistic children such as Leon and Jonathan, or if it represents primarily a delay in the normal sequence of language development which is seen in normal children as well. After all, two- or three-year-olds have not yet mastered prepositions and pronouns and many of the implicit rules of syntax. If it were possible to submit such normal children to the same kind of arduous training-testing as we have done with autistic children, we do not know if the results would be any different. But even if very young children were to perform the same as older autistic children, at some point normal children "catch on." What's the difference? If we knew that, we might better understand autistic children.

Through detailed and systematic scrutiny of the speech of higher-functioning autistic children such as Leon and Jonathan, we may well be looking at the defect—at the "something more" which is missing—which precludes language competence, regardless of how extensively and elegantly they may be trained to perform. In a stereotyped situation, almost any kind of direct responding can be shaped. But to use discriminative stimuli in two different dimensions or sets simultaneously appears to immediately exceed the linguistic competence of such children. We have seen no evidence that the underlying rules of grammar, spontaneously apprehended by the normal young child exposed to any language whatsoever, can be taught. And the autistic child, however good his speech, seems locked into a concrete world in which only one dimension can be responded to at a time.

REFERENCES

Bartak, L., & Rutter, M. The use of personal pronouns by autistic children. *Journal of Autism and Childhood Schizophrenia,* 1974, *4,* 217–222.

Bartak, L., Rutter, M., and Cox, A. A comparative study of infantile autism and specific developmental receptive language disorder. 1. The children. *British Journal of Psychiatry,* 1975, *126,* 127–145.

Bellugi, U., & Brown, R. The acquisition of language. *Society for Research in Child Development Monographs,* 1964, *29* (92).

Bryson, C. Systematic identification of perceptual disabilities in autistic children. *Perceptual and Motor Skills,* 1970, *31,* 239–246.

Churchill, D. The relation of infantile autism and early childhood schizophrenia to developmental language disorders of childhood. *Journal of Autism and Childhood Schizophrenia,* 1972, *2,* 182–197.

Churchill, D. An experimental nine-word language: Strategies and errors of autistic children. Paper presented at *Conference on Severe Psychopathologies in Childhood,* New York, NYU-Bellevue Medical Center, December 1973.

Cobrinik, L. Unusual reading ability in severely disturbed children: Clinical observation and a retrospective inquiry. *Journal of Autism and Childhood Schizophrenia,* 1974, *4,* 163–175.

DeMyer, M. Perceptual limitations in autistic children and their relation to social and intellectual deficits. In M. Rutter (Ed.), *Infantile autism: Concepts, characteristics and treatment.* London: Churchill, 1971.

DeMyer, M., Churchill, D., Pontius, W., & Gilkey, K. A comparison of five diagnostic systems for childhood schizophrenia and infantile autism. *Journal of Autism and Childhood Schizophrenia*, 1971, *1*, 175–189.

DeVilliers, J. G., & Naughton, J. M. Teaching a symbol language to autistic children. *Journal of Consulting and Clinical Psychology*, 1974, *42*, 111–117.

Ferster, C. Positive reinforcement and behavioral deficits of autistic children. *Child Development*, 1961, *32*, 437–456.

Frith, V., & Hermelin, B. The role of visual and motor cues for normal, subnormal and autistic children. *Journal of Child Psychology and Psychiatry*, 1969, *10*, 153–163.

Halpern, W. Psychosis of early childhood as a function of abnormal speech development. *Bulletin of the Mental Health Center, Rochester, N.Y.,* 1971, *3*, 5–14.

Hermelin, B. Recent psychological research. In J. K. Wing (Ed.), *Early childhood autism*. Oxford: Pergamon, 1966.

Hermelin, B. Recent experimental research. In P. J. Mittler (Ed.), *Aspects of autism*. London: British Psychological Society, 1968.

Hermelin, B., & Frith, V. Can autistic children make sense of what they see and hear? *Journal of Special Education*, 1971, *5*, 107–117.

Hermelin, B., & O'Connor, N. Effects of sensory impact and sensory dominance on severely disturbed autistic children and subnormal controls. *British Journal of Psychology*, 1964, *55*, 201–206.

Hingtgen, J., & Churchill, D. Identification of perceptual limitations in mute autistic children. *Archives of General Psychiatry*, 1969, *21*, 68–71.

Koegel, R., & Wilhelm, H. Selective responding to the components of multiple visual cues by autistic children. *Journal of Experimental Child Psychology*, 1973, *15*, 442–453.

Lenneberg, E. *Biological foundations of language*. New York: Wiley, 1967.

McNeill, D. Developmental psycholinguistics. In F. Smith & G. A. Miller (Eds.), *The genesis of language*. Cambridge, Mass.: M.I.T. Press, 1966.

Menyuk, P. The bases of language acquisition: Some questions. *Journal of Autism and Childhood Schizophrenia*, 1974, *4*, 325–345.

Ottinger, D., Sweeney, N., & Loew, L. Visual discrimination learning in schizophrenic and normal children. *Journal of Clinical Psychology*, 1965, *21*, 251–253.

Risley, T., & Wolf, M. Establishing functional speech in echolalic children. *Behaviour Research and Therapy*, 1967, *5*, 73–88.

Rutter, M. The influence of organic and emotional factors on the origins, nature and outcome of childhood psychosis. *Developmental Medicine and Child Neurology*, 1965, *7*, 518–528.

Rutter, M. Concepts of autism: A review of research. *Journal of Child Psychology and Psychiatry*, 1968, *9*, 1–25.

Rutter, M., Bartak, L., & Newman, S. Autism—a central disorder of cognition and language? In M. Rutter (Ed.), *Infantile autism: Concepts, characteristics and treatment*. London: Churchill-Livingstone, 1971. Pp. 148–171.

Schopler, E. Visual versus tactile receptor preference in normal and schizophrenic children. *Journal of Abnormal Psychology*, 1966, *71*, 108–114.

Weiss, H., & Born, B. Speech training or language acquisition? *American Journal of Orthopsychiatry*, 1967, *37*, 49–55.

Wing, L. The handicaps of autistic children—a comparative study. *Journal of Child Psychology and Psychiatry*, 1969, *10*, 1–40.

6

Language Disorder and Infantile Autism

MICHAEL RUTTER

Since Kanner (1943) first described the syndrome of infantile autism in 1943 there have been major changes in the concepts of the disorder and in methods of treatment. Perhaps the most striking shift of all has been the move from seeing autism as a condition involving social and emotional *withdrawal* to a view of autism as a disorder of development involving severe *cognitive* deficits which probably have their origin in some form of organic brain dysfunction (Rutter, 1974). The fact that autistic children do indeed have basic abnormalities in specific aspects of cognitive functioning is no longer in serious dispute. There is abundant evidence in support of this postulate. Research attention has now shifted to the further questions which arise from the observation that cognitive deficits exist. We need to go on to determine the nature and boundaries of the cognitive deficit, the mechanisms by which it is assoicated with social and emotional abnormalities shown by autistic children, and the biological basis of the cognitive dysfunction.

In considering the nature and boundaries of the cognitive deficit, attention has particularly focused on the possible importance of language impairment, and it is that issue which is the main focus of this chapter. There are many reasons why investigators have been concerned to study language difficulties in autistic children but the most obvious reason is that virtually all autistic children show serious disorders in the field of language. Thus, language abnormalities of various kinds were noted in all

MICHAEL RUTTER · Department of Child and Adolescent Psychiatry, Institute of Psychiatry, De Crespigny Park, Denmark Hill, London, SE5 8AF, England.

of the first 11 autistic children described by Kanner (1943) and in 1946, when he had observed 23 autistic children, he (1946) argued that "the peculiarities of language present an important and promising basis for investigation." The study of Maudsley Hospital cases (Rutter & Lockyer, 1967) confirmed that all autistic children showed a retarded development of spoken language (either through delay or regression); that half were still without speech at five years of age; and that of those who gained speech over three quarters showed echolalia or other abnormal features. The findings from other investigations are very similar (see Rutter, 1970, 1977a; Ornitz, 1973). The question naturally arises as to whether language impairment might be the basic defect which constitutes the central feature of infantile autism (Churchill, 1972; Pronovost, *et al.*, 1966; Rutter, 1966; Wing, 1969). This apparently simple issue actually involves several quite different questions which are best considered separately.

A TRUE INCAPACITY?

The first question asks whether there is a true inability or incapacity in the field of language, as distinct from an emotional or social block to the usage of language. There are three main pieces of evidence in this connection.

Type of Language Disorder

First, it is well established that the problem is *not* just that autistic children use little speech but rather that their language, when it develops, is abnormal in many respects. Echolalia and pronominal reversal are very frequent features, defects in the understanding of spoken language are widespread, and peculiarities in the form of neologisms or metaphorical usage are common (Shapiro *et al.*, 1970; Shapiro *et al.*, 1972; Cunningham & Dixon, 1961; Cunningham, 1968; Wolff & Chess, 1965; Tubbs, 1966; Rutter, 1966). In short, what has to be explained in autistic children is not only limited usage of speech but also a serious delay in speech acquisition and language which appears markedly deviant once it appears. This is quite different from the picture in elective mutism in which children fail to speak because of emotional disturbance (see Rutter, 1972, 1977b) or from that in which speech delay is due to a lack of adequate stimulation or appropriate experiences in early life (see Rutter, 1972, 1977b).

Pattern of IQ Scores

The second piece of evidence concerns autistic children's pattern of IQ scores. Figure 1 shows the findings on the Wechsler "performance"

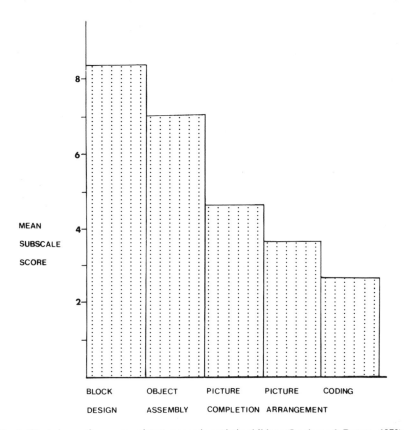

Fig. 1. Wechsler performance subtest means in autistic children (Lockyer & Rutter, 1970).

subtests for autistic children in the Maudsley Hospital follow-up study (Lockyer & Rutter, 1970). In the general population, mean scores are about the same on each of the five subtests but this was far from the case with autistic children. While they scored at a near normal level on block design and object assembly, their scores on picture arrangement and coding were in the mentally retarded range. Furthermore, this pattern was most striking in the autistic children with poor language skills. The finding is important because none of these Wechsler subtests require the child to use speech. All he has to do is to sort objects or write symbols. On the other hand, the tests do differ considerably in the extent to which they require verbal, sequencing, or abstracting skills. Thus, block design and object assembly both have very low loadings on the verbal factor (Cohen, 1959; Maxwell, 1959), whereas the other performance subtests have rather higher verbal loadings. It seems that autistic children perform badly on tests which require verbal or sequencing skills *even when the tests do not involve any use of speech.*

Experimental Studies

The third piece of evidence comes from the experimental studies undertaken by Hermelin, O'Connor, and their associates (Hermelin & O'Connor, 1970). By a series of ingenious experiments they showed that, in marked contrast to normal children, autistic children made relatively little use of meaning in their memory and thought processes. Autistic children showed a marked tendency to remember best what they heard last but only a slight tendency to remember sentences better than random words. The reverse was true for normal children. Other experiments showed that autistic children were not only impaired in their appreciation of the grammatical aspects of language but also they lacked an ability to associate words semantically. Their recall for words seemed to depend largely on the *sound* of the words rather than on their *meaning* or grammatical *usage*.

It is quite clear from these three sets of data that autistic children's failure to use speech adequately is *not* a motivational problem. Rather they show a serious cognitive deficit which involves language and language-related functions such as abstraction and sequencing.

IS THE COGNITIVE DEFICIT OF ANY BASIC IMPORTANCE?

The next question is whether this cognitive deficit is of any basic importance in infantile autism. The short answer is that obviously it is. The evidence may be considered generally in terms of intelligence and specifically with respect to language.

Intelligence

Table 1 uses data from a study of autistic children attending three different special units (Bartak & Rutter, 1973; Rutter & Bartak, 1973). The children's scholastic attainments were compared according to IQ scores at the start of the study. Whereas only a fifth of the autistic children with an IQ of 69 or less achieved all four basic arithmetic skills with double-figure numbers (i.e., addition, subtraction, multiplication, and division), the majority of those with an IQ of 70 or more did so (Bartak & Rutter, 1976). The findings on reading skills were much the same.

The Maudsley Hospital study (Rutter, 1970; Lockyer & Rutter, 1969; Rutter, *et al.*, 1967) confirmed the importance of intellectual level in a much longer follow-up into adult life. Table 2 shows that scarcely any of the children with an IQ below 70 were in regular paid employment. In marked contrast, about half of those with a nonverbal IQ in the normal range were either working or in higher education at follow-up. The biolog-

Table 1. Arithmetical Skills According to IQ Level (Three Units Study—Children Aged at Least 10 Years at Follow-Up)

	IQ (at start of study)		
Arithmetical skills	69 or less	70 or more	Total
No skills	13	0	13
Single skills only (e.g., counting)	12	4	16
All four basic skills (with double-figure numbers)	6	10	16
Total	31	14	45

$X^2 = 10.54; df = 2; p < 0.01.$

Table 2. Follow-Up Findings by IQ Level (Maudsley Hospital Follow-Up)

	Initial IQ		Follow-up IQ	
	< 70 ($n = 45$)	70 + ($n = 19$)	< 70 ($n = 47$)	70 + ($n = 17$)
Epileptic fits	35.6%	0.0%	31.9%	5.9%
Work/higher education	6.7%	36.8%	0.0%	58.8%

ical outcome also differed sharply according to IQ. Whereas a third of the mentally retarded autistic children developed epileptic seizures (usually during adolescence) very few of those of normal intelligence did so.

A variety of other studies have confirmed that the autistic child's IQ score in early childhood is a most important predictor of his later educational attainments and social adjustment (Lotter, 1974; Rees & Taylor, 1975; Rutter, 1970; DeMyer *et al.*, 1973).

Language

Language skills also predict outcome. In the Maudsley follow-up study, both a marked lack of response to sounds during the preschool years and the presence of useful speech at five years proved to be powerful predictors of later social adjustment. *All* of the children with a good social adjustment were speaking at five years of age and *none* had shown a marked lack of response to sounds (Rutter *et al.*, 1967). The only two individual measures with a strong prognostic implication were IQ and language. Lotter (1974) found exactly the same in his Middlesex follow-up. Speech and IQ in combination correlated 0.89 with outcome whereas no combination of factors that excluded speech and IQ produced multiple correlations greater than 0.4 with outcome. Evidence from other follow-up studies cited above points in the same direction.

Not only do IQ and language skills predict social outcome in infantile autism, but also they are themselves persistent characteristics. Thus, Lockyer and Rutter (1969) found that the autistic child's initial IQ level (at about age six) correlated 0.63 with the Wechsler full-scale IQ and 0.74 with the Vineland SQ obtained at follow-up a decade later. Intellectual retardation proved to be one of the most severe and persistent of handicaps in autistic children.

About half of the autistic children remained without useful speech in adult life. Even among those who achieved a normal or near-normal level of language competence, abnormalities in language usage and speech delivery often remained (Rutter, 1970). A monotonous flat delivery with little lability, change of emphasis, or emotional expression was characteristic of some children. In others speech was staccato and lacking in cadence and inflection. Frequently there was a formality of language and a lack of ease in the use of words which led to a pedantic mode of expression. Many autistic children tended to converse mainly through a series of obsessive questions. Difficulties with abstract concepts nearly always remained, even in adult life.

It may be concluded that there is ample evidence that the cognitive deficit in autism is of basic importance.

BOUNDARIES OF THE COGNITIVE DEFICIT

The third question refers to the boundaries of the cognitive deficit. In this connection, of course, the question is *not*: "What cognitive deficits *may* be associated with autism?" (for these are many and various) but rather: "What cognitive deficits *have* to be present for autism to develop?" For this purpose, there is much to be said for the strategy of studying autistic children of normal (nonverbal) intelligence, in whom autism occurs in a more "pure" form not associated with mental retardation or other handicaps. Even though these constitute a minority of autistic children they provide the best opportunity to delineate the cognitive deficits which are *specific* to autism.

A wide range of cognitive functions have been investigated through both experimental (Hermelin & O'Connor, 1970; Hermelin, 1976) and clinical studies (Rutter, 1974). From these various studies we may draw several conclusions.

First, the deficit in autism is certainly not just an abnormality of speech. Rather it involves a wide range of language and language-related functions which include impairments in verbal understanding, sequencing, and abstraction.

Second, it is certainly not just a question of a defect in language output. Autistic children also have serious impairments in their understanding of language, in their use of symbols in play, and in their use of meaning in memory processes.

Third, as our own studies demonstrated (Bartak *et al.,* 1975), the difficulties extend beyond spoken language. Autistic children are also impaired in their use and understanding of gesture and of written language.

Fourth, autistic children's problems in temporal sequencing are much worse than those in spatial sequencing (Hermelin, 1976).

Fifth, although the disabilities are not entirely confined to any one sensory modality, the associations tend to be with auditory rather than visual functions. Thus, Chess *et al.* (1971) found that autism was particularly frequent in children with congenital rubella, but that within the rubella group autism was especially common when mental retardation was accompanied by a hearing defect (see also Rutter, 1973). In contrast, a visual defect did not add to the risk of autism. In this connection, it is perhaps also relevant that O'Connor and Hermelin (1973; Hermelin, 1976) found that with respect to temporal and spatial ordering of visually presented digit sequences, autistic and deaf children behaved in a similar manner but one quite different from that shown by normal children.

Sixth, visuo-spatial perceptual defects do *not* play any essential role in the development of infantile autism. Mention has already been made of the finding that many autistic children have normal skills on the block design and object assembly subtests of the Wechsler scales. These are the two subtests with the highest loading on the visuo-spatial factor (Cohen, 1959; Maxwell, 1959). Our more recent studies with autistic children who have a nonverbal IQ in the normal range have also shown that they have somewhat *above*-average skills on Raven's Progressive Matrices (1965), a test which employs complex visual stimuli to assess reasoning skills. Clearly, autism can and does develop in children who have no discernible visuo-spatial perceptual defect.

Hermelin and O'Connor (1970) have concluded that the autistic child's language deficit is only one aspect of a more general inability to use signs and symbols; a disability which principally involves a deficit in the coding, extraction, or organization of incoming information (Hermelin, 1976). Their studies are much the best experimental investigations of autism which have been undertaken and their arguments are persuasive, but there is one important limitation and that is that in many of their studies the children were mentally retarded as well as autistic. This is important because mentally retarded autistic children and autistic children of normal intelligence differ in many respects (Bartak & Rutter, 1976). Furthermore, the cognitive deficit was wider and somewhat different in autistic children of low IQ (Hermelin & O'Connor, 1970). In order to try to sort out the issue of how far the cognitive deficit in autism extended beyond the field of language, we made a series of studies of autistic children all of whom had a nonverbal IQ of at least 70. In doing this we also focused on the fourth main question concerning the cognitive deficit in autism, namely, how far is the language deficit in autism the same as that in the most severe developmental language disorders?

AUTISM AND DEVELOPMENTAL RECEPTIVE DYSPHASIA

As several writers have noted (Churchill, 1972; Pronovost *et al.*, 1966; Rutter, 1966, 1968; Wing, 1969), autistic and developmental receptive dysphasic children have many problems in common. Furthermore, several investigators have reported cases in which these two conditions occurred in different children in the same family (Rutter *et al.*, 1971; Cohen *et al.*, 1976). It seemed that the two conditions might be linked.

Accordingly, we conducted a study of all the boys we could find with a severe developmental receptive language disorder (Bartak *et al.*, 1975). The group was then divided up on behavioral grounds into the 19 with the behavioral syndrome of infantile autism, the 23 clearly not autistic who seemed to have an uncomplicated developmental language disorder, and a much smaller mixed group of five boys who showed only some features of autism. This last group will be put on one side for the moment in order to focus on the comparisons between the two main subgroups of autistic and "dysphasic" children. The groups were well matched in terms of nonverbal intelligence (with a mean of 92–93) and language production. They were also comparable in age.

Obviously, the two groups differed in terms of behavior and social characteristics, as this was how the main group had been subdivided. There was some overlap but, as expected, autistic features such as lack of appropriate eye-to-eye gaze, lack of friends, limited group play, and ritualistic activities were all very much more common in the autistic group.

We then looked in detail at various language and language-related functions. As these had not been used in the differentiation into autistic and dysphasic subgroups there was no *a priori* reason why they should differentiate the groups. However, in fact, there were marked differences in language between the autistic and dysphasic children. In both groups many of the children had shown abnormal or diminished babble and many had been suspected of deafness because of their inconsistent response to sounds. However, an undue sensitivity to sounds was much more common in the autistic group. Table 3 shows that abnormalities of spoken language were all much more frequent in autistic children. These included

Table 3. *Abnormalities of Spoken Language in Autistic and "Dysphasic" Children*

	Group	
	Autistic	"Dysphasic"
Defects of articulation	10	21
Pronoun reversal (you-I) (ever)	11	4
Echolalia (ever)	19	6
Stereotyped utterances (ever)	12	2
Metaphorical language (ever)	7	0
Inappropriate remarks	6	0
Total	19	23

All differences statistically significant

pronoun reversal, echolalia, stereotyped utterances, metaphorical language, and inappropriate remarks. On the other hand, defects of articulation were more often evident in the dysphasic children. The autistic children showed less spontaneity in their use of spoken language, they talked less readily and they made less use of speech for social communication or chat. In addition, the autistic children made less use of gesture and showed less imaginative play (see Table 4).

Most of these findings derive from parental reports. However, systematic testing showed much the same picture. Bartak (Bartak *et al.*, 1975) devised a test for the understanding and use of gesture. Autistic children's scores on this test were significantly worse than those of dysphasic children. Cantwell *et al.* (1977) analyzed tape recordings of the children's speech at home and found that autistic children showed a much

Table 4. *Comparison of Other Language Features in Autistic and "Dysphasic" Children*

	Group	
	Autistic	"Dysphasic"
Chatted spontaneously regularly	5[a]	17
Regularly gives account of activities in answer to questions	7	18
Imaginative play	4	17
Use of gesture, other than pointing (ever)	2	13
Total children	19	23

All differences statistically significant
[a] Not known in one case.

Table 5.   Cognitive Test Scores in Autistic and "Dysphasic" Children

	WISC perf. scale	Matrices	S.Q.	Peabody IQ
Autistic	93.5	115.9	69.8	51.7
"Dysphasic"	91.8	104.1	92.6	71.4

higher proportion of inappropriate echolalia and nonsocialized utterances. On the other hand, the autistic and dysphasic children did not differ *at all* in any measures of syntactical errors or grammatical competence. Boucher (1976), using the Edinburgh Articulation Test, confirmed that articulation errors were more frequent with dysphasic children.

The next step was to examine the children's cognitive performance using standard psychometric tests. As shown in Table 5, the autistic and dysphasic groups did not differ in their mean scores on either the WISC performance scale or on Raven's Progressive Matrices. Thus, they were closely matched on nonverbal skills. On the other hand, the autistic children had a mean social quotient 23 points below that for the dysphasic group and a Peabody Picture Vocabulary score which was 20 points below. Also, far fewer autistic children were testable on the verbal section of the WISC scale. It may be concluded that the verbal deficit was much more severe in autistic children.

However, not only was the verbal impairment more severe, it was also *different* in the autistic group, as shown by the WISC subtest patterns. On both the verbal and the performance scales the scores of the dysphasic children were generally even. On the performance scale their scores were all roughly average with little variation between them. On the verbal scale the means were lower—being just over one standard deviation below average—but again with little variation between subtests. The pattern in the autistic group was quite different. On the performance scale their block design score (which was above average) was significantly superior to the scores on the other subtests. Subtest variation on the verbal scale was even more striking (as shown in Fig. 2). The mean for digit span was significantly higher than each of the others and was the only subtest on which the autistic children were superior to the dysphasic (although not significantly so). The autistic children's scores on arithmetic and information did not differ significantly from the dysphasic group but their scores on each of the other three subtests were significantly worse. Their poor performance was particularly striking on the comprehension subtest, where the autistic children scored very badly with a mean score well down into the retarded range.

The results clearly show that the verbal deficit in autistic children is both more severe and also different in pattern from that found in de-

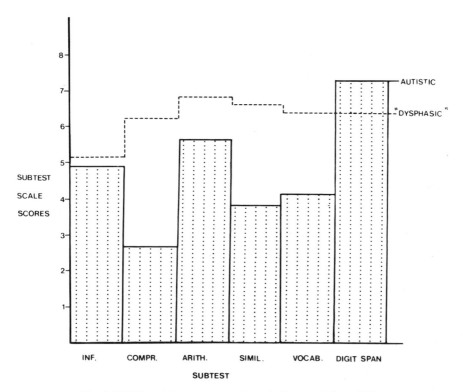

Fig. 2. WISC verbal scale pattern (Bartak, Rutter, & Cox, 1975).

velopmental receptive dysphasia. Two further sets of questions immediately arise. First, how different are the autistic and "dysphasic" groups? How much overlap is there between them? Is the difference sufficiently marked to allow diagnostic differentiation at the *individual* as well as at the group level? Second, how far does the differentiation on linguistic or cognitive features parallel that made on behavioral grounds? Would the *same* children be called autistic if diagnosis were made solely on the basis of language and intellectual function without reference to social relationships or behavior? To answer these two series of questions a discriminant functions analysis was carried out (Bartak *et al.*, 1977). The major variables on which there were substantial differences between the autistic and dysphasic children were divided into five groups: past and present behavior, past and present language, and scores on standardized tests of cognition and language. In each analysis a discriminant function was computed and used to calculate discriminant scores for each subject. For all analyses on items reflecting current performance, the groups were matched for age.

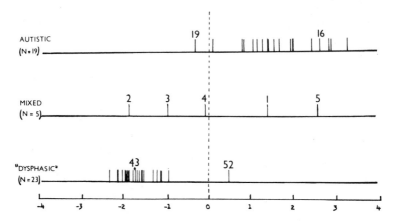

Fig. 3. Discriminant function analysis: behavior disturbance (past).

Figure 3 shows the pattern of discriminant scores for past behavior. As would be expected from the way in which autism was diagnosed, there was very little overlap between groups. Only one autistic child had a "dysphasic" discriminant score and only one "dysphasic" child had an autistic score. As already noted, the original selection of children with a severe developmental disorder of language comprehension included a "mixed" group of five children who could not be clearly placed in either an autistic or a nonautistic group. Their discriminant scores covered a wide range and reflected their "mixed" status.

The discriminant functions analysis findings for the children's current behavior and social relationships showed a perfect discrimination between the autistic and "dysphasic" groups, without any overlap.* But, again, the "mixed" group spread over the entire range of scores.

The third analysis, shown in Fig. 4, referred to the children's language in the past, as reported by parents. The discriminant score was based on the abnormalities of language previously found to be associated with autism—items such as undue sensitivity to loud noise, pronominal reversal, echolalia, and neologisms. Once more there was no overlap between the two main groups but the "mixed" group were scattered across the range. The fourth analysis was concerned with the children's current language performance in terms of both abnormal features and patterns of language usage for social communication. The results showed a quite sharp differentiation of the autistic and "dysphasic" groups with

*However, when misclassification is assessed on the basis of data used to derive the discriminant function (as in this case) the estimate of error tends to be too low (Hills, 1966; Lachenbruch & Mickey, 1968). Accordingly the true degree of overlap between groups is almost certainly higher.

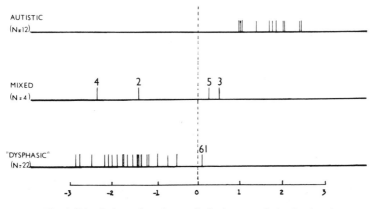

Fig. 4. Discriminant function analysis: language behavior (past).

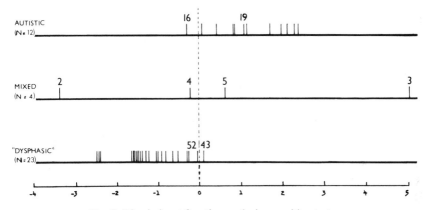

Fig. 5. Discriminant function analysis: cognitive tests.

no overlap at all between them. The findings are striking and the implications very important. The analyses show that the link between social behavior and language performance is so strong that autism can be diagnosed highly successfully simply by reference to the qualities of the child's language and the pattern of his language usage.

The fifth and final discriminant functions analysis (see Fig. 5) was concerned with the children's scores on standardized psychometric tests. These included the Reynell Developmental Language Scale, the various WISC verbal subtests, the Peabody Picture Vocabulary Test, the Columbia Mental Maturity Scale, the Vineland Social Maturity Scale, and three tests of gesture usage and understanding. Yet again, the analysis yielded a significant discrimination between groups with very little overlap. Only one autistic and one dysphasic child were misclassified. However, the "mixed" group were spread even more widely than before. It may be concluded that autism can be diagnosed almost as well on the basis of cognitive test performance as on behavioral or linguistic grounds.

When the five discriminant functions analyses were put together it was found that out of the total sample of 38 children (including "mixed" cases), there were only seven children (18%) who were not definitely classified into the *same* autistic or "dysphasic" group in each and every one of the five separate analyses. Four of these seven children were in the "mixed" group who were not classifiable on clinical criteria. It should be noted that *different* children were misclassified in different analyses. This indicates that children who are clearly autistic in terms of one kind of behavior may be clearly not autistic in terms of some other measure. This diagnostic ambiguity refers to less than a fifth of children but its presence has to be taken account of.

So far, the findings refer to comparisons between groups. It is now necessary to consider how far the various measures intercorrelate *within* groups, that is, how far children who were most autistic on one set of measures were also most autistic on another set. This was determined separately for the autistic and "dysphasic" groups by computing Kendell's W between rank orders of children in the five analyses (according to magnitude of discriminant scores). The overall W was +0.42 in the autistic group and +0.29 in the "dysphasic" group. The highest correlations in the autistic group were between past language features and either past behavior (rho = +0.82) or present behavior (rho = +0.64). The only other significant correlation was between past and present behavior (rho = +0.67). The correlations were rather lower in the "dysphasic" group and the only one to reach statistical significance was that between past language and past behavior (rho = +0.46).

Four main conclusions may be drawn from these various analyses comparing autistic and dysphasic children.

(1) First, the findings provide very strong support for the hypothesis that a cognitive deficit is an essential part of the syndrome of infantile autism. This is shown by the fact that linguistic and cognitive features differentiate autistic children from dysphasic children as sharply as do social and behavioral characteristics and by the fact that each set of items differentiates the *same* group of autistic children. The overlap between the various measures is so great that it would be fair to say that autism is as much a cognitive and linguistic disorder as it is a social and behavioral syndrome.

(2) Second, the findings provide important confirmatory evidence for the suggestion that abnormalities of language constitute a central feature of the cognitive deficit associated with autism. This is shown by the well-established observation that language disabilities are universal or

almost universal in autism and by the present finding that, even when a comparison group of children with severe language impairment was chosen, *all* the features which differentiated autistic children involved some form of verbal skill. There was no nonverbal measure on which autistic children scored significantly worse than dysphasic children. As no other study of autistic children without accompanying mental retardation has shown any necessary association between autism and any kind of nonverbal measure, it may be concluded that a disorder of language constitutes a crucial aspect of the cognitive deficit.

(3) Third, it is equally evident that this disorder of language is *not* the same as that in developmental receptive dysphasia. For the most part "dysphasia" simply constitutes a particularly severe variety of language *retardation* in which there are few *abnormalities* of language (although there are some). However, this is not the case in autism where abnormal features are as prominent as lack of language skills. The language differences between autistic and dysphasic children fall under four headings: (a) the autistic child's defect in the understanding of language is more *severe*; (b) the autistic language impairment is more *extensive* in that it extends well beyond spoken language to include gesture, written language, sequencing, and abstraction; (c) autism usually involves language *deviance* as well as language delay; and (d) the autistic disorder includes an *impaired usage* of both spoken language and gesture as well as reduced skills in these areas. This is an impressive list of differences which certainly outweighs the various similarities between the two disorders.

(4) Fourth, although the language disorders in autism and in "dysphasia" are clearly different in a number of crucial respects, there are important links between them. This is evident from four rather different sets of observations: (a) There is overlap between the two conditions so that no clear boundary can be drawn (a fifth of children with a severe developmental disorder of receptive language cannot be put unambiguously into either group); (b) within both autistic and "dysphasic" groups there is an association between language disabilities of an autistic type and the social/behavioral characteristics of autism (that is, it is meaningful to talk of degrees of autism and to relate this to degrees of language abnormality); (c) in both autism and "dysphasia" a family history of speech delay is reported in about a quarter of cases (Bartak *et al.*, 1975); and, most crucially of all, (d) as reported in Chapter 15 (see also Folstein & Rutter, 1977), the *non*autistic co-twin of the autistic member in discordant monozygotic twin pairs usually shows a cognitive deficit involving some level of language impairment. In short, although the language disorder in autism is very different from that in "dysphasia" there are important functional links between autism and at least some cases of "dysphasia."

QUESTIONS REMAINING

There are five main questions which remain unanswered. First, there is still the question of the nature of the cognitive deficit in autism. As we have seen, it fairly clearly involves a disorder of language, sequencing, and abstraction, but what does that mean? Language is a complex skill based on multiple cognitive processes operating in a social context (Hockett, 1960), so that it is not meaningful to ask whether autism is based on a pure language deficit. That merely leads to a fruitless debate on what is meant by language. However, it is important to ask which cognitive processes are disordered in order for the language disability in autism to arise. Hermelin (1976) has suggested that the "cognitive pathology seems to consist largely of an inability to reduce information through the appropriate extraction of crucial features such as rules and redundancies." Ricks and Wing (1976) emphasize abnormalities in the handling of symbols and in the development of language. Certainly, the disorder includes central coding processes, language, and communication but it has not proved easy to delineate the precise elements involved. Other workers have argued for the importance of memory deficits (Boucher & Warrington, 1976) or of various brain stem functions (MacCulloch & Williams, 1971; Des Lauriers & Carlson, 1969; Hutt *et al.*, 1964; Ornitz, 1970), but so far no convincing explanation has been proffered for how these could lead to language impairment and more particularly how they could account for the demonstrated links between autism and other forms of nonautistic language retardation.

Second, as follows from the first question, knowledge is still required on the biological basis of the cognitive deficit. Which brain systems are involved and what is the locus or loci of brain pathology?

Third, we still have very little notion of what is involved in the autistic child's relative failure to use speech for social communication. Is this simply a consequence of the severe disorder of language and of central coding processes or is some rather different disorder involved?

Fourth, obviously the cognitive deficit and the social abnormalities are closely associated but does one cause the other or are they both different facets of the same basic disability? Two rather different approaches are needed to investigate these alternatives. The possibility that the cognitive deficit may cause the autistic abnormalities of behavior and socialization is best investigated by means of experimental studies. The investigations by Koegel and Covert (1972) and by Churchill (1971) provide a good start in this connection. The possibility that the language disorder and the social abnormalities are both different aspects of the same deficit may best be studied by a more detailed analysis of what is involved in the social problem. Are there, for example, deficits in social perception as found with deaf children by Odom *et al*. (1973)?

Fifth, and last, given that autism is associated with a cognitive deficit involving language, is this a sufficient explanation of the genesis of autism? The need to ask this question arises from three different sets of observations. First, within families in which several members show a language impairment, usually only one is autistic (Rutter *et al*., 1971; Cohen *et al*., 1976; Folstein & Rutter, 1977). Why? What is different about the biological makeup or experiences of the autistic individual? Second, most investigators have found autism to be more frequent in middle-class or professional families. Why? What mechanisms could be involved? Third, various forms of environmental intervention seem to benefit autistic children (see Chapter 29). Does this imply that autism is more likely to develop in some sorts of environments than in others? So far, no psychosocial determinants of autism have been identified and various possibilities have been virtually ruled out (see Chapter 18). However, as indicated, there are still questions which remain to be answered.

CONCLUSION

In summary, there are four main conclusions that derive from the work so far into the associations between language disorder and autism. One, a cognitive deficit constitutes an essential part of the syndrome of autism. Two, abnormalities of language are a central feature in this cognitive deficit. Three, the disorder of language is not the same as that in "dysphasia" in that not only is the comprehension defect more severe but also it is more extensive and it involves both deviant characteristics and an impairment in social usage as well as language delay. Four, although the language disorders in autism and in "dysphasia" are clearly different in a number of crucial respects, nevertheless there are still important links between them. Many of the questions which remain are involved with an elucidation of the nature and meaning of these links.

REFERENCES

Bartak, L., & Rutter, M. Special educational treatment of autistic children: A comparative study. I. Design of study and characteristics of units. *Journal of Child Psychology and Psychiatry*, 1973, *14*, 161–179

Bartak, L., & Rutter, M. Differences between mentally retarded and normally intelligent autistic children. *Journal of Autism and Childhood Schizophrenia*, 1976, *6*, 109–120.

Bartak, L., Rutter, M., & Cox, A. A comparative study of infantile autism and specific developmental receptive language disorder. I. The children. *British Journal of Psychiatry,* 1975, *126,* 127–145.

Bartak, L., Rutter, M., & Cox, A. A comparative study of infantile autism and specific developmental receptive language disorder. III. Discriminant functions analysis. *Journal of Autism and Childhood Schizophrenia,* 1977, *7,* 383–396.

Boucher, J. Articulation in early childhood autism. *Journal of Autism and Childhood Schizophrenia,* 1976, *6,* 297–302.

Boucher, J., & Warrington, E. K. Memory deficits in early infantile autism: Some similarities to the amnesic syndrome. *British Journal of Psychology,* 1976, *67,* 73–87.

Cantwell, D., Baker, L., & Rutter, M. A comparative study of infantile autism and specific developmental receptive language disorder. IV. Syntactical and functional analysis of language. Submitted for publication, 1977.

Chess, S., Korn, S. J., & Fernandez, P. B. *Psychiatric disorders of children with congenital rubella.* New York: Brunner/Mazel, 1971.

Churchill, D. W. Effects of success and failure in psychotic children. *Archives of General Psychiatry,* 1971, *25,* 208–214.

Churchill, D. W. The relation of infantile autism and early childhood schizophrenia to developmental language disorders of childhood. *Journal of Autism and Childhood Schizophrenia,* 1972, 2, 182–197.

Cohen, J. The factorial structure of the WISC at age 7–6, 10–6 and 13–6. *Journal of Consulting Psychology,* 1959, *23,* 285–299.

Cohen, D. J., Caparulo, B., & Shaywitz, B. Primary childhood aphasia and childhood autism: Clinical, biological and conceptual observations. *Journal of the American Academy of Child Psychiatry,* 1976, in press.

Cunningham, M. A. A comparison of the language of psychotic and non-psychotic children who are mentally retarded. *Journal of Child Psychology and Psychiatry,* 1968, *9,* 229–244.

Cunningham, M. A., & Dixon, C. A study of the language of an autistic child. *Journal of Child Psychology and Psychiatry,* 1961, *2,* 193–202.

DeMyer, M. K., Barton, S., DeMyer, W. E., Norton, J. A., Allen, J., & Steele, R. Prognosis in autism: A follow-up study. *Journal of Autism and Childhood Schizophrenia,* 1973, *3,* 199–246.

DesLauriers, A. M., & Carlson, C. F. *Your child is asleep.* Homewood, Ill.: Dorsey, 1969.

Folstein, S., & Rutter, M. Genetic influences and infantile autism. *Nature,* 1977, *265,* 726–728.

Hermelin, B. Coding and the sense modalities. In L. Wing, (Ed.), *Early childhood autism: Clinical, educational and social aspects (2nd Ed.).* Oxford: Pergamon Press, 1976. Pp. 135–168.

Hermelin, B., & O'Connor, N. *Psychological experiments with autistic children.* Oxford: Pergamon Press, 1970.

Hills, M. Allocation rules and their error rates. *Journal of the Royal Statistical Society* (Series 3), 1966, *28,* 1–20.

Hockett, C. F. The origin of speech. *Scientific American,* 1960, *203,* 89–96.

Hutt, S. J., Hutt, C., Lee, D., & Ounsted, C. Arousal and childhood autism. *Nature,* 1964, *204,* 908–909.

Kanner, L. Autistic disturbance of affective contact. *Nervous Child,* 1943, *2,* 217–250.

Kanner, L. Irrelevant and metaphorical language in early infantile autism. *American Journal of Psychiatry,* 1946, *103,* 242–246.

Koegel, R. L., & Covert, A. The relationship of self-stimulation to learning in autistic children. *Journal of Applied Behavioural Analysis,* 1972, *5,* 381–387.

Lachenbruch, P. A., & Mickey, M. R. Estimation of error rates in discriminant analysis. *Technometrics,* 1968, *10,* 1–11.

Lockyer, L., & Rutter, M. A five-to-fifteen-year follow-up study of infantile psychosis. III. Psychological aspects. *British Journal of Psychiatry*, 1969, *115*, 865–882.

Lockyer, L., & Rutter, M. A five-to-fifteen-year follow-up study of infantile psychosis. IV. Patterns of cognitive ability. *British Journal of Sociology and Clinical Psychology*, 1970, *9*, 152–163.

Lotter, V. Factors related to outcome in autistic children. *Journal of Autism and Childhood Schizophrenia*, 1974, *4*, 263–277.

MacCulloch, M. J., & Williams, C. On the nature of infantile autism. *Acta Psychiatrica Scandinavica*, 1971, *47*, 295–314.

Maxwell, A. E. A factor analysis of the Wechsler Intelligence Scale for children. *British Journal of Educational Psychology*, 1959, *29*, 237–241.

O'Connor, N., & Hermelin, B. M. The spatial or temporal organisation of short-term memory. *Quarterly Journal of Experimental Psychology*, 1973, 25, 335–343.

Odom, P. B., Blanton, R. L., & Laukhuf, C. Facial expressions and interpretations of emotion-arousing situations in deaf and hearing children. *Journal of Abnormal Child Psychology*, 1973, *1*, 139–151.

Ornitz, E. M. Vestibular dysfunction in schizophrenia and childhood autism. *Comprehensive Psychiatry*, 1970, *11*, 159–173.

Ornitz, E. M. Childhood autism—a review of the clinical and experimental literature. *California Medicine*, 1973, *118*, 21–47.

Pronovost, W., Wakstein, M. P., & Wakstein, D. J. A longitudinal study of speech behaviour and language comprehension of fourteen children diagnosed as atypical or autistic. *Exceptional Children*, 1966, 33, 19–26.

Raven, J. C. *Guide to using the Coloured Progressive Matrices*. London: Lewis, 1965.

Rees, S. C., & Taylor, A. Prognostic antecedents and outcome in a follow-up study of children with a diagnosis of childhood psychosis. *Journal of Autism and Childhood Schizophrenia*, 1975, *5*, 209–322.

Ricks, D. M., & Wing, L. Language, communication and the use of symbols. In L. Wing, (Ed.), *Early childhood autism: Clinical, educational and social aspects* (2nd ed.). Oxford: Pergamon Press, 1976. Pp. 93–134.

Rutter, M. Behavioural and cognitive characteristics of a series of psychotic children. In J. K. Wing, (Ed.), *Early childhood autism*. Oxford: Pergamon Press, 1966. Pp. 51–81.

Rutter, M. Concepts of autism: A review of research. *Journal of Child Psychology and Psychiatry*, 1968, *9*, 1–25.

Rutter, M. Autistic children: Infancy to adulthood. *Seminars in Psychiatry*, 1970, *2*, 435–450.

Rutter, M. Psychiatric causes of language retardation. In M. Rutter & J. A. M. Martin, (Eds.), *The child with delayed speech*. Clinics in Developmental Medicine No. 43. London: SIMP/Heinemann, 1972. Pp. 147–160.

Rutter, M. Psychiatric disorders of children with congenital rubella by S. Chess, S. J. Korn, and P. B. Fernandez. *Journal of Autism and Childhood Schizophrenia*, 1973, *3*, 276–277.

Rutter, M. The development of infantile autism. *Psychological Medicine*, 1974, *4*, 147–163.

Rutter, M. Infantile autism and other child psychoses. In M. Rutter, & L. Hersov, (Eds.), *Child psychiatry: Modern approaches*. Oxford: Blackwell Scientific, 1977. Pp. 717–747.(a)

Rutter, M. Speech delay. In M. Rutter, & L. Hersov, (Eds.), *Child psychiatry: Modern approaches*. Oxford: Blackwell Scientific, 1977. Pp. 688–716.(b)

Rutter, M., & Bartak, L. Special educational treatment of autistic children: A comparative study. II. Follow-up findings and implications for services. *Journal of Child Psychology and Psychiatry*, 1973, *14*, 241–270.

Rutter, M., Bartak, L., & Newman, S. Autism—a central disorder of cognition and language? In M. Rutter, (Ed.), *Infantile autism: Concepts, characteristics and treatment.* London: Churchill Livingstone, 1971. Pp. 148–171.

Rutter, M., Greenfeld, D., & Lockyer, L. A five-to-fifteen-year follow-up study of infantile psychosis. II. Social and behavioural outcome. *British Journal of Psychiatry, 1967, 113,* 1183–1199.

Rutter, M., & Lockyer, L. A five-to-fifteen-year follow-up study of infantile psychosis. I. Description of sample. *British Journal of Psychiatry, 1967, 113,* 1169–1182.

Shapiro, T., Fish, B., & Ginsberg, G. L. The speech of a schizophrenic child from two to six. *American Journal of Psychiatry, 1972, 128,* 92–98.

Shapiro, T., Roberts, A., & Fish, B. Imitation and echoing in schizophrenic children. *Journal of American Academic Child Psychiatry, 1970, 9,* 548–567.

Tubbs, U. K. Types of linguistic disability in psychotic children. *Journal of Mental Deficiency Research, 1966, 10,* 230–240.

Wing, L. The handicaps of autistic children—a comparative study. *Journal of Child Psychology and Psychiatry, 1969, 10,* 1–40.

Wolff, S., & Chess, S. An analysis of the language of fourteen schizophrenic children. *Journal of Child Psychology and Psychiatry, 1965, 6,* 29–41.

7

Language: What's Wrong and Why

PAULA MENYUK

The papers presented by Churchill and Rutter contain a rich amount of data on the "language features" in the behavior of autistic children. From these data the authors have reached several conclusions about what these features are, the causes of their appearance, and some questions about the explanations that have been provided thus far. In this discussion I will summarize what appear to be agreements and disagreements between the two researchers in these areas, and then express my own views. Unlike Churchill and Rutter, I have not done extensive research with autistic children. Therefore, I am relying on observations from the literature about these children and my own work on the language development of normally developing children and children classified as aphasic or dysphasic to reach some tentative conclusions about what may be wrong with the language of autistic children and why.

LANGUAGE BEHAVIOR OF AUTISTIC CHILDREN

The language difficulties of autistic children involve both the acquisition of the system itself and the use of the system. That is, they have difficulty in acquiring the phonological, morphological, syntactic, and semantic rules of the language (the system) and, also, the pragmatic rules (use of the system). Thus, even those children who achieve the ability to communicate with oral language symbols apply this ability to limited domains of discourse and do not communicate appropriately about these

PAULA MENYUK · Applied Psycholinguistics Program, Boston University, Boston, Massachusetts.

domains. They do not take into account their listeners' state and responses (Ricks & Wing, 1975). The degree to which the system itself will be acquired varies in the population. There are some children who do not acquire oral language at all, others who acquire a system that is limited in structure and to particular contexts (stimulus-response chains), and still others who have made some generalizations and can retrieve from a limited repertoire a set of structures which are partially appropriate.

I would put into the category of stimulus-response chains echolalic behavior, metaphoric language, and neologisms. Although the behaviors appear, on the surface, to be quite different, they nevertheless all represent retrieval of unanalyzed wholes in response to a particular situation. The distinction between echolalic and delayed echolalic, metaphoric, and neologistic language appears to be the difference between online retrieval, that is, immediate memory reprocessing, and retrieval from long-term storage. Immediate echoing is neither appropriate nor inappropriate in that it is, presumably, merely a reiteration of some aspects of what has just previously been said. Delayed echolalia, use of metaphors, and neologisms can be either appropriate or inappropriate in that either these responses appear to the listener to fit the context or they do not, and probably reflect whether or not the child was attending to some of the situational cues that were also available to the listener. These, grossly, are the categories of language-production behavior observed in this population: mutism, echolalia, acquisition of certain aspects of the language system, and its use. The context and structure of these children's language comprehension and production behaviors have not been systematically described thus far. Only their gross performances have been studied.

COMPARISONS WITH OTHER LANGUAGE-HANDICAPPED POPULATIONS

Rutter highlights the issues of similarities and differences of autistic children and children diagnosed as having receptive developmental dysphasia. He raises (but rejects) the possibility of a continuum of severity of language defect along which the developmentally dysphasic and autistic populations fall. Most of the children in the two populations fall into two discrete categories in terms of language behavior; some few present a mixture of behaviors (i.e., dysphasic/autistic). The *degree* to which language is specifically impaired (i.e., abnormalities in spoken language *more* frequent, *less* spontaneity in speech, language comprehension *very* poor, etc.) is the degree to which a child will exhibit social/behavioral characteristics which lead to the categorization of the child as "autistic." That is, the less language the child has, and the more "bizarre" it appears, the more likely are the social/behavioral characteristics of the child said to be "autisticlike." There are, however, other characteristics which appear to be limited to autistic children: undue sensitivity to sound during infancy, few articulation problems, no use of gesture or imaginative play during a later period.

DeMyer (1976) carried out a longitudinal study comparing the linguistic and nonlinguistic performance of mentally retarded and autistic children. Standard intelligence tests were used to obtain base-line measures of intellectual competence. It was found that 94% of the autistic children studied had IQs of 67 or below, and 75% had below 51. Previous studies had indicated that one quarter to one third of autistic children in a sample population score in the normal range on standard intelligence tests (Rutter, 1968). The findings concerning the language performance as compared to the nonlanguage performance of these autistic children in this study also differ somewhat from that of other studies. That is, the autistic children in general performed more poorly than the mentally retarded children on both verbal and performance scales of the tests and so-called "splinter skills" were, in general, not observed. Thus, one might hypothesize that observed differences between the two populations might be due to differences in degree of retardation. However, it was also found that mentally retarded children, who had scores similar to those of high-scoring autistic children, nevertheless exceeded them in relating to other people, using speech communicatively, and demonstrating functional object use.

In summary, both in comparisons of autistic and dysphasic children and autistic and mentally retarded children, degree of language deficit and/or degree of intellectual retardation as measured on standard tests cannot wholly account for the differences in the communicative behaviors that are observed between these populations. The autistic child displays distinct difficulties in relating to other people and in using speech communicatively and he does not demonstrate functional object use, engage in symbolic play, or use gestures communicatively. A further distinct characteristic of autistic speech is inappropriate use of the suprasegmental features of language (intonation and stress). Rutter points to extension of language difficulties to the acquisition of reading and writing as a problem of autistic children. This seems entirely logical given the fact that these abilities are based on knowledge of language, and these difficulties are found in *all* the populations discussed. However, a distinct characteristic of autistic children may be that some few of them appear to be able to read and write at a certain level although they do not produce spontaneous speech.

In my own research with children categorized as receptive and/or expressive aphasics, I have observed some of the differences between them and autistic children already remarked upon. That is, they, unlike autistic children, communicate with peers as well as adults (Menyuk, in press) and they use objects not only functionally but also symbolically. Further, they use gesture and prosody (suprasegmental features) appropriately. In addition, however, and importantly, distinct differences within the population in the ability to process and produce language can be observed. It is not simply degree of language deficit that differs among the aphasic children but, rather, the nature of the involvement (phonological and/or syntactic and/or semantic) under varying processing conditions (amount and type of linguistic information; visual-auditory or motor-auditory or auditory-auditory) which distinguishes children within this population. Indeed, there are a sufficient number of differences in the language perception and production performances of these children to suggest that the neurological substrates of their behavior, like that found in adult aphasics, may be different (Menyuk, 1975a). The same question, of course, arises when observing the clear differences in the language behaviors within the autistic population. That is, distinctions in the language behavior of these children may, logically, be due to differences in the neurological substrates of this behavior. Therefore, simply *degree* of language difficulty does not seem to adequately describe the differences between autistic and aphasic children nor the difference among the children in the population. Further, acceptance of the notion that autistic children are simply more language deficient and/or more severely retarded than children in other language-handicapped populations may obscure those distinctions which would provide better explanations of their particular forms of behavior and, possibly, be helpful in designing interventional approaches for particular children.

COMPARISONS WITH NORMALLY DEVELOPING CHILDREN

I have suggested that it is inappropriate to describe these children's communicative behavior as simply more deficient. Can one describe their language behavior as simply severely delayed but following a normal course of development? Churchill raises the issue of whether or not the behaviors he has observed represent the normal sequence of language development but, obviously, severely retarded and arrested. He implies that there may be, grossly, three populations of autistic children: those who are not all conditionable vis-à-vis language, those who are conditionable but only up to a point (they level at some point in their acquisition of a nine-word language), and those who are able to make some but not all of the associations and generalizations deemed necessary for the acquisition of simple sentences. He suggests that "conditionability" is a prerequisite for language competence, but that language competence requires something more. It is this "something more" which autistic children appear to lack.

I would suggest, given the data of studies of normal language development, that these children lack something to begin with rather than only something more. The normally developing child over the first year of life uses vocalizations expressively and communicatively; that is, to indicate needs and feelings and to socialize. Turn-taking in vocalization interaction occurs at an extremely early age (approximately one to two months). In addition the infant is sensitive to those acoustic features (suprasegmental) which communicate intent, mark speech sound boundaries (segmental), and identify the speaker. He is also sensitive to the visual features which communicate direction of movement, mark object boundaries, and identify objects by the end of the first five months of life. Thus, both perceptually and productively, in the linguistic and visual-motor domain, the infant has categorized some criterial parameters and uses these categories to operate upon the environment. Further, and crucially, in attempting to compare the language behavior of normally developing and autistic children, the very young normally developing child both perceives and uses those linguistic, paralinguistic, and situational features which communicate intent: prosody, gesture, and actions and objects in the immediate environment (Menyuk, 1976a). This perception and use is exhibited *before* the child begins to acquire the standard lexicon of the language and continues to be used as the child produces one- and two-word utterances. By the time the child is four years old he/she has not only acquired a great deal of knowledge about the structure of language but also the rules for appropriate use of language with peers, younger children, and adults (Shatz & Gelman, 1973).

This ability to discriminate, categorize, and use appropriately linguistic and nonlinguistic classes and rules during infancy does not appear to be merely a function of conditionability. Along the same lines it has been stated that the echolalic behavior of the autistic child represents an arresting of normal development in that imitative linguistic behavior occurs in the normally developing child. This seems to me to be a gross misapprehension of the structure and function of imitative behavior in the normally developing child. The imitative behavior of the latter child is seldom, if ever, mere mimicry, disappears at the end of the second year of life, and serves as a means of testing hypotheses about both the structure of language and the state of affairs (Menyuk, 1976b).

In summary, the language behaviors of autistic children, both in structure and use, do not appear to be adequately described in terms of a continuum of degree of language impairment nor in terms of degree of delay. They appear, rather, to be a product of the unique processing strategies or incapacities of these children to begin with, which, in turn, are probably due to unique neurological substrates. This brings us to a discussion of the causes of the observed behavior presented by the two researchers.

CAUSES OF THE BEHAVIOR OF AUTISTIC CHILDREN

Churchill states that the language disturbance of autistic children is an explanation not only of the profound impairment found in the speech and gestures for communication of the children but also of their nonfunctional use of objects and difficulty in interpersonal relations. He suggests that differences in perception, memory, motor integration, and social history account for *differences* among these children. Thus, a language deficit is the cause of their developmental problems. Rutter states that these children suffer from a severe cognitive deficit which involves language and language-related functions, and is due to organic brain dysfunctions. One would assume, given the differences in functioning among these children, that the nature and/or extent of the dysfunction varies, and that these differences, in turn, lead to differences in perception and memory for both linguistic and nonlinguistic information. At this point both researchers would rule out or at least question psychosocial factors as being the principal cause of autistic development. Indeed both researchers suggest that the social/behavioral anomalies observed in autistic children may be due to their linguistic and/or cognitive incapacities.

If this is the case then both researchers reasonably question the nature of this linguistic and/or cognitive deficit. Rutter points to the results of intelligence testing and observes that these children show retardation in general and have particular difficulty in those subtests that have "high verbal loading," and involve temporal sequencing and coding. He quotes Hermelin (1976), "The cognitive pathology seems to consist largely of an inability to reduce information through the appropriate extraction of crucial features such as rules and redundancies." Churchill has, in his training of a group of these youngsters, attempted to do just that—reduce information so that crucial features can be abstracted, and to reinforce the appropriate responses so that they will continue the process with the set of stimuli used, and generalize to other stimuli.

Both researchers suggest that these children do not suffer from visuo-spatial defects in organization. Rutter reaches these conclusions from these children's scores on performance IQ subtests, and the Raven's Matrices. Churchill notes that use of a visual cue allowed one of the children examined to learn what verbal response was required (either the name of an object or its color) although he did not respond appropriately to the verbal cues of "color" or "object." It is not the case, however, that these children's cognitive deficit is principally in the auditory domain. Some of them, for example, do comparatively well on the digit-span subtest of an IQ test, and on the recall of lists of words. They do not do well, however, when recall is dependent on analysis (as in their poor performance on both verbal and performance subtests which call for such analyses) or when meaning, either situational or linguistic, needs to be employed in memory processes. I would assume this to be the case whether the input stimuli are acoustic or visual. That is, when analysis is called for rather than recall of wholes or gestalts these children should experience difficulty. Thus, sequential arrangements are more difficult than spatial arrangements.

Churchill goes on to describe some attempts to teach a child appropriate responses to verbal sequences that were composed of verb + noun + preposition + noun ("Put shoe in cup"). This child was the one who had difficulty in parcelling out the relation between a visual parameter and its verbal encoding from another in the task requiring either the name of an object (the whole) or an attribute (an aspect) of the object. In this training the child had difficulty in parcelling out the distinctive meanings of the prepositions ("beside" interpreted as "in") and in sequencing appropriately ("Put cup on shoe" interpreted as "Put shoe on cup"). In the given stimulus situation objects were manipulated but without analysis of either meaning of the preposition or the relation described by word order. Thus, the child appeared to understand the nouns and verbs in the sentence or at least that the objects named were to be manipulated (i.e., the situation), but this appeared to be the limit of his understanding.

Churchill describes this as a stereotyped response and this seems to be an appropriate description. With a "higher-functioning" child, who could carry out the above tasks appropriately, an attempt was made to teach him to respond appropriately to utterances such as "What do you/I have?" and "What are you/am I doing?" After the child was taught to respond to doing (sitting or standing) and having (yellow block or blue cup), the questions were tried. The child was unable during the same session to respond to *both* "having" or "doing" and to "you" or "I" having or doing. He could only respond to *one* parameter during a single session and could not switch during that session. Churchill refers to this as an inability to deal with and parcel out double classifications. Notice, however, that in *both* types of tasks discussed ("cross referencing" and "double classification") *analyses* of either the attributes of the objects per se or of the sentence were required for responding appropriately. In the first task, the child must classify the objects as an *X and* one that has the attribute Y. In the second task, each item must be so classified: subject = X or Y (you or I), verb = X_1 or Y_1 (have or do). The distinction between the two tasks is the number of possibilities available for response: in the first X or Y; in the second $X + X_1$, or $X + Y_1$, or $Y + X_1$, or $Y + Y_1$.

MEMORY PROCESSES IN AUTISTIC CHILDREN

Thus far, the linguistic/cognitive and related behavioral problems of the autistic child have been described as an inability to analyze linguistic and nonlinguistic data. The cause of this inability has been attributed to a neurophysiological dysfunction. Since analysis itself is dependent on having appropriate rules available for processing (Bruner, 1957) one might ask if the problem is due to an incapacity to store information (not enough computing space) or an incapacity to store this information in appropriate ways. In a study dealing with the paired associate learning of a group of retarded and echolalic children (Schmidt, 1976) it was found that mentally retarded children did much better with high associative pairs than they did with low associative pairs whereas no difference in learning occurred between the high and low associative pairs in the autistic population. (Similarly, autistic children recall lists of words and sentences with equal accuracy whereas in other populations sentence recall is better than word list recall.) On the other hand, the autistic children improved more markedly over trials with low associative pairs and did better with these pairs in transferring from association of pictures to association of objects than did the mentally retarded group. These results indicated to the researcher that the autistic children's difficulty with high associative pairs derived from having *too many* associations for a given stimulus.

To me these results indicate that (1) the autistic children did not have the same categorizations available to them that the mentally retarded had, but that during the experiment they were able to establish the associations required in the task (learned response), and (2) these new associations were maintained by them for a longer time than by the mentally retarded children (the transfer task). In the above study type of language behavior elicited by the experimenter in a play session and responses to questions on the verbal scale of the WISC were analyzed into categories of "no response," types of echolalia, and "adequate response." It was found that there was a significant positive correlation between paired associate learning and adequate responses to conversation and questions (.61) and significant negative correlation between paired associate learning and either "no response" a high proportion of the time ($-.75$) or repeating last words ($-.42$).

These data and the data presented by Rutter and Churchill indicate that perceptual processing deficits which prevent the child from discriminating, categorizing, and observing relations between categories of input stimuli, and, therefore, from storing appropriate categorizations and retrieving appropriate rules for analyses underlie both the linguistic and nonlinguistic problems of these children. Rutter states that a memory (capacity) deficit per se cannot account for these children's language impairment and I would agree. They display better memory *capacity* but not organization than do mentally retarded children. One is intrigued by the possibility, given Churchill's data, that the difference in the performance of the high-and low-functioning children is due to differences in *amount* of memory capacity for learned sets, but not, necessarily, differences in analytic and categorization abilities.

If one reexamines the linguistic and nonlinguistic behavior of normally developing children and children classified as autistic, this difficulty in discrimination and categorization of both types of input from the earliest stages on becomes clear. For example, the abnormal sensitivity to sound observed in autistic infants might be a reflection of the fact that these infants have not discriminated between and categorized speech versus nonspeech stimuli, and categories of these stimuli. Lack of functional object use and response to adult caretakers might be due to nondiscrimination and categorization of those visual or visual-motor features which identify objects and people. In the normally developing infant one observes habituation to familiar stimuli and arousal only to unique stimuli in studies of speech sound and sequence discrimination. Simultaneously, during this same period, studies of visual perception have indicated that infants are surprised at object disappearances and disconcerted when their mothers' faces appear at the same time in different locations. All these data indicate that by about five months of age the properties of visual and auditory input are perceived, discriminated, and identified (Menyuk, 1975b).

I am stressing the fact that both linguistic and nonlinguistic organizations are necessary to the development of communicative abilities. Two of these abilities, feature discrimination and categorization, appear to be part of the biological repertoire of the human infant (Eimas, 1974; Bower, 1971), while others are acquired in time, but both organizations, linguistic and nonlinguistic, are necessary. Further, the normally developing child uses objects and language first functionally and then representationally, and one would assume that representational use follows from functional use. Thus, not only are both organizations necessary, but there is, also, a sequence in the development of both organizations. Clearly children within the autistic population differ in initial capacities. Otherwise, the so-called "higher-functioning" autistic children would not achieve the higher level of attainment described by both researchers. However, even these children do not achieve communicative competence.

COMMUNICATIVE INTERACTION

If one analyzes any communicative interaction, several components of analysis and of operation are involved. Figure 1 is a graphic representation of some of these components and operations. The speaker-listener must take into account the external factors in the situation and the state of his/her addressee. Depending upon the intentions of the speaker he/she will select from the linguistic repertoire those structures in those forms (phonologically, suprasegmentally, and gesturally) which will most effectively communicate that intention given the state of the listener and the message being conveyed by the speaker. All of these factors *simultaneously* play a role in effective communicative interaction.

As the child matures, not only do the structures that are available in the linguistic repertoire increase, but, also, the ability to analyze and organize each of these components and operations simultaneously increases. The highest-functioning autistic child appears to be unable to take into account all of these factors although he/she is able to produce syntactically, semantically, and phonologically "correct" sentences. The normally developing child begins to deal with some aspects of *all* of these factors at a very early age.

In summary, in order to communicate effectively linguistic rules, rules of discourse and appropriate analysis of the situation and speaker are necessary. The ability to analyze and categorize linguistic data is needed for the acquisition of linguistic rules and rules of discourse. The ability to analyze and categorize situational cues and speaker state and response is also needed for the acquisition of rules of discourse. Thus both linguistic and nonlinguistic analyses are required in communicative

EXTERNAL ANALYSES		INTERNAL OPERATIONS	
Situation ↔	*Addressee* ↔	*Intentions* ↔	*Language Knowledge*
Place	State of	Convey Information	Structural Linguistic
Objects Available	Age		Paralinguistic
Actions Occurring	Amount of Shared Information	Express Needs and Feelings	Prosodic
		Elicit Information	

Fig. 1. Communication interaction.

interaction. It should be kept in mind that the ability to communicate effectively with oneself and others will, in turn, affect, respectively, performance in certain intellectual tasks *and* social adaptation. The problems of the autistic child lie in his/her inability to analyze and categorize both linguistic and nonlinguistic data. The level of achievement of communicative competence by subpopulations of these children seems to be dependent on both initial capacities to perceive, discriminate, and categorize in the linguistic and nonlinguistic domains, and then to observe relations between categories within and across domains.

SOME RESEARCH QUESTIONS

Rutter states that the biological bases of autistic behavior need to be determined. This is clear and unquestionable. However, a simultaneously pressing need seems to me to be a resolution of other types of questions raised so that effective intervention programs can be planned with individual children within this population. These questions concern the nature and boundaries of the children's linguistic and nonlinguistic deficits (as stated by Rutter). Or (as stated by Churchill) what is the difference between them and normally developing children? The research involved should not be limited to an assessment of what they can do but also include what they can learn to do. It is insufficient to only assess their intellectual, linguistic, and social performances on standard tests of these behaviors since we can merely prognosticize with these data. Taking as our model what has been found to be the *sequence* of development in both the linguistic and nonlinguistic domains in normally developing children, we can examine in detail the types of input these children can process and the conditions under which processing takes place. We can also begin to experiment with teaching those subcomponents which appear to lead to achievement in both domains and to communicative interaction. For example, it might be helpful to teach these youngsters to categorize the same group of objects along several different parameters and observe the consequences of these groupings before we ask them to respond differently to different parameters that are of a superordinate nature. I say "might be helpful" because clearly we are unsure of which subcomponents of the behavior we should break it down into, and in what order these subcomponents should be presented to these children, although we have a better idea with normally developing children. Indeed, although we know a great deal more about the functioning of these children than we did 20 years ago, we still have a great deal to learn.

REFERENCES

Bower, T. The object in the world of the infant. *Scientific American*, 1971, *225*, 30–38.

Bruner, J. S. On perceptual readiness. *Psychological Review*, 1957, *64*, 123–152.

DeMyer, M. K. The measured intelligence of autistic children. In E. Schopler & R. J. Reichler (Eds.), *Psychopathology and child development: Research and treatment*. New York: Plenum Press, 1976. Pp. 93–114.

Eimas, P. Linguistic processing of speech by young infants. In R. Schiefelbusch & L. Lloyd (Eds.), *Language perspectives: Acquisition, retardation, and intervention*. Baltimore: University Park Press, 1974. Pp. 55–74.

Hermelin, B. Coding and the sense modalities. In L. Wing (Ed.), *Early childhood autism: Clinical, educational and social aspects (2nd Ed.)*. Oxford: Pergamon Press, 1976, Pp. 135–168.

Menyuk, P. Children with language problems: What's the problem? *Georgetown University Round Table on Languages and Linguistics*. Georgetown University, 1975, 129–144.(a)

Menyuk, P. The language impaired child: Linguistic or cognitive impairment? *Annals of the New York Academy of Sciences*, 1975, *263*, 59–69. (b)

Menyuk, P. The bases of language acquisition: Some questions. In E. Schopler & R. J. Reichler (Eds.), *Psychopathology and child development: Research and treatment*. New York: Plenum Press, 1976. Pp. 145–165. (a)

Menyuk, P. That's the 'same', 'another', 'funny', 'awful' way of saying it. *Journal of Education*, Boston University, 1976, *158*, 25–38. (b)

Menyuk, P. Linguistic problems in children with developmental dysphasia. In M. Wyke (Ed.), *Developmental dysphasia*. London: Academic Press, in press.

Menyuk, P. Communication abilities of pre-school language disordered children. Forthcoming.

Ricks, D. M., & Wing, L. Language communication and the use of symbols in normal and autistic children. *Journal of Autism and Childhood Schizophrenia*, 1975, *5*, 191–220.

Rutter, M. Concepts of autism: A review of research. *Journal of Child Psychology and Psychiatry*, 1968, *9*, 1–25.

Schmidt, J. *Relations between paired-associate learning and utterance patterns in children with echolalia*. Unpublished doctoral dissertation. Boston University, School of Education, 1976.

Shatz, M., & Gelman, R. The development of communication skills: Modifications in the speech of young children as a function of listener. Monograph. Society for Research in Child Development, 1973, *38*, No. 5.

8

Neurophysiologic Studies

EDWARD M. ORNITZ

A striking deficiency in the physiologic modulation of sensory stimuli can be inferred from clinical observation of autistic behavior (Ornitz, 1969, 1970, 1973, 1974; Goldfarb, 1961, 1963; Bergman & Escalona, 1949). All sensory modalities are affected and the faulty modulation of sensory input may be manifest as either a lack of responsiveness or an exaggerated reaction to sensory stimuli (Goldfarb, 1961, 1963). This faulty modulation of sensory input is an intrinsic feature of the autistic behavioral syndrome. This is demonstrated by the responses of parents of 74 young (mean age 45.2 months) autistic children to a standard set of questions about their child's development. Tables 1 through 3 list some of the most predominant symptoms reported by the parents in the behavioral categories of autistic disturbances of relating, perception, and motility* and indicate the percent of this population of autistic children who were reported to exhibit each behavior. It can be seen from these tables that symptoms in the category of autistic disturbances of the modulation of sensory input occur at high frequencies as do symptoms in the category of autistic disturbances of relating. Furthermore, the occurrences of many of these symptoms are highly correlated. In particular, two of the more frequently occurring symptoms of faulty modulation of sensory input ("ignoring or failing to respond to sounds" and "staring into space as if

---

*These data and the methodology required for their proper acquisition will be described in greater detail in a subsequent publication (Ornitz et al., in press).

EDWARD M. ORNITZ · Mental Retardation and Child Psychiatry Division, Department of Psychiatry, and Brain Research Institute, School of Medicine, University of California, Los Angeles, California 90024. This work was supported in part by Grant MH 26798.

Table 1.   *Percentages of Autistic Children with Disturbances of Relating*

Disturbance[a]	Number of children[b]	Percent with symptom
*Seemed very hard "to reach" or to be "in a shell"	71	90
*Ignored people as if they did not exist	73	85
Avoided looking people directly in the eye	71	76
*Acquired things by directing another's hand	72	74
Looked through people as if they did not exist	69	65
*Ignored toys as if they did not exist	72	61
Responded to affection by ignoring it	73	53
Responded to being held by clinging without interest	70	51
*Became attached to an unusual object	73	48

[a]Entries preceded by an asterisk represent variables used for assigning autism scores.
[b]The number of children varies and is less than 74 since some of the parents did not answer some of the questions on the clinical history.

seeing something not there") correlate strongly with the most prominent symptom of the autistic disturbances of relating (seeming very hard "to reach" or being "in a shell") ($r = 0.46, P = 0.0002$, two-sided). A more systematic approach to these correlations *between* categories of autistic behavior utilizes a stepwise discriminant analysis (**Ornitz** *et al.*, in press)

Table 2.   *Percentages of Autistic Children with Disturbances of the Modulation of Sensory Input*

Disturbance[a]	Number of children[b]	Percent with symptom
*Ignored or failed to respond to sounds	70	71
*Excessively watched the motions of his hands or fingers	73	71
*Stared into space as if seeing something that was not there	73	64
Preoccupied with things that spin	73	57
Preoccupied with minor visual details	69	57
Preoccupied with the feel of things	72	53
Let objects fall out of hands as if they did not exist	70	53
Preoccupied with scratching surfaces and listening to the sound	72	50
*Agitated at being taken to new places	73	48
*Agitated by loud noises	71	42

[a]Entries preceded by an asterisk represent variables used for assigning autism scores.
[b]The number of children varies and is less than 74 since some of the parents did not answer some of the questions on the clinical history.

Table 3. Percentages of Autistic Children with Disturbances of Motility

Disturbance[a]	Number of children[b]	Percent with symptom
*Flapped arms or hands in repetitive way	72	76
*Whirled around without apparent reason	70	59
*Rocked head or body	74	51
Ran or walked on toes	73	40

[a]Entries preceded by an asterisk represent variables used for assigning autism scores.
[b]The number of children varies and is less than 74 since some of the parents did not answer some of the questions on the clinical history.

to remove the redundancy associated with strong correlations among symptoms *within* categories of autistic behavior. The stepwise discriminant analysis assigned *autism scores* in each category of behavior to each of the 74 autistic children. These scores are based on those symptoms (marked with asterisks in Tables 1 through 3) selected by the discriminant procedure. The scores for the category of autistic disturbances of the modulation of sensory input correlated highly with those in the category of autistic disturbances of relating ($r = .49$, $P = 0.0002$, two-sided). Consequently, the autistic disturbances of the modulation of sensory input must be considered an integral aspect of the autistic behavioral syndrome.

Disorders of the processing of sensory input in children are amenable to modern neurophysiologic investigation because a number of technical developments in recent years permit the recording of small responses of the nervous system to various types of sensory stimuli by noninvasive techniques.

NEUROPHYSIOLOGIC INVESTIGATIONS

Neurophysiologic investigations of autistic children have been carried out in several areas: (1) electroencephalographic (EEG) studies; (2) noncontingent sensory evoked response studies; (3) sensory evoked response studies involving contingencies; (4) autonomic responses; and (5) vestibular responses. This paper will review neurophysiologic studies of autistic children in these five areas of investigation. This review will be confined (except when otherwise stated) to those investigations in which there is some evidence that the subjects are autistic according to the following criteria: age of onset prior to 30 months of age; delayed or deviant language development; and the presence of autistic disturbances of relating, the modulation of sensory input, and motility (Ornitz, 1973).

Electroencephalographic (EEG) Studies

EEG studies have been carried out in both the waking state and during sleep. Several EEG studies in the waking state have been carried out in the context of testing a hypothesis of physiologic overarousal in autistic children. Results in the waking state are dependent upon the stimulus conditions which exist during the EEG recording since the background EEG is most sensitive to sensory input. Thus, two reports of unusually low-voltage EEGs suggestive of hyperarousal (Kolvin *et al.*, 1971; Hutt *et al.*, 1965) were not confirmed in two others (Creak & Pampiglione, 1969; Hermelin & O'Connor, 1968) when stimulus conditions were controlled. A recent carefully controlled study which minimized stimulus input and utilized quantitative EEG measures (voltage integration and auto- and cross-correlation) in seven autistic and seven normal children found significantly less session-to-session and transhemispheric variability in the autistic group (Small, 1975).

EEG studies during sleep are less influenced by sensory input and are easier to control. Early EEG investigations during sleep were motivated by analogies between the subjective experience of dreaming, a psychological state which occurs during rapid eye movement (REM) sleep, and the perceptual and conceptual processes of psychotic thought. It was thought, therefore, that autistic children might have a different REM-nonREM sleep cycle. However, it has been found that autistic children have a normal sleep cycle with normal amounts of rapid eye movement (REM) sleep (Ornitz *et al.*, 1965a,b; Onheiber *et al.*, 1965). However, the rapid eye movement activity of REM sleep is reduced in autistic children (Ornitz *et al.*, 1969a) and is similar to that found in normal infants (Ornitz *et al.*, 1971), suggesting a maturational defect (Ornitz, 1972). The reduced rapid eye movement activity and particularly the reduced organization of the rapid eye movements into "bursts" also suggest the possibility of vestibular dysfunction since the presence of the rapid eye movement bursts depends on the integrity of the vestibular nuclei (see discussion in Ornitz, 1970).

Noncontingent Sensory Evoked Response Studies

The study of sensory evoked responses originally appeared to be a natural approach, on a neurophysiologic level of investigation, to those facets of autistic behavior which suggested aberrant responses to sensory stimuli. These studies have also been carried out during both the waking and sleep states. It has proven difficult to gain the cooperation of these children; movement and inattentiveness seriously impair the recording of their responses in the waking state. Consequently, several attempts have been made to study sensory evoked responses during sleep in autistic children. The only reported study of evoked responses to flashes and clicks in waking autistic children is inconclusive since no statistical analysis of the data was presented (Small, 1971).

Studies of auditory evoked responses during sleep have revealed only marginal differences between autistic and normal children and have been characterized by great variability from subject to subject (Ornitz *et al.*, 1968, 1969b). During these investigations, it was found, however, that the normal inhibition of auditory evoked responses during the rapid eye movement bursts of REM sleep was significantly less likely to occur in autistic than in normal children (Ornitz *et al.*, 1968). As with the rapid eye movement bursts, the inhibition of sensory evoked responses during the burst epochs is dependent on the integrity of the vestibular nuclei. Thus, if replicated, this finding also points toward a vestibular dysfunction in autistic children (see discussion in Ornitz, 1970). Studies of the recovery cycle of the auditory evoked response during sleep demonstrated normal responses in autistic children (Ornitz *et al.*, 1972, 1973b, 1974c). In summary, the noncontingent sensory evoked response studies of autistic children have revealed inconclusive or minimal results. The failure of these studies to reflect clinical observations of aberrant responses to sensory stimuli may be due in part to the "averaging" technique used in these studies. The clinical observations are of abnormal responses to single stimuli, e.g., the absence of a startle response. These transient phenomena may be missed by a technique which depends on the averaging of responses to long trains of stimuli.

Sensory Evoked Response Studies Involving Contingencies

Unlike the noncontingent sensory evoked responses, the contingent sensory evoked responses are usually recorded by DC amplification to facilitate study of the long latency slow potentials which are elicited in response to the contingency which is established between two previously unrelated stimuli. Because slow potentials are recorded, these studies are complicated by artifact introduced by slow vertical eye movements which often accompany the response to the stimulus (Wasman *et al.*, 1970). This problem is illustrated in one noncontingent study of the transcephalic direct current potential. In that study, an unusually high percentage of DC "bursts" were found in six- to thirteen-year-old autistic children and one- to five-year-old normal children (Goodwin *et al.*, 1971). Unfortunately, adequate controls for eye movement activity were not reported so that this apparently interesting maturational finding may only reflect the tendency of older autistic and younger normal children to have more vertical eye movements. Returning to the contingent sensory evoked responses, in one study the contingent negative variation (CNV) occurred in autistic children when two stimuli were paired together, but the autistics, unlike normal controls, did not show a differential response to slides of familiar and strange faces (Small *et al.*, 1971). However, this does not seem to represent a neurophysiologic abnormality since the basic ability to establish a CNV is intact, but rather reflects the ordinary clinical experience that autistic children do not show adequate differential responses to strangers and do not participate in social interaction (Walter *et al.*, 1971).

Lelord *et al.* (1973) have reported the development of large slow negative or positive potentials in response to stimulus coupling in autistic children. These potentials were significantly larger than those found in normal controls and some measures were instituted to rule out the contribution of eye movement potentials. If replicated, this finding will be of great interest because these slow waves resembled those evoked by movement or its anticipation (Laffont *et al.*, 1971), while in this experimental situation they developed during sensorial conditioning in the absence of movement. The authors suggested, therefore, a diffuse motor component to the perceptual and associative processes of autistic children. This hypothesis is consistent with the suggestion that the spontaneous spinning and flicking of objects, the flapping and oscillating of their extremities, and the whirling and rocking of their bodies may be the autistic children's way of making sense out of the sensations in their environment, including their own bodies and their parts, through kinesthetic (sensorimotor) feedback (Ornitz, 1974). This notion of a strong motor component to the perceptual processes of autistic children finds further support in psychologic experiments (Hermelin & O'Connor, 1970; Frith & Hermelin, 1969) which indicate that autistic children learn through cues that are primarily manipulative, i.e., involving motor feedback. Thus, autistic children "seem to rely more on perceptual activity than on perceptual analysis" (Hermelin & O'Connor, 1970).

Autonomic Studies

Several investigations of autonomic responses to sensory stimuli have been attempted. An inadequate galvanic skin response to both auditory and visual stimuli has been demonstrated in autistic children (Bernal & Miller, 1970). This finding is consistent with the clinical observation that autistic children do not show an adequate startle response. It was not possible, however, to replicate the finding by using changes in heart rate as the experimental measure (Miller & Bernal, 1971).

Increased variability in the heart rate has been reported, but adequate control of activity level and stimulus conditions under which autistic and normal children were observed was not documented in one study (MacCulloch *et al.*, 1971). However, this finding has been very recently confirmed in a study which did control for activity level and also found that the greatest variance of beat-to-beat heart rate in autistic children occurred during their stereotyped behaviors (Hutt *et al.*, 1975).

It had been suggested by MacCulloch *et al.* (1971) that increased heart rate variability in autistic children might indicate vestibular system involvement since the nucleus of the tractus solitarius (vagal outflow) lies adjacent to the vestibular nuclei in the brain stem. The significant increase in heart rate variability during epochs of stereotyped behavior (Hutt *et al.*, 1975) provides a link between the clinical observations (referred to above) which relate the motility disturbances of autistic children to their perceptual processes and the hypothesis of vestibular system dysfunction. In this analysis of the behavior of autistic children, it is postulated that pathophysiologic central vestibular mechanisms drive the autistic disturbances of motility (stereotyped motor behaviors) which, in turn, provide sensorimotor feedback to compensate the faulty modulation of sensory input which, itself, is due to insufficient vestibular regulation of sensory processing. The experimental findings of reduced organization of rapid eye movements into bursts (motor output dysfunction) and reduced inhibition of auditory evoked responses (sensory input dysfunction) during the vestibularly mediated eye movement epochs of REM sleep (Ornitz *et al.*, 1968, 1969a) are also consistent with this hypothesis.

Review (Ornitz, 1970, 1974) of the neurophysiology of the vestibular system reveals that the vestibular nuclei either directly modulate or transmit modulating influences over motor output at the time of sensory input and over sensory input at the time of motor output. Thus, a dysfunction of a complex circuitry involving the central connections of the vestibular system may be responsible for the strange sensorimotor behavior observed in autistic children.

Vestibular Response Studies

I have made a speculative excursion into the possible vestibular origins of the sensorimotor dysfunction found in autistic children. Before reviewing the experimental studies of their vestibular responses, it seems prudent to first review the clinical observations which bear upon this theme.

Clinical observations during the past three decades have implicated a dysfunction of vestibular mechanisms in autistic children (Ornitz, 1970). Bergman and Escalona (1949) recorded disturbances of equilibrium among the "unusual sensitivities" found in young psychotic children whose behavior was similar to that of autistic children. It was observed, e.g., that one child "never tolerated any of the usual equilibrium stunts which most babies enjoy, like being swung around, lifted up high, etc. One of the fears that she developed in later years was that of being turned upside down." During their extensive clinical experience with "schizophrenic" children, some of whom actually manifested autistic behavior (Bender & Faretra, 1972; Bender, 1969, 1970), Bender and her colleagues recorded numerous examples of behavior symptomatic of vestibular dysfunction. These observations included "rotating and whirling motor play in all planes" and the tendency to seek "a dependable center of gravity" (Bender, 1947), fearful reactions to "rapid movements as in trains and elevators" (Bender & Freedman, 1952), an intolerance of "sensations arising from gravity, changes in spatial or position relationships, or from vestibular stimuli" and, with increasing age, a tendency to "respond with terror to antigravity play, to whirling or to flying fantasies," or by compensation to "become fascinated and fixated upon such perceptual experiences" (Bender, 1956). Some 20 to 22 out of 30 "schizophrenic" children showed abnormalities of posture and balance and spontaneous whirling during neuropsychiatric examinations (Bender & Helme, 1953). Colbert et al. (1959) have speculated about the vestibular mechanisms which might underlie the spontaneous whirling and toe walking (Colbert & Koegler, 1958) of these children. Developmental data recently obtained from parents of young autistic children reveal that 59% of a group of 74 children are reported to have "whirled around without apparent reason" (Ornitz et al., in press). Autistic children seem to induce labyrinthine stimulation by whirling, excessive head and body rocking, and head swaying and rolling. The absence of dizziness, vertigo, or loss of balance associated with whirling is a further indication of vestibular dysfunction. During the rotational tests described later in this paper, one autistic child was observed to repeatedly give himself Coriolis stimulation by tilting his head from side to side while rotating at 180 deg/sec. He could not be dissuaded from this behavior which he clearly enjoyed, and he never showed any sign of vertigo or nausea (personal observations). All of these observations suggest that autistic children are seeking out, overreacting to, or avoiding sensations induced by stimulation of vestibular mechanisms.

Vestibular dysfunction is also indirectly suggested by a history of delayed motor development (Rapin, 1974) which has been reported in schizophrenic children with the clinical features of early infantile autism (Fish, 1961). Similar data have recently been obtained from the parents of 74 young autistic children: a significant delay in walking was reported for 50% of the children (Ornitz et al., in press).

The preoccupation of many autistic children with spinning objects (57% of 74 young autistic children according to parental reports; Ornitz et *al.*, in press) may also be indicative of vestibular dysfunction. The prolonged observation of, e.g., spinning tops, may provide optokinetic stimulation which, in turn, may modify vestibular responsivity (Pfaltz & Ohtsuka, 1975; Young & Henn, 1974; Dichgans & Brandt, 1973; Young *et al.*, 1973).

The clinical observations which have been described are complemented by both indirect and direct evidence from experimental studies of autistic children. Since it is known that the rapid eye movement activity of REM sleep is mediated by the vestibular nuclei (Pompeiano, 1967; Pompeiano & Morrison, 1965), the effect of vestibular stimulation on the ocular activity of sleeping autistic and normal children was tested (Ornitz *et al.*, 1973a). The children slept on a special bed which provided a mild continuous sinusoidal acceleration and deceleration throughout the entire night. Under the influence of this stimulation, the duration and organization of the rapid eye movement bursts significantly increased during the course of the night in the normal children, but showed no response in the autistic children. This result suggested the possibility of an abnormal vestibularly mediated oculomotor response to an exogenous vestibular input.

A more direct approach to the problem of vestibular dysfunction in autistic children has been to elicit nystagmus by vestibular stimulation. The first such study of children who may have been autistic found minimal or absent postrotational nystagmus in seven out of 15 seven- to nine-year old patients; in contrast, adequate postrotational nystagmus occurred in six out of seven behavior disorder children and in all nine normal controls (Pollack & Krieger, 1958). This early study is difficult to interpret since the children were labeled "schizophrenic" and no behavioral description whatsoever was given. Several other difficulties in the conduct and reporting of vestibular experiments with autistic children are also illustrated by this investigation. The observation of postrotational nystagmus following ten revolutions of a Bárány chair in 20 seconds was apparently made in a lighted environment since electronystagmography was not used. The influence of alertness and visual fixation (Wendt, 1951; Marshall & Brown, 1967; Collins, 1966, 1968; Hood, 1970; Pfaltz & Piffko, 1970; Dix & Hood, 1969; Ornitz *et al.*, 1974a) was not evaluated and there was not quantification of the nystagmus which was merely reported as "minimal or absent" versus "present."

Considerable improvement occurred in a second study of vestibular nystagmus in "schizophrenic children" (Colbert *et al.*, 1959). Although the children were referred to as "schizophrenic," the clinical description which was provided permits the inference that this was essentially an autistic population. Forty of the 43 patients had shown major developmental disturbances within the first two years of life (Ornitz, 1973; Kolvin *et al.*, 1971; Rutter, 1967), and autistic disturbances of speech, play patterns, motility, and response to sounds were described. Vestibular nystagmus was elicited in these 43 patients, 18 "behavior problem" children and 32 normal children by both caloric and rotational stimulation. The "schizophrenic" children had a significant reduction of both caloric and postrotational nystagmus duration measured in lighted conditions by direct observation.

The third study (Ritvo *et al.*, 1969) of vestibular nystagmus in autistic children was carried out in the author's laboratory; the methodology permitted assessment of the influence of visual input. This is an important consideration since it is known that visual fixation suppresses nystagmus in all individuals. The 28 patients were clearly autistic, showing autistic disturbances of relating, motility, perception, and language by clinical history and direct examination (Ornitz & Ritvo, 1968; Ornitz, 1973). The children (28 autistic and 22 normal) received ten rotations in 20 seconds in a hand-rotated Bárány chair; the postrotational nystagmus was measured, following an abrupt braking deceleration, by electronystagmography. Half of the trials were carried out in lighted conditions which permitted visual fixation to occur and half of the trials were conducted in the dark so as to preclude visual fixation or any other aspect of visual input. The autistic children showed a highly significant reduction of postrotatory nystagmus duration in the lighted room, but not in the dark. This study eliminated some of the problems inherent in the earlier investigations. The patients were unequivocally autistic children. A quantitative measure of vestibular response, postrotatory nystagmus duration, was recorded by electronystagmography. The influence of visual input was considered and was found to be of critical importance.

In order to clarify the nature of the influence of visual input on the vestibular nystagmus, we compared the postrotatory nystagmus duration and the total number of nystagmus beats in 21 autistic children and 25 age-matched normal children under six different conditions of visual input (Ornitz *et al.*, 1974a,b). All of the autistic children showed autistic disturbances of relating, perception, and motility (Ornitz, 1973) prior to 24 months of age and were either mute or echolalic. Rotation was by the conventional Bárány procedure of ten revolutions in 20 seconds in complete darkness followed by an abrupt braking stop of an electromechanically driven and remotely controlled chair. Eye movements were recorded by electronystagmography. Controls were implemented to assure that the subjects remained in an alert state with eyes open to ensure optimal nystagmus (Tjernström, 1973) on all trials accepted for data analysis. The six different conditions of visual input which were maintained during the postrotational period included: (1) a standard visual field with standard illumination to determine the degree of suppression of nystagmus in light when fixation was possible but not encouraged; (2) a standard visual field with a fixation object added to determine if nystagmus suppression would be enhanced when deliberate fixation was encouraged; (3) a pinpoint of red light with minimal light value in complete darkness to study the effect of fixation in the absence of light; (4) goggles which admitted light without visual pattern to preclude fixation in order to study the effect of light without fixation; (5) same as (4), but with lower light intensity to study the effect of light intensity; (6) complete darkness for comparison with the various conditions of visual input. These six experimental conditions made it possible to distinguish the influence of visual fixation in the presence or absence of light from the influence of light in the absence of visual fixation. Figures 1 and 2 show the postrotatory nystagmus durations and total number of beats (mean of clockwise and counterclockwise rotations) under the different conditions of visual input for the 18 autistic and 22 normal children who had successful trials in all six experimental conditions. The major finding was that the inhibition of postrotatory nystagmus depends on the interaction between the vestibular stimulation *and* the visual input. The visual input may be either a stimulus to visual fixation in the absence of light or may be light in the absence of visual fixation. In the latter case, the nystagmus inhibition is greater if the light is brighter. These findings are important for an understanding of the pathophysiology of autism because they show that the

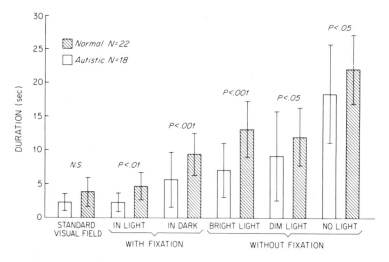

Fig. 1. Postrotatory nystagmus durations under six different conditions of visual input (mean ± one standard deviation). In this figure and Fig. 2, the probabilities that the means of the autistics are different from those of the normals are based on contrasts (tests on simple main effects) after an analysis of variance.

interaction of oculosensory input (light) with a vestibular stimulus (abrupt deceleration) is sufficient to suppress vestibularly induced nystagmus. An oculomotor action (visual fixation) is not necessary to obtain this result. This suggests that autistic children have not "learned" how to suppress vestibular responsiveness by fixating more than normal children do. Instead, the results suggest a more fundamental neurophysiologic interaction between the visual and vestibular systems in determining the deficient nystagmus response in autistic children.

Clinical Variability within the Autistic Population

Within a large population of autistic children, there is considerable variation in the degree to which the autistic behavior is manifest. Childhood autism clearly does *not* occur in an all-or-nothing manner. This is illustrated in respect to the direct psychiatric examination of 74 autistic children in Fig. 3 where it can be seen that, e.g., 5.5% of the children showed no autistic disturbances of perception, 15% showed minimal symptoms, and 11% showed maximal symptoms in this category. Another type of clinical characterization of the autistic behavior is based on a statistical approach to the historical data obtained from the parents of these same 74 children. The parental responses were compared to those obtained from parents of normal children who filled out the same written symptom inventory. The responses of all the parents were then subjected

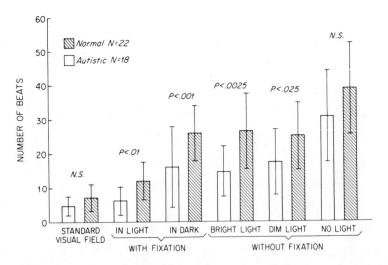

Fig. 2. Total number of postrotatory nystagmus beats under six different conditions of visual input (mean ± one standard deviation).

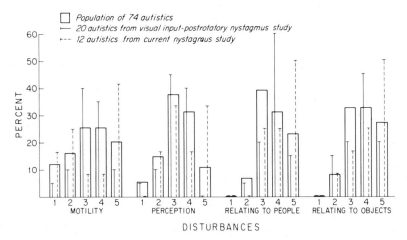

Fig. 3. The percentage of autistic disturbances of motility, perception, and relating observed during psychiatric examinations of a large population of autistic children and two smaller populations drawn in part from the larger population. The scales run from 1 = behavior not present at all to 5 = behavior unequivocally present.

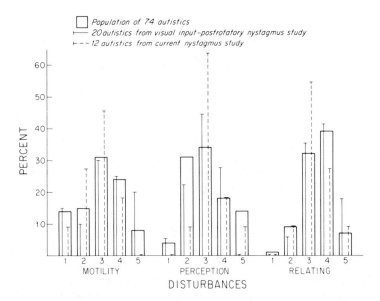

Fig. 4. The percentage of autistic children showing disturbances of motility, perception, and relating based on *autism scores* derived from a discriminant analysis of parental histories for the same populations of autistic children described in Fig. 3. The scales run from 1 = behaviors never occurred to 5 = behaviors almost always occurred.

Table 4. Correlation of Postrotatory Nystagmus Durations under Different Visual Input Conditions with Autistic Behavior Scores Derived from Parental Histories

		Autistic behavior disturbance scores		
Visual input condition		Perception	Motility	Relating
1. Standard visual field	r	−0.23	−0.45	−0.37
with standard	(N)[a]	(16)	(18)	(15)
illumination	P[b]	NS	<.05	NS
2. Standard visual field	r	−0.14	−0.47	−0.31
with fixation	(N)	(17)	(19)	(16)
encouraged	P	NS	<.025	NS
3. Pinpoint of red light	r	−0.16	−0.51	−0.27
to encourage fixation	(N)	(17)	(19)	(16)
in the absence of light	P	NS	<.025	NS
4. Bright light transmitted	r	−0.38	−0.50	−0.44
through frosted goggles	(N)	(16)	(18)	(15)
to preclude fixation	P	NS	<.025	≐.05
5. Same as condition 4 but	r	−0.10	−0.46	−0.23
with lower light	(N)	(13)	(15)	(12)
intensity	P	NS	<.05	NS
6. Complete darkness	r	−0.22	−0.39	−0.32
	(N)	(17)	(19)	(16)
	P	NS	≐.05	NS

[a] N (the number of subjects in each correlation) varies because of missing data.
[b] P is the probability that r reflects a significant decrease in postrotatory nystagmus duration as the disturbed behavior increases in severity.

to a linear discriminant analysis (Lachenbruch, 1975). The discriminant analysis assigned *autism scores* to each patient in each major category of behavior indicated in Tables 1 through 3. These scores can be thought of as weighted averages of those individual symptoms (in each category of behavior) which were selected by the discriminant analysis procedure as having value in classifying subjects as autistic.* The distribution of autism scores (or weighted averages of individual symptoms) is shown in Fig. 4 where the values one through five refer to the parents' estimates of the frequencies (1 = never . . . 5 = almost always) at which the behaviors were exhibited. Figures 3 and 4 show the distributions of autism scores for the smaller populations of autistic children who were subjects of the visuovestibular interaction study (Ornitz et al., 1974b) described above and the new vestibular nystagmus studies to be described below. Perusal

*A detailed account of these procedures will appear in Ornitz, Guthrie, and Farley (in preparation).

of both figures reveals a relatively normal distribution of behavioral characteristics for the large, relatively unselected population of 74 autistic children. In contrast, the behavioral characteristics of the smaller groups of autistic children selected for the nystagmus studies are less normally distributed. Thus, it can be seen that experimental results from vestibular studies, or any other experimental studies on small groups of autistic children may not necessarily be representative of the larger population of autistic children from which the experimental population is drawn.

With this caution in mind, we examined possible correlations between the clinical characteristics of the autistic children studied in the visuovestibular interaction study (Figs. 1 and 2) (Ornitz *et al.*, 1974b). Table 4 shows the correlations (r) between the durations of postrotatory nystagmus and the autism (discriminant analysis) scores derived from the parental histories. A large negative correlation indicates that short nystagmus durations are associated with high behavioral scores. It is of interest that, with one exception, the significant correlations are confined to the disturbances of motility. Since these include very stereotyped activities, e.g., hand flapping, which occur in patterns suggesting nonvolitional behavior (Sorosky *et al.*, 1968; Ritvo *et al.*, 1968; Ornitz *et al.*, 1970), it might be suggested that the reduced nystagmus is related to an aspect of the autistic syndrome that may involve lower brain centers which function in the regulation of motor output as suggested above. These correlations are not strong, and they are based on a small number of cases; they are presented to suggest avenues of investigation which might be opened by looking quantitatively at the variations within the autistic syndrome.

WORK IN PROGRESS: PRELIMINARY DATA ACQUIRED DURING A STUDY OF SECONDARY NYSTAGMUS IN AUTISTIC CHILDREN

In the nystagmus studies described above, the source of the vestibular stimulus was either caloric stimulation or an abrupt angular deceleration which followed 20 seconds after an acceleration of uncontrolled magnitude (the conventional Bárány procedure). Both types of stimulation have disadvantages. The caloric stimulation, while useful in the diagnosis of peripheral vestibular lesions, is a nonphysiologic stimulus. The conventional Bárány procedure confounds the nystagmus response to an acceleration with that to a deceleration occurring 20 seconds later. Consequently, the complete nystagmus response to a unidirectional angular acceleration cannot be studied. In particular, the *secondary nystagmus* as illustrated in Fig. 5 is not observed.

By stimulating subjects in such a way as to permit the development of

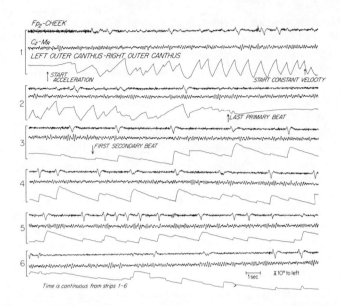

Fig. 5. 10 deg/sec² counterclockwise acceleration of a 25-month-old normal boy. Example of a successful recording. Note vigorous secondary nystagmus with eyes deviating in opposite direction to that of primary nystagmus. Blink artifacts (upper channel), EEG (middle channel), and DC oculogram (lower channel) are all consistent with a state of relaxed alertness.

secondary nystagmus (Wendt, 1951; Aschan & Bergstedt, 1955) as well as primary nystagmus, by quantifying specific parameters (e.g., slow and fast phase amplitude, duration, and velocity) of each separately, and by quantifying these response parameters in relation to graded stimulus intensity, information about vestibular *adaptation* processes may be obtained (Young & Oman, 1969; Malcolm & Jones, 1970). Such adaptation processes may reflect the activity of central nervous system mechanisms (Stockwell *et al.*, 1973). It should be noted, however, that although Wendt (1951), Aschan and Bergstedt (1955) and Collins (1968) have described the genesis of the secondary nystagmus in terms of *central* vestibular function, studies of the discharge rate of *peripheral* vestibular units have indicated prolonged adaptation effects, including secondary responses (Goldberg & Fernandez, 1971). These secondary responses of discharging units in primates have a time course which parallels that of the secondary nystagmus in man. Goldberg and Fernandez explain these late secondary adaptation effects in terms of the physiology of the hair cells and/or the afferent nerve terminals. However, they leave open the possibility that activation of efferents to the sensory epithelium by central processes is involved. Honrubia *et al.* (1973), in studies of induced nystagmus in unilateral labyrinthectomized cats, have provided evidence that both the postrotatory primary nystagmus and the secondary nystagmus are dependent on central nervous system influences.

Regardless of the ultimate origin of the secondary nystagmus, it can be considered a measure of vestibular adaptation, and the preliminary data presented here can be considered a measure of vestibular adaptation in autistic children.

The secondary nystagmus is a nystagmus of opposite direction to that of the primary nystagmus and is generated by sustained angular accelerations of sufficient magnitude. In normal subjects, there is a predictable relationship of the secondary nystagmus to the primary nystagmus which precedes it (Malcolm & Jones, 1970; Young & Oman, 1969). The secondary nystagmus in a normal child in response to an acceleration of 10 deg/sec^2 for 18 seconds is illustrated in Fig. 5. Using these parameters of stimulation in the dark with eyes open and with attention to the mental state, data have been analyzed from recordings of 19 normal children 22–139 months old (mean age 54.1 months) and 15 autistics (29–153 months old, mean age 57.6 months). The behavioral characteristics of 12 of these 15 autistic children are given in Figs. 3 and 4. The results of the nystagmus computations (averages of clockwise and counterclockwise trials) are shown in Fig. 6 where it can be seen that significant differences between the autistic and normal children are confined to one parameter (slow phase duration) of the secondary nystagmus.

We have tried on manipulation of these preliminary and very limited data. This approach is based on the notion that the secondary nystagmus is a measure of vestibular adaptation (Malcolm & Jones, 1970; Young & Oman, 1969) and that there is a quantitative relationship between the secondary nystagmus and the primary nystagmus (Malcolm & Jones, 1970). It was hypothesized that the autistic children would not have the expected quantity of secondary nystagmus given the amount of primary nystagmus elicited by the stimulus. The proportion of secondary to primary nystagmus for both the postacceleration and the total primary nystagmus was computed for each child and the resultant ratios of the 19 normal and 15 autistic children were compared. Figure 7 shows that, in respect to several of the nystagmus parameters, the autistics have significantly less secondary nystagmus relative to their primary nystagmus than do the normals. If these preliminary data based on a small number of cases should be confirmed for an autistic population of greater size, then we might postulate some pathologic involvement of vestibular adaptive mechanisms. It should be noted that if confirmation of significant differences between autistic and normal children should be obtained on these measures, then an additional comparison should be made between autistic children and nonautistic retarded children since the majority of autistic

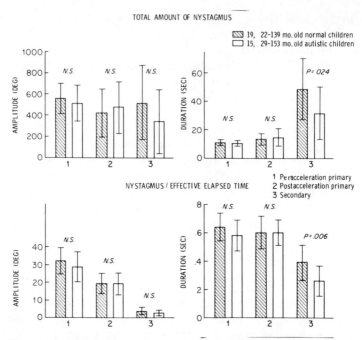

Fig. 6. Slow phase nystagmus in response to 10 deg/sec² acceleration for 18 seconds. Means (± one standard deviation) of total slow phase nystagmus (upper half) and slow phase nystagmus normalized for effective elapsed time (lower half) in normal and autistic children. Effective elapsed time is the total period of time (minus time obscured by artifact) during which the primary or secondary nystagmus occurred. Duration is the actual time during which the eyes moved during nystagmus beats.

children are developmentally retarded. The estimated developmental quotients of the 15 autistic children in this vestibular adaption study (based on the relative amounts of secondary and primary nystagmus) varied from 28 to 78 and no correlation with their nystagmus scores could be found. Nevertheless, our experimental program calls for replication of these experiments with a comparison of autistic and nonautistic retarded children of similar mental age.

SUMMARY AND CONCLUSIONS

Neurophysiologic investigation of autism is a particularly appropriate research approach to this disorder because a prominent feature of the clinical symptomatology appears to be a faulty modulation of sensory input. Neurophysiologic responses to sensory stimuli have been measured in the context of EEG studies, noncontingent and contingent sensory evoked response studies, autonomic response studies, and vestibular response studies. The occurrence of large slow potentials, resembling those evoked by movement, during a contingent sensory evoked response experiment and increased variability of heart rate during stereotyped motor behavior suggest a strong motor component to the sensory processes of autistic children. These findings, together with clinical observations of the strange sensorimotor behaviors of autistic children and their tendency to learn through motor feedback rather than normal perceptual processes, point toward a neurophysiologic dysfunction of some system which modulates the interaction of sensory and motor processes. The central connections of the vestibular system may subserve such a regulatory function, and some irregularities of sensory input and (oculo)motor output associated with the vestibularly mediated rapid eye movement bursts of REM sleep have been reported in autistic children. Experimental studies of vestibular nystagmus have implicated a disturbance of the interaction of two sensory systems, vestibular and visual, in autistic children, and the severity of this abnormality is correlated with the severity of

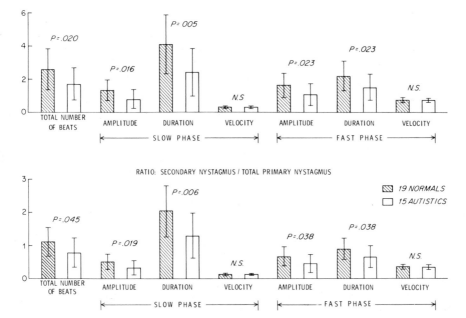

Fig. 7. The proportion of secondary to primary nystagmus in autistic and normal children. The values are means (± one standard deviation) of ratios of the secondary nystagmus to the postacceleration primary nystagmus (upper half) and to the total primary nystagmus (lower half).

the motility disturbances found in these children. Therefore, it can be suggested that the postulated vestibular dysfunction is intimately associated with the disturbances of perception (the modulation of sensory input) and motility (the modulation of motor output) found in autistic children. Finally, recent preliminary findings from a study of the secondary nystagmus, a measure of central vestibular mechanisms, indicate a disturbance of vestibular adaptation in autistic children.

REFERENCES

Aschan, G., & Bergstedt, M. The genesis of secondary nystagmus induced by vestibular stimuli. *Acta Societatis Medicorum Upsaliensis,* 1955, *60,* 113–122.

Bender, L. Childhood schizophrenia. Clinical study of one hundred schizophrenic children. *American Journal of Orothpsychiatry,* 1947, *17,* 40–56.

Bender, L. Schizophrenia in childhood—Its recognition, description, and treatment. *American Journal of Orthopsychiatry,* 1956, *26,* 499–506.

Bender, L. A longitudinal study of schizophrenic children with autism. *Hospital and Community Psychiatry,* 1969, *20,* 230–237.

Bender, L. The life course of children with autism and mental retardation. In F. J. Menolascino (Ed.), *Psychiatric approaches to mental retardation.* New York: Basic Books, Inc., 1970. Pp. 149–191.

Bender, L., & Faretra, G. The relationship between childhood schizophrenia and adult schizophrenia. In A. R. Kaplan (Ed.), *Genetic factors in "schizophrenia."* Springfield, Ill.: Charles C Thomas, 1972.

Bender, L., & Freedman, A. M. A study of the first 3 years in the maturation of schizophrenic children. *Quarterly Journal of Child Behavior,* 1952, *4,* 245–272.

Bender, L., & Helme, W. H. A quantitative test of theory and diagnostic indicators of childhood schizophrenia. *A.M.A. Archives of Neurology and Psychiatry,* 1953, *70,* 413–427.

Bergman, P., & Escalona, S. K. Unusual sensitivities in very young children. *Psychoanalytic Study of the Child,* 1949, *3–4,* 333–353.

Bernal, M. E., & Miller, W. H. Electrodermal and cardiac responses of schizophrenic children to sensory stimuli. *Psychophysiology,* 1970, *7,* 155–168.

Colbert, E., & Koegler, R. Toe walking in childhood schizophrenia. *Journal of Pediatrics,* 1958, *53,* 219–220.

Colbert, E. G., Koegler, R. R., & Markham, C. H. Vestibular dysfunction in childhood schizophrenia. *A.M.A. Archives of General Psychiatry,* 1959, *1,* 600–617.

Collins, W. E. Vestibular responses from figure skaters. *Aerospace Medicine,* 1966, *37,* 1098–1104.

Collins, W. E. Special effects of brief periods of visual fixation on nystagmus and sensations of turning. *Aerospace Medicine,* 1968, *39,* 257–266.

Creak, M., & Pampiglione, G. Clinical and EEG studies on a group of 35 psychotic children. *Developmental Medicine and Child Neurology,* 1969, *11,* 218–227.

Dichgans, J., & Brandt, T. Optokinetic motion sickness and pseudo-coriolis effects induced by moving visual stimuli. *Acta Oto-Laryngologica,* 1973, *76,* 339–348.

Dix, M. R., & Hood, J. D. Observations upon the nervous mechanism of vestibular habituation. *Acta Oto-Laryngologica,* 1969, *67,* 310–318.

Fish, B. The study of motor development in infancy and its relationship to psychological functioning. *American Journal of Psychiatry,* 1961, *117,* 1113–1118.

Frith, U., & Hermelin, B. The role of visual and motor cues for normal, subnormal and autistic children. *Journal of Child Psychology and Psychiatry,* 1969, *10,* 153–165.

Goldberg, J. M., & Fernandez, C. Physiology of peripheral neurons innervating semicircular canals of the squirrel monkey. I. Resting discharge and response to constant angular accelerations. *Journal of Neurophysiology,* 1971, *34,* 635–684.

Goldfarb, W. *Childhood schizophrenia.* Cambridge, Mass.: Harvard University Press, 1961.

Goldfarb, W. Self-awareness in schizophrenic children. *Archives of General Psychiatry,* 1963, *8,* 47–60.

Goodwin, M. S., Cowen, M. A., & Goodwin, T. C. Malabsorption and cerebral dysfunction: A multivariate and comparative study of autistic children. *Journal of Autism and Childhood Schizophrenia,* 1971, *1,* 48–62.

Hermelin, B., & O'Connor, N. Measures of the occipital alpha rhythm in normal, subnormal and autistic children. *British Journal of Psychiatry,* 1968, *114,* 603–610.

Hermelin, B., & O'Connor, N. *Psychological experiments with autistic children.* Oxford: Pergamon Press, 1970.

Honrubia, V., Strelioff, D., & Ward, P. H. Computer analysis of nystagmus induced by constant angular accelerations in normal and labyrinthectomized cats. *Advances in Oto-Rhino-Laryngology,* 1973, *19,* 254–265.

Hood, J. D. The clinical significance of vestibular habituation. *Advances in Oto-Rhino-Laryngology,* 1970, *17,* 149–157.

Hutt, C., Forrest, S. J., & Richer, J. Cardiac arrhythmia and behavior in autistic children. *Acta Psychiatrica Scandinavica,* 1975, *19,* 361–372.

Hutt, C., Hutt, S. J., Lee, D., & Ounsted, C. A behavioural and electroencephalographic study of autistic children. *Journal of Psychiatric Research,* 1965, *3,* 181–197.

Kolvin, I., Ounsted, C., & Roth, M. Six studies in the childhood psychoses. V. Cerebral dysfunction and childhood psychoses. *British Journal of Psychiatry,* 1971, *118,* 407–414.

Lachenbruch, P. A. *Discriminant analysis.* New York: Hafner Press, 1975.

Laffont, F., Sauvage, D., & Lelord, G. Activité lente corticale succédant au mouvement volontaire chez l'homme. *Comptes Rendus des Séances de la Société de Biologie et de Ses Filiales (Paris),* 1971, *165,* 660–665.

Lelord, G., Laffont, F., Jusseaume, P., & Stephant, J. L. Comparative study of conditioning of averaged evoked responses by coupling sound and light in normal and autistic children. *Psychophysiology,* 1973, *10,* 415–425.

MacCulloch, M. J., Williams, C., & Davies, P. Heart-rate variability in a group of cerebral palsied children. *Developmental Medicine and Child Neurology,* 1971, *13,* 645–650.

Malcolm, R., & Jones, G. M. A quantitative study of vestibular adaptation in humans. *Acta Oto-Laryngologica,* 1970, *70,* 126–135.

Marshall, J. E., & Brown, J. H. Visual-arousal interaction and specificity of nystagmic habituation. *Aerospace Medicine,* 1967, *38,* 597–599.

Miller, W. H., & Bernal, M. E. Measurement of the cardiac response in schizophrenic and normal children. *Psychophysiology,* 1971, *8,* 533–537.

Onheiber, P., White, P. T., DeMyer, M. K., & Ottinger, D. R. Sleep and dream patterns of child schizophrenics. *Archives of General Psychiatry,* 1965, *12,* 568–571.

Ornitz, E. M. The disorders of perception common to early infantile autism and schizophrenia. *Comprehensive Psychiatry,* 1969, *10,* 259–274.

Ornitz, E. M. Vestibular dysfunction in schizophrenia and childhood autism. *Comprehensive Psychiatry,* 1970, *11,* 159–173.

Ornitz, E. M. Development of sleep patterns in autistic children. In C. D. Clemente, D. Purpura, & F. Mayer (Eds.), *Sleep and the maturing nervous system.* New York: Academic Press, 1972.

Ornitz, E. M. Childhood autism: A review of the clinical and experimental literature. *California Medicine,* 1973, *118,* 21–47.

Ornitz, E. M. The modulation of sensory input and motor output in autistic children. *Journal of Autism and Childhood Schizophrenia,* 1974, *4,* 197–215.

Ornitz, E. M., Brown, M. B., Mason, A., & Putnam, N. H. The effect of visual input on postrotatory nystagmus in normal children. *Acta Oto-Laryngologica,* 1974, *77,* 418–425. (a)

Ornitz, E. M., Brown, M. B., Mason, A., & Putnam, N. H. The effect of visual input on vestibular nystagmus in autistic children. *Archives of General Psychiatry,* 1974, *31,* 369–375. (b)

Ornitz, E. M., Brown, M. B., Sorosky, A. D., Ritvo, E. R., & Dietrich, L. Environmental modification of autistic behavior. *Archives of General Psychiatry,* 1970, *22,* 560–565.

Ornitz, E. M., Forsythe, A. B., & de la Peña, A. The effect of vestibular and auditory stimulation on the REMs of REM sleep in autistic children. *Archives of General Psychiatry,* 1973, *29,* 786–791. (a)

Ornitz, E. M., Forsythe, A. B., Tanguay, P. E., Ritvo, E. R., de la Peña, A., & Ghahremani, J. The recovery cycle of the averaged auditory evoked response during sleep in autistic children. *Electroencephalography and Clinical Neurophysiology,* 1974, *37,* 173–174. (c)

Ornitz, E. M., Guthrie, D., & Farley, A. J. The early symptoms of childhood autism. Presented at Kittay Scientific Foundation Symposium, "Cognitive Defects in the Development of Mental Illness," in press.

Ornitz, E. M., & Ritvo, E. R. Perceptual inconstancy in early infantile autism. *Archives of General Psychiatry,* 1968, *18,* 76–98.

Ornitz, E. M., Ritvo, E. R., Brown, M. B., LaFranchi, S., Parmelee, T., & Walter, R. D. The EEG and rapid eye movements during REM sleep in normal and autistic children. *Electroencephalography and Clinical Neurophysiology,* 1969, *26,* 167–175. (a)

Ornitz, E. M., Ritvo, E. R., Lee, Y. H., Panman, L. M., Walter, R. D., & Mason, A. The auditory evoked response in babies during REM sleep. *Electroencephalography and Clinical Neurophysiology,* 1969, *27,* 195–198. (b)

Ornitz, E. M., Ritvo, E. R., Panman, L. E., Lee, Y. H., Carr, E. M., & Walter, R. D. The auditory evoked response in normal and autistic children during sleep. *Electroencephalography and Clinical Neurophysiology,* 1968, *25,* 221–230.

Ornitz, E. M., Ritvo, E. R., & Walter, R. D. Dreaming sleep in autistic and schizophrenic children. *American Journal of Psychiatry,* 1965, *122,* 419. (a)

Ornitz, E. M., Ritvo, E. R., & Walter, R. Dreaming sleep in autistic twins. *Archives of General Psychiatry,* 1965, *12,* 77–79. (b)

Ornitz, E. M., Tanguay, P. E., & Lee, J. C. M. The recovery function of the auditory evoked response during sleep in children. In U. J. Jovanovic (Ed.), *The nature of sleep.* Stuttgart: Gustav Fischer Verlag, 1973. (b)

Ornitz, E. M., Tanguay, P. E., Lee, J. C. M., Ritvo, E. R., Sivertsen, B., & Wilson, C. The effect of stimulus interval on the auditory evoked response during sleep in autistic children. *Journal of Autism and Childhood Schizophrenia,* 1972, *2,* 140–150.

Ornitz, E. M., Wechter, V., Hartman, D., Tanguay, P. E., Lee, J. C. M., Ritvo, E. R., & Walter, R. D. The EEG and rapid eye movements during REM sleep in babies. *Electroencephalography and Clinical Neurophysiology,* 1971, *30,* 350–353.

Pfaltz, C. R., & Ohtsuka, Y. The influence of optokinetic training upon vestibular habituation. *Acta Oto-Laryngologica,* 1975, *79,* 253–258.

Pfaltz, C. R., & Piffko, P. Studies on habituation of the human vestibular system. *Advances in Oto-Rhino-Laryngology,* 1970, *17,* 169–179.

Pollack, M., & Krieger, H. P. Oculomotor and postural patterns in schizophrenic children. *Archives of Neurology and Psychiatry,* 1958, *79,* 720–726.

Pompeiano, O. The neurophysiological mechanisms of the postural and motor events during desynchronized sleep. *Research Publications of the Association for Research in Nervous and Mental Disease,* 1967, *45,* 351–423.

Pompeiano, O., & Morrison, A. R. Vestibular influences during sleep. I. Abolition of the rapid eye movements during desynchronized sleep after vestibular lesions. *Archives Italiennes de Biologie,* 1965, *103,* 569–595.

Rapin, I. Hypoactive labyrinths and motor development. *Clinical Pediatrics,* 1974, *13,* 922–937.

Ritvo, E. R., Ornitz, E. M., Eviatar, A., Markham, C. H., Brown, M. B., & Mason, A. Decreased postrotatory nystagmus in early infantile autism. *Neurology,* 1969, *19,* 653–658.

Ritvo, E. R., Ornitz, E. M., & LaFranchi, S. Frequency of repetitive behaviors in early infantile autism and its variants. *Archives of General Psychiatry,* 1968, *19,* 341–347.

Rutter, M. Psychotic disorders in early childhood. In A. J. Coppen & A. Walk (Eds.), *Recent developments in schizophrenia—A symposium.* London: RMPA; Ashford, Kent: Headley Bros., 1967. Pp. 133–158.

Small, J. G. Sensory evoked responses of autistic children. In D. W. Churchill, G. D. Alpern, & M. DeMyer (Eds.), *Infantile autism: Proceedings of the Indiana University Colloquium.* Springfield, Ill.: Charles C Thomas, 1971. Pp. 224–242.

Small, J. G. EEG and neurophysiological studies of early infantile autism. *Biological Psychiatry,* 1975, *10,* 385–398.

Small, J. G., DeMyer, M. K., & Milstein, V. CNV responses of autistic and normal children. *Journal of Autism and Childhood Schizophrenia,* 1971, *1,* 215–231.

Sorosky, A. D., Ornitz, E. M., Brown, M. B., & Ritvo, E. R. Systematic observations of autistic behavior. *Archives of General Psychiatry,* 1968, *18,* 439–449.

Stockwell, C. W., Gilson, R. D., & Guedry, F. E., Jr. Adaptation of horizontal semicircular canal responses. *Acta Oto-Laryngologica,* 1973, *75,* 471–476.

Tjernström, Ö. Nystagmus inhibition as an effect of eye closure. *Acta Oto-Laryngologica,* 1973, *75,* 408–418.

Walter, D. O., Berkhout, J. I., Buchness, R., Kram, E., Rovner, L., & Adey, W. R. Digital computer analysis of neurophysiological data from Biosatellite 111. *Aerospace Medicine,* 1971, *42,* 314–321.

Wasman, M., Morehead, S. D., Lee, H. Y., & Rowland, V. Interaction of electroocular potentials with the contingent negative variation. *Psychophysiology,* 1970, *7,* 103–111.

Wendt, G. R. Vestibular functions. In S. S. Stevens (Ed.), *Handbook of experimental psychology.* New York: John Wiley & Sons, Inc., 1951. Pp. 1191–1223.

Young, L. R., Dichgans, J., Murphy, R., & Brandt, T. Interaction of optokinetic and vestibular stimuli in motion perception. *Acta Oto-Laryngologica,* 1973, *76,* 24–31.

Young, L. R., & Henn, V. S. Selective habituation of vestibular nystagmus by visual stimulation. *Acta Oto-Laryngologica,* 1974, *77,* 159–166.

Young, L. R., & Oman, C. M. Model for vestibular adaptation to horizontal rotation. *Aerospace Medicine,* 1969, *40,* 1076–1079.

9

Images and Language

BEATE HERMELIN

Whenever we are confronted with mental impairment in children, as for instance in autism, the most outstanding feature is that there is a general reduction of competence. However, as it is neither very illuminating nor profitable to come up repeatedly with the finding that children who are ill do less well than children who are not, the experimental psychologist is obliged to develop alternative approaches. A strategy which we have adopted is to look for communalities and dissimilarities in the cognitive processes of different diagnostic groups of children. Thus we do not adopt a distinct disease concept, but rather seek to establish how groups of children whose pathology is quite different will respond to situations and tasks which are defined in terms of certain specific underlying mental operations.

One additional advantage of such an approach for the study of autism is that it enables one to investigate cognitive processes independently of the cognitive level of functioning of an individual or a group. This does not imply that cognitive strategies are totally independent of level of function, but only that this correspondence is not inevitable. Not only may overall similar IQ scores be obtained by different patterns or processes, but similar cognitive strategies may also be used at different cognitive levels.

One method, which we have frequently adopted, is to use control groups of subnormal, intelligence-matched children in investigating child-

BEATE HERMELIN · MRC Developmental Psychology Unit, Drayton House, Gordon Street, London WC1H OAN, England.

hood autism. If the experimental results obtained from such groups should differ in spite of their similar intellectual level, then it would be justified to attribute the so-isolated process to something other than intelligence. Another way to deal with the same problem is to compare groups of children of very different levels of cognitive functioning. If, in spite of such differences their response patterns on certain tests were similar, one could conclude that this similarity must be due to some function which is not dependent on the level of intelligence. This is the line we have followed in the studies described below. All the experiments have been carried out jointly with my colleague N. O'Connor.

We have compared autistic children with subjects whose impairment is perceptual rather than cognitive, and we asked whether data from intact sensory sources would be handled similarly by blind, deaf, or autistic, when compared with normal children. We are, of course, fully aware of the differences in the nature of the language deficits of the deaf when compared with the autistic, and likewise the perceptual impairments of the blind are, of course, qualitatively different from those of autistic children. But the question is not whether the nature of the impairment is similar, but rather whether the consequences of sensory and central deficits are comparable. We are asking how children with various deficits and difficulties perceive the world around them. The following simple illustration might make the implications clear: If one imagines going into a completely dark room without switching on the light in order to get a book which one left lying on the table, one may guide oneself with outstretched hands along the way, and then grope on the tabletop in order to find the book. If one does this, one will in fact use information, i.e., touch, which is at this moment directly available. However, if the room is familiar, one will probably also use information, which though not present at this moment, is nevertheless present in one's mind, or stored in the brain. From previous visual experience one will be able to evoke a visual image of the room and of the position of the book on the table, i.e., one will see with the "mind's eye." If we turn from visual to auditory images, for instance while reading a poem or a musical score, one may similarly "hear" the sounds of the words or the music in one's head, though the actual input would be visual.

Such simple examples, familiar to all of us, illustrate that at any given moment we experience not only what we see or hear, but see or hear on the basis of what we have previously experienced. We do not only use that data which is directly available to our senses. In order to interpret the world around us, we also use data from previous experience which has been stored in our brain in the form of image codes. Normally, if one, for instance, shows a series of letters one after another on a screen and asks subjects to remember and recall them, most people would probably do this by "silently" saying them to themselves. In fact, experiments have clearly demonstrated that this is so, by showing that when people misremember, an incorrectly recalled letter does not look but sounds similar to the one for which it has been mistakenly substituted.

This tendency of the mind to change one code into another, more appropriate one, has however certain psychological consequences. The reason for this is that different sensory codes reflect the specific characteristics of the sense modality from which they are derived, and, therefore, these codes which are abstracted from the different senses are not equivalent. One can, for example, imagine a complete visual scene, which is present simultaneously before the "mind's eye." The visual image can provide one with information on the spatial location of all its components. Vision is a sense which organizes incoming data in space, and the visual image code tends to reflect this property and is a spatially organized code. Hearing on the other hand, and for that matter speaking, is a temporally organized affair. We tend to hear, say, and remember one sound after another. In fact, the human ear is notoriously bad in distinguishing between simultaneously heard sounds. One can only listen with difficulty to two or more voices talking at once, and even trained musicians must concentrate to specify all the simultaneously playing instruments in a piece of music. Of course we can also say *when* we saw something or *where* we heard a sound come from. But generally vision is primarily spatially organized, that is in terms of where things are, while hearing and speaking tend to be ordered in temporal sequences, according to when the items occurred.

Space and time, the framework for all our perceptions, are thus linked specifically to seeing and hearing. It follows from this that people without the experience of vision, or the ability to use visual images, might be impaired in their conceptualization of space, and people without hearing and with language deficits might have difficulties with temporal sequences.

Regardless of whether or not autistic children have developed some overt speech, their abnormalities in thinking have often been explained by an assumption that they lack "inner language." I would like to suggest that they may also lack "inner pictures," or to put it more generally that there may be in their case an absence of internal representation of external events, which normally serves as a code which is used to interpret the environment. According to the theory of child development put forward by Piaget and Inhelder (1969) there is a close association between different mental operations, which are all dependent on the symbolic function. These operations include not only language and play, but also the ability to construct mental images and to form stable emotional relationships. None of these operations are possible, Piaget suggests, without the ability to represent objects and events internally when they are physically absent.

In our context, the term "image" will be used in a somewhat different sense from that implied by Piaget. Nevertheless, one could utilize the Piagetian framework in suggesting that the reason why autistic children do not use language appropriately, and neither play imaginatively, nor form adequate relationships, may all be associated with an impaired development of symbolic functions. One might then predict that children with autism would function adequately when the directly available stimulation was sufficient to interpret a situation. However, when the situation was such, that in order to deal with it we would normally use other, additional, previously acquired information, which is internally represented in the form of an image code, then autistic children may be expected to behave either as if they did not have such representations available, or could not relate them to the present situation.

The autistic children in all the experiments which will now be described attended special schools and were between 10 and 14 years of age. As has already been pointed out, in these studies we were not concerned with the level of functioning, but with the qualitative analysis of the thought process. Therefore, the IQ range of the autistic subjects varied widely, from 30 to 70 points on the WISC verbal scale, and 40 to 100 points on the performance scale. Correspondingly, overall mental age ranged from 5 to 12 years. The minimum requirement for participation in the studies was the necessary degree of cooperation, and the ability to follow the instructions, although these were not dependent on language comprehension. All tasks were repeatedly demonstrated, and the children practiced them until the experimenters were satisfied that they were understood. A total population of 40 autistic children was available, and groups in the different experiments usually included 15 to 20 subjects.

Onset of autism in all cases was reported at birth or during the first 18 months. All children had been reported to have been aloof and emotionally remote at some stage of their lives, though this symptom had in many cases abated at the time of testing. All showed severe speech disabilities, were reported to be resistant to change, and exhibited various motor mannerisms and obsessive preoccupations with particular objects or activities. No overt symptoms of brain damage were present, and a psychiatric diagnosis of autism has been obtained in all cases.

Blind children were blind from birth, or onset occurred during the first 18 months. No child had useful pattern vision, though a degree of light vision was present in some. Onset of deafness in the deaf group was also at birth or before 18 months. No deaf child had any useful hearing in either ear over the speech frequency range. Blind and deaf children as well as normal control groups were between 10 to 14 years of age and of normal intelligence.

The first study was an immediate memory experiment with normal, deaf and autistic children (O'Connor & Hermelin, 1973). We presented the subjects visually with three letters, which were displayed in such a way that the successive-temporal order, and the left-to-right spatial order of the series did not coincide. The letters were displayed on a screen with three 4-cm windows placed horizontally from left to right at eye level. They appeared successively at these windows, one disappearing while the next one appeared. Each letter was visible for 0.66 sec and they never appeared in a left-to-right succession. If, for instance, the first-to-last presentation was in the order *A E M*, *A* would appear first, but in the right-hand window, *E* was shown second in the left-hand window, and *M* last, in the middle.

The children were simply asked to recall each set of three letters after they had appeared. Before the experiment, we had ascertained that the immediate memory space of all subjects exceeded three items, and that all could read and name letters. In the experimental conditions the child had two options to reproduce the order of presentation. He could either repeat the letters as they had been exposed in the order from first to last, or from left to right. He could, of course, also give a recall order which had not occurred, and these responses were termed "random."

There was no difference between the groups in the total number of three-letter sequences which were correctly recalled. However, the pattern of recall did differ. Normal children tended to recall the letters in the temporal-sequential order in which they had appeared, i.e., from first to last, while deaf and autistic children gave responses which were predominantly spatially organized, i.e., from left to right. This is illustrated in Table 1.

A statistical analysis confirmed that the pattern of these results was the same for deaf and autistic children, and that both groups differed from the normal controls. We interpreted these results as indicating that normal children had used an auditory-verbal, and therefore temporally ordered recall code, in spite of the visual presentation. On the other hand the left-to-right recall by the deaf and the autistic children seemed to indicate the use of a visual and therefore spatially organized code. What these children stored was presumably a complete "picture" of the display, and

Table 1. Total Number of Correctly Recalled Three-Letter Sequences

Subjects	Responses		
	Temporal	Spatial	Random
Normal	118	26	16
Deaf	16	132	12
Autistic	43	103	14

the "read off" from this image proceeded from left to right. While the normal children transposed the items from the visual presentation modality to an auditory-verbal storage code, the deaf and the autistic children did not.

One question which arises is whether the letters were either exclusively temporally or spatially ordered in memory or whether a dual code was available to the subjects. In order to test this we used a recognition procedure. For this the presentation conditions were identical to those just described. But following this, the subject was immediately presented with a card measuring 5 × 10 cm. On each card, two sets of three digits were printed. One set corresponded to either the temporal or spatial order of the preceding display. The other set consisted of the same three numbers, but ordered randomly, i.e., in a manner which did not correspond to either the temporal or spatial sequences. Thus if the temporal sequence of exposure had, for instance, been 3 5 9 and the left-to-right exposure 9 3 5, one card would show 3 5 9 together with 5 9 3, and another card would show 9 3 5 with 5 9 3. On half the cards the randomly ordered set was printed above and on half below the temporally or spatially ordered set. Each subject was tested for 32 trials, and for 16 of these the task was to recognize the temporal and for the other 16 the spatial order of the previously displayed digits, Instructions were: "You are going to see three numbers. After this you will see a card with two sets of numbers. Point to the one you have just seen."

The results of this recognition task are shown in Table 2. Although the scores were somewhat more evenly distributed than in the recall study, statistical analyses indicated that the normals recognized only the temporal displays successfully and tended to guess whenever spatially and randomly ordered digits were presented together. The converse was the case for the deaf and autistic children, who recognized the left-to-right order, but their temporal recognition scores did not differ significantly from chance.

The results from the recognition experiment confirmed that the material was remembered only in either its temporal or its spatial form, but not as a double storage code. This does not mean of course, that the children

Table 2. *Total Number of Correctly Recognized Three-Number Sequences*

Subjects	Responses		Responses	
	Temporal	Random	Spatial	Random
Normal	114	46	88	72
Deaf	94	66	130	30
Autistic	96	64	125	35

could not have retained either order under appropriate instruction. However, in the absence of such an instruction there were qualitative differences between the groups in the manner in which the displays were remembered and organized.

These results illustrate two points. The first is that an identical display can be remembered in terms of different memory codes, and these codes will determine the organization which is imposed on the material. The storage code can either reflect the modality of presentation, or storage occurs in another system, which has a better facility for handling this particular information. In the present instance, and this is the second point, normal children treated visually presented verbal material as if it were auditory, and consequently ordered it temporally. This probably reflected their tendency for implicit verbalization of the presented items, and their verbal rehearsal. Deaf children on the other hand do not tend to verbalize such visually presented material spontaneously, and in any case, in the absence of an auditory image code one would not expect them to opt for a sequential order. Instead they remembered in terms of a visual, spatially organized code. Autistic children, though they hear, nevertheless also treated the information in the manner in which it had been presented, i.e., as a visual display. Consequently they too recalled and recognized the letters and numbers not from first to last, but from left to right.

We will now turn to a comparison of autistic with blind and sighted normal children, which dealt with the question of the sufficiency of touch and movement for constructing spatial coordinates. It has been proposed for instance by von Senden (1960) that conceptualization of space may require the mapping of tactile and kinesthetic stimuli into some referent schema or image, derived from visual experience. Those born blind do not, of course, have such visually derived schema available, while the autistic children may be impaired in the construction and utilization of images and representations. The question is therefore whether space, experienced by touch and movement, differs for blind and blindfolded autistic children when compared with sighted blindfolded normals.

In the first of these studies to be described (Hermelin & O'Connor, 1971) blind, and blindfolded autistic and normal children were asked to extend two fingers of each hand, one in front of the other. They then learned a tactual-verbal association between one of the fingers which was touched by the experimenter and a word with which they had to respond. The words were "run," "sit," "walk," and "stand." When this association had been learned to criterion, the hands were reversed, so that, for instance, a finger previously in a position where "run" had been the response was now placed in the spatial location to which "walk" would be the appropriate association. The child could select one of two alternative response strategies. He could either respond to the finger touched, though this was now in a different spatial position, or he could retain the association between the verbal response and a fixed spatial location, though the finger stimulated at this location was now a different one. When we compared blindfolded autistic, blindfolded normal, and blind children on this task, the results were that the blind and also the autistic children responded in terms of a relative, movable, and the blindfolded normal according to a fixed absolute organization of space. Blind and autistic children primarily retained an association between a particular finger which was touched, and a previously learned verbal response. Normal children on the other hand tended to respond in terms of an association between a word, and a fixed spatial position. Thus though stimulation was through the sense of touch, the existence of a visually derived spatial framework determined the manner in which the normal blindfolded children responded, and this was qualitatively different from the way in which the autistic children and the blind treated the identical stimulus situation.

These are, of course, special situations in which it is possible to select either one of two alternative response strategies. There are, however, many tasks in which the available evidence is sufficient for interpretation, so that no additional internal representations from other sources need to be evoked. But in other operations a lack of such alternative memory representations may lead to an impaired performance. The next experiment illustrates this (O'Connor & Hermelin, 1975). The subjects in these studies were again blind, blindfolded normal, and blindfolded autistic children.

The question we asked was whether exploration by touch alone was sufficient to solve spatial rotation problems. Pairs of two-dimensional shapes were mounted alongside each other on a flat board. In one set, half of the cutout sections would fit directly into the gap in the squares while the rest would not. In the other set, the smaller piece would first have to be rotated through 180° before direct fitting was possible. However, as the two pieces were fixed and could not be moved, what was required was a "mental rotation." An illustration is shown in Fig. 1. The child felt around the shapes with the preferred hand and had to decide whether a square would result by fitting the shapes together. No difference between blind and blindfolded sighted normal or autistic children in this task was found. It seemed as if, after some practice, an appreciation of rotation-independent features for comparison and recognition had been developed. The children used sharp corners, curves, and other shape features rather than orientation, for making their decision. For this estimation of shape features, manual exploration was obviously efficient.

In another task we presented models of hands in various spatial

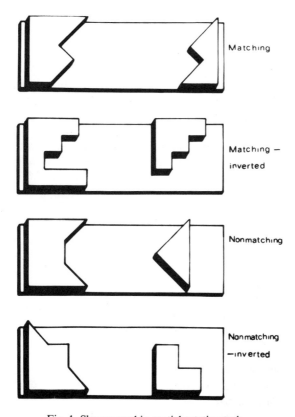

Fig. 1. Shapes used in spatial rotation task.

orientations and rotations, and the child had to decide in each trial whether he felt a right or a left hand. The experimenter sat opposite the subject and presented him with a right or a left three-dimensional model of a hand in one of six orientations, and in four axial rotations for each of these orientations. The subject was required to "feel" this hand and to decide whether it was a left or a right one. The six orientations were fingers pointing up vertically, down vertically, to the left, to the right, toward the experimenter, and away from him. The axial positions were in the four horizontal orientations of thumb up, thumb down, thumb left, or thumb right and in the two vertical orientations of thumb left, thumb right, thumb toward the experimenter, or thumb away from him. Hands have the same shape, and a right and a left hand are simply mirror images of each other. Thus, as the orientational features were the only distinguishing cues in this rotation task, the normal blindfolded children were both faster, and more often correct in deciding whether they felt a right or a left hand, than were the blind or the autistic. A spatial reference frame which included visual dimensions seems to have helped the blindfolded normal children in making a decision. This was in contrast to the congenitally blind and also to the autistic children. They relied on touch alone, while the normal sighted children were aided by their ability to relate information which was initially derived from another modality, i.e., from vision, to this tactual discrimination task.

Finally we compared blind and blindfolded children for their ability to reproduce different aspects of a limb movement (Hermelin & O'Connor, 1975).We asked whether kinesthetic data were sufficient for the derivation and storage of movement components in the absence of kinesthetic invariants. Held (1961) has expressed the view that movements of the body and the limbs are accompanied by a copy of the normally achieved visual results. The visual information resulting from movement is fed back and compared with the visual schema which was expected on the basis of previous movements. Attneave and Benson (1969) even suggested that the specific input modality of stimuli might be ignored and the data transferred to that modality system best able to process and store them. They concluded that information about spatial coordinates was primarily represented in visual terms, regardless of whether the data were derived through vision or through one or more of the other senses.

If a visual reference system is essential for coding aspects of limb movement, those who do not have or do not use such a reference system may be impaired in this respect. To investigate this we compared normal, sighted, blindfolded children with congenitally blind as well as with autistic children for their ability to remember and to reproduce location and distance features of an arm movement.

A picture of the apparatus which was used is shown in Fig. 2.

For practice trials, a lever was first moved vertically along a rod up to a stopper. The stopper was then removed, and the test movement begun from a starting point which differed from that used in the practice movement. For instance, the starting lever might have been set 5 cm from the bottom for the practice movements, and the stopper at a height of 15 cm. In the first condition, end position had to be reproduced, but from a different starting point. Thus in the test trials the starting lever might be set 10 cm from the bottom, while the movement from this point had to terminate, as in the practice trials, at 15 cm after the stopper had been removed. In this task the subject had to begin the movement from a different point than in the practice trials, but stop at the original end point.

In a second task, distance rather than end point of the previously practiced movement had to be reproduced. The child had to make an arm movement of the same extent as before, which, as he started from a different point, also had to terminate at a different position than the stopping point in the practice trials.

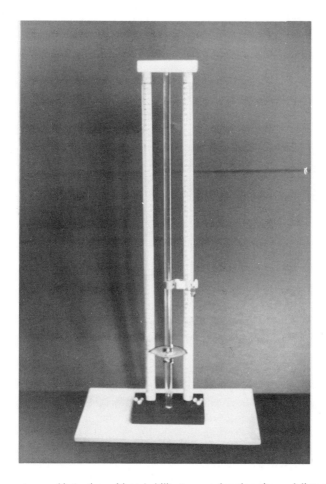

Fig. 2. Apparatus used in testing subjects' ability to reproduce location and distance features of an arm movement.

We found no significant differences between blind and blindfolded normal or autistic children in the first of these tasks, i.e., location reproduction. The position of the arm could apparently be recalled by all the children on the basis of kinesthetic information, and though there were errors they were similar in extent for all three groups.

However, while there are receptors in the muscles and joints which do signal limb position, there are no such receptors for signaling distance, and distance information seems not very precise, and perhaps even uncodable, when it is the only available information regarding limb movement. Thus, large differences between the groups in the accuracy of reproduction of distance were found, the normal blindfolded children being more correct than the autistic or the blind children. The mean errors in this task were 1.64 cm for the normals, 2.67 cm for the blind, and 2.64 cm for the autistic.

If movements of the body and the limbs are normally accompanied by a "copy" of the previously achieved visual results, as had been suggested, the normal children may have judged distance on this basis. Blind children, of course, could not use such visually derived distance estimates, and autistic children similarly seemed to have been unable to evoke the necessary visual images to supplement the information derived from movement.

These results may be interpreted as showing that the effects of absence of sight can be similar to those due to a failure to make cognitive inferences from a visual reference system. It seems that when spatial motor tasks can be solved by using directly available kinesthetic data, visually as well as cognitively impaired children can perform adequately. However, when it is necessary to make inferential judgments in terms of perceptual reference systems, those lacking such systems, or lacking the ability to use them, are at a disadvantage. The normal, human brain, to which a complete range of sensory information is available, does not process data solely in terms of the sense modality in which they are received. On the contrary, such information often seems to be transferred from the actually stimulated modality to that sensory system which has the greatest capacity for handling the particular data. As we have shown, language items tend to be processed by the auditory-verbal system, even when the actual input is visual, and such material therefore is remembered and recalled in a temporal order. Spatial information seems to be analyzed with the participation of a visual framework, independent of whether the input is visual or tactual.

This tendency for a flexible strategy of data analysis has consequences for the qualitative aspects of the codes which are used for storage and interpretation. Consequently, children with sensory impairments, who do not have the full range of possible codes available to them, may interpret information from intact, remaining sources differently than those who have no such deficits. Autistic children, though able to receive the complete range of sensory information, nevertheless seem to use only such data as are at the moment presented to them. This lack of mental mobility in handling information, and their tendency to remain stimulus-bound, can make them in fact operationally blind or deaf or even both. In view of this, it would be an oversimplification to attribute the cognitive pathology in autism primarily to an impaired language system. According to our results, such an interpretation fails to take account of the restricted access which autistic children seem to have also to other, nonlinguistic representations. It is also well to remember that it is not the possession of language which enables the child to think, but the ability to think and form symbols which makes language development possible. As Piaget has pointed out, the construction of images is the first symbolic act which is carried out by the child. It is only when objects and actions can be internally represented that stable emotional relationships are formed, imaginative play can develop, and language development can begin. Though in our context the term "image" is somewhat differently defined than it is by Piaget, nevertheless the overall conception of the role of internal representations remains relevant.

Some further points might be made. In some of the studies which were reported, we also tested subnormal children of similar age and IQ as the autistic. We found that some of these subjects, too, treated visually presented verbal input as pictures, and further analyses showed that it was those subnormals with low verbal IQs who responded in this manner. On the other hand the subnormals did not tend to behave like the blind or the autistic when confronted with spatial motor tasks. Instead, their response pattern was similar to that of the normals, which indicates that they were able to evoke visual images. Thus while some subnormal children can be said to behave in some situations as if functionally deaf, autistic children in addition also resemble the blind. A behavioral similarity with the deaf-blind has also been observed by Wing (1969).

Symbols and internal representations develop through an interaction between an individual and his environment. If environmental input is severely restricted, as in the deaf-blind, there is little on which these representations can be based. As the case of Helen Keller showed, it is only when some alternative input has become organized, thereby conveying information, that symbolic functions can develop.

The series of experiments which has been described seems to indicate to us that autistic children do not tend to integrate current experience with schemas and representations stored from previous sensory impressions. We had previously shown that these children deal with input purely in terms of an unclassified, immediate memory system, and that they do not use rules and structures in mental processing and remembering (Hermelin & O'Connor, 1970). This lack of integrated, rule-governed behavior was found not only in regard to language, but also, though to a lesser extent, in perceptual motor tasks. One must thus conclude that the basic cognitive deficit of autistic children, at least as far as the age group we studied is concerned, is not a language impairment per se, but rather that it affects those underlying processes which are necessary for the development of representational and symbolic systems.

However, it has also to be kept in mind that a simple developmental hypothesis, such as assuming that autistic children do not reach the stage where symbols become operative, will not be sufficient. Though it is true that the impairment in evoking internal representations is the most adequate description of the nature of the cognitive pathology of somewhat older autistic children, one should not forget that the condition is present from birth or early infancy, when such an explanation cannot apply. Autistic babies differ from normal ones long before the latter can handle symbols and images, so that at this level of development at least, the behavioral pathology must be due to other psychological mechanisms and processes. The nature of these processes must await detailed investigation of autistic infants, and this in turn is dependent on early identification of the syndrome.

REFERENCES

Attneave, F., & Benson, L. Spatial coding and tactual stimulation. *Journal of Experimental Psychology*, 1969, *81*, 216–222.

Held, R. Exposure history as a factor in maintaining stability of perception and coordination. *Journal of Nervous and Mental Disease*, 1961, *132*, 26–32.

Hermelin, B., & O'Connor, N. *Psychological experiment with autistic children*. Oxford: Pergamon, 1970.

Hermelin, B. & O'Connor, N. Spatial coding in normal, autistic and blind children. *Perceptual Motor Skills*, 1971, *33*, 127–132.

Hermelin, B., & O'Connor, N. Location and distance estimates of blind and sighted children. *Quarterly Journal of Experimental Psychology*, 1975, *27*, 295–301.

O'Connor, N., & Hermelin, B. The spatial or temporal organization of short-term memory. *Quarterly Journal of Experimental Psychology*, 1973, *25*, 335–343.

O'Connor, N., & Hermelin, B. Modality specific spatial coordinates. *Perception and Psychophysics*, 1975, *17*, 213–216.

Piaget, J., & Inhelder, B. *The psychology of the child*. London: Routledge & Kegan Paul, 1969.

Senden, M. von. *Space and sight*. London: Methuen, 1960.

Wing, L. C. The handicap of autistic children—a comparative study. *Journal of Child Psychology and Psychiatry*, 1969, *10*, 1–40.

10

Research Methodology: What Are the "Correct Controls"?

WILLIAM YULE

The history of the study of infantile autism is almost exactly contemporaneous with the development of clinical psychology as an applied science. Clinical psychology has been formally taught at degree level in Britain since 1947, and in these past 30 years there has been a very fruitful interaction between pure and applied research. In particular, the past three decades have witnessed the development of quite sophisticated and varied experimental methodologies for investigating an ever-increasing variety of clinical problems and phenomena.

While it would be inappropriate in the present context to attempt to delineate all these methodological developments, it is clear that many investigations of autistic children fail to use the most appropriate research designs. For optimal development, any applied science must both formulate researchable problems that are clinically meaningful and must use appropriate strategies in pursuing the investigations.

Kanner (1943) did autistic children and their families a great service when he so eloquently described the condition as a separate "syndrome." Good, objective description is one of the keystones to scientific progress. Experimental methodology is another. It is with the faulty application of experimental methodology that this commentary is concerned.

WILLIAM YULE · Departments of Psychology and Child and Adolescent Psychiatry, Institute of Psychiatry, De Crespigny Park, Denmark Hill, London SE5 8AF, England.

EXPERIMENTAL METHOD

Most established research workers in the field of infantile autism will have been brought up in an era when the classical experimental methodology of the social sciences involved a comparison between at least two groups of subjects on some critical variable. Null hypotheses were ritually formulated, although few experimenters may have fully understood the Popperian view of the philosophy of science which necessitated that the hypotheses be couched in negative terms. Data were solemnly gathered and subjected to statistical analyses, usually based on an analysis of variance model. This combination of group comparisons and statistical analyses has proved a very powerful tool in investigations in the social sciences.

However, such studies are based on the assumption that the variable on which the groups are being compared is a substantive (nontrivial) one, and that the groups do not substantially differ on any other relevant measure. In other words, studies are conducted to throw light on important hypotheses, and the comparison groups must be appropriately matched. (Of course, "matching" can sometimes be achieved statistically by partialling out covariance in a number of ways, but this usually requires larger groups than are to be found in most experimental studies of autistic children.) Another way of expressing this latter point is to demand that the experimenter has controlled all the relevant variables, or that he has selected "the correct controls." It is this search for "correct controls" which appears to have been widely misunderstood.

Currently, techniques developed from behavioral approaches are enjoying apparently great success in the treatment of autistic children. Professionals working within an applied behavior analysis framework have developed their own research methodology suited to investigating treatment questions with individual subjects (Baer *et al.*, 1968; Yule & Hemsley, 1977). Single-case experimental designs rely upon demonstrating a functional relationship between changes in treatment conditions and changes in the organism's overt behavior. If the child's behavior improves as treatment is applied but deteriorates when treatment is removed, then the treatment can be regarded as being useful.

In this sort of experimental paradigm, the experimental subject is used "as his own control." One is concerned with the behavior of the individual under different treatment regimes, and there is no need to have untreated controls to answer the question, "Does treatment make a difference?" However, many single-case enthusiasts have failed to realize that in order to answer questions about the relative effectiveness of treatments over sizable populations, group comparison studies of the classical experimental variety are still required.

As can be seen in our own home-based study (Chapter 26) different comparison groups are required to test different hypotheses. In our study, we needed one group of children matched with one experimental group on relevant variables who differed only in that they had not been offered systematic behavioral treatment. A comparison of the *children's* progress allows one to comment on the value overall of treatment versus no treatment. However, we also were interested in changes in parental behavior over time, so a separate control group of parents had to be formed. In other words, there is no such thing as *the* correct control group, different experimental controls are needed for different purposes, and within one study individual hypotheses may be tested using a variety of separate controls.

In passing, it is of interest to note an emerging paradox in the realm of applied behavior analysis. While within-subject designs are used almost exclusively, by so doing investigators focus on individuals at the expense of studying *individual differences*. As we pointed out a number of years ago (Yule & Berger, 1972), early studies of operant speech training with autistic children frequently ignored the fact that whereas some mute children learned useful labeling in a few trials, others remained mute after thousands of trials. Had investigators such as Lovaas (Chapter 25) provided adequate clinical descriptive data on each individual case, we might by now have been better able to predict which children would benefit most from treatment by which method. That is not to argue against the need to continue to evaluate intervention in each and every case treated. However, it is an argument for the broadening of the initial diagnostic work-up of each case.

THE STUDIES OF HERMELIN AND OF ORNITZ

My concern over experimental methodology in the investigation of autism was sparked off by comparing the two contrasting approaches of Hermelin and Ornitz (see Chapters 9 and 8). Both authors contributed to an international symposium held at the CIBA Foundation in London in 1970 (Rutter, 1971). It is of interest, first, to consider how their views on perception and perceptual process in autism have changed in the intervening six years.

In their 1970 presentation (O'Connor, 1971), Hermelin and O'Connor presented an overview of their many experiments with autistic children (see Hermelin & O'Connor, 1970). At that time, they had demonstrated that the autistic children whom they observed showed as much interest in people as in objects, and that the orienting behavior in autistic children was similar to that of severely subnormal (i.e., profoundly retarded) children of an equivalent mental age. The pattern of sensory dominance in autistic children differed from that in nonautistic but severely subnormal children. While both groups showed most preference for light stimuli, autistic children preferred touch to sound, while subnormal children preferred sound to touch. The authors concluded that there was "a lack of normal development of a hierarchical pattern for sensory input in the case of autistic children. Presumably this was because of their relatively poor auditory or auditory-semantic development."

Already in 1970, Hermelin and O'Connor were showing an interest in the cognitive processes whereby information obtained in one sensory channel is coded and stored in the form appropriate for another sensory modality. As reflected in Chapter 9, this has been one of the main shifts of emphasis for their research. The interpretation of sensory input depends both on the immediate sense data and on the capacity to integrate these data with past data stored in other channels. The demonstration that visually presented material may be interpreted within an auditory channel is both as fascinating as it is ingenious. When it is then realized that the inability of autistic children to utilize this channel of sensory integration renders them functionally deaf, a whole new way of conceptualizing the cognitive deficits of autistic children is opened up.

Interest in cross-modal integration has a long tradition in experimental psychology and the more recent field of developmental neuropsychology. The work of Birch and Belmont (1964) on auditory-visual integration in poor readers, and of Bakker (1971) on temporal order perception in poor readers originally seemed to indicate central defects of sensory integration, particularly dealing with sequential material. However, these early studies failed to investigate within-modality difficulty and as Bryant (1974) points out, without this essential control, the findings are open to ambiguous interpretation. Hermelin and O'Connor carried out one of the earliest investigations in this area to include the appropriate controls, and their careful experimental approach has been continued into their present work. Their interest in cross-modal and within-modal problems has developed steadily, and the past six years have seen a shift from the study of perceptual processes per se to the study of the mechanisms integrating perceptual processes. Although the emphasis is rightly on the examination of psychological processes, Hermelin acknowledges that defects in such processes may well reflect some neurophysiological dysfunction or defect.

In his CIBA Symposium paper six years ago, Ornitz (1971) concluded his discussion of theoretical issues by suggesting that "childhood autism may be viewed as a disorder of sensorimotor integration" and later that "early infantile autism is fundamentally a specific type of maturational delay in sensorimotor integration and that the disturbances of relating and language are secondary events."

Thus, there appears to be a good deal of agreement between Ornitz and Hermelin. However, Hermelin continues to examine psychological processes and to seek psychological explanations when she moves from the observation of a defect in sensorimotor integration to an examination of linguistic codes which underlie such integration. In contrast, Ornitz seeks to explain psychological deficits by reference to neurophysiological deficits and concludes that language disorders are secondary to the neurophysiological dysfunction.

Ornitz was severely criticized by DeMyer (1971) for the view which he held in 1970 on "perceptual inconstancy." It is interesting to note that this concept is no longer invoked. Instead, he has continued to develop the hypothesis (which he himself labels as a "speculative excursion") that the neurophysiological basis for the deficit lies in the vestibular system.

His work gives rise to feelings of ambivalence. On the one hand there has been great ingenuity in devising technically complex experiments and in the scientific detective work involved in the long chains of deduction from observation to interpretation. On the other hand, doubts are raised by premature neurologizing. The etiology of autism is likely to be found eventually in some neurophysiological system, but we do not yet know enough about either autism or developmental neurology to be able to implicate one subsystem rather than another. Moreover, the conclusions which can be drawn from the experimental findings are necessarily constrained by the limitations which stem from the types of control groups used.

CONTROLS FOR MENTAL AGE

While both Hermelin and Ornitz have used experimental approaches, they have employed widely different control groups. Hermelin points out that "it is neither very illuminating nor profitable to come up repeatedly with the finding that children who are ill do less well than children who are not. . . ." The experimental psychologist is more interested in examining psychological processes which may underlie the observed gross deficits. To do this adequately requires the selection of appropriate control groups. Let us examine this more carefully.

A group of autistic children may be compared with a group of nonautistic children of similar chronological age in terms of their scores on a particular test. If the autistic children score significantly lower than the normal controls, the experimenter cannot conclude very much. After all, the autistic children will, as a group, be of significantly lower intelligence. The question then is whether their poor scores are a function of their low intelligence or rather of their autism? When autism is the focus of attention, one needs to be able to say what deficits are *specific* to the autistic group. Thus, at the very least, one requires an additional comparison group of mentally subnormal (or profoundly retarded) children who do *not* show the behavioral syndrome of autism.

The point is best illustrated by specific examples. Ornitz refers to Goldfarb's (1961) finding that autistic children showed more postural whirling than normal children. But is this specific to autistic children, or is it characteristic of children of low mental age? Rachman and Berger (1963) conducted a study including the relevant control group of mildly retarded children. They showed that whirling was indeed specifically associated with autism even after controlling for mental age. On the other hand, this was not found by Lockyer (1967) with a much older group of autistic children, so it may be that whirling is specific to autism but only in those who are also young and mentally retarded.

In this example, the addition of a missing control group helped to clarify the data and indicated that the phenomenon of whirling did indeed appear to be relatively specific to autistic children. Another example illustrates the opposite sort of clarification, and probably serves as a more salutary reminder of the need for appropriate control groups.

In a number of studies, mainly carried out in Lovaas's laboratory in UCLA, it had been claimed that autistic children differed from normal, retarded, and aphasic children in that they were "overselective" or showed an abnormally narrow focus of attention. This finding of a cognitive deficit which appeared to be specific to autistic children was regarded as both underlying many of the characteristic behaviors of the group and as having implications for educational techniques. However, as Schover and Newsom (1976) point out, "A major problem in interpreting all of the above findings is that there were no controls for mental age." When they added in the appropriate controls, they found that the autistic children behaved much the same as normal children matched for mental rather than chronological age. Thus, a set of potentially misleading findings were placed in a proper, developmental context, and different implications were spelled out for educational intervention.

It remains to be determined whether Ornitz's results will be found to be specific to autism or merely a reflection of general immaturity and low cognitive level. So far his technologically sophisticated studies have generally lacked appropriate control groups. This, together with the fact that few of the findings have yet been replicated, means that Ornitz's argument that the vestibular system is involved in the genesis of autism must be regarded as "not proven."

IMPLICATIONS FOR TREATMENT

We must also ask what are the implications of the findings from these detailed experimental studies of psychological processes for the treatment of autistic children. Often there are none but there are implications from the Hermelin and Ornitz investigations. However, I am uncertain how to use the data to devise treatment approaches. Both sets of studies lead to the conclusion that autistic children have difficulty integrating sensory data from different channels. This implies that teachers must be very cautious in utilizing multisensory approaches to teaching. As Hermelin concludes, "for the autistic child, sense training is certainly not enough." But she does not speculate what alternative teaching strategies might be relevant.

DeMyer (1971) suggested that the addition of kinesthetic cues might assist autistic children to learn tasks which normally rely on visual memory. By contrast, Schreibman (1975) found that the addition of cues from another stimulus dimension was usually more confusing than helpful to autistic children. Instead, they benefitted when the extra cue took the form of an exaggeration of the relevant stimulus dimension. Schover and Newsom's (1976) findings in turn suggest that, at least in discrimination learning, autistic children should be treated as if they were like younger, normal children. Clearly, the apparently contradictory suggestions coming from all these studies need to be synthesized. In part, the authors are each talking of results tied to particular techniques in particular studies, and it is dangerous to generalize straight from laboratory findings to classroom practices. Any hypotheses about desirable teaching approaches must be framed in an empirically testable manner. Even so, it should be possible to produce firmer guidelines for positive action to help autistic children either to overcome or to bypass the sort of cognitive deficits which have been graphically illustrated in the papers presented in this volume.

REFERENCES

Baer, D. M., Wolf, M. M., & Risley, T. R. Some current dimensions of applied behavior analysis. *Journal of Applied Behavior Analysis*, 1968, *1*, 91–97.

Bakker, D. J. *Temporal order in disturbed reading*. Rotterdam: University Press, 1971.

Birch, H. G., & Belmont, L. Auditory-visual integration in normal and retarded readers. *American Journal of Orthopsychiatry*, 1964, *34*, 852–861.

Bryant, P. *Perception and understanding in young children*. London: Methuen, 1974.

DeMyer, M. K. Perceptual limitations in autistic children and their relation to social and intellectual deficits. In M. Rutter (Ed.), *Infantile autism: Concepts, characteristics and treatment*. London: Churchill, 1971.

Goldfarb, W. *Childhood schizophrenia*. Cambridge, Mass.: Harvard University Press, 1961.

Hermelin, B., & O'Connor, N. *Psychological experiments with autistic children*. Oxford: Pergamon Press, 1970.

Kanner, L. Autistic disturbances of affective contact. *Nervous Child*, 1943, *2*, 217–250.

Lockyer, L. A Psychological Follow-up Study of Psychotic Children. Unpublished Ph.D. thesis, University of London, 1967.

O'Connor, N. Visual perception in autistic childhood. In M. Rutter (Ed.), *Infantile autism: Concepts, characteristics and treatment*. London: Churchill, 1971.

Ornitz, E. M. Childhood autism: A disorder of sensorimotor integration. In M. Rutter (Ed.), *Infantile autism: Concepts, characteristics and treatment*. London: Churchill, 1971.

Rachman, S., & Berger, M. Whirling and postural control in schizophrenic children. *Journal of Child Psychology and Psychiatry*, 1963, *4*, 137–155.

Rutter, M. (Ed.), *Infantile autism: Concepts, characteristics and treatment*. London: Churchill-Livingstone, 1971.

Schover, L. R., & Newsom, C. D. Overselectivity, developmental level, and overtraining in autistic and normal children. *Journal of Abnormal Child Psychology*, 1976, *4*, 289–298.

Schreibman, L. Effects of within-stimulus and extra-stimulus prompting on discrimination learning in autistic children. *Journal of Applied Behavior Analysis*, 1975, *8*, 91–112.

Yule, W., & Berger, M. Behaviour modification principles and speech delay. In M. Rutter and J. A. M. Martin (Eds.), *The child with delayed speech*. London: Heinemann/SIMP, 1972.

Yule, W., & Hemsley, D. Single-case methodology in medical psychology. In S. Rachman (Ed.), *Contributions to medical psychology*, Vol. 1. Oxford: Pergamon Press, 1977.

11

Biochemical and Hematologic Studies: A Critical Review

EDWARD R. RITVO, KAREN RABIN,
ARTHUR YUWILER, B. J. FREEMAN,
and EDWARD GELLER

Biological studies of psychiatric disease generally rely on examination of cerebrospinal fluid, blood, urine, or other tissues far removed from the brain. Indeed, metabolism in the brain is frequently so rapid that even autopsied brains, when available, shed little light on any but the most gross psychiatric or neurological disorders. Cerebrospinal fluid is physically closest to the brain. Even here, however, evidence suggests that the bulk of its composition of transmitter and transmitter metabolites comes from the local spinal cord rather than the more distant brain. The composition of blood is even more complex since it contains nutrients for and metabolites from the entire body. Similarly, urine is the repository of the entire body's metabolic wastes and the fraction attributable to the brain is uncertain and usually indistinguishable. Due to these factors it is particularly difficult to design experiments to find significant correlations between biological measures, specific brain functions, and psychiatric syndromes. Even more difficult is the task of critically analyzing such research. One must constantly bear in mind that any differences or correlations noted may be only incidental, or due to factors unrelated to the psychiatric diagnostic categories studied.

As recently as the 1930s, it was generally accepted that "nerve im-

EDWARD R. RITVO *ET AL.* · Mental Retardation and Child Psychiatry Program, Department of Psychiatry, School of Medicine, University of California, Los Angeles, California 90024.

pulses" were electrical in nature and "jumped" from one cell to "excite" the next. Neurobiochemical research during the last two decades, however, has proven that such direct electrical couplings between nerve cells, or neurons, are rare exceptions rather than the rule. Instead nerve cells have been shown to communicate mainly by the stimulus-coupled release of chemicals called neurotransmitters from one neuron (the presynaptic neuron) into the narrow space (the synaptic cleft) between neurons, with resultant stimulation of the adjacent neuron (the postsynaptic neuron). Different groups of cells produce, store, and release different transmitters and two systems have been extensively studied (Cohen & Young, in press). First is the catecholaminergic system which consists of the neurotransmitters norepinephrine and dopamine. Second is the indoleamine system in which serotonin is the transmitter.

INDOLEAMINES

The indoleamine serotonin (5-HT, 5-hydroxytryptamine) is the neurotransmitter in a group of CNS neurons located in several nuclei of the Raphé system. Their terminals richly innervate such important brain regions as the hypothalamus, amygdala, septum, and caudate, as well as the cortex and spinal cord. Some serotoninergic fibers appear to innervate the cell bodies of other transmitter systems and, in turn, some of these appear to have terminals on or near serotoninergic cell bodies. Thus, there are rich and complex interactions between neurotransmitter systems and it is likely that an abnormality in one system can also influence other systems.

Serotonin is synthesized, utilized and degraded within the central nervous system. It is also synthesized in the enterochromaffin cells of the intestine where it is taken up by blood platelets and stored in vesicles similar to those of serotoninergic nerve cells. It is degraded and excreted in the urine as 5-hydroxyindolacetic acid (5-HIAA). These similarities have given impetus to many *in vitro* studies because of the availability of whole blood and platelets. However, while there are many similarities in the systems controlling serotonin biosynthesis in the brain and in the periphery, some distinct and important differences make it impossible to simply generalize from studies of one system to the other.

Serotonin Levels in the Blood

The first study of serotonin levels in the blood of children diagnosed as having "autism" and "other forms of mental retardation" was reported by Schain and Freedman in 1961. They studied 23 children labeled "autistic" by a set of criteria based on those of Kanner (Schain & Yannet, 1960).

Interpretation of their results was complicated by the fact that "some children were receiving phenobarbital, Dilantin, or chlorpromazine at the time of the study," and three on reserpine who had no detectable blood serotonin were omitted from data analysis. They concluded that "consistent, unusual elevations of blood serotonin were found only in children with the diagnosis of autism, although the mean blood 5-HT level of the other severely retarded children was higher than that of the mildly retarded group." They went on to analyze the histories of the six autistic children with the highest blood 5-HT levels. Their presenting symptomatology did not differ from that of the other autistic children who had normal 5-HT levels. The only feature that appeared possibly noteworthy was the absence of seizure disorders in the six autistics with elevated serotonin since frequent occurrence of seizures was one of the most significant findings of their earlier clinical studies (Schain & Yannet, 1960). This observation is interesting in the light of later studies by Rutter (1970) indicating that some autistic children develop seizures in later life.

This pioneering study was hampered by limitations of the available assay techniques and by a lack of specificity of diagnostic criteria. They did not subdivide "autistics" by IQ. Thus, since their nonautistic severely retarded comparison group also had elevated levels of serotonin, it is possible that the elevations of serotonin may have been primarily related to the degree of retardation and not to autism.

The second major study of serotonin levels in whole blood was reported in 1970 by Ritvo *et al.*, who surveyed serotonin and platelet levels in 24 "autistic" patients and 36 comparison subjects all of whom were off medication of any type for at least three months. Detailed clinical descriptions of each patient were provided, and all demonstrated unequivocal evidence of the syndrome of perceptual inconstancy both on history and when examined immediately prior to testing (Ornitz & Ritvo, 1968; Ritvo *et al.*, 1970).

Our results indicated: (1) both blood serotonin and platelet levels were inversely related to age in the normal population; (2) mean serotonin levels and platelet counts of age-matched groups of autistics were significantly higher than comparison cases; and (3) mean serotonin per platelet values were not significantly different between age-matched groups of autistics and comparison subjects. However, a significantly greater variability of serotonin per platelet was found within the youngest group of age-matched patients compared with controls (24 to 47 months). This appeared to be related to a significantly greater variability of platelet levels in the autistic patients.

This study is important for three reasons. First, a highly reliable method of determining whole blood serotonin levels was developed and utilized (Yuwiler *et al.*, 1970). Second, uniform blood-drawing techniques were employed and attention paid to assure that both hospitalized patients and a comparison group received the same diet. Third, normative data were obtained and developmental effects were taken into account. However, while detailed clinical descriptions of patients were given and previously published diagnostic criteria used, no objective basis for defining or rating severity of symptoms was utilized.

A third study of serotonin and platelets in autistic children was reported in 1970 by Boullin *et al.* It was designed primarily to measure the efflux of serotonin from platelets and this aspect of their data will be discussed in a later section. Six autistic children, ages 4 to 14, diagnosed by using the Rimland E-2 "parental rating" questionnaire (Rimland, 1964) were studied along with six "normal" (not otherwise specified) children "of comparable ages." Serotonin determinations were made using a laboratory assay procedure on quickly frozen blood which involves protein precipitation with zinc and sodium hydroxide. (In our hands this technique has given inconsistent results.)

Their results indicated no differences between mean platelet size "but the number of platelets per milliliter of plasma was significantly increased in the autistic patients. No concomitant increase in serotonin per platelet was noted, rather the concentration was slightly diminished."

This study confirmed previous findings (Ritvo *et al.*, 1970) of elevated platelet counts in autistic children, but several methodological problems existed. The patient population was defined only by use of a parental questionnaire, the scoring techniques of which had not and, as yet, have not been published. (A detailed critique of the questionnaire approach to diagnosis has been reviewed elsewhere by DeMyer *et al.*, 1971.) Also, no clinical descriptions of subjects were supplied so that cross-comparison is not possible on an anecdotal basis.

In 1971, we (Yuwiler *et al.*, 1971) reported on circadian rhythmicity of blood serotonin levels and platelet counts in seven autistic and four nonautistic inpatients at the UCLA Neuropsychiatric Institute. The autistics were diagnosed as described by Ritvo *et al.* (1970) and serotonin was determined by the method of Yuwiler *et al.* (1970). Blood samples were drawn every 4 hours over a 24-hour cycle.

These experiments were based on the possibility that serotonin concentration and blood platelet counts might undergo a systematic daily variation and that autistic and nonautistic children might differ in the phasing of this circadian rhythm. In such an event, a single determination at a constant time of day might show differences between populations because of circadian phase differences.

The results failed to demonstrate any clear circadian variation in platelet counts, blood serotonin concentration, nor any differences in variation between populations. As in their previous study, both serotonin and platelets were higher among autistics but the number of subjects was too small to permit meaningful statistical comparisons.

Campbell *et al.* (1974) studied blood serotonin in "psychotic and brain damaged children '' (11 patients, nine boys and two girls attending the inpatient service at Bellevue Hospital). "Seven were diagnosed as 'childhood schizophrenic,' two diagnosed as 'early infantile autism' based on criteria of the British Working party as delineated by Creak *et al*. (1961) and two 'a withdrawing reaction' (one had Klinefelter's syndrome).'' All the patients had symptoms prior to the age of two years and intellectual functioning ranged from average to severely retarded. None had a history of seizure disorder. Six inpatient comparison cases, four boys and two girls (ages 3 to 8 years) were also studied.

Their results indicated that serotonin concentration in the platelet-rich plasma of the 11 patients was not significantly different from their comparison cases. "Low intellectual functioning was the only parameter which seemed to be clearly associated with higher serotonin levels." Unfortunately the large number of diagnostic categories in both the psychotic and comparison groups make this study difficult to evaluate. For example, of the two autistics, one had a very high serotonin level and one had a low level. Also, since serotonin was calculated per volume of platelet-rich plasma of undetermined platelet concentration comparison with other studies is impossible. Regarding the conclusion that low intellectual functioning was associated with elevated serotonin levels, it must be noted that a variety of intelligence tests were used to determine "levels of functioning." Using different tests introduces severe methodological problems which makes subject comparison difficult (Freeman & Ritvo, 1976) and casts doubt on any relationship noted between intellectual function and other variables. However, these results do confirm those of Schain and Freedman (1961) whose nonautistic, severely retarded comparison groups also had elevated levels of serotonin.

In 1975 Campbell *et al*. reported a second study of serotonin levels. Twenty-three patients (childhood schizophrenic—having at least 7 of Creak's 9 points) and 16 comparison age- and sex-matched children were surveyed. As before, they also rated the patients by IQ and DQ (developmental quotient).

The results indicated that serotonin levels were higher in the patient group than in the controls, although the difference did not reach significance. However, serotonin levels were significantly higher in a subgroup of patients with "florid psychosis, and those with lower IQs than in patients in remission or partial remission or higher IQs."

The next study to be discussed in this section was done in our laboratory (Ritvo *et al*., 1971). We wished to determine if whole blood serotonin could be lowered in autistic children by the administration of L-Dopa and, if so, to assess possible concomitant clinical or physiological changes. Following a 17-day placebo period, four hospitalized autistic boys (ages 3, 4, 9, and 13 years) were given 300 to 500 mgs of L-Dopa per day for six months.

The results indicated a significant decrease in blood serotonin concentrations in the three youngest patients, a significant increase in platelet counts in the youngest patient, and a similar trend in the others. Urinary excretion of 5-HIAA decreased significantly in one patient and a similar trend was found in all the others. No changes were observed in the clinical course of the autistics, nor in their amount of motility disturbances, REM sleep, or endocrine levels (FSH and LH).

The findings of this study were limited by two main factors. First, only four autistic patients were studied, far too few a number from which to draw definite conclusions. Second, while the investigators were concerned with changes in the clinical status of the patients, only subjective measures of itemized change were used. We discussed these and other methodological difficulties which limited this study, as well as theoretical problems involved in relating clinical state to blood levels of serotonin at a given point in a disease process.

A second study of L-Dopa administration was reported by Campbell et al. in 1976. They administered from 900 to 2500 mg per day of L-Dopa and from 3.5 to 42 mg per day of levoamphetamine in a cross-over type experimental paradigm. Their patients were 11 childhood schizophrenics (diagnosed as previously cited). Clinical rating scales were developed and applied in a blind manner (Fish, 1968).

Their results confirmed those of Ritvo et al. (1971) in that L-Dopa administration did not appear to alter the course of the disease. It did, however, appear to be "stimulating" to the patients and they suggest this effect warrants further study.

The final study to be reviewed in this section is on monoamine oxidase (MAO) and was reported by Campbell et al. (in press). MAO activity was measured in blood platelets of 21 schizophrenic children (criteria as previously cited) and 18 normal comparison children matched for age and sex.

No significant differences were found between patient and comparison groups. However, they conclude that because of the small number of patients and their heterogeneity, and the difficulties in assay techniques, this study needs to be extended to other populations.

Platelet Uptake and Efflux of Serotonin

Four groups of investigators have studied the uptake and efflux of serotonin from the platelets of autistic children. The first report by Siva-Sankar (1970) was on 200 inpatient boys, ages 5 to 15, whose diagnoses "varied from childhood schizophrenia (undifferentiated; autistic), disorders of behavior in character (termed primary behavior disorders or non-schizophrenic), psychosis (due to known or unknown causes, with or without brain damage, and with or without mental deficiency), and mentally defective." All subjects were withdrawn from medication for at least six weeks prior to the study. Uptake was measured by incubating platelets with radioactively labeled serotonin creatinine sulfate. Experiments were also done on the uptake of norepinephrine by platelets, the uptake of serotonin by red blood cells, and the uptake of norepinephrine by red blood cells. Unfortunately, platelets were separated by centrifugation, a drastic procedure likely to rupture the platelets.

The results indicated that autistic, schizophrenic, nonautistic, and other patient groups did not differ with respect to the uptake of norepinephrine by platelets, the uptake of serotonin by red blood cells, nor the uptake of norepinephrine by red blood cells. However, a significantly lower rate of uptake of serotonin by platelets was found in the autistic group, when compared to the schizophrenic, nonautistic and other patient groups. Furthermore, analysis of patients by age groups indicated a relationship between decreased uptake of serotonin and chronological age. The autistic patients had lower levels similar to those found in the younger age groups of other patient populations. These developmentally related findings are consistent with ours (Ritvo *et al.*, 1970) on serotonin and platelet levels, and previous theoretical suggestions of Bender (1953) that a maturational lag exists in the syndromes of autism and childhood schizophrenia.

Again, it is difficult to evaluate the results of this study because neither specific diagnostic criteria nor descriptions of individual patients were provided. Furthermore, there has been much controversy in the literature as to the relationship between the diagnosis of childhood schizophrenia and autism (Ornitz & Ritvo, 1977), with some investigators separating autism and childhood schizophrenia into separate syndromes (Rutter, 1972) and others viewing them as different points on the continuum of a single disease process. Siva-Sankar (1970) apparently conceived the diagnostic categories as being on a continuum and separated out subgroups of autistic nonschizophrenics for special study, but did not specify the criteria for making these clinical divisions.

The second study of serotonin uptake by platelets of normal children and "childhood schizophrenics" was reported by Lucas *et al.* (1971). They failed to observe any difference between "normals" and patients. The method they used was a less drastic one than the multiple-centrifugation procedure of Siva-Sankar (and Boullin, see below) and was more likely to preserve platelet integrity during the assay. It should also be noted that clinical diagnostic criteria were not specified in this report.

A third series of studies on uptake and efflux was begun by Boullin and his associates in 1970. In the first study (Boullin *et al.*, 1970), they incubated platelet-rich plasma with radioactive serotonin and separated the platelets by centrifugation. Uptake was estimated from the loss of radioactivity in the supernatant solutions and efflux from the increase in radioactivity of the supernatant solution after resuspension of the platelets in fresh plasma. Their subjects (six children, sex unspecified, ages 4 to 14) were diagnosed by the Rimland E-2 parental rating questionnaire. Six "normal children" (otherwise unspecified) of comparable ages were used for comparative purposes.

They concluded from this uptake study that "autistic platelets accumulated serotonin to a slightly greater extent than normals, 2.6 times the endogenous content, compared with 2.0 times for normal platelets. The serotonin uptake activity was unchanged or increased in platelets from the autistic subjects." They also found that mean platelet size was not different, but that the concentration of platelets was significantly greater in the autistics. They noted no increase in serotonin levels, but rather that the concentration of serotonin per platelet was slightly diminished.

Results of the efflux part of this study indicated that platelets from normal and autistic children lost a similar percentage of the total amount of radioactivity during the resuspension procedure, before reincubation for 90 minutes. This indicated that platelets from autistic children were not preferentially affected by the experimental procedure. However, twofold increase in the efflux of radioactive serotonin during the incubation from the autistics represented a "definite abnormality." This apparently did not arise from lack of ATP (resulting in diminished binding of serotonin) since the ATP content of platelets was the same for autistics and normals.

As they noted, their procedure did not differentiate between efflux of serotonin from storage vesicles and efflux of serotonin degradation products (5-HIAA, 5-hydroxytrytophol). The increased efflux they observed is different from that in Down's syndrome where it appears to be due to defective binding associated with decreased ATP. Thus, they caution that increased efflux of platelet serotonin *per se* may not be confined to autism.

In 1971, Boullin *et al.* reported a second study to determine if increased serotonin efflux could be used to predict the diagnosis of autism as determined by the Rimland E-2 parental questionnaire. Platelets from ten children were examined without knowledge of their E-2 ratings. They developed an "efflux ratio," obtained by dividing the efflux from an autistic child by that obtained from a control. It is unclear, however, how the pairs were assigned. Using this ratio, they predicted that six children were autistic and four were "non-autistic psychotics." The E-2 scores labeled seven children as autistic and three as nonautistic psychotics. Neither platelet ATP nor serotonin uptake by platelets, which were also examined in this study, showed significant differences between the autistic and comparison cases.

In 1972 Boullin and O'Brien reported a third study using labeled dopamine rather than serotonin. By analogy with vesicular storage in the central nervous system, dopamine may compete with serotonin for storage sites. Platelets were obtained from five autistic children, four boys and one girl, ages 9 to 19. Normal age-matched children were used as controls. Only two of these patients were diagnosed by the Rimland E-2 questionnaire as having autism, and the criteria for diagnosis of autism of the other three children were not specified.

Their results indicated no significant differences of uptake or loss of dopamine, but there was much greater variation in individual values within the autistic group. In one subject, both dopamine uptake and loss were abnormally high, which they stated, "may be indicative of a body defect for dopamine in platelets of autistic children."

The problems inherent in this study with regard to patient-selection procedures and methods have been discussed previously and the greater variation among subjects noted could be due to patient heterogeneity. Since dopamine may occupy the same binding sites as serotonin, the failure to observe increased dopamine efflux from platelets in the autistic children suggests that the increased serotonin efflux previously reported may be due to altered serotonin metabolism rather than altered binding. Methodologically, the efflux procedure was complicated by repeated centrifugations. Because of platelet fragility, such centrifugation may have led to unintended selection of platelet populations and contributed to the variability observed.

The final studies to be reviewed on serotonin binding by platelets were reported by us (Yuwiler *et al.*, 1975). We attempted to replicate the previous work of Boullin *et al.* (1970) using methods to minimize platelet damage.

The autistic patients were diagnosed according to our criteria (see above). The comparison groups were 15 inpatients (seven girls and eight boys) who received the same clinical and laboratory evaluations as the autistics, had the same diet and ward conditions, and were diagnosed as having mild to severe mental retardation and developmental delays without symptoms of autism, and also normal children.

Our results failed to show significant differences in either uptake or efflux rates between the autistic and the hospitalized comparison groups or normals. Methodological considerations which could possibly account for the failure of this study to confirm previous findings were discussed in detail. In particular, the authors explain how differences in subject selection as well as biochemical assay techniques may have accounted for our inability to confirm previous findings of group difference.

Bufotenin

The next group of studies that we shall review deals with the excretion of bufotenin, the N-dimethylated derivative of serotonin. It is known to be mildly hallucinogenic in humans. A number of studies have reported that bufotenin is present in the urine of adult schizophrenic patients (Himwich *et al.*, 1972) and N-dimethylation of serotonin can be carried out by biological materials *in vitro* and possibly *in vivo* as well. The N-methylating enzyme carrying out this reaction, for example, is found in the lungs.

Himwich *et al.* (1972) investigated whether patients diagnosed as autistic excreted N,N-dimethyltryptamine, 5 methoxy-N-dimethyl-tryptamine, and bufotenin. Their subjects were ten boys and two girls, ages 8 to 21 years. Based on a review of their records, they identified five boys and one girl as having "a possible diagnosis of early infantile autism."

In the first part of the study, 24-hour urine samples were collected from all 12 subjects while they were on different diets in their respective homes. In the second part of the study the six children "suspected of autism" were used as subjects. Twenty-four hour urine collections were made on 14 consecutive days while dietary intake was controlled. The presence of bufotenin was assessed by thin-layer chromatography of urine samples.

The results indicated that five of the six patients who were "suspected of having infantile autism" excreted bufotenin. The lack of diagnostic specificity complicates interpretation of this finding. While three diagnostic "systems" were used, their criteria were not specified and the brief clinical descriptions suggest that widely differing criteria must have been employed by each "system." Also, these descriptions indicate that specific CNS dysfunction probably existed in four of the "possible autistic patients." Thus, although related to the author's categories, excretion of bufotenin may be quite unrelated to autism. Furthermore, the entire significance of bufotenin excretion is still being debated in the literature, and no consensus has been reached in the years since this study was published.

In 1975 Narasimachari and Himwich summarized their further investigations. Parents of the five "autistic" children who had been positive for bufotenin in the previous study were administered psychological tests and their urine studied. The behavioral data did not reveal significant evidence of psychopathology and one or both of each pair of parents also excreted bufotenin. They also referred to their other studies, including one in which only 6 of 18 "autistic children" were found to excrete bufotenin. (Reference to this study was not specified.)

While this research regarding bufotenin is hardly definitive, it is conceivable that its excretion may be genetically linked, and their data suggest that psychiatric disturbances are not always manifested in bufotenin excretors, although such excretion may define a biological subpopulation. Further work is obviously required to assess the meaning of these results.

TRYPTOPHAN METABOLISM

A number of recent studies have identified inborn errors of amino acid metabolism in patients with developmental disabilities and mental retardation. Phenylketonuria and branched-chain ketonuria are examples of such diseases. Similar studies have also been undertaken to search for possible errors in amino acid metabolism in autism. The first was reported by Sutton and Read in 1958, on an 18-month-old girl with seizures, whom they diagnosed as autistic. Urinary chromatograms were run before and after dietary administration of an oral tryptophan load.

Their results indicated that during tryptophan loading the "autistic" excreted decreased amounts of 5-hydroxyindoleacetic acid, indoleacetic acid, and tryptophan. They hypothesize that "the mental aberration" in autism may be the result of "an altered ability to maintain normal brain serotonin levels" (Coleman, 1973a, b).

Since the authors failed to provide a detailed description of their diagnostic criteria, one could question whether the subject studied indeed had the syndrome of autism as we now understand it. Also, this study suffers from the problems inherent in single-subject research, an issue the investigators readily acknowledged.

Shaw *et al.* (1959) also reported on the effects of tryptophan loading on indole excretion. Their 21 subjects included 11 with "childhood schizophrenia" who received 3 g of L-tryptophan orally.

Their results revealed that all but three subjects (two schizophrenics and one nonschizophrenic) showed increased excretion of 5-HIAA over their control periods. No other significant differences were observed between the schizophrenic and nonschizophrenic groups with respect to either basal excretion or to altered excretion following tryptophan loading. Failure to specify diagnostic criteria for both the schizophrenic and nonschizophrenic patients makes it difficult to interpret these results. However, they also reported the following interesting anecdotal observation which bears further study: "Side effects from the tryptophan were observed in several of the children. These were more prominent in the schizophrenic children, and consisted of exaggerated bizarre behavior with apparent subjective effects as evidenced by the increased activity, fearfulness, motility, and resistance to control."

In 1965 Heeley and Roberts reported a study of tryptophan loading and amino acid excretion in 16 "psychotic" children. Their ages ranged from two to seven (14 boys, two with histories of convulsions). The authors do not specify how their urine samples were analyzed but referred to an unpublished paper. Their results indicated that "some of the psychotic children had increased excretion of their tryptophan load." They concluded that "the abnormality appears to be associated with those children who have always shown deviant behavior, rather than the other group in whom psychosis developed later." However, they do not describe their statistical procedures or further define their clinical parameters.

In summary, both Heeley and Roberts (1965) and Shaw *et al.* (1959) reported ambiguous results. At this time it has not yet been ascertained whether abnormal responses to tryptophan loading occur in certain developmentally disabled children. Certainly this area of investigation needs to be pursued in order to follow up these suggestive observations.

In 1970 Jorgensen *et al.* reported a survey of amino acid excretion in the urine of 178 children (ages 3 to 12 years) in three Danish hospitals. The following diagnostic categories were used: "(1) psychosis, (2) neurosis, (3) character disorder and (4) mental deficiency and other functional disturbances."

Urines were tested for total protein, phenylpyruvic acid, glucose, reducing carbohydrates, and two-dimensional amino acid chromatography.

Results of this study "indicated that there was no case of specific hyperaminoaciduria found. In the patient group as a whole, valine excretion was significantly lower in the three- and four-year-old children than in the 'control material' (not specified). The three- to six-year-old psychotic children excreted significantly lower amounts of tryptophan than the 'controls.' In the same age group, valine was excreted in lower amounts in the neurotic children and in the children with character disorders. The one- to twelve-year-old patient group did not differ from the "control group" in the amino acid excretion pattern."

Interpretation of this study is difficult. First, the authors did not specify the diagnostic criteria used. Second, biochemical results are confounded by the fact that no controls for diet, medication, or motor activity were made. Since these factors are known to affect amino acid metabolism, they must be studied as independent variables before relating results of excretion levels to clinical categories. Finally, the authors did not make it clear as to how groups were arranged for comparison, i.e., which groups were designated as "controls" for statistical purposes.

CATECHOLAMINES

In 1974 Cohen *et al.* reported a study of homovanillic acid (HVA, the end product of brain dopamine metabolism) and 5-hydroxyindolacetic acid (5-HIAA) in the cerebrospinal fluid of 20 children (9 "autistic," 11 "atypical," and 10 with epilepsy requiring constant use of anticonvulsant medications). Five other boys with severe disorders of motor control were also studied.

Their results revealed significant differences in cerebrospinal fluid HVA and 5-HIAA levels between the psychotic and epileptic patients. HVA and 5-HIAA levels were also significantly correlated with each other. No differences were found within the group of "psychotic children" between those labeled "autistic" and those labeled "atypical psychotic but non autistic."

This pioneering study of neurotransmitters in cerebrospinal fluid metabolites used diagnostic labels in an idiosyncratic fashion. For example, their use of the term "atypical" does not overlap with previous uses in the literature (Rank, 1955); however, their careful clinical descriptions allow one to take this into account. As the authors noted, there also are methodological difficulties inherent in any study of cerebrospinal fluid in children, and these become compounded when attempting to collect data from groups of children without specific medical indications for lumbar puncture. Indeed, such medical indications (questionable meningitis, etc.) would probably make such patients unacceptable for research purposes. Despite these difficulties, this study points the way to extend this interesting area of investigation. Cerebrospinal fluid is at least physically and perhaps biochemically one step closer to the site of action of brain neurotransmitters because a considerable fraction of its monoamine products likely come from the cord itself, rather than the brain.

Pink Spot

The urinary excretion product 3, 4-dimethoxyphenylethylamine (DMPEA) has been identified in the urine of some adult schizophrenic patients. It has been labeled the "pink spot" since it turns pink when exposed to a modified Ehrlich's reagent. Widelitz and Feldman (1964) reported a study designed to examine the urine of 18 children (6 to 15 years) for the presence of "pink spot." Twelve were diagnosed as "childhood schizophrenic" and six were "normal." No diagnostic criteria were cited, nor were the normals categorized or further described. For the isolation of the material causing pink spots, a two-dimensional thin-layer chromatography technique was employed.

Their results indicated that eight of the twelve schizophrenics had pink spots of 5 μg or more. Five of the six normals also had pink spots, but their total amount per aliquot was less than 2 μg. The authors stressed that although the incidence of pink spots in both groups was similar, the intensity was greater in the schizophrenics. Further research is necessary to replicate this study and to explore the possible association between pink spot and psychiatric disturbances in children.

HEMATOLOGIC STUDIES

Fowle (1968) reported a study of leukocyte patterns in three groups of children (26 "schizophrenics," 30 normals, and 33 siblings of the schizophrenic patients). She concluded that schizophrenic children have "an atypical leukocyte pattern which is characterized by higher frequency of several morphologically distinct types of lymphocytes and plasmacytes than are found in the blood of their brothers and sisters and normal control children."

This study is important because of the possible association among different types of leukocyte patterns and diagnostic categories. However, it was not clear what criteria were used to diagnose patients. One important aspect of this study which warrants further investigation is that leukocyte patterns of older "schizophrenic children" resembled those of normal two-year-olds. This could possibly be further evidence of a maturational lag in "schizophrenic" patients (Ritvo *et al.*, 1970).

A study of blood magnesium levels was reported in 1969 by Gittelman and Cleeman. Their subjects were 61 inpatients (35 "psychotic" and 26 "who did not manifest overt psychiatric disturbances").

Their results indicated no significant differences between the mean serum magnesium levels of the "psychotic" and "control" groups. Results of this study are difficult to interpret for several reasons. First, the authors did not specify how the "psychotic" children were diagnosed nor was an adequate description of the control group given. Second, the subjects' age range (3 to 15 years) was so wide that developmental factors should also have been assessed.

In 1969, Saladino and Siva-Sankar reported a study on the erythrocyte magnesium and potassium levels in 78 hospitalized boys, 6 to 15 years, variously diagnosed as schizophrenic-autistics, psychotics, and primary behavior disorders. Forty-four were administered the Wechsler Intelligence Scale for Children and all had been removed from medication one month prior to study. Overnight fasting bloods were obtained and red blood cells analyzed for magnesium, potassium, and hemoglobin.

Their results indicated that the "schizophrenic-autistic" patients showed "the lowest levels [of magnesium and potassium] and the least variation with chronological growth." In ascending order, the groups were ranked: schizophrenic-autistic, schizophrenic-nonautistic, psychotic, and then the primary behavior disorders. The latter group had blood levels of magnesium and potassium per red blood cell "closely approximating the adult levels reported in standard manuals of clinical chemistry.''

No further specification of diagnostic criteria was given and no clinical descriptions were provided. While the authors conclude that their results support the hypothesis that a ''maturational lag'' exists in the ''schizophrenic-autistic patient,'' their data are not sufficient to support this statement.

In 1962 Siva-Sankar *et al.* reported a study of urine of schizophrenic boys, ages 5 to 16. They concluded that there ''seems to be more indolic substances in the urine of 'schizophrenic children.' '' In addition, they noted ''an unidentified Ehrlich positive substance that occurs more frequently in the chromatograms of schizophrenic urine preparations,'' and ''there was more inorganic phosphate in the plasma of the erythrocytes hemolysates of schizophrenic children.''

The data from this study are presented in a confusing manner. The selection criteria for the patients and controls are not specified, and while developmental and sex factors are acknowledged to be important, they were not studied separately.

In 1971 Siva-Sankar, in a study primarily focused on blood platelets, also reported data on the leukocyte chromosome breakage in ''schizophrenic'' children. He concluded ''in the case of autistic-schizophrenic children, there was a higher incidence of chromosome breakage in leukocytes in culture.'' In the discussion he attempts to relate this to other biological ''defects'' which are ''often found in childhood autistic-schizophrenics.'' (See *Serotonin Levels in the Blood* for additional comments on this study.)

In 1971 Goodwin *et al.* reported on malabsorption and cerebral dysfunction in children diagnosed as ''autistic.'' Their subjects were 15 patients (ages 6 to 13 years) selected from 23 families. Of these, ''8 families reported 2 mentally handicapped children, including 6 pairs that were autistic. This latter group yielded 7 with observed coincidence of gastrointestinal disorders. All mothers experienced complications in pregnancy and prenatal stresses. Episodes of colic, diarrhea and dehydration, and intolerance of milk or other foods were reported for all autistic subjects. Signs of neurological dysfunction were also reported in each autistic child.'' On fast days, subjects were placed on a gliadin free diet, and then given a gliadin challenge or a placebo. One hour later blood was drawn for eosinophil counts and plasma cortisol determination. Behavior was observed and some electrophysiological recordings made.

The results indicated that ''there was no significant difference between the groups indicative of any increased stress or altered hormonal stress responses in autistic subjects.'' Histamine Wheal Tests revealed no significant difference between groups. In Transcephalic Direct Current

(TCDC) studies "the differences between the autistic children, their siblings, or a group of control children and adults of various ages were 'not outstanding.' " The authors also noted that "elevated blood cortisol levels were associated with decreased eosinophil counts in the autistic children and their siblings." Gliadin markedly decreased blood cortisol levels and abolished "the normal circadian cortisol rhythm." While it did not significantly alter frontal TCDC baseline activities in normals or siblings of autistics, it significantly increased frontal potential generation in schizophrenics, and significantly inhibited frontal voltage generation.

Unfortunately, the results of this extensive study are quite inconclusive. The "autistic" population was extremely heterogeneous, subgroups were not defined precisely, nor clinical descriptions provided to permit comparison with other studies. No conclusions can be drawn with regard to the electrophysiological and behavioral reactions since validity and reliability data were not presented. Most unfortunately, their data also shed no light on the interesting question of the possible relationship between autism and celiac disease. Our clinical experience and reports of others (Coleman, personal communication) indicate that there may well be a significant number of children with the syndrome of autism who also have malabsorption problems, celiac disease, and multiple allergic responses. Hopefully, carefully controlled studies in the future will document the incidence of these multiple occurrences.

Lucas *et al.* reported in 1971 a study of plasma and R.B.C. cholinesterase activity in 16 childhood schizophrenic boys (ages 8 to 15 years) and 16 age-matched nonpsychotic hospitalized comparison patients. All of the children were in good physical health and received approximately the same diet each day. The schizophrenic children were diagnosed "according to the criteria of Bender (1953)."

Their results indicated significant differences between groups for R.B.C. cholinesterase activity. In a substudy, they noted a significant inverse correlation between acetylcholinesterase activity in R.B.C.'s and serotonin uptake by platelets in schizophrenic patients and nonschizophrenics. This negative correlation was less significant in the schizophrenic patients.

Several methodological problems are present in this study. First, the authors did not specify how patients were selected nor the criteria for assigning them to the various groups. Second, all but one of the schizophrenic patients and three of the controls had been receiving phenothiazines up to three days prior to the study and the results may have been confounded by the presence of the phenothiazines.

DeMyer *et al.* (1968, 1971a, b) have reported a series of studies on macronutrients. They began with the observation that autistics and schizophrenics had decreased food intake when compared to normal children, although the proportion of carbohydrate, protein, and fat was the same. They also noted greater variability in caloric intake in the schizophrenic group and suggested this may be related to greater variability of plasma-free-fatty-acids.

The results of their studies revealed greater variability in psychotic than in control children of free-fatty-acids and plasma glucose levels, but normal response to insulin administration, high carbohydrate diet, and glucose administration. Mean levels of plasma-free-fatty-acids, however, were higher in the psychotic group. They conclude "that a possible reason for free-fatty-acid variability in psychotic children may be a deficiency in the regulatory feedback mechanism at a neurogenic or cellular level."

These studies investigating free-fatty-acid responses in autistic and normal children are of great interest. Unfortunately, they have not been extended or replicated, but hopefully this will be done.

In a recent study, Cohen *et al.* (1976) measured blood lead levels in 34 children diagnosed as having "childhood autism" and "atypical development." Ten normal siblings of these patients were also studied. Using the Hollingshead-Redlich two-factor index, social position was found to be as follows: Class I, eight families; Class II, six families; Class III, eight families; and Class IV, ten families. There were no families in Class V, which is the lowest socioeconomic group. None of the children had a history of acute lead exposure nor clinical evidence to suggest they may have had elevated blood levels. They also measured a broad range of blood factors and obtained electroencephalograms, skull Xrays, and other Xrays when indicated.

Their results indicated that the group of "autistics" had significantly higher mean levels of blood lead than the "atypical" children, and the normal sibling group. However, even though the authors report a significant difference between the autistic and the nonautistic groups, there was a great deal of variability in their data and the distributions of all three groups were overlapping. Thus, the significant differences found between the autistic group and the other two groups may be due to the fact that three of the autistic children had very high levels, which raised their group means.

CONCLUSIONS

Even a cursory analysis of the studies we have reviewed reveals difficulties caused by lack of diagnostic specificity and objective clinical ratings. The only hope we have at our present "state of the art" is to strive to mitigate these difficulties. To this end, future studies must contain careful clinical descriptions of patient and comparison groups as well as specific references to the diagnostic criteria employed. We, and others, have begun to develop objective behavioral rating scales. Unfortunately, they are not yet perfected to the point where they can be applied by different investigators.

An overview of the two decades of research summarized in this paper indicates that certain results warrant replication and others point to areas for future investigation. For example, surveys of serotonin levels and platelet counts need to be replicated in different patient populations. The patients must be carefully defined, age-matched to comparison cases, racial and family factors studied, and attempts made to subdivide according to both clinical (e.g., specific symptoms, IQ) and biological (e.g., high vs. low levels) parameters.

On the basis of serotonin platelet uptake and efflux studies, it appears important to reexamine separately high and low serotonin subpopulations for differences in symptom pattern, prognosis, or clinical history. We may also be able to determine if platelets differ between those subpopulations with regard to age and size distribution, binding characteristics (perhaps using binding competitors), uptake and efflux properties, and platelet serotonin metabolism. We believe it is particularly important to determine if reported differences are due to differences in platelet factors or serotonin metabolism.

On the assumption that serotonin metabolism may be disturbed in autism it seems important to relate changes of blood levels to changes in clinical states. Dietary manipulations (increasing or decreasing tryptophan intake) and administering specific drugs which could alter serotonin levels, and also presumably bufotenin, should be attempted. The clinical efficacy of serotonin reuptake inhibitors or receptor blockers might be examined. The patient's age at the time of such experimental manipulations may be a crucial factor, just as it is in phenylketonuria, and should be examined as a separate parameter.

In a similar vein, there is a need to replicate and extend studies of hematologic and CSF factors. In particular, normative data need to be collected so that meaningful comparisons among patient groups can be conducted.

Hopefully, studies will be undertaken in many countries so that genetic and sociocultural variables can also be assessed.

Finally, there is an obvious need for longitudinal studies aimed at determining if biological measures covary with clinical course and/or age.

In conclusion, let us hope that in the future, creative researchers will be able to pinpoint specific pathological processes which will allow us to develop rational therapeutic techniques based upon ameliorating etiologic factors.

ACKNOWLEDGMENTS

Research by the authors referred to in this chapter was supported by the Division of Mental Retardation and Child Psychiatry, Department of Psychiatry, UCLA Center for the Health Sciences and the Neurobiochemistry Laboratory, Veterans Administration, Brentwood Hospital, Los Angeles, California; and in part by NIH grants: HD-04612, Mental Retardation Research Center, UCLA; and MCH-927. Interdisciplinary Training Project, UCLA; and HD-08575 to Dr. Yuwiler.

The authors are indebted to Mrs. Margaret Campbell and Mrs. Rose Weisler for their editorial assistance in preparing this manuscript.

REFERENCES

Bender, L. Childhood schizophrenia. *Psychiatric Quarterly*, 1953, *27*, 663–681.

Boullin, D., Coleman, M., & O'Brien, A. Abnormalities in platelet 5-hydroxytryptamine efflux in patients with infantile autism. *Nature,* 1970, *226*, 371–372.

Boullin, D., Coleman, M., O'Brien, R., & Rimland, B. Laboratory prediction of infantile autism based on 5-hydroxytryptamine efflux from blood platelets and their correlation with the Rimland E-2 score. *Journal of Autism and Childhood Schizophrenica*, 1971, *1*, 63–71

Boullin, D., & O'Brian, R.A. Uptake and loss of 14 C-dopamine by platelets from children with infantile autism. *Journal of Autism and Childhood Schizophrenia*, 1972, *2*, 67–74.

Campbell, M., Friedman, E., Devito, E., Greenspan, L., & Collins, P. Blood serotonin in psychotic and brain damaged children. *Journal of Autism and Childhood Schizophrenia,* 1974, *2*, 33–41.

Campbell, M., Friedman, E., Green, W., Collins, P., Small, A., & Breuer, H. Blood serotonin in schizophrenic children. *International Pharmacopsychology*, 1975, *10*, 213–221.

Campbell, M., Friedman, E., Green, W., Small, A., & Burdock, E. Blood platelets monoamine oxidase activity is schizophrenic children and their families: A preliminary study. *Neuropsychobiology*, in press.

Campbell, M., Small, A., Collins, P., Friedman, E., David, R., & Genieser, N. Levodopa and levoamphetamine: A crossover study in young schizophrenic children. *Current Therapeutic Research*, 1976, *19*, 70–86.

Cohen, D., Johnson, W., & Caparulo, B. Pica and elevated blood lead level in autistic and atypical children. *American Journal of Diseases of Children*, 1976, *130*, 47–48.

Cohen, D., Shaywitz, A., Johnson, W., & Bowen, M. Biogenic amine in autistic and atypical children: Cerebrospinal fluid measures of homovanillic acid and 5-hydroxyindoleacetic acid. *Archives of General Psychiatry*, 1974, *31*, 845–853.

Cohen, D., & Young, G. *Neurochemistry and child psychiatry,* in press.

Coleman, M. Serotonin and central nervous system syndromes of childhood: A review. *Journal of Autism and Childhood Schizophrenia*, 1973, *3*, 127–135. (a)

Coleman, M. (Ed.). *Serotonin in Down's Syndrome*. New York: North Holland, 1973. (b)

Coleman, M. Personal Communication, June, 1975.

Creak, M., Cameron, K., Cowie, V., Ini, S., MacKeith, R., Mitchell, G., O'Gorman, G., Orford, F., Rogers, W., Shapiro, A., Stone, F., Stroh, G., & Yudkin, S. Schizophrenic syndrome in children. *British Medical Journal*, 1961, *2*, 889–890.

DeMyer, M., Churchill, D., Pontius, W., & Gilkey, K. A comparison of five diagnostic systems for childhood schizophrenia and infantile autism. *Journal of Autism and Childhood Schizophrenia*, 1971, *1*, 175–189. (a)

DeMyer, M., Schwier, H., Bryson, C., Solow, E., & Roeske, N. Free fatty acid response to insulin and glucose stimulation in schizophrenic, autistic and emotionally disturbed children. *Journal of Autism and Childhood Schizophrenia*, 1971, *4*, 436–452. (b)

DeMyer, M., Ward, S., & Lintzenich, J. Comparison of macronutrients in the diets of psychotic and normal children. *Archives of General Psychiatry*, 1968, *18*, 584–590.

Fish, B. Methodology in child psychopharmacology. In D. Efron *et al.* (Eds.), *Psychopharmacology: Review of progress*. Washington, D.C.: Public Health Service, 1968.

Fowle, A. Atypical leukocyte patterns of schizophrenic children. *Archives of General Psychiatry*, 1968, *8*, 666–680.

Freeman, B., & Ritvo, E. Cognitive assessment. In E. Ritvo, B. Freeman, E. Ornitz, & P. Tanguay (eds.), *Autism: Diagnosis, current research and management*. Holliswood, N.Y.: Spectrum Publications, 1976.

Gittelman, M., & Cleeman, J. Serum magnesium level in psychotic and normal children. *Behavioral Neuropsychiatry*, 1969, *8*, 51–52.

Goodwin, M., Cowen, M., & Goodwin, T. Malabsorption and cerebral dysfunction: A multivariate and comparative study of autistic children. *Journal of Autism and Childhood Schizophrenia*, 1971, *1*, 148–162.

Heeley, A., & Roberts, G. Tryptophan metabolism in psychotic children. *Developmental Medicine and Child Neurology*, 1965, *7*, 46–49.

Himwich, H., Jenkins, R., Fujimori, M., & Narasimachari, N. A biochemical study of early infantile autism. *Journal of Autism and Childhood Schizophrenia*, 1972, *2*, 114–126.

Jorgensen, S., Mellerup, E., & Rafaelsen, O. Amino acid excretion in urine of children with various psychiatric diseases: A thin layer chromatographic study. *Danish Medical Bulletin*, 1970, *6*, 166–170.

Lucas, A., Krause, R., & Domino, E. Biological studies in childhood schizophrenia: Plasma and RBC cholinestrase activity. *Journal of Autism and Childhood Schizophrenia*, 1971, *1*, 172–181.

Narasimachari, N., & Himwich, H. Biochemical study in early infantile autism. *Biological Psychiatry*, 1975, *10*, 425–432.

Ornitz, E., & Ritvo, E. Perceptual inconstancy in early infantile autism. *Archives of General Psychiatry*, 1968, *18*, 79–98.

Ornitz, E., & Ritvo, E. The medical diagnosis. In E. Ritvo, B. Freeman, E. Ornitz, & P. Tanguay (Eds.), *Autism: Diagnosis, current research and management*. Holliswood, N.Y.: Spectrum Publications, 1976.

Ornitz, E., & Ritvo, E. The syndrome of autism: A critical review. *American Journal of Psychiatry*, 1977, *133*, 609–621.

Rank, B. Intensive study and treatment of preschool children who show marked personality deviations of "atypical development" and their parents. In G. Caplan (Ed.), *Emotional problems of early childhood*. New York: Basic Books, 1955.

Rimland, B. *Infantile autism*. New York: Appleton-Century-Crofts, 1964.

Ritvo, E., Yuwiler, A., Geller, E., Kales, A., Rashkins, S., Schico, A., Plotkin, S., Axelrod, R., & Howard, C. Effects of L-dopa in autism. *Journal of Autism and Childhood Schizophrenia*, 1971, *1*, 190–205.

Ritvo, E., Yuwiler, A., Geller, E., Ornitz, E., Saeger, K., & Plotkin, S. Increased blood serotonin and platelets in early infantile autism. *Archives of General Psychiatry*, 1970, *23*, 566–572.

Rutter, M. Autistic children-infancy to adulthood. *Seminars in Psychiatry*, 1970, *2*, 435–450.

Rutter, M. Childhood schizophrenia reconsidered. *Journal of Autism and Childhood Schizophrenia* 1972, *2*, 315–337.

Saladino, C., & Siva-Sankar, D. Studies on erythrocyte magnesium and potassium levels in childhood schizophrenia and growth. *Behavioral Neuropsychiatry*, 1969, *1*, 24–28.

Schain, R., & Freedman, D. Studies on 5-hydroxyindole metabolism in autistic and other mentally retarded children. *Journal of Pediatrics*, 1961, *58*, 315–320.

Schain, R., & Yannet, H. Infantile autism: An analysis of 50 cases and a consideration of certain relevant neurophysiologic concepts. *Journal of Pediatrics*, 1960, *57*, 560–567.

Shaw, C., Lucas, J., & Rabinovitch, R. Metabolic studies in childhood schizophrenia. Effects of tryptophan loading on indole excretion. *Archives of General Psychiatry*, 1959, *1*, 366–370.

Siva-Sankar, D. Biogenic amine uptake by blood platelets and RBC in childhood schizophrenia. *Acta Paedopsychiatrica*, 1970, *37*, 174–182.

Siva-Sankar, D. Studies on blood platelets, blood enzymes and leucocyte chromosome breakage in childhood schizophrenia. *Behavioral Neuropsychiatry*, 1971, *2*, 11–12.

Siva-Sankar, D., Gold, E., Phipps, E., & Sankar, B. General metabolic studies on schizophrenic children. *Annals of the New York Academy of Sciences*, 1962, *96*, 392–398.

Sutton, E., & Read, J. Abnormal amino acid metabolism in a case suggesting autism. *Journal of Diseases of Children*, 1958, *96*, 23–28.

Yuwiler, A., Plotkin, J., Geller, E., & Ritvo, E. A rapid accurate procedure for the determination of serotonin in whole human blood. *Biochemical Medicine*, 1970, *3*, 426–436.

Yuwiler, A., Ritvo, E., Bald, D., Kipper, D., & Koper, A. Examination of circadian rhythmicity of blood serotonin and platelets in autistic and non-autistic children. *Journal of Autism and Childhood Schizophrenia*, 1971, *1*, 421–435.

Yuwiler, A., Ritvo, E., Geller, E., Glossman, R., Schneiderman, G., & Matsuno, D. Uptake and efflux of serotonin from platelets of autistic and nonautistic children. *Journal of Autism and Childhood Schizophrenia*, 1975, *5*, 83–98.

Widelitz, M., & Feldman, W. Pink spot in childhood schizophrenia. *Behavioral Neuropsychiatry*, 1964, *1*, 29–30.

12

A Report on the Autistic Syndromes

MARY COLEMAN

What do we know about autism? What hard scientific facts can we all agree are correct? Do we know if it is a single disease or a syndrome? Is it caused by emotional stress; or by exogenous agents, such as viruses; or by endogenous factors, such as biochemical abnormalities?

What do we really know about the etiology of autism? At this time, we know very little indeed. The striking lack of consensus regarding the pathogenesis of autistic symptoms suggests that the right questions regarding the anatomical, electrophysiological, and biochemical substratum of the symptoms have not yet been asked. The study reported here is an initial attempt to formulate such questions.

In the spring of 1974, Dr. Bernard Rimland and other members of the United States National Society for Autistic Children approached the Children's Brain Research Clinic of Washington, D.C., to discuss the possibility of the Clinic studying their autistic children at the time of the annual meeting to be held that June in that city. The possibility of a large number of autistic individuals congregating in a single city created an opportunity to perform studies that might have results with statistical significance. Such a study was undertaken in June of 1974 and is the basis of this report.

MARY COLEMAN · 2525 Belmont Rd., N.W., Washington, D.C. 20008. The work reported on in this paper was done in collaboration with Hemmige N. Bhagavan, Ph.D., David J. Boullin, Ph.D., Philip L. Calcagno, M.D., Menek Goldstein, Ph.D., Brenda Herzberg, M.B., M.R.C.Psych., Albert R. Landgrebe, Ph.D., Marilyn Landgrebe, Ph.D., Jovita Lee, B.S., Dersh Mahanand, M.A., Robert O'Brien, Ph.D., Michael Peterson, M.D., Bernard Rimland, Ph.D., G. Semenuk, Ph.D., Sidney Spector, M.D., Jean Symmes, Ph.D., E. Fuller Torrey, M.D., Harry A. Walker, Ph.D., Lucille Weidel, A.A., Mary Kelleher Wypych, B.A., M.B.H. Youdim, Ph.D.

For each autistic child included in our study, we asked that the Society for Autistic Children locate and provide a normal child of the same age and sex. Relatives of autistic children were excluded as controls. Twenty investigators from around the world were invited to participate in the study.

We recognized that the central nervous system of a young child is limited in its ability to react to injury and that such reactions tend to fall into one of several major constellations of symptoms, such as the infantile spasms syndrome or nonspecific retardation. Another such syndrome, it seemed to us, was the autistic syndrome.

Each investigator participating in this study formulated his own research questions. These were mostly based on the assumption that infantile autism is not a single disease entity but rather is an organic syndrome with a number of different etiologies.

Patients in our research clinic were given the diagnosis of the primary autistic syndrome Class I (formerly called Kanner's infantile autism) if they met five criteria: (i) An early age of onset of clinical symptoms; (ii) a profound inability to relate to other people—Kanner's extreme autistic aloneness or Chess's lack of affective human contact; (iii) language retardation, including impaired comprehension of language appropriate for age and unusual use of language (such as pronominal reversals); together with at least two out of three further criteria: (iv) ritualistic and compulsive behavior; (v) disturbances of motility and appearance of stereotypies; and (vi) abnormal perceptual responses to stimuli in auditory, visual, and tactile modalities (the child who had relative indifference to pain but normal processing of the other sensory systems would not be included under this criterion). The olfactory system—the only sensory modality not processed through the thalamus—appears to be intact in most autistics.

A seventh criterion (the presence of an "area of excellence" or an isolated normal or superior level of mental functioning in a child otherwise performing at a retarded level) was used for extra confirmation.

Patients with the other two autistic syndromes that we define in our Clinic (Class II—the childhood schizophrenic syndrome with autistic symptoms and Class III—the neurologically damaged autistic patients) were excluded from this study.

Examples of patients with Class II autism are children who appear well until past 30 months of age and then slowly develop bizarre behavior disorders which include some of the symptoms seen in autistic patients. Occasionally these are mistaken for Class I autistic patients. Examples of Class III patients with the neurologically impaired autistic syndrome are seen in the presently defined etiologies of autism. On neurological examination these patients often have abnormalities or they have EEG abnormalities or a seizure disorder which starts prior to adolescence. (In our system of categorization, Class I patients are not excluded from that category if a seizure disorder starts at adolescence or later.) Examples of Class III patients are patients with phenylketonuria (Friedman, 1969) mistaken for autistic children or patients with the rubella syndrome (Chess *et al.*, 1971) who have marked autistic symptoms.

Seventy-eight children who met our criteria of Class I and who met the other criteria of participation (such as collecting a 24-hour urine specimen) participated in the study. Because of the large number of patients and the relatively large number of investigators who studied these children, a number of new findings were reported and now await further investigation by other research groups. The full details and methodologies of these studies are reported in a separate monograph (Coleman, 1976).

In this paper we will summarize the results found and discuss some of the interesting new questions that they have raised.

SUBJECTS

Seventy-eight patients came to the Clinic from all over the Eastern seaboard of the United States. A few children were brought from as far away as California and Michigan. There were 64 males and 14 females, a ratio of 4.6 to 1. A previous study of autism by Rutt and Oxford in 1971 showed an overall male to female ratio of 3.8 to 1 and a study by Spence *et al.* (1973) showed an overall sex ratio of 4.8 to 1. These ratios are in marked contrast to the sex ratio in childhood schizophrenia which does not essentially differ from a 1 to 1 ratio (Kallmann & Ritvo, 1956; Mayer-Gross, 1955). The median age of the children was nine years (range 3 to 23 years).

The parents of both autistic and control children filled out an elaborate six-page questionnaire providing retrospective historical data on the child. The value of such retrospective questionnaires is quite limited. Nevertheless, we were unable to identify any case-control differences regarding the gestational period. Our only two interesting new findings in the questionnaire concerned data from the preconception history: the parents of autistic children had a greater exposure to chemicals (20 out of 78 compared with only one control family) and a higher incidence of hypothyroidism than control parents (five fathers and 11 mothers of autistic children as against one father and three mothers in the control group). These data are difficult to interpret because it is possible that only "chemically minded" parents of autistic children would participate in a study of this kind designed to study organic factors in autism. However, this finding cannot be dismissed since it is possible that chemical toxins could affect genetic materials prior to conception. Clearly, this is an area where more prospective research is needed. The higher incidence of hypothyroidism in parents of autistic children was intriguing. In view of current speculation about the role of autoimmunity in the etiology of autism, it is relevant that one form of hypothyroidism can be identified in families with autoimmunity.

The height, weight, and cranial circumference of each child and each mother were recorded. The autistic childrens' mean height and weight showed only a nonsignificant trend toward lower values. Two of the autistic patients and four of the control children had cranial circumferences of above the 98th percentile. However, there was no significant difference in mean cranial circumference between the groups of children or between the two groups of mothers.

One of the universal problems in studying autistic children is the obtaining of suitable controls from families of comparable economic and social strata. In a study of the epidemiology of infantile autism, Treffert (1970) reported that 47% of the parents of autistic patients had completed college and a number had gone on to master's or doctoral work. This was in contrast to only 19% of the parents of other psychotic and mentally retarded children. Therefore, we were fortunate to obtain a control population that appeared to have, if anything, higher educational and occupational levels than the cases. The median yearly income for the autistic families (as reported on the anonymous questionnaires) was $15,000–20,000 whereas it was $20,000–25,000 for the controls. (The median average family income in the United States for 1973 was $12,051 as reported by the U.S. Census Bureau.) In our sample, only 10% of the autistic families had an income over $25,000/year while 25% of the control families were in this higher income range. Also, 44% of the autistic families had incomes below $15,000/year while only 18% of the control families had an average income in this range.

Twenty percent of the mothers of autistic children and 21% of control mothers went to graduate school, indicating that both groups differed markedly from the general population. However, among the remaining mothers, twice as many control mothers as mothers of autistic children finished college, thus indicating a somewhat higher level of education in control families.

INDIVIDUAL STUDIES

A Study of Infectious Agents (Drs. Peterson and Torrey)

An immunologic study of the participants and their mothers included antibodies to herpes simplex virus 1 and 2, cytomegalic inclusion virus, rubella, rubeola, and toxoplasmosis gondii; hepatitis B antigen, IgA, and IgM titers as well as pharyngeal swabs and urine for rubella and cytomegali inclusion virus. There was an increased incidence in serum antibodies to herpes simplex virus 1 and 2 (21 cases vs. nine controls) and to toxoplasma gondii (six autistic children vs. two controls). Serologic evidence from the biological mothers was negative for toxoplasmosis; this showed that the toxoplasmosis infections in the children had developed after birth.

The patients in this study had no neurological abnormalities. Whether the increased incidence in serum antibodies to herpes simplex virus 1 and 2 and to toxoplasmosis gondii will be in any way relevant to the central nervous system symptoms in autistic children remains to be determined. There is a special predilection of herpes virus for the temporal lobe area of the brain, an area found abnormal in some autistic patients (see Chapter 14). As antibodies to HSV-2 are not ordinarily found in prepubertal children, it is of particular interest that these were present in a small subgroup of autistic children.

A Study of Serotonin and Catecholamine Metabolic Pathways
(Drs. Goldstein, O'Brien, Semenuk, Spector, Boullin, Bhagavan, Youdim, Coleman, A. R. Landgrebe, M. Landgrebe, D. Mahanand, and J. Lee)

The role of tryptophan and tyrosine pathways (specifically their amines, serotonin, dopamine, and norepinephrine) is under investigation in all the major psychiatric disease entities. In the case of autism, serotonin (5HT) abnormalities have been reported in blood studies of autistic patients. The first report by Schain and Freedman (1961) described elevated levels of whole blood 5HT in six out of 23 patients with "autism." (A more detailed report of these patients is given by Schain and Yannet, 1960, who describe what appears to be a group of withdrawn, disturbed children with a variety of neurological and psychiatric diagnoses, such as hemiparesis, hydrocephalus, and prematurity.)

Ritvo et al. (1970) also reported increased 5HT levels with increased platelet counts in a group of children "variously diagnosed as early infantile autism, atypical ego development, symbiotic psychosis, and certain cases of childhood schizophrenia"; they point out that the serotonin per platelet level is not increased in their patient group. Circadian rhythmicity of serotonin and platelets in autistic children did not account for their results (Yuwiler et al., 1970). In a recent study of endogenous levels of whole blood serotonin in 11 psychotic and brain-damaged children, Campbell et al., (1974) noted a tendency toward higher serotonin levels in the patients but the difference did not reach statistical significance.

Siva Sankar et al., (1963) studied the mechanisms underlying 5HT abnormalities reported in the blood of disturbed children. They measured the uptake of 5HT by platelets (the cell in the blood containing most of the 5HT) and found the uptake reduced. In 1970 our research group studied a number of patients with Kanner's "primary" autism as well as retarded psychotic children with autistic features. We found that a patient with clinical criteria of Kanner's "primary" autism may have a high, normal, or low endogenous level by the whole blood method. However, biochemical identification of a "primary" autistic child may be possible by a study of the binding of 5HT in platelets compared to controls (Boullin et al., 1970). In a related study, the efflux ratio determined on a blind basis was used to distinguish primary autistic patients from children with other psychiatric diagnoses in a mixed group of disturbed children. There was a 90% correlation in that study between the efflux ratio prediction of autism and the clinical diagnosis of Kanner's primary autism (Boullin et al., 1971). A recent study by Yuwiler et al. (1975) using similar but not identical efflux methods failed to confirm a difference between autistics and controls. We suspect the difference between the Boullin and Yuwiler studies may be in patient selection, rather than a methodological difference.

The Yale group of investigators headed by Cohen has published a study of cerebral spinal fluid levels of the end product of serotonin, 5-hydroxyindole acetic acid (5HIAA), using the probenecid method in autistic children. Compared to epileptic children, autistic children were reported to have higher levels of 5HIAA (Cohen 1974).

High endogenous levels of serotonin in the blood have been reported in a number of patients with mental retardation. These patient groups include the infantile spasm syndrome, infant hypothyroidism, maternal rubella, kernicterus syndrome, Schilder's disease and other leukodystrophies, hydranencephaly, postencephalitic and postmeningitic retarded children, and in encephalocele and meningomyelocele associated with retardation (Tu & Partington, 1972). Most of the studies describing high levels in retarded children were done in institutions for severely retarded children. The mechanisms resulting in the 5HT abnormality appear to differ in each group of diseases. Sometimes they are related to factors in the metabolic pathways and sometimes to factors of binding of the amine. Some of these patient groups, particularly the infantile spasm syndrome and the leukodystrophies, have children with fragments of the autistic syndrome or partial autistic features. However, their lack of the full features of the syndromes plus the neurological findings makes differentiation quite possible.

In our present June 1974 study we again confirmed that whole blood levels of serotonin (total 5-hydroxyindole method) were significantly higher ($t = 2.67, p > .01$) in the autistic patients (mean 86.2 ng/ml) than in the controls (mean 73.2 ng/ml) and had a greater variability.

Investigation of the role of biogenic amines in the catecholamine pathways is a relatively new area in autism. An earlier investigation of DBH levels in a mixed group of autistic and schizophrenic children had found no difference compared with age-matched controls (Coleman et al.,

1974). In our present study we elected to repeat this study with this larger more homogeneous sample of Class I autistic patients and compared them with age-sex matched controls. Dopamine-beta-hydroxylase (DBH) is the enzyme responsible for the final step of norepinephrine biosynthesis. The serum DBH levels were significantly lower ($t = 2.40, p < .02$) in the autistic patients (mean of 22.0) than in the controls (mean of 32.9). This result was surprising because serum DBH has not so far been found to be of value in differentiating adult psychiatric patients from normal individuals. However, a confirmation that there may be some relevance to a block at the dopamine-beta-hydroxylase level in autism was found in a separate study using thin-layer chromatography to study the urine in our patient group. An unusual compound (3,4-dihydroxyphenylacetic acid or homoprotocatechuic acid) was found in the urine of 88% of the autistic children. This acid is a minor metabolite in the tyrosine pathway and would be expected to be found when there is some block at the dopamine-beta-hydroxylase level. The same thin-layer chromatographic study of urine also found parahydroxymandelic acid in 80% of the autistic group and epinephrine in 73% of the autistic children. Urinary catecholamine screening was not undertaken with controls.

We also studied catechol-O-methyl transferase, another enzyme in the catecholamine pathway that is used in the degradation of the biogenic amine. Enzyme activity was determined in the red blood cells. There was no difference in overall enzyme activity between the two groups. However, when the red cell homogenates were divided into particulate and soluble fractions, some differences (falling short of statistical significance) were observed. The COMT activity in the soluble fraction of cells from autistic males was less than in control males whereas that in autistic females was higher than control levels. The other degradative enzyme, monoamine oxidase, was measured in the platelets of a small subgroup of nine autistic subjects and compared with nine normal subjects of comparable age and weight. There was no significant difference in the rates of metabolism of the substrates between autistic and control groups.

Zinc and Copper Studies (Dr. Calcagno, D. Mahanand, and Mary Wypych)

A study of the zinc and copper levels in the serum of autistic and control patients was undertaken. No difference between the two groups was found in the serum copper levels but serum zinc levels were significantly elevated in the autistic group. The control mean was 112.5 plus or minus 30.28 ng/%, while the autistic mean was 171.05 plus or minus 134.44 ng/% ($t = 3.52, p < .001$). This was a startling result since at the present time there are no human diseases known to be associated with elevated zinc levels. The problem is currently under further investigation with studies of total zinc metabolism in the hair, saliva, urine, and serum of autistic children and controls.

Studies of Physical Stigmata (Dr. Walker)

A study of minor physical anomalies (cranial circumference, fine electric hair, epicanthus, hypertelorism, seating of ears, adherence of ear lobes, asymmetry or malformation of ears, high palate, furrowed tongue, fifth-finger curvature, stubbing of fifth finger, syndactyly of toes, large gap between first and second toes, comparative length of third and second toes, and transverse crease on hand) was completed in both autistic children and controls. Clusters of stigmata that might be associated with known chromosomal disorders could not be identified. However, stigmata scores were significantly higher in the autistic group (5.8 vs. 3.5; $t = 5.95$; $p < 0.001$). Low seating of ears, hypertelorism, and minor syndactyly were particularly more frequent among autistic than among normal subjects.

Both schizophrenic (Goldfarb, 1967) and hyperactive (Quinn & Rapoport, 1974) children have been reported to have a significantly greater number of anomalies than normal children. This study now adds another major population among disturbed children to the category of stigmatized individuals. All of these studies are suggestive of some congenital factor playing a role in at least the risk factors in any individual patient. Among the beautiful children in this study, there were four autistic subjects who displayed absolutely no stigmata or any other distortion of physical form. Most of the children, including the majority with stigmata scores, had the attractive features described in most patients with the Class I autistic syndrome.

Dermatoglyphic Studies (Dr. Walker)

Dermatoglyphic patterns of the participants in the study also were completed. Analysis of dermal ridge patterns and ridge counts resulted in significant differences from normal expectation in the autistic children. These were most apparent in the reduced numbers of whorls (19% vs. 30% in controls) and increased numbers of arches (18% vs. 11% in controls) coupled with less distinctness in formation of triradii, and in lowered ridge counts (mean of 53.4 compared with 68.2 for controls). The number of dermal ridges on autistic hands was much lower for each finger as well as for the total hand.

Fig. 1. Autistic boy hand flapping at elbow level, the most common adventitious movement in autistic patients. (From Coleman, 1976.)

Again, abnormalities in dermatoglyphics are suggestive of embryological insults and lead us back to reconsideration of congenital factors in autism.

Studies of Adventitious Movements (Drs. Walker and Coleman)

Motion picture recordings were made of 55 children participating in this study. The most common adventitious movements observed were elbow-level hand flapping in 35% of the patients (Fig. 1) and head-level hand flapping in 27% of the patients. Other frequent behaviors were hand clasping and patting and hand posturing. Facial grimacing, choreic and athetotic postures, jumping and rocking, and the behaviors focusing on the ears and mouth were also observed. One child had myoclonic jerking of the head. Both flapping and clapping occurred at the rate of .28 seconds per cycle with the mean length of a flapping episode at 3.51 seconds. The correlation between duration and cycling rate of activity was significant. The mean duration of posture holding was 3.67 seconds. This suggested that some of these postures may be suppressions of flapping. Double adventitious movements were described in four of the patients.

Family History Studies (B. Herzberg)

Brief (10-minute) interviews and questionnaires were used to study family patterns of illness in both groups. Mental illness occurred with equal frequency in the families of autistic children and the families of controls. An unusual number of Down's syndrome relatives were identified (four in the autistic group vs. one in the controls), and this finding is interesting because increased parental hypothyroidism has been found in Down's syndrome families. There was little to suggest that siblings of autistic children could be distinguished from other children, with the possible exception of minor delay in maturation of the central nervous system. However, there appears to be an increase of gastrointestinal problems in autistic families.

Psychological Testing (Dr. Symmes)

Apparently lawful relationships between the levels of receptive language, expressive language, the ability to classify objects, and visual motor function occur in our sample of autistic children. Autistic girls seem to be at a lower level developmentally than autistic boys, or more likely, to be unresponsive to any task presented. A visual-matching-to-sample task was devised in an attempt to utilize commonly occurring behaviors of autistic children and thus maximize responsiveness. Children in the sample who were six years of age or younger responded to this task with far greater frequency than to any of the standard psychological tasks (see Fig. 2). The VMS task seems to have promise as a framework for training and testing autistic children and is being developed further.

AUTISTIC SUBGROUPS

Finally, when we put all of our results together it appeared to us that we had identified several subgroups of autistic patients that needed further investigation. It should be emphasized that these are all subgroups within the Class I syndrome and do not include patients in the other classes of autistic syndromes.

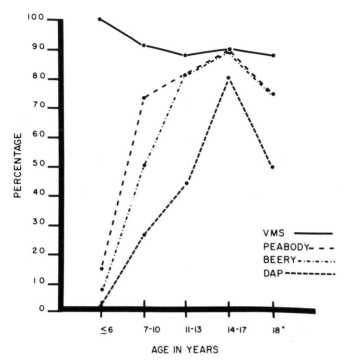

Fig. 2. Percentage of autistic children's response to task by age: a comparison of the new visual-matching-to-sample task (VMS), the Peabody Picture Vocabulary Test, the Beery Visual Motor Integration Test, and the Harris-Goodenough drawing of a person (DAP).

Familial Autism (Drs. Rimland and Coleman)

The first group appeared to have a familial form of autism. Of the 78 autistic patients in the study, six (8%) had a relative who also met our criteria for the Class I autistic syndrome (see Fig. 3). In this small family study of the 12 autistic children including each proband and his/her autistic relative, ten of the familial autistic patients were male and eight were Jewish. Six families is too small to definitely establish any genetic pattern. However, our present results suggest that the genetic pattern of familial autism may be autosomal recessive.

Hyperuricosuria in Autism (Drs. A. R. Landgrebe, M. Landgrebe, and Coleman)

Another subgroup of patients were classified as having a purine abnormality. Serum and 24-hour urinary studies of uric acid in the autistic and control subjects disclosed that no participant in the study had increased uric acid in the blood or hyperuricemia. However, 3% of the

CASE 95

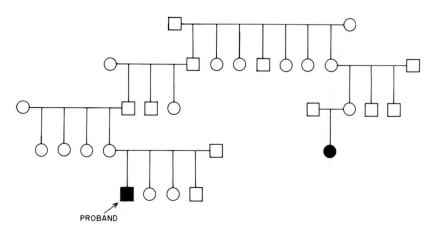

PROBAND

Fig. 3. An example of the pedigree found in patients with familial autism. Circles represent female family members; squares represent males. (From Coleman, 1976.)

controls and 22% of the patients had hyperuricosuria, that is, increased levels of uric acid in the urine. It is interesting that these results were most clearly noted in preadolescents (Fig. 4). The possibility of a research diagnosis of purine autism in some patients is raised by these data. Identifying such an abnormality in the turnover of purines, however, is just the first step in trying to identify a classifiable metabolic illness in such patients. The differential diagnosis of increased purine turnover is extremely long and has to take into account renal mechanisms as well as many other well-defined metabolic errors in the biosynthesis of purines. Each patient in the study is undergoing an extensive evaluation to see if the cause of the increased uric acid excretion can be specifically identified. The possibility of modifying purine turnover by diet or drugs, such as allopurinol, is raised by these research findings.

Celiac Disease and Autism (Drs. A. R. Landgrebe, M. Landgrebe, and Coleman)

There also appeared to be a small percentage of autistic children who had celiac disease. A diagnosis of celiac disease had been made on an unusually high number (10%) of the patients who participated in our project. Two of the patients were on a gluten-free diet treatment for celiac disease and these patients had normal levels of calcium in the serum and urine. The other six untreated celiac patients had a depression of calcium level. Celiac disease is a specific diagnostic entity that can be determined by jejunal biopsy. Most of the pediatricians who had made the diagnosis prior to this study had not performed a jejunal biopsy. Therefore, we currently have an ongoing research project studying malabsorption in these patients. One such study of jejunal biopsy has not confirmed celiac disease in several patients but the study has not yet been completed. Other leads to studying malabsorption in these patients may be more promising. It is possible that we are dealing with another form of malabsorption, possibly wheat-sensitive, rather than classical celiac disease in these autistic patients.

An additional 11 patients who did not have a diagnosis of celiac disease had a depression of urinary calcium for no reason that could be identified. None of the controls had an abnormality of calcium in either serum or urine. In our Clinic, we have found an unusually large variation in calcium levels in patients with developmental aphasia. So, the relationship between the calcium levels and the failure of language development in autistic patients is of great interest. Again, a research problem has been set up that needs further investigation.

CONCLUSION

There is evidence that the parents of autistic children may function at a higher intellectual or social and economic level than parents of schizophrenic or retarded children. This is indeed an unusual finding in the his-

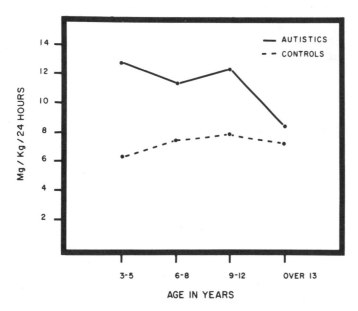

Fig. 4. Autistic and control uric acid levels by age groups in this study, from 24-hour urine specimens. (From Coleman, 1976.)

tory of pediatrics. Such gifted parents can help future research in autism; certainly already they are responsible for the existence of the study we have reported here. Most inborn errors of metabolism produce deleterious effects. However, the purine inborn error of metabolism that produces gout is sometimes associated with higher intelligence in the patients. Since we now have found that some autistic children are over-producers of the purines, a look at purine metabolism in the parents of autistic patients might be an interesting future project. Another new area needing investigation is a study of the small subgroup of autism in females. In our project, we found them less stigmatized but more obese and more seriously impaired by their symptoms than male autistics.

Our understanding of the pathogenesis of autistic symptoms is in its infancy. This study has provided no final answers but it has raised a number of questions that we continue to study and that we hope will be explored by other research teams.

REFERENCES

Boullin, D. J., Coleman, M., & O'Brien, R. A. Abnormalities in platelet 5-hydroxytryptamine efflux in patients with infantile autism. *Nature (London),* 1970, *226,* 371–372.

Boullin, D. J., Coleman, M., O'Brien, R. A., & Rimland, B. Laboratory predictions of infantile autism based on 5-hydroxytryptamine efflux from blood platelets and their correlation with the Rimland E-2 score. *Journal of Autism and Childhood Schizophrenia,* 1971, *1,* 63–71.

Campbell, M., Friedman, E., DeVito, E., Greenspan, L., & Collins, P. J. Blood serotonin in psychotic and brain damaged children. *Journal of Autism and Childhood Schizophrenia,* 1974, *4,* 33–41.

Chess, S., Korn, S. H., & Fernandez, P. B. *Psychiatric disorders of children with congenital rubella.* New York: Brunner/Mazel, 1971.

Cohen, D. J., Shaywitz, B. A., Johnson, W. T., & Bowers, M. Biogenic amines in autistic and atypical children. *Archives of General Psychiatry,* 1974, *31,* 845–853.

Coleman, M. (Ed.). *The autistic syndromes.* New York: Elsevier, 1976.

Coleman, M., Campbell, M., Freedman, L. S., Roffman, M., Ebstein, R. P., & Goldstein, M. Serum dopamine-beta-hydroxylase levels in Down's syndrome. *Clinical Genetics,* 1974, *5,* 312–315.

Friedman, E., The "autistic syndrome" and phenylketonuria. *Schizophrenia,* 1969, *1,* 249–261.

Goldfarb, W. Factors in the development of schizophrenic children: An approach to sub-classification. In J. Romano (Ed.), *The origins of schizophrenia.* Amsterdam: Excerpta Medica, 1967.

Kallmann, F. J., & Ritvo, B. Genetic aspects of pre-adolescent schizophrenia. *American Journal of Psychiatry,* 1956, *112,* 599–606.

Mayer-Gross, W. Clinical research in psychiatric—retrospect and prospect: President's address. *Proceedings of the Royal Society of Medicine,* 1955, *48,* 217–223.

Quinn, P. O., & Rapoport, J. L. Minor physical anomalies and neurologic status in hyperactive boys. *Pediatrics,* 1974, *53,* 742–747.

Ritvo, E. R., Yuwiler, A., Geller, E., Ornitz, E. M., Saeger, K., & Plotkin, S. Increased blood serotonin and platelets in early infantile autism. *Archives of General Psychiatry*, 1970, 223, 566–572.

Rutt, C. N., & Oxford, D. R. Prenatal and perinatal complications in childhood schizophrenics and their siblings. *Journal of Nervous and Mental Diseases*, 1971. *152*, 324–331.

Schain, R. J., & Freedman, D. X., Studies on 5-hydroxyindole metabolism in autistic and other mentally retarded children. *Journal of Pediatrics*, 1961, *58*, 315–320.

Schain, R. J., & Yannet, H. Infantile autism. *Journal of Pediatrics*, 1960, *57*, 560–567.

Siva Sankar, D. V., Cates, N., Broer, H., & Sankar, D. B. Biochemical parameters of childhood schizophrenia (autism) and growth. In (J. Wortis (Ed.), *Recent advances in biochemical psychiatry*. New York: Plenum Press, 1963.

Spence, M. A., Simmons, J. Q., Brown, N. A., & Wikler, L. Sex ratios in families of autistic children. *American Journal of Mental Deficiency*, 1973, *77*, 405–407.

Treffert, D. A. Epidemiology of infantile autism. *Archives of General Psychiatry* (Chic.), 1970, *22*, 431–438.

Tu, J., & Partington, M. 5-hydroxyindole levels in the blood and CSF in Down's syndrome, phenylketonuria and severe mental retardation. *Developmental Medicine and Child Neurology*, 1972, *14*, 457–466.

Yuwiler, A., Plotkin, S., Geller, E., & Ritvo, E. R. A rapid accurate procedure for the determination of serotonin in whole human blood. *Biochemical Medicine*, 1970, *3*, 426–431.

Yuwiler, A., Ritvo, E., Geller, E., Glousman, R., Scheiderman, G., & Matsuno, D. Uptake and efflux of serotonin from platelets of autistic and non-autistic children. *Journal of Autism and Childhood Schizophrenia*, 1975, *5*, 83–98.

13

Biochemical Strategies and Concepts

R. RODNIGHT

There are critical problems of data interpretation in the field of biochemical research in autism. The excellent chapters by Coleman (Chapter 12) and Ritvo (Chapter 11) illustrate these problems very graphically. They also bring out the obvious fact that it is even more difficult to obtain meaningful results in this area when studying immature subjects than when working with adult populations. Compared with adults, body chemistry in children is more labile, more subject to variation from endogenous as well as exogenous factors, and, of course, more age-dependent. The question of controls is therefore crucial, but for ethical reasons even more difficult to solve satisfactorily than when investigating mature subjects. Further, since autism is not a disease entity but almost certainly a syndrome with different etiologies, the results of comparisons between different groups of children with behavioral abnormalities can become very difficult to interpret. Yet one feels certain that here, as in the major psychoses of adult life, metabolic factors may play some role in etiology and could be revealed if only one knew how best to display them.

Indeed, it is now generally recognized that in any normal population individual biochemical differences play some part, however small, in influencing behavior patterns. Except for identical twins, no two individuals have exactly the same constellation of genes, and since genes determine enzymes and enzyme activity is a major factor in determining body fluid composition, considerable variation for the latter may be expected in the normal population. Therefore, when particular biochemical results are

R. RODNIGHT · Department of Biochemistry, Institute of Psychiatry, De Crespigny Park, Denmark Hill, London SE5 8AF, England.

obtained in sick children one may need to compare these with a large number of results from normal individuals before making any conclusions. In other words, the range of genetically determined biological variation for many parameters is still relatively unexplored in pediatric research, and unless the deviation from the available normative data is very large (as for example in the typical inborn errors of metabolism) it is often difficult to come to any conclusions.

Of course, ultimately there is value in recording all good data on body fluid composition in mental illness. But, to be of value they must be obtained by sound reproducible methodology under well-controlled conditions, so that exogenous influences such as diet and drugs are excluded, or at least recorded. It is likely that in autism, as in many instances of psychotic illness, the biological contribution to the phenotype may be multifactorial, in which case the pattern or profile of body fluid metabolites may bear more meaningful relationship to clinical abnormality than will any single measurement.

ABNORMALITY IN THE BRAIN OR IN BODY CHEMISTRY?

We also have to recognize that the study of the chemical composition of the peripheral body fluids, however detailed and sopisticated, may yet not give us any clues as to the etiology of autism. On present evidence it seems very unlikely that autism, or any subgroup within the syndrome, any more than schizophrenia, is due to a defined inborn error in a metabolic pathway generally represented in body tissues. Thus autism is in no way analogous to phenylketonuria or to the Lesch-Nyhan syndrome. One suspects that it may well arise from pathology confined to the brain, in which chemical factors are involved but not expressed in the peripheral body fluids. Two examples from neurology illustrate this point.

Parkinson's disease and Huntington's Chorea are both due to imbalance of neurotransmitter function in the central nervous system resulting from degenerative lesions. However, in neither of these conditions are there any specific consequences for the biochemistry of the peripheral body fluids and it was not until postmortem samples of brain tissue were examined that the natures of the lesions were discovered (Hornykiewicz, 1972). In autism, and perhaps in similar disturbances of childhood, neurotransmitter imbalance could conceivably arise from developmental failure of certain neuronal pathways rather than degeneration. Such failure could be related to neonatal or postnatal infection, vascular anomalies or even dietary factors. Accordingly, it is good that a start has been made in studying the composition of cerebral spinal fluid in autism. There are obviously ethical problems here, but few would doubt that once our knowledge of basic neurochemistry has proceeded to a stage where reasonable hypotheses can be set up to explain abnormal behavior in children, the sampling of this fluid would be justified.

PLATELET STUDIES

There have been very extensive studies on serotonin uptake, storage and efflux in blood platelets in autism. Unfortunately, as Ritvo recognizes (see Chapter 11), these studies are extremely difficult to interpret because of the many factors involved. Although there is evidence from studies with drugs like reserpine and the tricyclics that serotonin is taken up and stored in platelets by a similar mechanism to that operating in the CNS (Pletscher *et al.*, 1969), I am frankly skeptical about the extent to which information from these blood cells can be extrapolated to the brain. Coleman and others have argued cogently for the view that the platelet may be considered as a model, albeit imperfect, of a serotoninergic neurone. On this argument it is presumably suggested that, if an abnormally high efflux of serotonin from platelets is observed (possibly as a result of a binding defect) in a group of patients, there is likely to be faulty binding of serotonin in the CNS. The reasoning assumes that platelet binding in the two locations not only involves a common mechanism (for which there is much evidence; Pletscher *et al.*, 1969) but also is influenced and controlled by the same factors. That seems to be very unlikely since the blood platelet is much more directly exposed to the environment than are the brain cells. This applies both to endogenous and exogenous influences.

Thus, the blood platelets are carried in the bloodstream through all the tissues of the body and are inevitably exposed to a whole variety of tissue metabolites and dietary components which never impinge on the brain cells. Further, it seems likely that in our present stage of knowledge there are important differences in the way in which serotonin is handled in the bloodstream and the brain. In particular the efflux or release of serotonin from platelets cannot be compared to transmitter release at synapses. For instance, bacterial endotoxins are known to stimulate serotonin release from platelets, as are proteolytic enzymes, antigens, fatty acids, and glycogen (Hawiger *et al.*, 1975). These agents are far more likely to be encountered by the platelet than the synapse, protected as it is behind the blood brain barrier. It may be argued that efflux studies are usually performed on washed cells in media of defined composition, but it cannot be excluded that exposure of the platelets *in vivo* to one or other of the above factors may have modified the integrity of the cell membrane. This is not to doubt the validity of the correlations which many workers have shown between clinical symptoms and platelet serotonin; the fashion in which platelet serotonin is sometimes normalized as a result of treatment (as in PKU) is particularly striking (Closs, 1966). And to give another example, Coleman (1971) has observed that some hyperactive disturbed children exhibit abnormal platelet serotonin concentrations which tend to return to normal, when, as a result of treatment, the patients become quiet and more manageable. However, rather than extrapolate this finding to the nervous system, I would prefer to suggest that it is more likely that platelet function is modified as a result of some unknown peripheral consequences of overactivity.

ZINC METABOLISM

Two other observations require comment. One is the very striking difference in serum zinc found in Coleman's study (Chapter 13) between the controls and autistic children. The significance of this finding, however, would be more obvious if the autistic children exhibited a zinc deficiency. Zinc is widespread in the diet and, while not much is known about the factors which control its uptake from the alimentary tract, its absorption is influenced by other minerals in the diet such as calcium. Also, zinc deficiency may be rectified by administering $ZnCO_3$. It will be very important to replicate these findings on groups of subjects consuming exactly the same diet. The same comments apply to the occasional finding of hyperuric aciduria in autism, since much of the urinary uric acid is derived from purines in the diet.

RESEARCH STRATEGIES

Two strategies mentioned by Ritvo (Chapter II) may be of particular value in biochemical research in autism. One is the need to do more longitudinal studies with children. There are problems in taking repetitive samples from children but replication of unusual findings is essential in individual patients as well as in different groups of subjects. In this approach the patient to some extent acts as his own control. If the abnormality persists over a period of time regardless of clinical state, then it is more likely to have a constitutional basis and perhaps be related to the disease process than if it only appears during exacerbation of the condition. Secondly, the approach of studying body biochemistry after a loading dose of a precursor of some compound thought to be of significance in a particular pathway may sometimes reveal an abnormality which does not show up in the basal state. For instance, to investigate Coleman's interesting finding of low values of dopamine-β-hydroxylase in autism, one might feed a loading dose of amino acid tyrosine and measure the rate at which it is converted to noradrenaline and its metabolites in subjects with normal and low levels of the enzyme in their blood.

One final comment: Although biochemical studies of autism (and indeed of all abnormal behavior states in man) have often proved negative and when positive are fraught with problems of interpretation, I have no doubt that the effort being expended will eventually prove worthwhile. As I see it, the main obstacle to progress at present is our relative ignorance of the chemical basis of normal behavior in man. As knowledge of the ways in which biochemical events in the brain modulate normal behavior increases, so it will become more feasible to devise testable hypotheses to explain autistic behavior.

REFERENCES

Coleman, M. Serotonin concentrations in whole blood of hyperactive children. *Journal of Pediatrics*, 1971, *78*, 985–990.

Closs, K. The biochemistry of phenylketonuria. In O. Walaas (Ed.), *Molecular basis of some aspects of mental activity, I*. New York: Academic Press, 1966. Pp. 231–247.

Hawiger, J., Hawiger, A., & Timmons, S. Endotoxin-sensitive component of human platelets. *Nature (London)*, 1975, 256, 125–127.

Hornykiewicz, O. Neurochemistry of parkinsonism. In A. Lajtha (Ed.), *Handbook of neurochemistry, VII*. New York: Plenum Press, 1972. Pp. 465–501.

Pletscher, A., da Prada, M., & Tranzer, J. P. Transfer and storage of biogenic monoamines in subcellular organelles of blood platelets. *Progress in Brain Research*, 1969, *31*, 47–52.

14

A Neuropsychologic Interpretation of Infantile Autism

G. ROBERT DeLONG

Infantile autism is a behavioral syndrome, almost certainly having various etiologies. The present paper focuses on one group of autistic children, which might be designated the neurologically impaired group. In the course of investigating a number of children exhibiting features of early infantile autism, we have been impressed by the finding of consistent changes on pneumoencephalography, apparently reflecting a specific anatomical pathology (Hauser *et al.*, 1975). These changes have centered on the left medial temporal lobe and have been seen as dilatation of the temporal horn of the left lateral ventricle. The present chapter represents our attempt to relate these anatomical abnormalities to the known clinical manifestations of this disorder.

Seventeen cases were selected retrospectively from a group of 105 children referred for neurological evaluation of retarded language development and/or infantile autism. The sole criterion for selection was the presence of a pneumoencephalogram as part of the diagnostic study. Identifiable metabolic disorders were absent in all cases. None of the children were blind or deaf. Somatic abnormalities were present in only two cases (one dolichocephalic, one with synostosis of the coronal suture, surgically repaired).

The primary language disorder, remarkably uniform throughout the group, may be characterized as a failure to develop expressive speech.

G. ROBERT DeLONG · Pediatric Neurology Unit, Massachusetts General Hospital, Boston, Massachusetts.

Comprehension of spoken language, although also delayed for age, was somewhat less deficient in 14 cases. Thus, when last seen by us, five cases had failed to develop any speech, four were using single words only (range of vocabulary one to 50 words) and the remaining eight had developed four–five-word sentences. In the last group, acquisition of sentences was accompanied by hyperlalia and echolalia in five cases and was used as a form of self-amusement in addition to communication. Although fluent, words were poorly articulated, improperly accented and often used with incorrect pronouns. Regression was present in only three cases; one lost all expressive language following a prolonged episode of status epilepticus; another used only one word at a time, losing a previously spoken word with each new acquisition; and a third had subsequently lost the six or seven words spoken at 15 months. Particularly curious is one, whose early words included "soaked," "soggy," and "tasty." With only three exceptions the entire group was able to understand and respond appropriately to simple commands. Thirteen could point to common objects and/or parts of the body when requested to do so. Thus, the language disorder observed in these children can only be interpreted as both receptive and expressive in nature. The expressive deficit may have predominated in most cases.

All our cases (with one exception) exhibited, during some stage in their development, profound disturbances in relationships with people. Twelve were frankly autistic, refusing to be held, establishing no eye contact, and usually described by family as in a world of their own. Five others might be classified as symbiotic, ignoring all people except for one or two close family members, with whom they were overaffectionate and demanding of attention. These two groups were not separable by other clinical criteria and hence are considered together.

Behavior disorders represented a prominent clinical feature in 16 cases. The disturbances noted were quite similar to those usually accepted as characteristic of the autistic syndrome. They included (a) hyperactivity (in 13/17); (b) self-stimulatory behavior, i.e., rocking, spinning, banging (12/17); (c) violent episodes of severe temper tantrums (9/17); (d) unusual preoccupation with objects, often water, dust in the sunlight, things that spin (8/17); (e) apparent insensitivity to pain (7/17); (f) rituals and obsessions (6/17); (g) abnormal eating habits, usually hyperphagia and refusal to change a single item of diet (6/17); (h) perseverative activity (4/17); (i) use of proximal more than distal senses (4/17); (j) hyperacusis (4/17); (k) precocious skills in musical or mechanical areas (4/17); (l) unusual fears and anxiety (3/17); (m) tendency to preserve sameness in environment (3/17); (n) excessive masturbation (2/17); and (o) waxy rigidity (2/17).

Fifteen of our 17 cases satisfy at least five of the criteria set forth by Creak (1963) for the diagnosis of infantile autism. Sixteen satisfy the two criteria which seem to us most central to the diagnosis, that is, the failure of normal development of communicative language and gross and sustained impairment in emotional relationships.

There was reasonable evidence in support of a prenatal insult in five of the 17 cases; two children were adopted shortly after birth and little is known of their pregnancy or birth history or of their biological family. One child was premature at 36 weeks, another postmature at 42 weeks. All 17 were thought to be normal at birth and no perinatal problems arose with the exception of a forceps-induced seventh nerve palsy in one.

With a single exception, all these children did well during their first year of life, feeding well, cooing, and babbling normally and making progress in motor development. One child spent his first two years in a minimum-care nursery where he was malnourished and repeatedly ill. By the age of 18 months he was unable to sit up, but, within one year of leaving the institution, his motor milestones had become normal for his age.

In only three cases was there reasonable evidence that the child possessed social responsiveness during the first year of life which was subsequently lost. In the other 14 regression of previously attained social interaction was not part of the clinical picture.

Gross motor milestones were reported as normal (walking by 15 months) in seven, as mildly delayed (walking by 18 months) in five, and as delayed (not walking by 18 months) in five. While many in the group were described by their families as clumsy, none were considered at any time to have obvious motor disorders.

Major illnesses predated or accompanied the clinical problem in three cases. One was hospitalized several times during his first two years of life for malnutrition and for fever of unknown origin (investigations negative). One had premature closure of the coronal sutures and had developed mild ataxia which was cured by surgical correction of the synostosis. Another was developmentally normal at age 3 years and her symptoms appeared only after an episode of prolonged status epilepticus. Two other cases had a history of middle-ear infections, thought at the time to be uncomplicated. In addition, a history of seizures was obtained in five cases.

There was a significant history of familial neuropsychopathology in five of the 15 cases whose natural families were known. One had three older brothers, all of low intelligence (IQ 70), all doing poorly in school, and two demonstrating sociopathic behavior; an uncle had been in prison for robbery. Another had an older brother who was a late walker and a younger sister described as hyperactive and destructive. The mother of another was alcoholic. Both siblings of another were late to speak (but of normal intelligence) and a second cousin was mentally retarded. The father of one, as a child, had been a toe-walker and a head-banger.

Six children had head circumferences greater than the 97th percentile for age. Five had definite focal findings on examination, including: an upgoing right plantar response; slight hyperreflexia in the left arm; neglect of the right visual field; duplication of knee-jerks, ankle clonus, and bilateral adductor spasticity; and bilateral ankle clonus with a left upgoing plantar response.

Of the 17 children, eight were left-handed and three had failed to establish preferred handedness when last seen.

EEGs demonstrated unequivocal paroxysmal activity in eight cases, with spiking in the right temporal area (3), in the left temporal area (1), bitemporally (2), generalized (1), or suggestive of atypical petit mal. Four other cases had probably abnormal EEGs due to excessive slow activity (3) or fast activity for age (1). Only five EEGs were considered to be entirely normal.

Psychological testing was performed on 12 of 17 cases. The other five cases, although uncooperative and therefore not tested, would have been expected to perform only on a moderately to severely retarded level. Five cases had normal intelligence as tested by nonverbal means. Two cases were mildly retarded. The remaining ten cases were significantly retarded in nonverbal and verbal skills.

The most consistent abnormality found in the laboratory investigations was in the pneumoencephalograms. Fifteen cases demonstrated some enlargement of the left lateral ventricle (with respect to the right) and, particularly, enlargement of the left temporal horn (in one case the temporal horns were not visualized). The widening of the temporal horn seemed to vary independently of the overall left ventricular enlargement; the increased width of the temporal horn reflected primarily flattening and atrophy of the hippocampal contours which form the medial wall of the horn, bulging prominently into the ventricle.

In all our cases cited as abnormal, the left coronal cleft is both (a) 4 mm or greater in width, and (b) at least 2 mm wider than its counterpart on the right. The single exception to this was one patient for whom only (a) was true.

In all cases except two, in which only the temporal horn was involved, the entire left ventricle, but particularly the left temporal horn, was enlarged. Five of these 15 patients had, in addition, some dilatation of the right ventricular system, either of the lateral ventricle, the temporal horn or of both. In these cases of bitemporal horn enlargement, the widening was greater on the left in all cases except one. Other abnormalities observed by PEG included: cortical atrophy in three; gross central atrophy (lateral ventricular span of 40 mm or greater) in three; atrophy at the level of the third ventricle (anterior width 7 mm or greater) in two; widening of the aqueduct (caliber wider than 3 mm) in two; and increased height of the fourth ventricle (beyond 20 mm) in three.

Maximum allowable values were adopted from data collected by Taveras and Wood (1964) and by Nielsen *et al.* (1966). For a complete review of pneumoencephalographic measuring methods in children, the reader is referred to Melchior (1961). It should also be noted that the left lateral ventricle may normally be slightly larger than its counterpart on the right (Heinrich, 1939; Wolff & Brinkmann, 1940; Melchior, 1961).

As controls, we measured the temporal horns in ten normal pneumoencephalograms of children (ages 3–17 years). In one of these cases, the left temporal horn measured 4 mm; in all others it measured 1–3 mm. One right temporal horn measured 3 mm and one 2.5 mm; in the others it measured 1–2 mm. The greatest difference between left and right temporal horn measurements was 1.5 mm (left greater than right) in two cases.

The second control group consisted of 12 cases of children (ages 3 months–17 years) with abnormal PEGs, excluding those with mass lesions or obstructive hydrocephalus. Of the twelve, two had normal temporal horn measurements. Three had central nervous system malformations (one neurofibromatosis, one tuberous sclerosis, one undefined) in association with large temporal horns; the structural abnormalities make interpretation of the temporal horn data difficult. Three further cases showed mild bilateral enlargement of temporal horns (range 4–6.5 mm with no more than 1 mm difference between sides); one of these had pseudotumor cerebri, and two had spastic cerebral palsy with generalized central atrophy by PEG. Two further cases had mild right temporal horn enlargement (5 mm) with normal left temporal horn (2.5 and 3 mm); one of these had adrenoleukodystrophy, and the other unspecified dementia. Finally, two of the abnormal control cases had enlarged left temporal horns (6 and 5 mm) with normal right temporal horns (2 and 2.5 mm); one of these children had a left temporal EEG focus, an aggressive behavior disorder, and poor school performance, but was not autistic; the other child, age 11, investigated because of sexual precocity, displayed many features suggesting autistic behavior: very limited social relationships, learning disabilities, hyperactivity, abnormal fears, obsessions, and compulsions, and periodic rocking activity. Interpretation of this heterogeneous abnormal group is complicated, but overall lends substance to the findings recorded in the group of autistic children.

In summary, the PEGs demonstrated enlargement of the left ventricular system and especially of the left temporal horns in 13 cases and isolated widening of the left temporal horn in two others. These were isolated findings in five cases and in the ten others were associated with a variety of other findings. The two patients with no temporal horn enlargement (one with a normal PEG, and one with left hemispheral central atrophy) each had prominent electroencephalographic abnormalities focused on the temporal lobe: one had constant left temporal bursts of sharp and slow wave activity, and the other had slow wave phase reversals bitemporally. Thus all of our patients had either PEG or EEG evidence implicating the temporal lobe, either on the left side or bilaterally.

Some precedent for these observations is found in the literature. Boesen and Aarkrog (1967) reviewed PEGs from 117 children with a wide variety of psychiatric syndromes, reporting that "in the material there are cases of local particularly severe dilations, e.g. in the temporal region, and a smaller number of asymmetrical dilatations." However, they go on to say that "specific psychiatric clinical pictures corresponding to these locations were not observed." Aarkrog (1968) expanded this study to include 46 children with infantile autism or "borderline" autism and 'found that 25 (54%) had abnormalities by PEG. Recently, Dalby (1975) has presented pneumoencephalographic findings in a large group of children characterized by developmental language retardation. Forty of 87 (46%) of these children had dilatation of the left temporal horn, which in 26 (30%) was limited solely to the left temporal horn. A control group of children with petit mal showed only 15% with dilatation of left temporal horn. Dalby's group was not characterized behaviorally except for language retardation, but in personal communication he has noted that many of them had more extensive behavioral abnormalities. It seems likely that they overlap considerably with the group of patients we have described.

Having reviewed this evidence of medial temporal lobe anatomic abnormalities, especially left-sided, in some children with features of autistic behavior, these anatomical findings need to be related to the clinical and behavioral manifestations in this condition. It should be noted that autism presumably has several etiologies, and clearly not all autistic children would necessarily be expected to show the same anatomic changes—nor indeed any evident anatomic change. Nevertheless a clinical-anatomic correlation demonstrated in one subgroup of patients might be useful in clarifying neuropsychological understanding of the disorder.

At this point it may be useful to review the known effects of temporal lobe lesions; first, the more dramatic consequences of bilateral lesions, then unilateral. Two prominent syndromes have been associated with bilateral temporal (particularly medial temporal) lesions in adult man, and both may be briefly reviewed.

The Klüver-Bucy syndrome (1939), first studied in monkeys, has been described in man, first by Terzian and Delle Ore (1955) after bilateral neurosurgical ablations of medial temporal lobes, and more recently as a consequence of herpes simplex encephalitis (Marlowe *et al.*, 1975), which has a known predilection for the medial temporal areas. In monkeys, this syndrome is characterized by "psychic blindness." This refers to the inability of the lesioned animal to recognize the significance—the "emotional" significance if you will—of objects seen visually or handled; or to remember their significance, once determined by mouthing, taste, or smell. Other deficits in the Klüver-Bucy monkey are seemingly purposeless hyperactive, hyperexploratory behavior; hypersexuality in adult males, and most interestingly, a profound incapacity for social behavior. Monkeys after bilateral amygdalectomies—reproducing part of the Klüver-Bucy syndrome—are incapable of functioning in their social group and remain apart from the group (Dicks *et al.*, 1969; Kling, 1972).

Deficits in the Klüver-Bucy syndrome in man have been described as an incapacity for adaptive social behavior, and a loss of recognition of the significance of persons and events. Such patients show an empty blandness, an absence of emotion or concern for family or other persons, and pursue no sustained, purposive activity. For one seeing such patients, the analogy to infantile autism is striking and compelling.

Korsakoff's psychosis is the second well-recognized syndrome in man, which may result from bilateral hippocampal lesions—or indeed from lesions at other points in the Papez memory circuit. The syndrome is common in adult man. But curiously enough, apparently it has not been described or commented upon in a developmental context in young children (Geschwind, 1972; but see Boucher & Warrington, 1976). This may be because in the young child it has been called autism, or mental retardation, without the anatomical or neuropsychological analogies to adult Korsakoff psychosis being recognized. Korsakoff's psychosis is commonly thought of as an amnestic syndrome—a failure to register or recall recent experience beyond the immediate span of attention—but in full-blown form it has much greater importance even than that obvious deficit. A patient who cannot retain one momentary segment of experience and relate it to the next has no continuity of mental life, no real unity of experience, and no sustained purposive or goal-oriented behavior (Estes, 1977). As a consequence of this, or independently, these patients also are notoriously shallow in affective responses, and profoundly lacking in motivation. Importantly, also, they do not recognize or take an interest in even those persons who were previously most important and most meaningful to them; everyone is approached as a total stranger at every successive encounter.

It is evident there is considerable room for overlap between this syndrome and the Klüver-Bucy syndrome, and also that a powerful analogy can be drawn between both and the infantile autism syndrome. Let us look closely at this suggested analogy by considering two related questions. The first question has to do with possible differences between an amnestic syndrome appearing in the adult and in the young child. The adult has a store of skills (including language facility) and remote memories, both of which are preserved after he sustains his Korsakoff lesion. Such patients may score normally on a standard IQ test, for instance, though totally incapable of managing even the simplest affairs of daily life. The situation of the child having a similar lesion from early life might be far different: He would lack the apparatus for at least certain kinds of learning and memory from the outset, and thus could not fill the stores that the adult patient can draw on. Thus, the resulting clinical picture might be expected to be quite different. This is an obvious consideration, but one for which no pertinent data exist, and no further comment seems possible at present.

The second question consists of two parts and has to do with the nature of the learning deficit produced by hippocampal lesions, and the question of unilaterality vs. bilaterality of the lesion. First, it is generally thought that bilateral lesions are necessary to produce Korsakoff's psychosis; unilateral lesions to either side in adults typically do not produce salient permanent deficits.

Patients with Korsakoff's psychosis (and presumably bilateral lesions) do retain a capacity for nonverbal, visuo-motor learning, as demonstrated by Milner (1972). The learning deficit is greatest in language and social areas. Superficially, at least, we might seize on this to say: Yes, this is like autistic children, who learn poorly in social and language contexts, but may learn well in visuo-motor, object-oriented contexts.

However, Milner and her associates (1970, 1972, 1974) have also demonstrated clearly that unilateral hippocampal lesions do produce clear-cut and characteristic learning deficits in adults. Left medial temporal lesions impair verbal learning, but do not affect memory for places, faces, melodies, or nonsense patterns; whereas right medial temporal lesions have the opposite effect. Here obviously, we are presented with a very attractive hypothesis. A unilateral left medial temporal lesion in the young child might be expected to impair verbal learning at a time when the acquisition of such information is exceedingly crucial; while learning of places, faces, melodies, and nonsense patterns is not impaired. A possible analogy to autism becomes quite compelling: Memory for places and music is often excellent in autistic children, a fact often commented upon by parents. In some cases, at least, face recognition is strikingly good despite generally constricted social skills. And most interestingly, Hermelin and O'Connor (1970) in their excellent studies have shown that autistic children remember nonsense material as well as sense, in contrast to controls who remember sense material with much greater facility.

The above notes suggest a hypothesis that a significant lesion of the left medial temporal lobe in early life may impair learning of material ordinarily processed by the left hemisphere, notably relating to language and social interaction, and that nonverbal, right hemispheral material is normally learned and retained.

The hypothesis is tenable only if there is no sufficient interhemispheral transfer (from left to right), or at least if this transfer is delayed and not totally efficient. Some such transfer might occur, accounting for delayed improvement in autism, especially in those patients with more limited lesions.

Milner (1974) has obtained results relating to the lateralization and interhemispheric transfer of language function that are critical to our postulations about autism. Her data, obtained both from patients with early left-hemisphere injury, and from patients undergoing cortical excisions for the treatment of epilepsy, indicate that lesions outside the specific speech areas have no effect on the lateralization of speech, whereas early damage to the left hemispheral primary speech areas causes the transfer of speech development to the right hemisphere. She specifically states that damage outside the primary speech areas is associated with highly specific defects which reflect the specialization of the hemisphere and the site of the lesion. In particular, she states that "regardless of whether the lesion occurs early or late in life, left anterior temporal damage is likely to cause a specific impairment of verbal memory but no general intellectual loss."

Here then we have a potentially quite satisfactory answer to a perplexing question, and the answer may solve a persistent riddle about autism. The riddle may be stated thus: It has long been suspected that autism involves, particularly, functions related to the left, dominant hemisphere; yet in other cases with left hemisphere lesions in early life, these functions are taken over by the right hemisphere, and the result is far different from autism. The proposed answer to the riddle is, then: Transfer of function to the right hemisphere occurs only when the lesion involves specific left hemispheral cortical areas. In autism, if we are correct that a left medial temporal lesion is critically important, the result would be a failure of verbal memory (and other specific left hemisphere functions) without transfer of those cortical functions to the nondominant side. The resulting clinical picture would be quite different from a left hemispheral destruction. Specifically, the deficit in left hemisphere functions (because of failure of learning or memory of these functions) could be profound, and uncompensated by shift of function to the right hemisphere; and right hemisphere functions would be entirely normal. This accords with what is commonly found in autism. In contrast, with left hemispheral neocortical lesions resulting in transfer of functions to right hemisphere, there is a general intellectual deficiency; verbal functions tend to develop at the expense of nonverbal ones; and thus nonverbal, right hemispheral skills tend to be low.

The overall consequence of the above is that in autism, one finds the most profound differences between left and right hemisphere functions: for instance, profoundly defective language development alongside remarkably good, even superior, skills in music, mechanical, and visuospatial areas. And this accords very well with the facts.

Two other considerations should be discussed. It seems very likely that a further refinement will indicate there may be two more-or-less separable autistic syndromes: left hemispheral and bilateral. The left medial temporal syndrome would be as described above and might well be partial, less severe, associated with areas of preserved normal function, and associated with higher IQ. The bilateral syndrome, also recognizable as autism, would be associated with profound retardation in all cortical functions, probably a more dense failure of social and emotional contact, and a poor prognosis. It is evident, of course, that the causative lesion might also affect other areas besides the medial temporal lobes, thus compounding the deficits.

The hypothesis developed above seems satisfactory with respect to the language deficit in autism. The question of social interactions is more difficult, for several reasons: We do not possess any satisfactory neurological understanding of social interactions in general, as we do for language (though certainly language is a very important part of social interaction). We may suggest that social interaction in general has at least two major components: cognitive and affective. The cognitive component includes a number of specific cortical functions: language, recognition of gestures, recognition of faces, recognition of speech tones, etc.—as well as the expressive counterpart of each of these, and the memory facility for each that makes them efficacious. The affective components include emotions, mood, motivation, and the capacity to express each of these. Obviously the cognitive and affective components are, normally, closely linked.

Can we dissect which of these components is at fault? At present, we cannot. It may be either, both, or the vital connecting link. Speculatively one might suppose that a pure hippocampal lesion, causing solely an amnestic deficit, might damage only the cognitive component. On the contrary, a lesion restricted to the amygdala bilaterally might damage only the affective contribution. We do not have the knowledge at present to sort these out, though it promises to be an interesting and worthwhile challenge. Both can be encompassed nicely within the concept of a medial temporal lesion. There is, however, the possibility of further refinement in this context. It may be that a unilateral left hippocampal lesion can cause a significant cognitive defect in social communicative function, as it certainly does for language. However, it is probable that a bilateral and extensive amygdalar lesion is necessary to produce important deficits in affective or social behavior. Thus unilateral autism, if such exists, may reflect cognitive failure, and bilateral autism may also involve, at least in some cases, an affective deficit.

In summary, there is evidence that some cases of autism show anatomic defects, by pneumoencephalography, involving medial temporal lobe structures, particularly on the left. Such lesions have been speculatively related to the known clinical characteristics of autistic children. Two known syndromes were suggested to correlate with salient features of autistic behavior: the Klüver-Bucy syndrome and the Korsakoff amnestic syndrome, with the recognition that these are not mutually exclusive and are probably overlapping to various degrees. A particularly attractive hypothesis is drawn from the work of Milner, who has found that the memory function of the temporal lobe is specific for each hemisphere and that specific cortical functions are not transferred to the contralateral hemisphere unless there is a lesion in the cortical area involved. This leads naturally to the suggestion that in autism the function of the left hemisphere may be uniquely impaired by the failure of the hemispheral memory or integrative learning function, without takeover of the critical functions by the opposite hemisphere. Such a formulation can explain in a satisfying fashion some of the most striking phenomena of the autistic syndrome. Finally, a suggestion was made that there may be unilateral and bilateral autism, the first being in general milder, more reversible, and associated with preserved islands of normal function.

Whether right or wrong, these ideas, and challenges to them, may help to focus attention on the neuropsychology of social, cognitive, and emotional development of the infant and young child.

REFERENCES

Aarkrog, T. Organic factors in infantile psychoses and borderline psychoses. Retrospective study of 46 cases subjected to pneumoencephalography. *Danish Medical Bulletin*, 1968, *15*, 283–288.

Boesen, V., & Aarkrog, T. Pneumoencephalography of patients in a child psychiatric department. *Danish Medical Bulletin*, 1967, *14*, 210–218.

Boucher, J., & Warrington, E. K. Memory deficits in early infantile autism: Some similarities to the amnesic syndrome. *British Journal of Psychology*, 1976, *67*, 73–88.

Creak, E. M. Schizophrenic syndrome in childhood: Progress report of a working party. *Cerebral Palsy Bulletin*, 1963, *3*, 501–503.

Dalby, M. Air studies of speech-retarded children; evidence of early lateralization of language function. Abstract, First International Congress of Child Neurology, Toronto, 1975.

Dicks, D., Meyer, R. E., & Kling, A. Uncus and amygdala lesions: Effects on social behavior in the free-ranging monkey. *Science*, 1969, *165*, 69.

Estes, W. K. The structure of human memory. Science and the Future. Chicago: *Encyclopedia Brittanica Year Book*, 1977. Pp. 60–74.

Geschwind, N. Disorders of higher cortical function in children. *Clinical Proceedings: Children's Hospital National Medical Center* (Washington, D.C.), 1972, *28*, 261–272.

Hauser, S. L., DeLong, G. R., & Rosman, N. P. Pneumographic findings in the infantile autism syndrome: A correlation with temporal lobe disease. *Brain*, 1975, *98*, 667–688.

Heinrich, A. Das normale Enzephalogramm in seiner Abhangigkeit vom Libinsalter. *Zeitschrift fur Alternsforschung*, 1939, *1*, 345–354.

Hermelin, B., & O'Connor, N. *Psychological experiments with autistic children*. Oxford: Pergamon Press, 1970.

Kling, A. Effects of amygdalectomy on social-affective behavior in non-human primates. In B. Eleftheriou (Ed.), *The neurobiology of the amygdala*. New York: Plenum Press, 1972, Pp. 511–536.

Klüver, H., & Bucy, P. Preliminary analysis of functions of the temporal lobes in monkeys. *Archives of Neurology and Psychiatry*, 1939, *42*, 979–1000.

Marlowe, W. B., Mancall, E. L., & Thomas, J. J. Complete Klüver-Bucy syndrome in man. *Cortex*, 1975, *11*, 53–59.

Melchior, J. C. Pneumoencephalography in atrophic brain lesions in infancy and childhood. *Acta Paediatrica* (Stockholm), Supplement, 1961, *128*, 1–320.

Milner, B. Memory and the medial temporal regions of the brain. In K. H. Pribram & D. E. Broadbent (Eds.), *Biology of memory*. New York: Academic Press, 1970. Pp. 29–50.

Milner, B. Disorders of learning and memory after temporal-lobe lesions in man. *Clinical Neurosurgery*, 1972, *19*, 421–446.

Milner, B. Hemispheric specialization: Scope and limits. In F. O. Schmitt & F. C. Worden (Eds.), *The neurosciences third study program*. Cambridge, Massachusetts: M.I.T. Press, 1974.

Nielsen, R., Petersen, O., Thygesen, P., & Willanger, R. Encephalographic ventricular atrophy. Relationships between size of ventricular system and intellectual impairment. *Acta Radiologica: Diagnosis* (Stockholm), 1966, *4*, 240–256.

Taveras, J. M., & Wood, E. M., *Diagnostic neuroradiology*. Baltimore: Williams and Wilkins, 1964.

Terzian, H., & Delle Ore, G. Syndrome of Klüver and Bucy reproduced in man by bilateral removal of the temporal lobes. *Neurology* (Minneapolis), 1955, *3*, 373–380.

Wolff, H., & Brinkmann, L. Das "normale" Enzephalogramm. *Deutsche Zeitschrift fur Nervenheilkunde*, 1940, *151*, 1–25.

15

A Twin Study of Individuals with Infantile Autism

SUSAN FOLSTEIN and MICHAEL RUTTER

In his original description of the syndrome of infantile autism, Kanner (1943) noted that the condition was distinctive in that in most cases the children's behavior had appeared abnormal right from early infancy. He suggested the presence of an inborn defect of presumably constitutional origin. Since then there have been numerous hypotheses concerning the possible nature and origins of this defect (see Ornitz, 1973; Rutter, 1974). However, in spite of the supposition that the disorder is inborn there have been surprisingly few attempts to investigate possible genetic influences (see Rutter, 1967; Hanson & Gottesman, 1976).

The first set of evidence comes from family studies. There is no recorded case of an autistic child having an overtly autistic parent and it is decidedly unusual for a family to contain more than one autistic child, although such cases have been reported (Seidel & Graf, 1966; Verhees, 1976). The usually negative family history for autism seems to be out of keeping with genetic determination. However, this line of reasoning is fallacious. First, it is extremely rare for autistic persons to marry (Rutter, 1970) and there is only a single published report of one having given birth to a child (Kanner & Eisenberg, 1955). This fact alone invalidates the usual assumptions about the meaning of a family history. Second, autism

SUSAN FOLSTEIN · Division of Child Psychiatry, Johns Hopkins Hospital, Baltimore, Maryland. MICHAEL RUTTER · Department of Child and Adolescent Psychiatry, Institute of Psychiatry, De Crespigny Park, Denmark Hill, London, SE5 8AF, England.

is a very uncommon disorder occurring in only about two to four children out of every 10,000 (Brask, 1967; Lotter, 1966; Wing et al., 1976). If the population frequency is very low, the rate in relatives will also be low even in conditions with a high heritability (Smith, 1974; Curnow & Smith, 1975). On both these grounds a strong family history would not be expected even if autism were largely genetically determined.

Moreover, there are two positive findings from family history studies which do suggest possible hereditary influences. First, although the best available estimate indicates that only about 2% of the siblings of autistic children suffer from the same condition (Rutter, 1967), this rate is 50 times that in the general population. Second, although a family history of autism is very rare, a family history of speech delay is much more common, being present in about a quarter of cases (Bartak et al., 1975; Rutter et al., 1971). This last observation raises the possibility that it is not autism as such which is inherited but rather that the genetic influence concerns some broader linguistic or cognitive impairment of which autism is but one part.

The second set of evidence comes from twin studies. These were reviewed ten years ago (Rutter, 1967) with the conclusion that no valid inferences could be drawn. There have since been several further reports (McQuaid, 1975; Kotsopoulos, 1976; Kean, 1975), but the conclusion remains the same (Hanson & Gottesman, 1976). The problems in interpretation are twofold. First, the reports of monozygotic pairs far outnumber those of dizygotic pairs (22 compared with 10). As dizygotic pairs are twice as frequent in the general population, it is clear that there must have been serious selective biases in reporting.* This is sufficient in itself to disregard the findings. Second, excluding two pairs where the autism is associated with an overt physical disorder (Kallman et al., 1940; Keeler, 1958) only five papers reporting same-sexed pairs include both an adequate clinical description and evidence of zygosity (Bakwin, 1954; Kamp, 1964; McQuaid, 1975; Ward & Hoddinott, 1962; Vaillant, 1963). For what it's worth these show two out of three concordance for monozygotic and one out of two concordance for dizygotic twin pairs. In addition there are two opposite-sexed pairs, one concordant (Kotsopoulous, 1976), and one discordant (Böök et al., 1963). The great majority of the remainder report concordance in monozygotic pairs but the papers lack either clinical details or evidence of zygosity and many are no more than passing references in publications on other topics (Chapman, 1957, 1960; Creak & Ini, 1960; Ornitz et al., 1965; Polan & Spencer, 1959; Sherwin, 1953; Bruch, 1959; Keeler, 1957, 1960; Lovaas et al., 1965; Lehman et al., 1957; Brown, 1963; Weber, 1966; Stutte, 1960). The same

---

*Unless MZ twins were peculiarly liable to autism which seems implausible.

problems apply to reports of twins with childhood schizophrenia (Havel-kova, 1967; Cline, 1972). O'Gorman (1970) has described two monozygotic pairs concordant for "pseudoschizophrenia" but the criteria for zygosity were not specified.

In studying genetic factors it is necessary to bear in mind that autism is probably a behavioral syndrome with multiple etiologies (Rutter, 1974). Certainly it is known that the syndrome can develop in association with medical conditions as pathologically diverse as congenital rubella (Chess *et al.*, 1971) and infantile spasms (Taft & Cohen, 1971). Accordingly, the investigation of possible hereditary factors must take account of etiological heterogeneity.

The need was apparent for a systematic and detailed study of a representative sample of twin pairs containing an autistic child. Because of the possibility that the genetic factor might apply to a broader range of disorders than autism *per se*, it would be essential to obtain detailed assessments of social, emotional, cognitive, and linguistic functions in the nonautistic as well as the autistic twins. This demanded a personal study of the twins. Because twins are especially liable to suffer perinatal complications and because such complications have been thought to play a part in the etiology of autism, it would also be necessary to obtain obstetric and neonatal data in order to check whether the concordance findings were a consequence of physical environmental factors rather than heredity. This is what we set out to do and this chapter reports the main findings.

METHODS

Subject Selection

The first task was to obtain a complete and unbiased sample of same-sexed twin pairs which included an autistic child. Opposite-sexed pairs were excluded in view of the well-established finding that autism is very much commoner in boys. A list of autistic twin pairs collected over the years by the late Dr. M. Carter provided the start. Then we sought, using multiple sources of information, to obtain information on all school-age autistic twin pairs in Great Britain. Letters and personal approaches were made to psychiatric and pediatric colleagues known to have a special interest in autism or who were consultants to special schools including autistic children. A request for cases was also made to all members of the British Child Psychiatry Research Club. Through the Association of Head Teachers of Schools/Classes for autistic children approaches were made to those running special schools or units for autistic children in Britain. Mrs. Monica White kindly searched the records of all children known to the National Society for Autistic Children to identify all who were twins. A request for cases was also published in the *Society Newsletter*. Finally, a personal search was made using the twin registers at the Maudsley Hospital and at the Hospital for Sick Children, London.

In this way 33 possible pairs were identified and a detailed scrutiny was made of all available case notes and other clinical information. The sample was restricted to cases which might meet the clinical diagnostic criteria for autism outlined by Kanner (1943) and further delineated by Rutter (1971, 1977), namely a serious *impairment in the development of social relationships* of the type characteristic of autism (that is with limited eye-to-eye gaze, poor social responsiveness, impaired selective bonding, a relative failure to go to parents for comfort, and when older a lack of empathy, a lack of personal friendships, and little group interaction); together with *delayed and deviant language development* with some of the specific features associated with autism (namely poor language comprehension, little use of gesture, echolalia, pronominal reversal, limited social usage of language, repetitive utterances, flat or staccato speech, and very restricted imaginative play); and also *stereotyped, repetitive, or ritualistic play and interests* (as indicated by an abnormal attachment to objects, marked resistance to change, rituals, repetitive behavior, unusual preoccupations, and restricted interest patterns). Cases with an onset after age 5 years were excluded but no further restriction was placed in terms of age of onset. Because this was a genetic study, children whose autism was associated with a known diagnosable neurological disorder (such as tuberose sclerosis or cerebral palsy) were also ruled out.

On the basis of information in case notes, eight twin pairs given the diagnosis of some form of child psychosis were excluded because they did not meet our criteria for infantile autism. This left a sample of 25 twin pairs to be studied in detail. After the children and parents had been seen and interviewed by one of us (SF), diagnoses were then made using all available data. At that final stage, a further four pairs were excluded because they failed to meet our diagnostic criteria, leaving a sample of 21 same-sexed pairs ranging in age from 5 to 23 years (six aged 5-9 years, eight aged 10-14 years, and seven aged 15+ years). Table 1 gives the sources of selection for these 21 pairs, which constitute the basis of this paper. In a third of cases the names were available from just one source but most cases were notified by several different sources. It is clear that no one source would have been adequate.

Zygosity

Zygosity was determined by physical appearance, finger prints, and blood grouping. Attention was paid to such detailed physical characteris-

Table 1. Source of Cases

	Sole source	Joint source
Dr. Carter's list	1	7
National Society	0	9
Schools	0	4
Maudsley Hospital Register	2	3
Hospital for Sick Children Register	0	3
Individual psychiatrists	3	2
Newsletter advertisement	1	0

tics as eye color and pattern of iris; hair color, texture, and curliness; and shape of nose, ears, and hands as well as general appearance (Gedda, 1961). In eight of the pairs the differences between the twins were sufficiently marked to be sure of dizygosity without the need for further testing. In two pairs a designation of monozygosity was made on the basis of very close physical similarity plus the results of finger print analysis using the ridge count method described by Holt (1961). Blood group testing was undertaken for 12 pairs,* nine of which proved to be monozygotic. Thus, the sample consists of 11 monozygotic and 10 dizygotic pairs.

Data Collection

In all cases an attempt was made to interview the parents using a standardized interview and also to interview and examine both twins. Complete information was obtained on 19 pairs. In one case the parents and the autistic twin were interviewed but the normal twin was not seen; in a second, no interview was undertaken. However, in both cases where personal interviewing was incomplete the children had been previously studied very extensively and detailed descriptions, findings, and photographs were made available to us.

Topics covered in detail by the parental interview included a systematic account of the children's social, emotional, and behavioral development and present status; language development, competence, and characteristics; early history and developmental milestones; account of pregnancy, labor, and perinatal period; illnesses and separations; family characteristics and social circumstances; and family history of psychiatric and neurological disorder. Vineland Social Maturity Scale (Doll, 1947) and Mecham Language Scale (Mecham, 1958) assessments were also undertaken.

The children were closely observed and interviewed at home or in hospital and all were given a detailed neurodevelopmental examination. If

---

*In two pairs the blood tests showed that the parents' view of zygosity was wrong.

systematic psychological test findings were not already available, further testing of cognition and language was undertaken using the Wechsler (1949), Merrill-Palmer (Stutsman, 1948), and Reynell (1969) scales.

Pediatric and psychiatric case records were obtained and studied for all hospital admissions and attendance. Finally, hospital obstetric records were examined for all but one of the 17 twin pairs born in hospital.

Diagnosis of Autism

Systematic biases readily arise in twin research through the possibility of the psychiatric diagnosis of one twin being influenced by knowledge of his co-twin and of the zygosity of the pair. Accordingly, rigorous precautions were taken in the study to insure that such diagnostic contamination could not occur. The procedure was as follows: First, one of us (SF) prepared a detailed separate summary of all available psychiatric and developmental information for each of the twin children included in the study. These summaries were then carefully scrutinized to insure that all possible identifying information (such as family characteristics) were deleted. As a further precaution the age of the child was given only in terms of a five-year grouping. The case histories were then put into random order and given a new case number so that it was no longer possible to sort by pairs. These randomized case histories without identifying information were then given to the other investigator (MR) for diagnosis, made "blind" both to pair and to zygosity. His diagnoses are those used for the purpose of all analyses. Autism was diagnosed on the basis of the strict criteria already outlined. This provided a sample of 21 pairs (11 MZ and 10 DZ) including 25 autistic children.

A separate diagnosis of cognitive/linguistic impairment was made on the basis of at least one of the following features: lack of phrase speech by 30 months, a verbal IQ or social quotient of 70 or below, grossly abnormal articulation persisting to age 5 years or older, and scholastic difficulties of such severity as to require special schooling. All 25 autistic children met at least two of these criteria and a further six nonautistic children also did so.

Finally, a psychiatric assessment was made with respect to any nonautistic disorders which were present. In view of its possible connection with infantile autism, particular attention was paid to the possible presence of so-called "autistic psychopathy"—meaning a condition characterized by gross social impairment, obsessive preoccupations or circumscribed interest patterns and poor coordination but normal general intelligence (van Krevelen & Kuipers, 1962; van Krevelen, 1963).

Table 2.   Characteristics of Autistic Twins

	MZ (n = 15)	DZ (n = 10)
Mean lang. abn. score[a]	4.7	4.8
Mean social abn. score[b]	3.6	4.1
Mean repetitive behav. score[c]	3.2	3.3
Mean IQ	54.4	60.0
Mean age (years)	11.7	14.3

[a] Based on six items: lack of use of gesture, echolalia, stereotyped speech, repetitive speech, lack of social use of speech, abnormal mode of delivery (nine nonspeaking children omitted from this analysis).
[b] Based on five items: lack of social smiling in first year, lack of eye-to-eye gaze, lack of attachment to parents, abnormal relationship with peers, lack of empathy.
[c] Based on five items: abnormal attachment to objects, resistance to change, stereotyped play or interests, rituals, repetitive movements.

DESCRIPTION OF CASES

Table 2 summarizes the main features of the 25 autistic twins. Six of the twins were aged 5–9, eight 10–14; and seven 15+. Both the MZ and DZ twins were alike in showing the characteristic deviant language, social impairment, and repetitive behavior peculiar to autism. They had a mean IQ in the mildly retarded range. It is clear that the behavioral features were closely comparable to nontwin samples. Table 3 shows how other features compare with a study of singletons previously undertaken by one of us (Rutter *et al.*, 1967).

Table 3.   Comparison of Sex, IQ, and Social Class in Autistic Twins and Singletons

	Twins (This study) (n = 25)	Singletons (Rutter, Greenfeld, & Lockyer, 1967) (n = 63)
Sex ratio:	3.2:1	4.25:1
IQ		
<50	48.0%	43.0%
50–69	20.0%	28.5%
70+	32.0%	28.5%
Social class		
I and II	57.0%	55.5%
III	28.5%	41.3%
IV and V	14.5%	3.2%

Of the 25 autistic children, 19 were male giving a male:female ratio of 3.2 to 1 which is similar to most other studies. The parents came from all social strata but were predominantly middle class, which is in line with most other series. Other family characteristics were also much as expected. Thus none of the parents suffered from schizophrenia and only one of the 36 sibs was autistic (a rate of 2.8%). However, in three of the 21 families (14%) either a parent or sib had experienced a severe delay in the acquisition of spoken language.

About half the autistic children were severely retarded, but nearly a third had an IQ in the normal range on nonverbal tests, which again is closely comparable to other findings. By definition, none of the children had a diagnosable neurological condition. However, two-thirds showed impairment on developmental functions such as motor coordination or had isolated minor signs such as strabismus or choreiform movements. Four of the autistic children had developed epileptic fits during adolescence. In 11 cases EEGs had been reported as abnormal, but in most cases the abnormalities were of a nonspecific nature. Air encephalograms had been undertaken in three children; these showed left-sided cortical atrophy in one case, slight dilatation of the right lateral ventrical in a second case, and no abnormality in a third. It may be concluded that, apart from the fact that they are twins, the 25 autistic children in the sample are closely similar to the autistic children described in nontwin populations.

RESULTS

Concordance for Autism

Of the 10 dizygotic twin pairs, *none* was concordant for autism (see Table 4) whereas *four* of the 11 monozygotic pairs were concordant (Exact test; p =0.055). This gives a 36% concordance rate by pair or a 53% concordance rate by proband for MZ pairs and in each case a 0%

Table 4. *Pairwise Concordance for Autism and for Cognitive Disorder by Zygosity*

	MZ pairs ($n = 11$)	DZ pairs ($n = 10$)	MZ-DZ difference (Exact test)
Concordance for autism	36%	0%	$p = 0.055$
Concordance for cognitive disorder (including autism)	82%	10%	$p = 0.0015$

concordance for dizygotic pairs (see Gottesman & Shields, 1976 for a discussion of concordance by pair or by proband).

Two of the concordant MZ pairs were closely similar in all respects (see Table 5). In each case the twins were severely retarded and the

Table 5. Characteristics of Concordant Monozygotic Pairs

Age and sex	IQ[a]	Description
22 years male	(i) <25	No babble, speech, or gesture. Marked lack of social responsiveness. Attachment to a skittle. Repetitive play. Finger-flicking mannerisms.
	(ii) <25	Normal babble but no speech or gesture. Marked lack of social responsiveness. Very severe food fads. Upset by changes. Puts objects in rows. Finger-flicking mannerisms. Waves string in front of eyes.
10 years male	(i) FS 56, VS 50, PS 75	Marked lack of social responsiveness and severe speech delay. Echolalia, pronominal reversal, and stereotyped utterances. Fixed routines. Stereotyped drawings. Finger-flicking mannerisms.
	(ii) FS 57, VS 60, PS 57	Marked lack of social responsiveness when younger but now friendly. Moderate speech delay. Echolalia and stereotyped phrases. Stereotyped play and drawing. Finger-flicking mannerisms.
14 years male	(i) <25	Marked lack of social responsiveness. Odd words only; echoing. Whirls self and spins knives. Food fads. Attracted to shiny objects. Wrist biting.
	(ii) <25	Lack of differentiation people to 5 years. No friendships but normal eye-to-eye gaze. Odd words only; echoing. Upset by changes. Spins knives. Food fads. Attracted to shiny objects.
8 years male	(i) Leiter IQ 90	Normal development to 3 years when speech deteriorated to monosyllables. After 6 months gradual improvement. Doesn't play with other children who regard him as odd. Affectionate with parents. Speech complex but stereotyped and echolalic with pronominal reversal and lack of abstract concepts. Upset by changes. Attached to cardboard buses. Repetitive drawing.
	(ii) Merrill Palmer IQ 51	Moderate lack of social responsiveness. No friendships. Few words only; echolalic. Dislikes new situations. Several attachments to odd objects. Swings string in front of face. Repetitive stereotyped play.

[a] All tests specifying VS and PS are either the Wechsler Intelligence Scale for Children or the Wechsler Adult Intelligence Scale, depending on the age of the child.

autism was somewhat atypical in terms of the limited evidence of ritualistic features. However, in both the other two pairs there were important differences between the twins in spite of concordance for autism. In one there was an 18-point difference in nonverbal IQ and a 24-point difference in Peabody language quotient. The twin with a lower nonverbal IQ but higher verbal IQ made much more progress in both social relationships and use of language. In the fourth pair there was a 39-point IQ difference; in this case the more intelligent twin was less severely autistic although the type of behavior was closely similar in both. It is also notable that the more intelligent twin did not develop autism until 3 years of age, although apart from the late onset the clinical picture was typical of autism.

Concordance for Cognitive or Social Impairment

The next question is: What is inherited? Is it autism as such or is it some broader form of disability of which autism is but one part? To answer this it is necessary to examine the pattern of disabilities in the nonautistic co-twins and to determine the concordance in MZ and DZ pairs for these disabilities.

In addition to the 25 autistic children, another six showed some form of cognitive impairment (see Table 6). In all cases this involved some kind of speech or language deficit but the type of deficit varied. Three of the six children had been markedly delayed in early speech development, not using phrase speech until 3 years or later. One of these was also mildly retarded and attended a special school. A further child had markedly abnormal articulation to age 7 years although she had not been delayed in early speech development. Another child with an SQ of 70 had been generally mildly retarded in development and did not use phrase speech until 28 months. The sixth child had a verbal IQ 21 points below the nonverbal and attended an ESN school but there had been no speech delay.

Five of the six children with cognitive impairment were in MZ pairs. Thus, five of the seven nonautistic children in MZ pairs had cognitive abnormalities compared with only one of the 10 nonautistic children in DZ pairs (Exact test; $p = 0.0175$). Since all the 25 autistic children also met the criteria for cognitive abnormality, the concordance rates may be recalculated for all forms of cognitive impairment, both autistic and nonautistic (see Table 4).

The results are striking. Nine of the 11 MZ pairs were concordant for some kind of cognitive disability, usually involving language, whereas this was so for only one out of the 10 DZ pairs (Exact test; $p = 0.0015$).

Only one child, included in the six just mentioned, had social or behavioral problems at all reminiscent of autism, and he was diagnosed as

*Table 6.   Cognitive/Emotional Disorder in Nonautistic Twins in
         Discordant Pairs*

Age, sex, and zygosity	IQ[a]	Language/cognition	Social/emotional
15-year female MZ	FS 51, VS 61, PS 48	Phrases from 3 years; now normal usage. Attends special school.	Shy and sensitive but affectionate and sympathetic.
9-year male MZ	VS 79, PS 100	Normal use language but mild articulation defect. Attends special school.	Sociable and friendly but had severe dog phobia when younger.
12-year female MZ	SQ 104	Marked articulation defect to age 7 years.	Very sympathetic, talkative, sociable. Worrier; cries daily over imagined slights.
10-year male MZ	VS 108, PS 122	Phrases from 3 years; still limited social speech. Special tutoring for severe reading/ spelling difficulties.	Socially awkward; only recently started to play with peers. Circumscribed interests.
5-year female MZ	SQ 70	Normal usage speech but poor articulation.	Friendly and responsive.
5-year male DZ	Merrill Palmer IQ 74	Phrases from 3 years; some echoing.	Friendly and sociable.
18-year male DZ	VS 108, PS 131	Normal speech. Above average school career.	From age 16 years school dropout. Lies in bed reading cybernetics and listening to music.

[a]All tests specifying VS and PS are either the Wechsler Intelligence Scale for Children or the Wechsler Adult Intelligence Scale, depending on the age of the child. SQ is the Vineland Social Maturity Scale.

showing autistic psychopathy on the basis of little social usage of speech, circumscribed interest patterns, and a lack of social relationships.

Three of the other children with cognitive impairments, however, also showed some kind of social or emotional disability. One child was painfully sensitive and self-conscious, crying over imagined slights; another, although friendly and sure of himself, had had a severe and disabling dog phobia when younger; and a third was rather shy, sensitive, and lacking in confidence. A fourth child without cognitive impairment developed a psychiatric disorder of uncertain nature at age 17 years. Because of the overlap with cognitive impairment the concordance in terms of social/emotional difficulties (including autism) is similar: eight out of 11 MZ pairs compared with two out of 10 DZ pairs.

Biological Hazards and Concordance

The major difference in concordance between MZ and DZ pairs strongly suggests the importance of hereditary influences in the etiology of autism. However, before drawing that conclusion it is necessary first to check whether the concordance patterns are explicable in terms of biological hazards associated with the birth process. We identified five features known to be associated with brain damage (and hence likely to predispose to autism); severe hemolytic disease (Gerver & Day, 1950), a delay in breathing of at least 5 minutes after birth (Drage & Berendes, 1966; Hunter, 1968), neonatal convulsions (Rose & Lombroso, 1970), a second birth which was delayed by at least 30 minutes following the birth of the first twin (Dunn, 1965; Kurtz *et al.*, 1955), and multiple congenital anomalies. Such features were present in 11 out of 42 children.

Table 7 shows the concordance for autism in terms of biological hazards. In only two pairs did both children experience biological hazards and both these pairs were *dis*cordant for autism. It may be concluded that the concordance is likely to be due to genetic factors and certainly is not explicable in terms of perinatal complications on which we had data and the same applies to the concordance on cognitive impairment. In *none* of the six pairs concordant for cognitive impairment but not autism were biological hazards present in both twins.

Biological Hazards and Discordance

The next question is why only some of the children with a cognitive impairment showed the syndrome of autism. The possible importance of biological hazards in this connection was reexamined by focusing on the 17 pairs discordant for autism. In six of these pairs one, but only one, of the twins had experienced one of the five specified biological hazards. In *all six* cases it was the autistic twin who was affected (see left-hand side of Table 8). However, there were a further 11 cases (see right-hand side of Table 8) in which the biological hazards affected neither twin or both twins, and so did not account for the discordance.

In order to further examine these 11 discordant cases, a wider definition of biological hazard, in terms of a marked difference between the twins, was employed. This included a birth weight at least a pound less than the other twin (three cases) (Willerman & Churchill, 1967), a pathologically narrow umbilical cord (one case), a more severe hemolytic anemia associated with neonatal apnea (two cases) and a severe febrile

Table 7.   Concordance/Discordance for Autism in Twin Pairs by Presence of Biological Hazards

	Biological hazards		
	Both twins	One twin only	Neither twin
Concordant	0	1	3
Discordant	2	6	9

Table 8. Twins with Biological Hazards and Discordant for Autism

Biological hazard			
Autistic twin only	Other twin only	Both twins	Neither twin
6	0	2	9

Table 9.   Biological Differences and Discordance for Autism

Autistic twin worse	Other twin worse	No difference
6	0	5

illness possibly involving encephalitis (one case). This differentiated a further six cases (see Table 9) and again it identified the autistic one each time. It may be concluded that some form of biological impairment, usually in the perinatal period, strongly predisposed to the development of autism. The pattern of findings is summarized in Table 10.

Did the same biological hazards explain the presence of a cognitive deficit? To examine this question we compared the six nonautistic twins who showed cognitive impairment with the 11 nonautistic twins without a cognitive deficit. The only two children (out of these 17) who had experienced a biological hazard were both *without* a cognitive disability. Clearly, biological hazards did *not* account for the presence of cognitive abnormalities.

Psychosocial Influences

The final issue was whether psychosocial environmental influences were associated with discordance in terms of either autism or cognitive impairment. Because both were evident from early life it was necessary to focus on possible factors in the infancy period, which meant that our data were necessarily retrospective in large part and often crude. All pairs had been reared together during infancy, although in one case the autistic

Table 10. Summary of Biological Hazards in Discordant Pairs (MZ pairs, top; DZ pairs, bottom)

Hazard	Autistic twin	Nonautistic twin
Definite	Multiple congenital anomalies	—
	Neonatal convulsions	—
Possible	Severe febrile illness	—
	Pathologically narrow cord	—
None	—	—
	—	—
	—	—
Definite	Apnea	—
	Delay second birth	—
	Delay second birth	—
	Delay second birth	—
Possible	Severe hemolytic disease	Delay second birth
or	+ apnea	
difference	Severe hemolytic disease	Mild hemolytic disease
in severity	+ apnea	
	Birth weight 1.3/4 pounds lower	—
	Birth weight 1.3/4 pounds lower	—
None	—	—
	—	—

child was often in hospital during the first year. In this case the severe early lack of responsiveness was followed by maternal rejection. There were no differences between the autistic and nonautistic children in experiences other than those which were associated with the greater frequency of neonatal biological hazards. Thus, out of the nine cases in which there was a difference in time before discharge home after birth, in seven it was the autistic child who stayed in hospital longer. In some instances this involved periods in an incubator or some kind of intensive care.

DISCUSSION

Sampling and Selection

Before discussing the meaning of the findings it is necessary to consider the adequacy of our sampling as the rest of the results hinge on that. In order to obtain as complete a sample as possible we used an unusually large number of sources of diverse kinds. As a result most of the children were reported by several different agencies. This in itself provides some indication of the efficiency of our sampling techniques. However, two better checks are available. First, there is the monozygotic-dizygotic pair ratio. For same-sexed pairs surviving the first year the ratio should be approximately 6 : 7 (Slater & Cowie, 1971) which is very close to our observed ratio of 11 : 10. Second, there is the number of autistic twins found. By using the Registrar General's population data for the number of same-sexed twin pairs born (see Slater & Cowie, 1971) together with the one-year survival figures for twins (Gittelsohn & Milham, 1965) and the prevalence estimates for autism (Wing *et al.*, 1976; Treffert, 1970), it is possible to calculate that we should have found about 19 to 27 autistic twins born between 1958 and 1970 (see Folstein & Rutter, 1977, for details of calculation). In fact we obtained 20 autistic children born in these years—which is very close to the expected number. It may be concluded that there is every reason to believe that our sample of autistic twins was about as complete as it could be.

It is also necessary to consider whether the choice of a twin sample introduced any particular biases. The most obvious possibility concerns the frequency of perinatal complications. These tend to be rather commoner in multiple births than in single births (Dunn, 1965) and this may have increased the likelihood of our finding an association between birth hazards and autism. On the other hand, studies of singletons have also suggested that perinatal complications tend to be somewhat commoner in autistic children than in other children (see, e.g., Lotter, 1967; Whittam *et al.*, 1966; Hinton, 1963; Moore, 1972; Knobloch & Pasamanick, 1974; Torrey *et al.*, 1975), although not usually to the extent found in this twin sample. It should be noted, however, that our sample did not have particularly low birth weights. Thirty of the 42 children had a birth weight of over 5 pounds and none had a birth weight under 3 pounds. We may conclude that our choice of a twin sample probably increased the likelihood of finding an association between perinatal complications and autism, but similar associations of lesser degree have been noted in singletons.

Similarly, it is well known that delayed acquisition of speech is commoner in twins than singletons. It might be suggested that this is why so many of the nonautistic twins showed impaired language. However, were this simply due to twinning, it would be expected to occur with equal frequency in the MZ and DZ pairs (Mittler, 1971). In fact we found that abnormalities of language were much more frequent in the MZ than in the DZ twins. Moreover, the abnormalities we found did not consist of just speech delay but rather involved a wider range of cognitive functions.

Finally, there is the question of sample size. How much confidence can be placed on the MZ-DZ differences in concordance in view of the relatively small sample size of 21 pairs? Obviously some caution is needed before drawing too sweeping conclusions and clearly replication is required. Nevertheless, as already indicated, there are good reasons for supposing that this twin study has avoided the serious biases which plague twin research. Moreover, although the sample is small, the MZ-DZ differences were large and statistically significant. It seems likely that the concordance differences are true ones.

Hereditary Influences

The MZ-DZ difference in concordance for autism and the much larger difference in concordance for cognitive disorder clearly points strongly to the importance of genetic factors in the etiology of autism. Indeed the size of the MZ-DZ difference together with the population frequency of autism indicate a very high heritability or coefficient of genetic determination (Smith, 1974; Curnow & Smith, 1975). The finding that concordance is strongly associated with the zygosity of the twin pairs and not at all with the presence of physical environmental hazards indicates that the concordance truly represents an hereditary influence rather than biological damage during the birth process. In this connection it should be noted that there are *greater* intrauterine environment differences in MZ than in DZ pairs, as reflected, for example, in the greater mean difference in birth weights in MZ pairs (Mittler, 1971).

What Is Inherited?

The findings clearly point to the conclusion that the hereditary influences are concerned with a variety of cognitive abnormalities and not just with autism. In other words, autism is genetically linked with a broader range of cognitive disorders. The results also show that the cognitive deficits linked with autism usually involve delays or disorders in the acquisition of spoken language. Thus, of the six pairs concordant for cognitive impairment, in three the nonautistic twin was not using phrase speech until after his third birthday. One of the remaining three showed a lesser degree of speech delay, a second had verbal skills much inferior to visuospatial, and the third had very abnormal articulation. It may be inferred that language difficulties of some kind are generally part of the problem. The conclusion is in keeping with the extensive evidence for the importance of abnormalities in language and symbolization in autism (Rutter, 1974).

On the other hand, in most of the nonautistic children it was not usually a straightforward isolated developmental delay in language acquisition. First, two of the six children also had some general intellectual impairment. Second, in one case the language delay involved echolalia and in another it involved a lack of social usage comparable to that found in infantile autism. It seems that a language deficit may be a part of the cognitive impairment in most cases but it is not usually a "pure" or isolated delay in the acquisition of spoken language. Of course, it is not suggested that all forms of language impairment are genetically linked to autism. Indeed, in most respects, the language characteristics of autistic children are very different from those of children with a developmental language disorder (Bartak *et al.*, 1975, 1977). However, it seems that *some* cases of language abnormality are genetically linked with autism. Unfortunately, knowledge is lacking on how to tell which these are.

It is also necessary to ask whether the social and emotional difficulties which were present in most of the children with a cognitive deficit are also part of what is inherited. For several reasons, no firm conclusions are possible on this point. In the first place, social difficulties and emotional disturbance are quite common in any group of children with language delay (Rutter, 1972; Stevenson & Richman, 1977), and with a sample as small as ours it was not possible to determine whether difficulties were more common in this group. In the second place, only one of the six children with cognitive impairment had social difficulties of a kind at all similar to those shown by autistic children. It may be that the shyness, fears, and sensitivity are part of what is inherited or it may be that, as in other children with language delay, they are merely temporary secondary emotional reactions to cognitive and communication difficulties. The present data do not allow a choice between these two possibilities.

A twin study could provide the opportunity to examine possible links between autism and schizophrenia. However, very few of the twins in this sample were old enough to determine whether autism and schizophrenia ever occur together in monozygotic pairs. None of the monozygotic twins had a disorder with any resemblance to schizophrenia. But there was one nonautistic dizygotic twin who showed social withdrawal at age 17 years. The possibility of schizophrenia clearly arises, but there was no evidence of thought disorder, delusions, hallucinations, or any other first-rank symptoms. Further follow-up is needed to make a diagnosis. Nevertheless, it should be added that in spite of a large number of twin studies of schizophrenia no case has ever been reported of infantile autism occurring in a nonschizophrenic co-twin.

Mode of Inheritance

It is obvious from the low rate of disorder in the sibs that autism is not a disease inherited in clear-cut Mendelian fashion. However, many factors (e.g., phenocopies, genetic heterogeneity, incomplete penetrance, high mutation rate, etc.) may distort the simple Mendelian ratios. In practice, it is extremely difficult on the basis of family data to differentiate between monogenic inheritance with incomplete penetrance and poly-genic or multifactorial effects (Curnow & Smith, 1975). In the case of autism, the sorting out of mode of inheritance is much complicated by the fact that autistic children rarely marry and have children. One crucial piece of information which is needed is what happens to the offspring of nonautistic sibs or twins with cognitive impairment. Unfortunately, no information is available on that point and until this is known, genetic model building seems premature.

Environmental Influences

Our findings clearly indicate that, in addition to hereditary factors, environmental hazards involving the risk of brain damage also play an important part in the etiology of autism. Out of the 17 pairs discordant for autism, there were 12 in which autism was associated with some kind of biological hazard or difference which affected the autistic child and not his co-twin. In this series, with one exception, the biological features were all perinatal in origin. However, it is clear from studies of nontwin samples that autism may arise on the basis of quite diverse forms of brain pathology, including congenital rubella (Chess et al., 1971) and infantile spasms (Taft & Cohen, 1971).

Although both hereditary and environmental influences play an im-portant part in the genesis of autism, the findings from this study suggest that they work in rather different ways. The MZ-DZ concordance differ-ences showed that the hereditary factor(s) was concerned with the genesis of cognitive/linguistic abnormalities rather than with just autism as such. But this was not the case with the biological hazards at all. They were completely unassociated with nonautistic cognitive deficits in spite of a strong association with autism.

Genetic-Environmental Interactions

That difference raises the question of how far hereditary and en-vironmental influences cause different cases of autism and how far they act in conjunction as part of a multifactorial determination. Our data do not allow any firm conclusions on this point but they suggest that both occur.

The four MZ pairs concordant for autism suggest that in some cases genetic factors may be sufficient to cause autism. Only one of the eight autistic children in these four pairs suffered a hazard at all likely to lead to brain injury—his disorder was more severe than that of his co-twin.

On the other hand, it appears that brain injury alone may also be a sufficient cause of autism. This is suggested by the fact that biological hazards occurred with much the same frequency in MZ and DZ pairs. It is also indicated by the finding from other studies that the rate of autism in children with particular forms of brain pathology, such as caused by congenital rubella (Chess *et al.*, 1971) is considerably higher than that in the sibs of autistic children.

Nevertheless, many cases of autism appear to result from a combination of brain damage and in inherited cognitive abnormality. This is suggested by the finding that out of the seven MZ pairs discordant for autism, in four cases the autistic child but not his nonautistic co-twin had experienced some form of biological hazard liable to cause brain damage. In three of these four cases the nonautistic child had a cognitive deficit, suggesting that it may have been brain injury that converted the deficit into a full-blown autistic syndrome.

In this regard, it is interesting that over a decade ago van Krevelen (1963) suggested that autism might result from the combination of an inherited personality deficit plus organic brain damage. The present results are in accord with that general hypothesis, but the deficit found involved cognitive/linguistic abnormalities rather than the "autistic psychopathy" syndrome postulated by van Krevelen.

In summary, we may conclude that this systematic study of 21 same-sexed twin pairs in which at least one twin showed the syndrome of infantile autism indicates the importance of a genetic factor which probably concerns a cognitive deficit involving language. It also indicates the importance of biological hazards in the perinatal period which may operate either on their own or in combination with a genetic predisposition. However, uncertainty remains on both the mode of inheritance and exactly what it is which is inherited.

ACKNOWLEDGMENTS

We are deeply indebted to the many colleagues whose help made this study possible. These are listed more fully in "Infantile autism: A genetic study of 21 twin pairs," published in the *Journal of Child Psychology and Psychiatry*, 1977, upon which this chapter is based.

This study was supported in part by a grant from the Medical Research Council.

REFERENCES

Bakwin, H. Early infantile autism. *Journal of Pediatrics*, 1954, *45*, 492–497.
Bartak, L., Rutter, M., & Cox, A. A comparative study of infantile autism and specific developmental receptive language disorder. I. The children. *British Journal of Psychiatry*, 1975, *126*, 127–145.

Bartak, L., Rutter, M., and Cox, A. A comparative study of infantile autism and specific developmental receptive language disorders. III. Discriminant functions analysis. *Journal of Autism and Childhood Schizophrenia*, 1977, 7, 383–396.

Böök, J. A., Nichtern, S., & Gruenberg, E. Cytogenetical investigations in childhood schizophrenia. *Acta Psychiatrica Scandinavica*, 1963, 39, 309–323.

Brask, B. H. The need for hospital beds for psychiatric children: An analysis based on a prevalence investigation in the County of Arkus. *Ugeskrift fur Laeger*, 1967, 129, 1559-1570.

Brown, J. L. Follow-up of children with atypical development (infantile psychosis). *American Journal of Orthopsychiatry*, 1963, 33, 855–861.

Bruch, H. Studies in schizophrenia. *Acta Psychiatrica Neurologica Scandinavica*, 1959, Supplement 130.

Chapman, A. H. Early infantile autism in identical twins. *AMA Archives in Neurology and Psychiatry*, 1957, 78, 621–623.

Chapman, A. H. Early infantile autism. *AMA Journal of Diseases in Childhood*, 1960, 99, 783–786.

Chess, S., Korn, S. J., & Fernandez, P. B. *Psychiatric disorders of children with congenital rubella*. New York: Brunner/Mazel, 1971.

Cline, D. W. Videotape documentation of behavioural change in children. *American Journal of Orthopsychiatry*, 1972, 42, 40–47.

Creak, M., & Ini, S. Families of psychiatric children. *Journal of Child Psychology and Psychiatry*, 1960, 1, 156–175.

Curnow, R. N., & Smith, C. Multifactorial models for familial diseases in man. *Journal of the Royal Statistical Society of America*, 1975, 138 (Pt. 2), 131–169.

Doll, E. A. *Vineland Social Maturity Scale*. Minneapolis: Educational Test Bureau, 1947.

Drage, J. S., & Berendes, H. Apgar scores and outcome of the newborn. *Pediatric Clinics of North America*, 1966, 13, 637–643.

Dunn, P. M. Some perinatal observations in twins. *Developmental Medicine and Child Neurology*, 1965, 7, 121–134.

Folstein, S., & Rutter, M. Infantile autism: A genetic study of 21 twin pairs. *Journal of Child Psychology and Psychiatry*, 1977, 18, 297–321.

Gedda, L. *History and science*. Springfield, Ill.: Charles C Thomas, 1961.

Gerver, J. M., & Day, R. Intelligence quotient of children who have recovered from erythroblastosis fetalis. *Journal of Pediatrics*, 1950, 36, 342–348.

Gittelsohn, A. M., & Milham, S. Observations on twinning in New York State. *British Journal of Preventative and Social Medicine*, 1965, 19, 8–17.

Gottesman, I. I., & Shields, J. A critical review of recent adoption, twin, and family studies of schizophrenia: Behavioural genetic perspectives. *Schizophrenia Bulletin*, 1976, 2, 360–401.

Hanson, D. R., & Gottesman, I. I. The genetics, if any, of infantile autism and childhood schizophrenia. *Journal of Autism and Childhood Schizophrenia*, 1976, 6, 209–234.

Havelkova, M. Abnormalities in siblings of schizophrenic children. *Canadian Psychiatric Association Journal*, 1967, 12, 363–369.

Hinton, G. G. Childhood psychosis or mental retardation: A diagnostic dilemma. II. Pediatric and neurological aspects. *Canadian Medical Association Journal*, 1963, 89, 1020–1024.

Holt, S. The inheritance of dermal ridge patterns. In L. Penrose (Ed.), *Recent advances in genetics*. Boston: Little Brown & Co., 1961. Pp. 101–119.

Hunter, J. A. Perinatal events and permanent neurological sequelae. *New Zealand Medical Journal*, 1968, 68, 108–113.

Kallmann, F. J., Barrera, S. E., & Metzger, H. The association of hereditary microphthalmia with mental deficiency. *American Journal of Mental Deficiency*, 1940, 45, 25–36.

Kamp, L. N. J. Autistic syndrome in one of a pair of monozygotic twins. *Folia Psychiatria, Neurologia, Neurochirurgia, Neerlandt,* 1964, *67,* 143–147.

Kanner, L. Autistic disturbances of affective contact. *Nervous Child,* 1943, *2,* 217–250.

Kanner, L., & Eisenberg, L. Notes on the follow-up studies of autistic children. In P. H. Hoch & J. Zubin (Eds.), *Psychopathology of childhood.* New York: Grune and Stratton, 1955. Pp. 227–239.

Kean, J. M. The development of social skills in autistic twins. *New Zealand Medical Journal,* 1975, *81,* 204–207.

Keeler, W. R. Stress, experimental psychology, child psychiatry. Discussion of paper presented by Leo Kanner. In *APA Psychiatry Research Reports,* 1957, No. 7, 66–76.

Keeler, W. R. Autistic patterns and defective communication in blind children with retrolental fibroplasia. In P. H. Hoch & S. Zubin, (Eds.), *Psychopathology of communication.* New York and London: Grune and Stratton, 1958. Pp. 64–83.

Keeler, W. R. Personal communication. 1960. Cited by Rimland, 1964.

Knobloch, H., & Pasamanick, B. (Eds.) *Gesell and Amatruda's developmental diagnosis* (3rd ed.). Hagerstown, Maryland: Harper and Row, 1974.

Kotsopoulos, S. Infantile autism in DZ twins: A case report. *Journal of Autism and Childhood Schizophrenia,* 1976, *6,* 133–138.

Kurtz, G. R., Keating, W. J., & Loftus, J. B. Twin pregnancy and delivery. Analysis of 500 twin pregnancies. *Obstetrics and Gynaecology,* 1955, *6,* 370–378.

Lehman, E., Haber, K., & Lesser, S. R. The use of reserpine in autistic children. *Journal of Nervous and Mental Disease,* 1957, *125,* 351–356.

Lotter, V. Epidemiology of autistic conditions in young children. I. Prevalence. *Social Psychiatry,* 1966, *1,* 124–137.

Lotter, V. Epidemiology of autistic conditions in young children. II. Some characteristics of the parents and children. *Social Psychiatry,* 1967, *1,* 163–173.

Lovaas, O. I., Schaeffer, B., & Simmons, J. Q. Building social behaviour in autistic children by use of electric shock. *Journal of Experimental Research and Personality,* 1965, *1,* 99–109.

McQuaid, P. E. Infantile autism in twins. *British Journal of Psychiatry,* 1975, *127,* 530–534.

Mecham, M. J. *Verbal Language Development Scale.* California: Western Psychological Services, 1958.

Mittler, P. *The study of twins.* Harmondsworth: Penguin, 1971.

Moore, M. A study of the aetiology of autism from a study of birth and family characteristics. *Journal of the Irish Medical Association,* 1972, *65,* 114–120.

O'Gorman, G. *The nature of childhood autism* (2nd ed.). London: Butterworth: 1970.

Ornitz, E. M. Childhood autism: A review of the clinical and experimental literature. *California Medicine,* 1973, *118,* 21–47.

Ornitz, E. M., Ritvo, E. R., & Walter, R. D. Dreaming sleep in autistic twins. *Archives of General Psychiatry,* 1965, *12,* 73–79.

Polan, C. G., & Spencer, B. L. A checklist of symptoms of autism in early life. *West Virginia Medical Journal,* 1959, *55,* 198–204.

Reynell, J. *Reynell Developmental Language Scales.* Slough: N.F.E.R., 1969.

Rimland, B. *Infantile autism.* New York: Appleton-Century-Crofts, 1964.

Rose, A. L., & Lombroso, C. T. Neonatal seizure states: A study of clinical, pathological and electroencephalographic features in 137 full-term babies with a long-term follow up. *Paediatrics,* 1970, *45,* 404–425.

Rutter, M. Psychotic disorders in early childhood. In A. Coppen & A. Walk (Eds.), *Recent developments in schizophrenia. British Journal of Psychiatry,* Special Publication No. 1, 1967. Ashford, Kent: Headley Bros. Pp. 133–158.

Rutter, M. Autistic children: Infancy to adulthood. *Seminars in Psychiatry,* 1970, *2,* 435–450.

Rutter, M. The description and classification of infantile autism. In D. W. Churchill, G. D. Alpern, & M. K. DeMyer (Eds.), *Infantile autism*. Springfield, Ill.: Charles C Thomas, 1971. Pp. 8–28.

Rutter, M. The effects of language delay on development. In M. Rutter and J. A. M. Martin (Eds.), *The child with delayed speech*. London: SIMP/Heinemann Medical, 1972. Pp. 176–188.

Rutter, M. The development of infantile autism. *Psychological Medicine,* 1974, *4,* 147–163.

Rutter, M. Infantile autism and other child psychoses. In M. Rutter & L. Hersov (Eds.), *Child psychiatry: Modern approaches*. Oxford: Blackwell Scientific, 1977. Pp. 717–747.

Rutter, M., Bartak, L., & Newman, S. Autism—a central disorder of cognition and language? In M. Rutter (Ed.), *Infantile autism: Concepts, characteristics and treatment*. London: Churchill-Livingstone, 1971. Pp. 148–171.

Rutter, M., Greenfield, D., & Lockyer, L. A five to fifteen year follow-up of infantile psychosis. II. Social and behavioural outcome. *British Journal of Psychiatry,* 1967, *113,* 1183–1199.

Seidel, U. P., & Graf, K. A. Autism in two brothers. *Medical Officer,* 1966, *115,* 227–229.

Sherwin, A. C. Reactions to music of autistic (schizophrenic) children. *American Journal of Psychiatry,* 1953, *109,* 823–831.

Slater, E., & Cowie, V. *The genetics of mental disorders*. London: Oxford University Press, 1971.

Smith, C. Concordance in twins: Methods and interpretation. *American Journal of Human Genetics,* 1974, *26,* 454–466.

Stevenson, J., & Richman, N. Behavior, language and development in three-year-old children. *Journal of Autism and Childhood Schizophrenia,* 1977, in press.

Stutsman, R. *Guide for administering the Merrill–Palmer Scale of Mental Tests*. New York: World Books, 1948.

Stutte, H. Kinder und Jugend psychiatrie. In H. W. Gruhle & W. Mayer-Gross (Eds.), *Psychiatrie der gegenwort Bd. II*. Berlin-Gottingen-Heidelberg: Springer, 1960.

Taft, L. T., & Cohen, H. J. Hypsarrhythmia and infantile autism: A clinical report. *Journal of Autism and Childhood Schizophrenia,* 1971, *1,* 327–336.

Torrey, E. F., Hersh, S. P., & McCabe, K. D. Early childhood psychosis and bleeding during pregnancy: A prospective study of gravid women and their offspring. *Journal of Autism and Childhood Schizophrenia,* 1975, *5,* 287–298.

Treffert, D. A. Epidemiology of infantile autism. *Archives of General Psychiatry,* 1970, *22,* 431–438.

Vaillant, G. E. Twins discordant for early infantile autism. *Archives of General Psychiatry,* 1963, *9,* 163–167.

van Krevelen, D. A. On the relationship between early infantile autism and autistic psychopathy. *Acta Paedopsychiatrica,* 1963, *30,* 303–323.

van Krevelen, D. A., & Kuipers, C. The psychopathology of autistic psychopathy. *Acta Paedopsychiatrica,* 1962, *29,* 22–31.

Verhees, B. A pair of classically early infantile autistic siblings. *Journal of Autism and Childhood Schizophrenia,* 1976, *6,* 53–60.

Ward, T. F., & Hoddinott, B. A. Early infantile autism in fraternal twins. *Canadian Psychological Association Journal,* 1962, *7,* 191–195.

Weber, D. Zur Atiologie autistischer syndrome des kindesalters. *Praxis der kinder psychologie und kinder psychiatrie,* 1966, *15,* 12–18.

Wechsler, D. *Wechsler Intelligence Scale for Children*. New York: The Psychological Corporation, 1949.

Whittam, H., Simon, G. B., & Mittler, P. J. The early development of psychotic children and their sibs. *Developmental Medicine and Child Neurology,* 1966, *8,* 552–560.

Willerman, L., & Churchill, J. A. Intelligence and birth weight in identical twins. *Child Development*, 1967, *38*, 623–629.

Wing, L., Yeates, S. R., Brierley, L. M., & Gould, J. The prevalence of early childhood autism: comparison of administrative and epidemiological studies. *Psychological Medicine*, 1976, *6*, 89–100.

16

Biological Homogeneity or Heterogeneity?

EDWARD M. ORNITZ

The diagnosis of infantile autism is sometimes used as if it refers to a specific disease (e.g., Rimland, 1964) which might have a single identifiable etiology both in psychophysiological and pathological terms. However, it appears unlikely that autism will prove to be a biologically homogeneous condition. Already there is a long list of organic brain syndromes which have been reported to occur in association with autism. These include the sequelae of prenatal and perinatal complications (Taft & Goldfarb, 1964), neonatal conditions such as retrolental fibroplasia which may be associated with brain damage (Keeler, 1958), infantile spasms (Creak, 1963; Kolvin et al., 1971; Menolascino, 1965; Schain & Yannet, 1960; Taft & Cohen, 1971), cerebral lipoidosis (Creak, 1963), and metabolic conditions such as phenylketonuria (Wing, 1966; Sorosky et al., 1968; Anthony, 1962), Addison's disease (Money et al., 1971), and celiac disease (Goodwin & Goodwin, 1969; Goodwin et al., 1971), as well as infectious conditions such as congenital rubella (Wing, 1969; Chess, 1971).

In a recent survey of 74 autistic children (Ornitz et al., 1977), we found that 23% of the children suffered from a major concomitant organic brain condition and an additional 17.6% suffered from a minor concomitant condition. The major associated syndromes included four cases of

EDWARD M. ORNITZ · Mental Retardation and Child Psychiatry Division, Department of Psychiatry, and Brain Research Institute, School of Medicine, University of California, Los Angeles, California 90024.

neonatal respiratory distress, three cases of congenital rubella, one case of infantile spasms and one later seizure disorder, two prominent EEG abnormalities, two cases of cerebral palsy, one case of failure to thrive, one case of microencephaly, one Moebius syndrome, and one case of oculocutaneous albinism.

DeLong (see Chapter 14) has demonstrated another type of association between the behavioral syndrome of childhood autism and other diagnosed conditions. He found that a certain number of autistic children have associated medial left temporal lobe pathology as demonstrated by abnormal pneumoencephalograms. The possibility of an association, in certain patients, between childhood autism and temporal lobe pathology has been suggested before. For example, 16 years ago, Sarvis and Garcia (1961) reported an association between autistic behavior and psychomotor seizures. The concurrence of childhood autism with temporal lobe pathology adds yet another organic brain syndrome to a long list of those which may be associated with autism.

When childhood autism was first recognized as a behavioral syndrome, attempts were made to differentiate it from mental retardation (Kanner, 1949). Terms such as pseudoretardation were devised on the assumption that the retardation seen in autistic children was only apparent and due to the child's unwillingness rather than inability to perform on developmental or intellectual tests (Bender, 1947, 1956). However, greater clinical experience with autistic children and recent follow-up studies have demonstrated that low scores or untestability on developmental tests in the early years of life may be predictive of retarded functioning in later life (Rutter *et al.*, 1967; Lockyer & Rutter, 1969). Autistic children not only will not, but actually cannot perform many tasks (Alpern, 1967). The notion that autistic children have a primary affective deficiency (Anthony, 1962) and good cognitive potential (Kanner & Lesser, 1958) has given way to the recognition that the cognitive deficiency in children with autism is every bit as real as in mental retardation (Rutter *et al.*, 1971; Rutter & Bartak, 1971), and approximately 75% of autistic children can be expected to perform throughout life at a retarded level (Rutter, 1970). Thus, mental retardation and childhood autism often coexist (Goldberg & Soper, 1963). Folstein and Rutter's twin study (see Chapter 15) provides additional evidence of this association.

The evidence from these many studies indicates that autism is often the behavioral manifestation of organic brain pathology. However, the occurrence of autism in association with so many diffuse, disparate, and different organic brain syndromes makes it unlikely that autism can be considered related to a single etiology. Thus, the premise put forward by DeLong (Chapter 14) that there may be a unique impairment of the function of the left hemisphere, particularly that part of the left hemisphere involved with memory function (the temporal lobe), cannot be readily generalized to all cases of childhood autism. To do this it would be necessary to demonstrate that in the many different types of organic brain syndromes with which childhood autism is associated there is significant damage to the left temporal lobe. The association of a relatively high percent of cases of childhood autism (23% in our recent experience) with organic brain syndromes also raises questions about the possible genetic etiology of childhood autism.

If the term genetic is used in the sense of an inheritable condition transmitted from parents to children through the genes then we would have to assume that a considerable number of quite different organic brain conditions can involve those same parts of the central nervous system which are involved by the hypothesized inherited defect. If, however, the genetic hypothesis is broadened to include genetic damage, that is pathologic changes in genetic structure, such damage could occur early enough in the course of a pregnancy to involve both members of a monozygotic twin pair. A specific type of genetic damage might then result in autism. Such a mechanism could be replicated by a number of different medical conditions in which there is not a uniform pathology. Congenital rubella provides such an example. The central nervous system pathology accompanying infection by the rubella virus may vary from individual to individual. By an almost random process those functional systems of the brain involved in the generation of autistic behavior might be affected in some individuals. It could be postulated that these same functional systems are those which are also controlled by genes liable to be damaged early in the course of a pregnancy.

Thus, while it is not tenable to think in terms of a single etiology for all cases of autism, it is plausible that a single pathologic mechanism involving particular functional brain systems could be affected by multiple disease processes. These postulated multiple etiologic agents may be infectious, traumatic, anoxic, metabolic, hormonal, or electrophysiologic.

Folstein and Rutter (Chapter 15) have postulated that among the possible etiologies of the autistic behavioral syndrome there is a significant hereditary influence. There are, however, some problems with this hypothesis. First, one of the most striking results of this study is the relatively low concordance rate for autism in the monozygotic twins. In fact, the difference between the concordance rate of 36% for the MZ twins and the 0% for DZ twins just barely reaches statistical significance. Rutter (1967) has pioneered the critical evaluation of studies of twins in autism; the current data emphasize again that evidence from twin research must be used very cautiously in evaluating genetic influences in childhood autism (see also review in Ornitz, 1973). Second, the low rate of autism in nontwin siblings argues against an inherited defect. The issue is clouded, of course, by the fact that autistic individuals do not tend to marry and reproduce. Third, as Folstein and Rutter show, many cases of childhood autism are associated with perinatal pathology of types known to predispose to brain damage. They conclude that some form of biological impairment, usually in the perinatal period, predisposes to the development of autism.

If the genetic concept is understood not in terms of an inheritable condition but rather of damage to the genes, it is then necessary to consider the possibility of a significant early prenatal as well as a perinatal fetal insult. A biological insult to the fetus occurring early enough in the course of pregnancy may certainly cause genetic damage, particularly damage which could affect both members of a monozygotic twin pair and result in concordance for autism. With the current state of knowledge, it would seem prudent to consider a hierarchy of possible etiologies, beginning with relatively rare cases of truly inherited autism, followed by more common occurrences of genetically induced autism (related to very early intrauterine insult), and finally a significant number of cases in which perinatal and perhaps on rare occasions postnatal insult has replicated the genetic defect in the first and second group of cases.

Folstein and Rutter have convincingly demonstrated that a cognitive-linguistic defect is significantly associated with monozygosity in their sample. In the context of the above discussion, it can be said that there is a clear genetic influence underlying the high rate of concordance in the monozygotic twins for this cognitive impairment, but it cannot be said that this is a characteristic which is necessarily inherited. It should also be noted that this cognitive-linguistic impairment does not necessarily define any particular special type of cognitive and/or language-related impairment. Furthermore, while 25 autistic children showed this cognitive linguistic defect, six nonautistic children also showed the impairment. Thus, for every three autistic children who showed the impairment, there was also one nonautistic child who showed it. Therefore, while this particular impairment has been shown to be subject to a genetic influence, it is not so clear that this form of impairment is specifically related to childhood autism.

DeLong (Chapter 14) has suggested that in childhood autism the function of the left hemisphere may be uniquely impaired by the failure of its critical memory function (there being no takeover by the opposite hemisphere). It is necessary, however, to ask whether there is evidence for a lesion of the left temporal lobe which is unique to childhood autism and which occurs in all cases of childhood autism. DeLong and his colleagues (Hauser *et al.*, 1975) have presented case studies on 18 children, 17 of whom had successful pneumoencephalograms in which the lateral cleft of the left temporal horn was visualized. Of the 17 cases, 15 showed an enlarged left temporal horn lateral cleft, and of the 17 cases, 16 were felt to be autistic. Now in order to postulate a specific etiologic agent such as left temporal horn damage, one of two conditions must be met. Either all of the cases which have been diagnosed as autistic must have the pathologic condition or, in the absence of such a universal incidence, then a statistically significant proportion of the autistic children must show the pathologic condition compared to those cases who are not autistic and show it. In this situation, scrutiny of the distribution of cases reveals that 14 of the 17 children who were felt to be autistic and the one child who was not diagnosed autistic had enlarged temporal horn lateral clefts. The remaining two children who were diagnosed as autistic did not have enlarged temporal horn lateral clefts. This distribution could occur by chance.

If autistic disturbances of relating to people (Ornitz, 1973; Ornitz & Ritvo, 1976) are the *sine qua non* for the diagnosis of childhood autism, then scrutiny of the case descriptions of the 17 children with successful pneumoencephalograms reveals that nine out of the 17 clearly showed this phenomenon and an addition four children showed some evidence of it (together with other autistic symptoms such as abnormal mobility and abnormal sensory sensitivity). This makes a total of 13 out of the 17 who met these criteria for the diagnosis of childhood autism.

The criteria for enlargement of the temporal horn lateral cleft present an additional problem. Hauser *et al.*, (1975) provide a table of "normal pneumoencephalograms" based on ten cases. The width of the left temporal horn lateral cleft in these cases ranges from 1 mm to 4 mm with three cases showing widths of 3 mm. In the absence of good data on the range of "normal," it is prudent to insist that an "abnormal" temporal horn lateral cleft width exceed the greatest width which is considered part of the normal series. On this basis, only 13 out of the 17 cases can be said to clearly show pathologic enlargement of the temporal horn lateral cleft on the left. By using this conservative approach to the available data, both in respect to the pneumoencephalographic findings and the criteria for the diagnosis of autism, the distribution of cases becomes ten autistic children and three nonautistic children with temporal horn lateral clefts exceeding 4 mm and three autistic children and one nonautistic child with left temporal horn lateral clefts equal to or less than 4 mm. This distribution could also occur by chance.

The above analysis, while pointing to some of the difficult problems in carrying out this type of research, does not detract from the importance of DeLong's findings. He has demonstrated that at least ten autistic children certainly do show a particular lesion in the left temporal horn lateral cleft which can be visualized by pneumoencephalography. This provides an interesting lead which needs to be followed through the use of new techniques. Obviously routine pneumoencephalography in young autistic children is not desirable but the development of computerized axial-tomography may make it possible to study the same phenomena by noninvasive techniques. It remains to be seen, however, whether this procedure will give the visual resolution necessary to make sufficiently accurate measurements of the temporal horn lateral clefts.

Lessons should also be drawn from DeLong's very important case of a four-year-old child who suffered a herpes simplex encephalitis and displayed a fully reversible episode of childhood autism (as shown by the child using other people's hands to carry out her wants, avoiding eye contact, and showing no recognition or response to persons, including her mother). Full recovery was reported following the recovery from the infectious disease process. The case is important because we have thought of childhood autism as a chronic, usually permanent, irreversible process rather than as a transient condition. We might, however, remember that childhood autism occurs in a significant number of cases of congenital rubella and that it has also been demonstrated that the rubella virus may remain active through many months of postnatal life in some congenital rubella patients.

Finally, it is necessary to consider the possible significance of the language delay found in the patients studied by Folstein and Rutter and by DeLong. Rutter and his colleagues have described the retardation and distortion of language development which usually occurs in autistic children (see Chapter 6) and in our own studies, too, we have found that some 75% of preschool autistic children fail to use phrase speech (Ornitz et al., 1977). Folstein and Rutter (see Chapter 15) have now shown in their sample of autistic twin pairs that a language deficit showed a very high concordance in MZ pairs and a very low concordance in DZ pairs. This highly significant result clearly indicates a genetic defect underlying the language deficit. DeLong (Chapter 14) used analogies with adult syndromes in which language memory deficits occur with temporal lobe pathology underlying the linguistic deficits in his patients. He has accumulated pneumoencephalographic data on a series of children with a severe impairment of language development. Excluding one child whose language development was normal until a prolonged episode of status epilepticus, and excluding the child whose pneumoencephalogram was unsatisfactory, 14 of 16 children with developmental language delays had enlarged left temporal horn lateral clefts. Many, but not all, of those children were autistic, using the diagnostic criterion of autistic disturbances of social relationships. Putting the evidence from these studies together, might we not postulate a genetically (not to be confused with "inheritable") induced left temporal lobe pathology underlying a particular type of language impairment (a primarily expressive language delay) in a certain group of developmentally disabled children, some of whom also show an associated autistic behavioral syndrome?

REFERENCES

Alpern, G. D. Measurement of "untestable" autistic children. *Journal of Abnormal Psychology*, 1967, *72*, 478–486.

Anthony, E. J. Low-grade psychosis in childhood. In B. W. Richards (Ed.), *Proceedings of the London Conference on the Scientific Study of Mental Deficiency* (Vol. 2). London: May & Baker, 1962. Pp. 398–410.

Bender, L. Childhood schizophrenia. Clinical study of one hundred schizophrenic children. *American Journal of Orthopsychiatry*, 1947, *17*, 40–56.

Bender, L. Schizophrenia in childhood—its recognition, description, and treatment. *American Journal of Orthopsychiatry*, 1956, *26*, 499–506.

Chess, S. Autism in children with congenital rubella. *Journal of Autism and Childhood Schizophrenia*, 1971, *1*, 33–47.

Creak, E. M. Childhood psychosis. *British Journal of Psychiatry*, 1963, *109*, 84–89.

Goldberg, B., & Soper, H. H. Childhood psychosis or mental retardation—A diagnostic dilemma—1. Psychiatric and psychological aspects. *Canadian Medical Association Journal*, 1963, *89*, 1015–1019.

Goodwin, M. S., Cowen, M. A., & Goodwin, T. C. Malabsorption and cerebral dysfunction: A multivariate and comparative study of autistic children. *Journal of Autism and Childhood Schizophrenia*, 1971, *1*, 48–62.

Goodwin, M. S., & Goodwin, T. C. In a dark mirror, *Mental Hygiene*, 1969, *53*, 550–563.

Hauser, S. L., DeLong, G. R., & Rosman, N. P. Pneumographic findings in the infantile autism syndrome: A correlation with temporal lobe disease. *Brain*, 1975, *98*, 667–688.

Kanner, L. Problems of nosology and psychodynamics of early infantile autism. *American Journal of Orthopsychiatry*, 1949, *19*, 416–426.

Kanner, L., & Lesser, L. I. Early infantile autism. *Pediatric Clinics of North America*, 1958, *5*, 711–730.

Keeler, W. R. Autistic patterns and defective communication in blind children with retrolental fibroplasia. In P. H. Hoch & J. Zubin (Eds.), *Psychopathology of communication*. New York: Grune and Stratton, 1958. Pp. 64–83.

Kolvin, I., Ounsted, C., Humphrey, M., McNay, A., Richardson, L. M., Garside, R. F., Kidd, J. S. H., & Roth, M. Six studies in the childhood psychoses. *British Journal of Psychiatry*, 1971, *118*, 381–419.

Lockyer, L., & Rutter, M. A five- to fifteen-year follow-up study of infantile psychosis—III. Psychological aspects. *British Journal of Psychiatry*, 1969, *115*, 865–882.

Menolascino, F. J. Autistic reactions in early childhood: Differential diagnosis consideration. *Journal of Child Psychology and Psychiatry*, 1965, *6*, 203–218.

Money, J., Borrow, N. A., & Clarke, F. C. Autism and autoimmune disease—A family study. *Journal of Autism and Childhood Schizophrenia*, 1971, *1*, 146–160.

Ornitz, E. M. Childhood autism—A review of the clinical and experimental literature (Medical Progress). *California Medicine*, 1973, *118*, 21–47.

Ornitz, E. M., Guthrie, D., & Farley, A. J. The early development of autistic children. *Journal of Autism and Childhood Schizophrenia*, 1977, *7*, 207–229.

Ornitz, E. M., & Ritvo, E. R. The syndrome of autism: A critical review. *American Journal of Psychiatry*, 1976, *133*, 609–621.

Rimland, B. *Infantile autism.* New York: Appleton-Century-Crofts, 1964.

Rutter, M. Psychotic disorders in early childhood. In A. J. Coppen & A. Walk (Eds.), *Recent developments in schizophrenia: A symposium.* London: R.M.P.A., Ashford, Kent: Headley Bros., 1967. Pp. 135–158.

Rutter, M. Autistic children—Infancy to adulthood. *Seminars in Psychiatry,* 1970, *2,* 435–450.

Rutter, M., & Bartak, L. Causes of infantile autism—Some considerations from recent research. *Journal of Autism and Childhood Schizophrenia,* 1971, *1,* 20–32.

Rutter, M., Bartak, L., & Newman, S. Autism—A central disorder of cognition and language? In M. Rutter (Ed.), *Infantile autism: Concepts, characteristics and treatment.* London: Churchill-Livingstone, 1971. Pp. 148–172.

Rutter, M., Greenfeld, D., & Lockyer, L. A five- to fifteen-year follow-up study of infantile psychosis—II. Social and behavioral outcome. *British Journal of Psychiatry,* 1967, *113,* 1183–1200.

Sarvis, M. A., & Garcia, B. Etiological variables in autism. *Psychiatry,* 1961, *24,* 307–317.

Schain, R. J., & Yannet, H. Infantile autism. *Journal of Pediatrics,* 1960, *57,* 560–567.

Sorosky, A. D., Ornitz, E. M., Brown, M. B., & Ritvo, E. R. Systematic observations of autistic behavior. *Archives of General Psychiatry,* 1968, *18,* 439–449.

Taft, L. T., & Cohen, H. J. Hypsarrhythmia and infantile autism: A clinical report. *Journal of Autism and Childhood Schizophrenia,* 1971, *1,* 327–330.

Taft, L., & Goldfarb, W. Prenatal and perinatal factors in childhood schizophrenia. *Developmental Medicine and Child Neurology,* 1964, *6,* 32–43.

Wing, J. K. Diagnosis, epidemiology, aetiology. In J. K. Wing (Ed.), *Early childhood autism.* Oxford: Pergamon Press, 1966. Pp. 3–49.

Wing, L. The handicaps of autistic children—A comparative study. *Journal of Child Psychology and Psychiatry,* 1969, *10,* 1–40.

17

Personality Characteristics of Parents

WM. GEORGE McADOO and
MARIAN K. DeMYER

The relationship between parental personality characteristics and infantile autism has been viewed in various ways. When demonstrable personality characteristics or psychopathology are found in parents of autistic children, one interpretation of these results is that autism in the child is the result of the parents' deviant personality characteristics. Within this interpretation, there are a number of different hypotheses. An extreme version is that deviant parents have created autism in a biologically normal child as a result of their isolation, coldness, rage, psychosis, etc. Goldfarb (1961) felt that this conceptualization was appropriate for only one subgroup of autistic children. Neurological impairment was seen as the primary etiological agent in the other subgroup. From Goldfarb's point of view, deviant parents are only associated with a portion of the autistic children. A different parental-biological interaction hypothesis states that deviant parents exacerbate psychological abnormalities in children with special biological vulnerabilities. These parents may have failed to provide adequate support for the biologically vulnerable infant because of inadequate child-rearing practices. Thus, demonstrable personality characteristics have been interpreted as the cause of autistic development in the child.

An alternative interpretation of these same results is that the deviancy in parents of autistic children develops as a response to the child's

WM. GEORGE McADOO and MARIAN K. DeMYER · Indiana University School of Medicine, Indianapolis, Indiana.

abnormalities. Uncertainty and confusion about how to respond to the child's severe social withdrawal, unusual object use, and lack of communication are likely to produce intense stress for these parents. This stress, acting over a period of time, is likely to result in personality changes for the parents of autistic children. Thus, demonstrable personality characteristics have also been interpreted as the result of a parent's response to severe pathology in the child.

A lack of demonstrable differences in personality characteristics of parents of autistic and other diagnostic groups would be expected if there were no consistent relationship between parental personality characteristics and autism. A wide range of personality characteristics would be expected in any given group of parents. These characteristics would be expected to result in only minor symptom variation in the autistic child because, according to these theorists, the autistic child has basic biological abnormalities. Therefore, if demonstrable personality characteristics were found in a particular study, then these results would likely be due to chance selection.

Kanner was the earliest investigator of the role of parental factors in the etiology of infantile autism. His work is frequently cited as evidence for a psychogenic cause. Kanner (1949) described parents as intelligent, obsessive, perfectionistic, and humorless individuals who used set rules as a substitute for life's enjoyment. As parents, they were reputed to be cold, formal, and sometimes extremely anxious (Eisenberg & Kanner, 1956). As spouses and friends, they were supposedly undemonstrative and mechanistic. They seemed uncomfortable socially, preferring solitary activities. The fathers were deemed similar to the mothers (Eisenberg, 1957). Instead of specific personality characteristics, other investigators (Esman et al., 1959; Kaufman et al., 1960; Block, 1969) found more general psychopathology in these parents than in control parents.

Goldfarb (1961) hypothesized two groups of psychotic children based on the child's level of neurological integrity: (a) those with evidence of neurological impairment ("organic") and (b) those without such evidence ("nonorganic"). Using ratings of family interaction, Goldfarb found that families of organic psychotic children did not differ in terms of adequacy from families of normal children, but the families of organic psychotic children were rated as more adequate and had a lower proportion of psychotic mothers than families of nonorganic children (Goldfarb, 1961; Meyers & Goldfarb, 1962). Goldfarb (1970) concluded that the evidence supported the general inference that families of nonorganic psychotic children are more deviant than families of organic psychotic children.

Most of the evidence for the parental reaction or parental stress hypothesis has been gathered from the failure of child-rearing and parental attitude scales to discriminate between the parents of children in various diagnostic categories while, at the same time, indicating differences between these groups and parents of normal children. Anthony (1958) reported no differences between mothers of neurotic and psychotic children in terms of their feelings of how well they handled various child-rearing behaviors. Fewer pathological attitudes were found in the mothers of psychotic children than in the mothers of neurologically disabled children (Klebanoff, 1959). However, both these maternal groups showed higher pathological scores than mothers of normal children. After comparing the attitudes of mothers of normal, psychotic, and Down's syndrome children, Pitfield and Oppenheim (1964) contended that the mother's attitude was influenced by her child's condition rather than the other way around. When mothers of autistic and aphasic children were compared, Rutter *et al.* (1971) found that 50% of both groups of mothers had been medically treated for neurotic or depressive disorders. While these two groups of mothers did not differ on a measure of neurotic disposition, they were both above the general norms. Rutter *et al.* interpreted their results as indicating a parental response to caring for a young, handicapped child.

If the parents of autistic children are deviant in either severity of psychopathology or in intensity of personality traits, then these deviancies should be reflected in specific parental practices during the formative period of the illness of the autistic children, namely the first two years of life. Furthermore, parents of autistic children should show more coldness, rage, isolation, or psychopathology. DeMyer *et al.* (1972) interviewed parents of autistic, normal, and nonpsychotic subnormal children about the nature of parental practices in the formative years of the index children's lives. Neither the psychogenic nor the parental-biological positions were supported. It was not possible to identify any specific or cumulative defects in the parents of autistic children in the dimensions of infantile acceptance, nurturing warmth, feeding, tactile and general stimulation.

The failure to find differences between groups in this study could be criticized from the point of view that in the interview situation, the parents may say all the right things but that this may not be reflected in their actual behavior. A retrospective study of child-rearing practices is likely to involve some memory distortion, especially in parents of disturbed children. This study also did not directly assess parental psychopathology or personality traits.

The purpose of the present study was fourfold. First, parents of autistic children were compared with parents who were being seen in an adult outpatient psychiatric clinic. If the parents of autistic children were severely disturbed, then they should resemble the adult outpatient parents. Second, parents of autistic children were compared with a random sample of parents of disturbed, child-guidance children to determine whether significant differences in personality existed between the two groups of parents. A random sample was used since we were interested in results that would provide some generalizations to the entire group of child-guidance parents seen at our clinic. Third, a group of parents of nonpsychotic child-guidance children were matched with the parents of autistic children to evaluate the parental stress hypothesis. Autistic children cause severe and prolonged stress in their families which should be reflected in personality differences. Age, education, number of children in the family, social class and other demographic variables described more fully below were controlled because we wished to determine whether the degree of the child's impairment would in itself be sufficient to cause significant personality differences in the parents. Fourth, the autistic children were divided into two groups based on the child's level of neurological integrity using Goldfarb's criteria. As a result of Goldfarb's work, parents of nonorganic autistic children were expected to have more deviant personality characteristics than parents of organic autistic children.

METHOD

The Minnesota Multiphasic Personality Inventory (MMPI) was individually administered to parents of children who were undergoing a preadmission evaluation at the Clinical Research Center for Early Childhood Schizophrenia. Each child was examined by two staff psychiatrists at the Clinical Research Center. For a diagnosis of autism, the child needed to exhibit all three of the following symptoms at the time of the evaluation: (a) severe social withdrawal, (b) noncommunicative speech or muteness, and (c) nonfunctional object use. As a result of the evaluation, 39 children were diagnosed as autistic. This represented four years of consecutively diagnosed autistic children. In addition to the behavioral observations and interview impressions, neurological, laboratory, and x-ray data on the autistic children were examined by a staff psychiatrist. Using Goldfarb's criteria (1961) for organicity, each child was categorized as either organic or nonorganic. The interrater agreement between the two psychiatrists on this classification was .83.

A random sample of 100 adult outpatient mothers and 100 adult outpatient fathers was obtained from a prior study by McAdoo and Connolly (1975). The adult outpatient mothers were diagnosed as follows: Psychosis 6%, Neurosis 22%, Personality Disorders 36%, Transient Situational Disturbance 5%, Psychophysiological Disorder 1%, Marital Maladjustment 24%, No Mental Disorder 2%, and Diagnosis Deferred 4%. The diagnoses on the adult outpatient fathers were as follows:

Psychosis 4%, Neurosis 7%, Personality Disorders 36%, Transient Situational Disturbance 1%, Marital Maladjustment 47%, No Mental Disorder 2%, and Diagnosis Deferred 3%.

A random sample of 100 fathers and 100 mothers of children seen at the Riley Child Guidance Clinic was also obtained from this same study (McAdoo & Connolly, 1975). In a recent survey of children attending this clinic, the following diagnoses were obtained after the completion of the diagnostic evaluation: Psychosis 4%, Neurosis 13%, Behavioral Disorder 43%, Specific Learning Disabilities 4%, Mental Retardation 4%, Transient Situational Disturbance 2%, No Mental Disorder 17%, and Diagnosis Deferred 13%. There were neurological problems reported in 19% of these children. As part of this evaluation, the MMPI is also routinely administered to the parents of the children. From a population of 489 families seen at this clinic, parents were matched with parents of the autistic children on marital status, race, relationship to the child, as well as on the age and educational level of both parents. Following the initial selection of the potential match, the child's clinical data were carefully examined. If the child was psychotic, the case was excluded from consideration. The parents of two of the autistic children could not be successfully matched. This resulted in a total of 37 cases which were matched on the demographic characteristics of the parents but which differed on the intensity of the child's disturbance. In addition to the variables on which the parents were matched, the two groups did not differ significantly with respect to the number of children in the family, social class, or sex of the child. There was a difference, as might be expected, in the age of the child, with the autistic children being younger than the child-guidance children. Thus, the two groups of parents seemed to be reasonably well matched on demographic variables, as shown in Table 1.

RESULTS

The MMPI data for the comparisons between the various groups of parents were analyzed by means of a multivariate analysis of variance. The vector of three validity scales and the vector of ten clinical scales were analyzed separately. If it was possible to reject the null hypothesis of equal population mean vectors, then the univariate tests were subsequently carried out on each MMPI scale in that particular vector of scores.

Figure 1 contains the MMPI profiles of mothers of autistic and child-guidance children as well as the adult outpatient mothers. The MMPI profiles of mothers of child-guidance and autistic children were

Table 1.   Means and Percentages for Demographic Data on Parents of
           Autistic and Child-Guidance Children

Variables	Autistic children mean	Child-guidance children mean
Mother's age	31.62	31.62
Father's age	34.67	34.39
Mother's education	12.32	12.16
Father's education	12.97	13.02
Child's age	5.38	7.81*
Number of children in family	2.81	2.73
Social class	Percentage	Percentage
I	11	8
II	11	5
III	11	22
IV	54	46
V	13	19

*$p < .001$

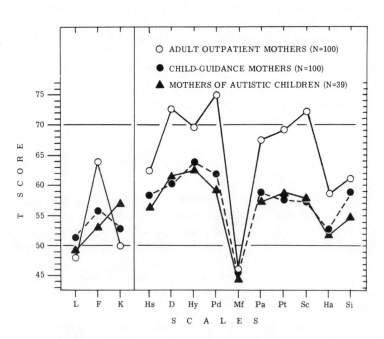

Fig. 1. MMPI profiles of three groups of mothers.

very similar in terms of the shape and elevation; however, both these groups of mothers were strikingly different from the adult outpatient mothers. When the MMPI scores of the mothers of autistic children and the adult outpatient mothers were compared, it was possible to reject the null hypothesis for both the vector of validity scales (F' (3,100) = 13.12, p < .01) and the vector of clinical scales (F' (10,97) = 6.91, p < .01). Significant differences were found on 11 of 13 MMPI scales. The adult outpatient mothers scored higher on *Hs, D, Hy, Pd, Pa, Pt, Sc, Ma, Si,* and *F*; but lower on *K*. All the probability levels were less than .001 except for *Hy, Ma,* and *Si,* which were less than .01; and *Hs,* which was less than .05. The mothers of the autistic children were clearly quite different from mothers who were seeking psychiatric help for themselves in an adult outpatient clinic.

In spite of the similarity between mothers of child-guidance and autistic children, the null hypothesis for the clinical scales was rejected (F' (10,66) = 2.29, p < .05). The Social Introversion (*Si*) scale was the only clinical scale on which significant differences were found (t (137) = 2.14, p < .05). The child-guidance mothers had higher scores, indicating that they are likely to be more uncomfortable in interpersonal situations. The null hypothesis for the validity scales was also rejected (F (3,135) = 4.28, p < .01). Significant differences occurred on one of the validity scales (t (137) = 2.31, p < .05), with the mothers of autistic children having had higher *K* scores. This would suggest that the autistic mothers are likely to be seen as having a more positive self-image.

The MMPI profile of each of the mothers was examined in terms of the similarity of her MMPI profile to that of the adult outpatient mothers (McAdoo, Note 1). Twelve of the 39 mothers of autistic children (31%) had MMPI profiles which were similar to the MMPI profiles of adult outpatient mothers. There were 21 child-guidance mothers (21%) who had this same type of MMPI profile. There was no significant difference in the proportion of mothers in each group with this type of MMPI profile (z = 1.42). It is interesting to note that the base rate for 294 child-guidance mothers with this type of MMPI profile was 26%. In summary, mothers of autistic children had MMPI profiles which, for the most part, were very similar to those of the child-guidance mothers but very different from mothers in an adult outpatient clinic.

The MMPI profiles of the fathers of autistic and child-guidance children together with the adult outpatient fathers are contained in Fig. 2. The MMPI profiles of the two groups of fathers of disturbed children were almost identical. As was the case with the mothers, the fathers of disturbed children had MMPI profiles which were remarkably different from those of the adult outpatient fathers. When the fathers of the autistic children were compared with the adult outpatient fathers, the null

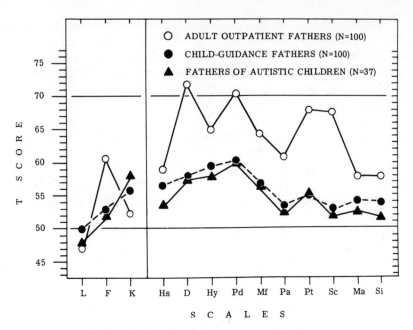

Fig. 2. MMPI profiles of three groups of fathers.

hypothesis was rejected for the vector of validity scales (F' (3,121) = 13.31, $p < .01$) and for the vector of clinical scales (F' (10,120) = 6.88, $p < .01$). Again, significant differences were found on 11 of the 13 MMPI scales. The fathers in the adult outpatient clinic scored higher on D, Hy, Pd, Mf, Pa, Pt, Sc, Ma, Si, and F; but lower on K. All the probability levels were less than .001 except for Hy, Ma, and Si, which were less than .01. The fathers of autistic children were thus strikingly different from the adult outpatient fathers.

When the fathers of child-guidance children were compared with the fathers of the autistic children, neither the null hypothesis for the clinical scales (F' (10,82) = 1.50) nor the null hypothesis for the validity scales (F (3,133) = 2.03) was rejected. Thus, the MMPI profiles of the fathers of autistic and child-guidance children were very similar.

When each of the fathers of the autistic and child-guidance children was compared in terms of the similarity of his MMPI profile to the adult outpatient fathers (McAdoo, Note 1), five of the 37 fathers of autistic children (14%) and 18 of 100 child-guidance fathers (18%) had MMPI profiles similar to fathers who were seeking help in an adult outpatient clinic. The difference in the proportion of fathers in each of the two groups with this type of MMPI profile was not significant ($z = .37$). Thus, there were no significant differences between the fathers of disturbed children but there were dramatic differences between these fathers and fathers who were seen in an adult outpatient clinic.

The results of the matching procedure will be examined next. The MMPI profiles of mothers of autistic and child-guidance children which were used to evaluate the parental stress hypothesis can be seen in Fig. 3. There was very little difference in the two MMPI profile configurations even though there were differences in the intensity of the child's disturbance. It was not possible to reject the null hypothesis for either the vector of validity scales (F (3,34) = 2.71) or the vector of clinical scales (F (10,27) = .98). Thus, in spite of major differences in the children which were expected to cause differential amounts of stress, the mothers of autistic and child-guidance children did not seem to differ. This lack of differential responding between the mothers of disturbed children does not provide much support for a parental stress hypothesis.

A similar comparison for the fathers of autistic and child-guidance children is portrayed in Fig. 4. The fathers of child-guidance children had higher MMPI profile elevations; however, the null hypothesis was not rejected for the clinical scales (F (10,25) = 1.38). For the validity scales, the null hypothesis was rejected (F (3,32) = 3.78, $p <$ .01). Significant differences occurred on only the F scale (t (34) = $-3.38, p <$ .01) with the child-guidance fathers having a higher mean score. In contrast to the results which would be expected from a parental stress hypothesis, the

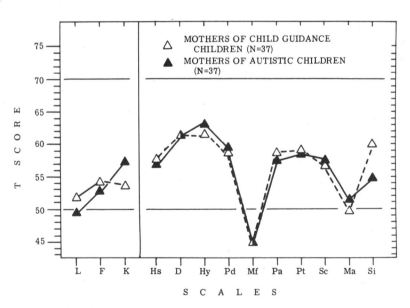

Fig. 3. MMPI profiles of mothers of autistic and child-guidance children.

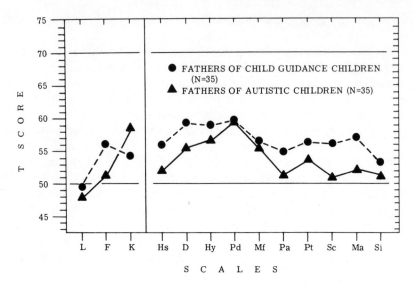

Fig. 4. MMPI profiles of fathers of autistic and child-guidance children.

child-guidance fathers had the more elevated MMPI scores; however, for the most part, the differences in profile elevation were not significant.

Using Goldfarb's (1961) criteria for organicity, 24 of the autistic children (62%) were categorized as organic and 15 were categorized as nonorganic (38%). This resulted in a ratio of 1.6 to 1. Figure 5 contains the MMPI profiles of the mothers of these two groups of autistic children. Contrary to Goldfarb's report, the mothers of the autistic organic children had slightly higher MMPI profile elevations. Neither the null hypothesis for the vector of clinical scales (F' (10,28) = .89) nor the null hypothesis for the vector of validity scales (F (3,35) = .61) was rejected. When the individual MMPI profiles were examined, seven of 24 mothers of autistic organic children (29%) and five of 15 mothers of autistic nonorganic children (33%) had MMPI profiles which were similar to mothers in an adult outpatient clinic. Thus, there appears to be little difference in the two groups of mothers in terms of their MMPI profiles.

The MMPI profiles of the fathers of autistic children can be seen in Fig. 6. The fathers of the autistic organic children had MMPI profiles which were similar in shape to the fathers of the autistic nonorganic children but their profiles were generally more elevated. However, while the null hypothesis for the vector of clinical scales was not rejected (F' (10,26) = .80), the null hypothesis for the vector of validity scales was rejected (F (3,33) = 4.66, p < .01). There were differences between the fathers on two of the three scales: Lie (L) scale [t (35) = 2.38, p < .05) and

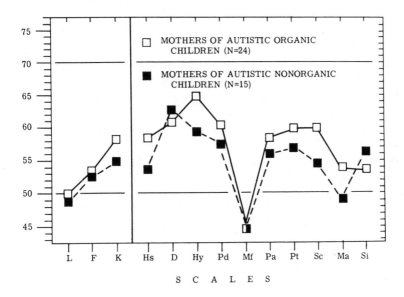

Fig. 5. MMPI profiles of mothers of organic and nonorganic autistic children.

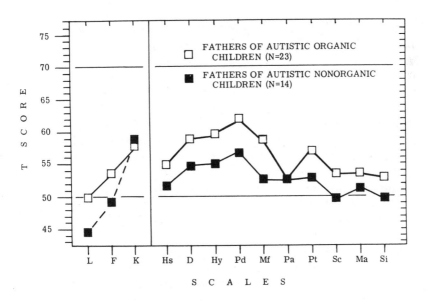

Fig. 6. MMPI profiles of fathers of organic and nonorganic autistic children.

F scale (t (35) = 2.59, p < .05]. The fathers of nonorganic autistic children scored lower on the L scale, indicating that they were inclined to be overly truthful and to shun social approval. They also scored lower on the F scale, suggesting that they were also more likely to be described as unassuming, unpretentious, and dependable. Four of 23 fathers of autistic organic children (17%) had MMPI profiles similar to fathers in an adult outpatient clinic. Only one of 14 fathers of autistic nonorganic children (7%) had this type of MMPI profile. Thus, the fathers of the organic autistic children had higher MMPI profile elevations but these differences were generally not significant. The findings for the parents of nonorganic autistic children were contrary to the direction expected from Goldfarb's data.

DISCUSSION

Families of autistic children have been described as having high rates of parental psychopathology. Werry (1972) has stated that, in general, the studies that have found an increased frequency of psychopathology were methodologically poor in that they relied on unstructured clinical interviews without a control group. In the present study, a random sample of child-guidance parents was utilized as a control group. The parents of autistic and child-guidance children had higher elevated MMPI profiles than the norm group on which this test was constructed. However, the MMPI profiles of both these groups of parents were not as elevated as the MMPI profiles of a sample of parents in an adult outpatient clinic. One of the objections raised by Rutter (1968) to the psychogenic hypothesis of autism was that this position would require very severe parental psychopathology to cause such a profound disorder as autism so early in the child's life. The MMPI profiles of the parents of autistic children obviously do not provide support for a psychogenic position.

There is, of course, variability among the MMPI profiles of parents of disturbed children. The proportion of parents of autistic children who might be described as disturbed did not differ from the proportion of parents of child-guidance children. Neither was there a difference between the fathers and mothers in each of these two groups. Thus, parents of autistic children were remarkably similar to parents of disturbed children in general. It seems likely that with small samples of autistic children, the proportion of parents that would be considered to be disturbed would fluctuate widely and could generate conflicting results in the literature, especially in the absence of a control group.

Various personality characteristics have also been attributed to the parents of autistic children. In terms of personality variables as measured by the MMPI, parents of disturbed children appear to be quite similar. For the fathers, there were no significant differences between the two groups. For the mothers, differences were found on only two scales: Social Introversion (*Si*) and *K* scale. In contrast to the stereotyped descriptions of the mothers of autistic children, it was the child-guidance mothers who had higher scores on the *Si* scale. As a group, the child-guidance mothers were likely to display more discomfort in social and interpersonal situations than mothers of autistic children. The *K* scale is a measure of test-taking attitudes as well as a measure of personality characteristics. Individuals who score low on this scale have a poor self-concept and are lacking either in the self-managing techniques or the interpersonal skills required to deal effectively with their situation in such a way as to improve their lot in the world (Dahlstrom *et al.*, 1972). On the other hand, high scores suggest defensiveness about personal inadequacies, as well as serious limitations in personal insight and understanding. Thus, the difference between these two groups of mothers on this scale could be due either to more child-guidance mothers scoring low or to more of the mothers of autistic children scoring high. When the scores on the *K* scale were analyzed according to the number of mothers scoring high (> 65) or low (< 44), the first possibility was apparently confirmed since 24 of the 100 child-guidance mothers had low scores as compared with two of the 39 mothers of autistic children ($z = 2.32$, $p < .05$).

Since *K* scores in the high average range (57–64) are generally earned by individuals in the upper-middle and lower-upper social classes, it is possible that more parents of autistic children are in this range of social class which resulted in higher but not necessarily elevated *K* scores. Less than 25% of the parents of autistic children were in social classes I and II as determined by Hollingshead (Note 2). Surprisingly, more than 50% of the parents of autistic children were in social class IV. Thus, it seems that the difference in *K* scores was not related to social class and that therefore, the child-guidance mothers had poorer self-concepts and were more self-critical and self-dissatisfied.

In summary, parents of disturbed children have similar personality characteristics in spite of differences in the severity of their child's problems; however, more of the mothers of child-guidance children have poor self-concepts and discomfort in interpersonal and social situations.

It seems likely that every disturbed child has some stressful impact on his family. The parental stress hypothesis implies that the child's behavior has resulted in an increase in parental distress or maladjustment. This possibility is usually invoked when several conditions occur: (a) the parents of disturbed children differ from the parents of normal children on a measure of interest and (b) there is a similarity among the parents of various groups of disturbed children on this same measure. For example, Erickson (1968) found that parents of young emotionally disturbed children and parents of young organically retarded children had similar MMPI profiles. Both these groups of parents had elevated MMPI profiles when compared with the Minnesota norm group. Erickson pointed out that some brain-damaged children may exhibit behavior that is grossly deviant and therefore as stressful as that of many psychotic children. Klebanoff (1959) also agreed with this position, pointing out that while the brain-injured and retarded children in his study differed in diagnosis from the psychotic children, they had comparable severity of disorganization and behavior. Therefore, the parental response should vary with the intensity of the disturbance. In the present study, when the matched groups of parents were compared, there were marked differences in the two groups of children, with the autistic children being younger as well as more severely disturbed. In addition, the two settings were likely to involve a differential amount of stress, with the child-guidance clinic being an outpatient clinic and the clinical research center being an inpatient treatment center. These factors would suggest a differential response, with parents of autistic children being under greater distress and therefore having a more elevated MMPI profile than the parents of the child-guidance children. This clearly was not the case. Furthermore, in contrast to the expected results, the child-guidance fathers had somewhat higher MMPI profiles. The results of this analysis are in agreement with the results of the comparison with the random control group. There seems to be little difference in personality characteristics of parents of disturbed children, even in spite of major differences in the severity of the child's problems.

The MMPI is probably not the best instrument to measure situational distress since the MMPI is a trait rather than a state measure. However, it seems likely that higher levels of state anxiety or distress, which continue for a period of time, could result in some personality changes which might be measured by the MMPI. This seems to be implied in the parental stress hypothesis; that is, the child's behavior is a continuing problem for the parents, which in turn results in certain personality characteristics. Prospective studies which follow the parents of autistic children may find personality changes as a function of time and provide some support for a parental stress hypothesis.

The lack of differences between the parents of autistic and child-guidance children might be expected if there were two subgroups of autistic children. Goldfarb (1970) has stated that virtually all the families of nonorganic psychotic children were deviant and contributed markedly to the disorders in their children. If most of the parents of nonorganic autistic children were disturbed and if most of the parents of organic autistic children were like parents of normal children, then combining these two groups of parents into a single group might obscure and minimize dramatic differences in parental psychopathology. This procedure might create the impression that the parents of autistic children were not very different from the parents of disturbed children. Therefore, the parents of autistic organic and nonorganic children (as defined by Goldfarb, 1961)* were compared to assess whether there were differences in parental personality characteristics and psychopathology in these two groups of autistic children.

Contrary to Goldfarb's hypothesis, the parents of autistic organic children had higher MMPI profile elevations than the parents of autistic nonorganic children. In general, these differences were not large enough to be significant with the exception of the vector of difference on the validity scales. The fathers of the nonorganic autistic children had a lower mean *L* score, while the fathers of organic autistic children had a higher mean *F* score. The frequencies of high and low scores on these scales were not large enough to permit a meaningful analysis. In spite of these differences, there was very little difference between the two groups of parents of autistic children.

Meyers and Goldfarb (1962) found that almost half of the mothers of the nonorganic subgroup were psychotic as compared with about 20% of the mothers in the organic subgroup. In the present study the numbers of mothers and fathers who might be described as disturbed enough to seek help for their own problems were not significantly different for the two groups of autistic children. Thus, there seems to be little evidence from the present results for Goldfarb's contention that virtually all the families of nonorganic psychotic children were deviant. The study also provides little support for two subgroups of autistic children which differ in parental psychopathology.

It is interesting to note that the parental stress hypothesis seems to best fit these data. Goldfarb (1970) has stated that the most severely impaired children were virtually all in the organic subgroup. Therefore, this would suggest that the parents of the organic autistic children would have more elevated MMPI profiles since impairment is likely to be related to parental stress. While the direction of the results was consistent with the parental stress hypothesis, the differences were not significant. This hypothesis might be better evaluated by classifying each autistic child into either a high or low functioning group, regardless of organic-nonorganic distinction. A lower functioning autistic child would be expected to cause greater parental distress than a higher functioning autistic child. The data will be further analyzed to examine this possibility.

*The clinical research center staff used more measures than Goldfarb's and a resulting neurological dysfunction index to define the relative amounts of overtly discernible brain dysfunction in autistic children.

ACKNOWLEDGMENTS

The authors would like to express their appreciation to Gerald Alpern, Ellen Smith Yong, and Lois Hendrickson Loew for collecting the MMPIs on the parents of autistic children, and to Sandy Barton, Vera McAdoo, and especially Victoria Ruckman for their assistance with this study. Computer service was provided by IUPUI computing facilities.

REFERENCES*

Anthony, J. An experimental approach to the psychopathology of childhood: Autism. *British Journal of Medical Psychology*, 1958, *31*, 211–225.

Block, J. Parents of schizophrenic, neurotic, asthmatic, and congenitally ill children: A comparative study. *Archives of General Psychiatry*, 1969, *20*, 659–674.

Dahlstrom, W. G., Welsh, G. S., & Dahlstrom, L. E. *An MMPI handbook* (Vol. 1). *Clinical interpretation* (Rev. ed.). Minneapolis: University of Minnesota Press, 1972.

DeMyer, M. K., Pontius, W., Norton, J. A., Barton, S., Allen, J., & Steele, R. Parental practices and innate activity in normal, autistic, and brain-damaged infants. *Journal of Autism and Childhood Schizophrenia*, 1972, *2*, 49–66.

Eisenberg, L. The fathers of autistic children. *American Journal of Orthopsychiatry*, 1957, *27*, 715–724.

Eisenberg, L., & Kanner, L. Early infantile autism 1943–55. *American Journal of Orthopsychiatry*, 1956, *26*, 556–566.

Erickson, M. T. MMPI comparisons between parents of young emotionally disturbed and organically retarded children. *Journal of Consulting and Clinical Psychology*, 1968, *32*, 701–706.

Esman, A. H., Kohn, M., & Nyman, L. Parents of schizophrenic children: Workshop, 1958. 2: The family of the "schizophrenic" child. *American Journal of Orthopsychiatry*, 1959, *29*, 455–459.

Goldfarb, W. *Childhood schizophrenia*. Cambridge, Mass.: Harvard University Press, 1961.

Goldfarb, W. Childhood psychosis. In. P. H. Mussen (Ed.), *Carmichael's manual of child psychology* (Vol. 2). New York: Wiley, 1970. Pp. 765–830.

Kanner, L. Problems of nosology and psychodynamics of early infantile autism. *American Journal of Orthopsychiatry*, 1949, *19*, 416–426.

Kaufman, I., Frank, T., Heims, L., Herrick, J., Rusir, D., & Willer, L. Treatment implications of a new classification of parents of schizophrenic children. *American Journal of Psychiatry*, 1960, *116*, 920–924.

Klebanoff, L. B. I. Parents of schizophrenic children: Workshop, 1958: Parental attitudes of mothers of schizophrenic, brain-injured and retarded, and normal children. *American Journal of Orthopsychiatry*, 1959, *29*, 445–454.

Reference Notes: (1) McAdoo, W. G. *Measures of parental psychopathology and their correlates.* Unpublished manuscript, 1976. (2) Hollingshead, A. B. *Two-factor index of social position.* Unpublished manuscript, 1957. (Available from August B. Hollingshead, 1965, Yale Station, New Haven, Connecticut.)

McAdoo, W. G., & Connolly, F. J. MMPIs of parents in dysfunctional families. *Journal of Consulting and Clinical Psychology*, 1975, *43*, 270.

Meyers, D. I., & Goldfarb, W. Psychiatric appraisal of parents and siblings of schizophrenic children. *American Journal of Psychiatry,* 1962, *118,* 902–915.

Pitfield, M., & Oppenheim, A. N. Child rearing attitudes of mothers of psychotic children. *Journal of Child Psychology and Psychiatry and Allied Disciplines*, 1964, *5*, 51–57.

Rutter, M. Concepts of autism: A review of research. *Journal of Child Psychology and Psychiatry*, 1968, *9*, 1–25.

Rutter, M., Bartak, L., & Newman, S. Autism—a central disorder of cognition and language? In M. Rutter (Ed.), *Infantile autism: Concepts, characteristics and treatment.* London: Churchill-Livingstone, 1971. Pp. 148–151.

Werry, J. S. Childhood psychosis. In H. C. Quay & J. S. Werry (Eds.), *Psychopathological disorders of childhood.* New York: Wiley, 1972. Pp. 173–233.

18

Family Factors

DENNIS P. CANTWELL, LORIAN BAKER, and
MICHAEL RUTTER

In his original description of the syndrome of infantile autism, Kanner (1943) commented that the parents tended to be highly intelligent obsessive people interested in abstractions and lacking real human warmth. He also observed that: "the children's aloneness from the beginning of life makes it difficult to attribute the whole picture exclusively to the type of the early parental relations." In the years that followed, numerous other writers drew attention to supposed abnormalities in this or that aspect of family functioning or early life experiences. A wide variety of theoretical formulations were proposed to suggest how psychogenic factors might have caused autism. At first, there were many suggestions that autism usually arises entirely on a psychogenic basis. However, evidence accumulated on the presence of biological abnormalities of various kinds in many autistic children, and few people today maintain an exclusively psychogenic position.

Of course, demonstration of an "organic" deficit does not mean that environmental influences play no part in the causation or course of autism. Indeed, the research findings which show that autistic children's development may be influenced both by the quality of schooling they receive (see Chapter 29) and by behavioral interventions in the home (see Chapter 26) emphasize the importance of environmental influences.

DENNIS P. CANTWELL and LORIAN BAKER · The Neuropsychiatric Institute, Center for the Health Sciences, University of California, Los Angeles, California 90024.   MICHAEL RUTTER · Department of Child and Adolescent Psychiatry, Institute of Psychiatry, De Crespigny Park, Denmark Hill, London SE5 8AF, England.

Such influences could operate in several very different ways. First, although some varieties of autism may be organically determined, others might be due to psychogenic factors (e.g., Goldfarb, 1961a). Second, it could be that, although a biological deficit is a necessary factor in the causation of autism, it is not a sufficient factor—an interaction with some psychogenic stress is also required for autism to develop (e.g., O'Gorman, 1970). Third, it is possible that, while autism is not linked with any abnormal type of environment, nevertheless the autistic child's biological deficit makes him liable to be damaged by ordinary family influences which are not hazardous to other children (e.g., Tinbergen & Tinbergen, 1972; see also Chapter 3). Fourth, the possibility remains that psychosocial influences play no part in the causation of autism but yet they are still important as determinants of prognosis and the course of development.

We consider these various possibilities later in this chapter but first it is necessary to outline the main psychogenic hypotheses which have been put forward and to review the empirical evidence which is relevant to them. Although views on the nature of family factors involved in the genesis of autism are legion, they may be roughly grouped under four broad headings: (1) severe stress and traumatic events early in the child's life, particularly in the first two years; (2) parental psychiatric disorder or deviant personality characteristics; (3) parental IQ and social class; (4) deviant parent-child interaction.

SEVERE EARLY STRESS

It has been proposed that early separation from the parents, parental rejection in the infancy period, maternal depression, and family conflict and disharmony may all be involved in the development of autism.

Specific traumatic events in early life (including births of siblings, separation from the parents, and physical illnesses) have been stressed by Rank and others in the Putnam group (Putnam, 1955; Rank, 1949, 1955). Szurek and his colleagues (Szurek, 1956; Boatman & Szurek, 1960) have suggested that severe, internalized unconscious conflicts in the parents over repressed and distorted libidinal impulses result in marital discord and the use of the child to resolve conflict. The stresses on the child distort his early personality development. These views derive from intensive psychotherapeutic work with children and their families.

Despert (1951) described emotional detachment and over intellectualization in the mothers of autistic children and indicated that the children were subjected to severe early stress and early rejection by the mother. However, the same author (Despert, 1938) had previously described the mothers of psychotic children as aggressive, overanxious and oversolicitous.

A number of clinicians have suggested that parental depression, early in the child's life, is another predisposing factor to the development of autism (Reiser, 1963; Ruttenberg, 1971; Szurek & Berlin, 1973; Tustin, 1972; Sarvis & Garcia, 1967). Sarvis and Garcia feel that autism develops during a period (6 months to 3 years) when the mother is the primary object to the child. Thus any event punishing to the child is likely to be perceived as deriving from the mother and leads to a paranoid reaction. In addition to maternal depression, they list a number of other family psychodynamic factors which promote autism. These include: unconscious needs in the parents, mothers who are unable to empathize with their children, paternal compulsivity, marital difficulties, illness in the parents, and overstimulation of the children which is perceived as an assault on the child.

These stresses, conflicts, and traumatic events are common in any group of children and the hypotheses are unclear on why autism only rarely develops and what is different about the circumstances when it does.

Bettelheim, on the other hand, has suggested a much more specific hypothesis (Bettelheim, 1967) that during infancy parents direct extremely negative feelings specifically and only to the child who is to become autistic. He then perceives the world as a dangerous place and so withdraws. This occurs during "critical periods" in the child's development; namely during the first 6 months of life when object relationships begin, at 6 to 9 months when language is beginning, and at 18 to 24 months when the child starts to have some influence on his relationships with the environment. Bettelheim's views lack empirical support and have been criticized on a number of grounds by several investigators (Wing, 1968; Rutter, 1968).

Because these hypotheses all suppose the existence of some stress *in the past* they are difficult to test directly. Nevertheless, they may be examined in several different ways. First, retrospective inquiry may be made about circumstances in the child's early life. Thus, Williams and Harper (1973) reported that 11 out of 97 psychotic children (diagnosed on Creak's nine points) had been locked in a room for long periods before the onset of the psychosis. This is a high proportion but conclusions are difficult in the absence of information on the circumstances of the locking, on the type of psychosis, and on the age of onset. Altogether, a quarter of the children had experienced severe environmental trauma. There was no control group so that it remains uncertain whether this is above or below expectation. Curiously, only seven of the 97 mothers were said to have been depressed which appears an unusually low proportion.

Systematic comparisons with a control group are provided by the Maudsley Hospital study (Cox *et al.*, 1975). Nineteen boys with infantile autism were compared with 23 boys with developmental receptive dysphasia (henceforth referred to as "dysphasia") in terms of their family background. The boys were aged 4 years 6 months to 9 years 11 months and all were of normal nonverbal intelligence and without overt neurological disorder or peripheral deafness. Information was systematically sought from parents on death, divorce, separation from the family for more than 4 weeks, hospital admission for more than a week and family stress (ncluding problems of finance, housing, health, and interpersonal relationships) during the child's first two years. No significant difference between the groups was found on any of these measures.

Special attention was paid to the parents' mental state during the first two years of the child's life. Five mothers of autistic children and seven mothers of aphasic children had been depressed during the first two years of their child's life. These differences were not statistically significant.

Folstein and Rutter (see Chapter 15) studied stress events in the early life of autistic and nonautistic children in twin pairs. Except that a higher proportion had experienced perinatal complications of various sorts, the autistic children did not differ from their nonautistic twins in terms of psychosocial stresses experienced before the onset of autism.

Lowe (1966) found that only 11% of autistic children came from broken homes, a rate far below the 50% for emotionally disturbed children. Rutter and Lockyer (1967) found that 9% of autistic children were in broken homes compared with 22% of children with other forms of psychiatric disorders.

Similarly, Bender and Grugett (1956) found that both broken homes and a poor "emotional climate" were much less frequent in the background of young "schizophrenic" children than of children with other psychiatric disorders.

Secondly, the possibility of early parental rejection may be assessed through inquiry on parental attitudes and child-rearing practices. DeMyer *et al.* (1972) used a semistructured interview to compare the parenting of autistic and brain-damaged children. Few differences were found and the findings provided no support for the notion of parental rejection.

Pitfield and Oppenheim's (1964) study also failed to show rejection. They administered a 50-item attitude survey on current child-rearing practices to the mothers of 100 psychotic children between the ages of 3 and 15 years and also to the mothers of 100 normal children and 100 children with Down's syndrome. There were no differences in "overprotection," "democracy," "acceptance," or "rejection." On the other hand, Holroyd and McArthur (1976) found that mothers of autistic children had more negative attitudes to their children that did mothers of Down's syndrome children.

Klebanoff (1959) used the Parental Attitude Research Instrument (PARI) to study maternal attitudes to "schizophrenic," brain-damaged, and normal children. The attitudes of mothers of schizophrenic children were less pathological than those of mothers of brain-damaged youngsters, but both these groups showed more pathological attitudes than the mothers of normal children. Only six of the 15 "schizophrenic" children had on onset before 3 years so few can have been autistic. The author suggested that the child's disorder shaped the maternal attitude rather than the other way round.

Anthony (1958), using a Shoben-type child-rearing questionnaire, found that the mothers of children with primary autism tended to be "self-righteous" whereas those with secondary autism tended to be "perplexed." The questionnaire findings failed to differentiate "organic" autism from "environmental" autism. Gonzales *et al.* (1977), using the Maryland Parental Attitude Scale, found no differences between the parents of autistic and subnormal children.

The third approach is to take a group of children known to have experienced early psychosocial traumata and then to examine their development. Numerous studies of this kind have been undertaken in relation to separation experiences (Rutter, 1971), hospital admission (Douglas, 1975; Quinton & Rutter, 1976), institutional upbringing (Tizard & Rees, 1976; Wolkind, 1974), and other stresses or hazards (Rutter, 1972). The evidence is decisive in indicating that early stresses do lead to a substantial increase in later psychiatric disorder. Equally, however, the same studies show that such disorders only very rarely take the form of autism.

Altogether, the evidence runs counter to hypotheses which suggest the importance of early family stresses in etiology. Clearly, such stresses are the exception rather than the rule and comparative studies indicate that stresses are no more frequent in the early lives of autistic children than in those of children with other forms of handicap. Inevitably, of course, the data are largely retrospective (and hence open to bias) and it is possible that there have been important stresses of a transient nature. Also, it is possible that early stresses may be crucial in certain individual cases but it seems highly unlikely that they play a central role in the majority.

PARENTAL DEVIANCE AND DISORDER

Numerous pejorative epithets have been applied to the parents of autistic children. They have been described as cold, undemonstrative, formal, introverted, and obsessive (Cooper, 1964; Singer & Wynne, 1963; Rank, 1955; Eisenberg, 1957; Bene, 1958; Kanner, 1949). Eisenberg

(1957) noted that these characteristics mainly applied to fathers. The parents have also been considered overprotective, symbiotic, indecisive, lacking dominance, and showing "perplexity" or psychic "paralysis" (Kasanan et al., 1934; Alanen, 1960; Cooper, 1964; Rank, 1955; Bowen et al., 1959; Meyers & Goldfarb, 1961; Goldfarb, 1961a,b). It should be added, however, that other clinicians (e.g., Creak & Ini, 1960) have found the parents to be quite varied and generally unremarkable in personality.

Meyers and Goldfarb (1961) have placed most weight on "parental perplexity" as the main etiological influence of a psychogenic nature. The family atmosphere associated with parental perplexity is characterized by extreme parental indecisiveness; a lack of parental spontaneity and of empathy with the child; an inability on the part of the parents to grasp their child's needs and to gratify them at a suitable time; and an unusual absence of control and authority (Goldfarb, 1967). The Putnam group (Rank, 1949, 1955; Putnam, 1955) have described the mothers as incapable of performing their maternal function or of experiencing genuine gratification in being mothers. The mothers are thought to show emotional immaturity and narcissism and to be devoid of genuine maternal drive and capacity for spontaneous affection. The fathers were also considered important in that they seemed extremely passive in their relationships and failed to counteract the negative influence of the mother.

Kaufman and his colleagues (Kaufman et al., 1959, 1960) describe four types of parents of "schizophrenic" children: pseudoneurotic, somatic, pseudodelinquent, and psychotic. These descriptions were based on data obtained from long-term psychotherapy with 82 schizophrenic children and both parents, diagnostic evaluations of the parents, psychological testing, and direct observations of parent-child interaction. According to Kaufman and his colleagues the children become "schizophrenic" due to their intense pathological relationship with the parents. This forces the child to become psychotic because of the parents' emotional needs and defenses.

In his review of the family environment of schizophrenic patients Alanen (1960) described four aspects of inadequate ego development: (1) defective establishment of initial important object relationships; (2) inadequate separation from symbiotic object relationships; (3) lack of healthy and useful identification patterns within the family; (4) disturbances in the process of role taking and identity formation in the family. Four different types of psychosis were thought to result when ego development becomes arrested at each of these four phases. Arrest at phase one leads to autism; arrest at phase two to symbiotic psychosis; arrest at phase three to the child's identification with abnormal (especially psychotic) parents; and arrest at phase four to an adult-type psychotic picture.

Other writers have placed the main emphasis on mental illness in the parents. Thus, Goldfarb (1967) and Bender (Bender & Grugett, 1956) have reported very high rates of schizophrenia in the parents of psychotic children. Fish (1975) and others have also noted serious developmental disorders in the offspring of schizophrenic adults—although these disorders have not been autistic in type.

Many different studies have been undertaken to test these various hypotheses about parental deviance. First, several have used projective tests. Meyer and Karon (1967) noted more "pathogenic" stories in the TAT responses of mothers of six schizophrenic children compared with the mothers of six normal children. Ogdon *et al.* (1968) found more Rorschach responses indicating relationship difficulties (but fewer indicating obsessiveness) in the parents of 12 autistic or symbiotic children than in the parents of 12 normal volunteers. However, in both these studies the control groups were unsatisfactory. Bene (1958) used the Rorschach test to compare the mothers of children with primary and with secondary autism. This distinction had been made by Anthony (1958) who hypothesized that in primary autism the child fails to emerge from an autistic state having been unresponsive from the time of birth. In secondary autism the child regresses to an autistic state after having developed normally for some period of time. Bene (1958) showed that mothers of children with primary infantile autism were more removed in social and emotional relationships than the mothers of children with secondary infantile autism. On the other hand, contrary to prediction, there were no differences between the groups on the number of morbid responses on the Rorschach.

Singer and Wynne (1963) studied the parents of 20 childhood schizophrenics and 20 late onset young adult schizophrenics, ten withdrawn neurotics and ten neurotics with conduct disorders. Using TAT and Rorschach protocols of the parents of each child, Singer was able to correctly predict the patient's diagnosis as either childhood schizophrenia or neurosis with a very high degree of accuracy. She described the parents of psychotic children as cynical, embittered, passive, apathetic, obsessive, and distant.

Rice *et al.* (1966) used ratings to assess 'communicative impact' between ten mothers and their psychotic children (including three autistic) and that between ten mothers and their behaviorally disordered children. The mothers of the psychotics were rated less likable and more pathogenic. However, little weight can be attached to the findings, not only because they were based on only 5 minutes of interaction but also because of the subjectivity of the ratings and the heterogeneity of the psychotic group.

Block (1969), using a Q sort based on Rorschach, TAT, and MMPI protocols, was unable to differentiate the parents of schizophrenic and neurotic children although both differed from the parents of asthmatic and congenitally ill children.

The Eysenck Personality Inventory was used to measure extroversion and neuroticism in the parents of autistic children by Kolvin *et al.* (1971a), Netley *et al.* (1975), and Cantwell *et al.* (1977b). The last study showed no difference between the parents of autistic and "dysphasic" children and the first two found no differences from the published general population norms. Neurotic tendencies have also been assessed through the MMPI by Kolvin *et al.* (1971a) and by McAdoo (see Chapter 17). In both studies the parents of autistic children proved unremarkable. Cox *et al.* (1975) found no significant difference between the parents of autistic and "dysphasic" children with respect to their scores on the psychiatric section of the Cornell Medical Index, or their scores on the Leyton Obsessional Inventory.

Gonzales *et al.* (1977) used Hogan's empathy scale and Wallin's sociability scale to compare the parents of autistic, deaf, and subnormal children. The parents of autistic children tended to be somewhat more empathic and more sociable. Their scores on Cattell's 16 PF were generally unremarkable.

In the Maudsley Hospital study, systematic interview techniques (Brown & Rutter, 1966; Rutter & Brown, 1966) were used to assess parental warmth to the handicapped child in groups of autistic and "dysphasic" boys. No significant differences were found between the groups (Cox *et al.*, 1975). However, there was a slight tendency for mothers to show less warmth to their autistic children than to the nonautistic sibs in the same family. Interview measures were specially devised to assess parental demonstrativeness and responsiveness using data on verbal and physical expressions of sympathy, pleasure, and enthusiasm in various specified situations. Again no parental differences were found between groups of autistic and "dysphasic" children. Sociability was assessed on the basis of the parents' range and frequency of contacts with relatives and with friends. The quality of friendships was judged from their duration and the degree to which confidences were exchanged. Again there were few differences between the groups although the parents of autistic children were more sociable on a few of the measures. Most of the parents appeared to have normal personalities but there was a clinical impression (not reflected in test findings) that marked oddities of various kinds, although present in only a minority of cases, were more frequent among the parents of autistic children (Cantwell *et al.*, 1977b).

It has also been suggested that the parents of autistic children may show thought disorder of a type characteristic of schizophrenic patients, and possibly also their relatives. Schopler and Loftin (1969a, b) found that abnormal scores on the Goldstein-Scheerer Object Sorting Test were present in 41% of the parents of psychotic children in conjoint family therapy, 30% of untreated parents (of psychotic children) tested in a nonstress situation, 32% of the parents of retardates, and 12% of the parents of normals. Only the difference between the first and last groups reached statistical significance and Schopler and Loftin (1969b) concluded that the OST measured situationally circumscribed anxiety rather than thought disorder. Lennox (1973) found that the parents of autistic children had more abnormal scores on the OST than controls but the OST scores were unrelated to anxiety. Netley *et al.* (1975) used a different test of thought disorder (the Bannister-Fransella Grid Test) to compare the parents of autistic children with the parents of nonautistic psychotic children. No difference was found for fathers but five of the seven mothers of autistic children had abnormal scores compared with only five of the 19 mothers of nonautistic psychotic children.

Lennox *et al.*, (1977) sought to resolve the contradictions by using *both* tests of thought disorder and also measures of state and trait anxiety with the parents of autistic and normal children. It was found that there was a striking lack of agreement between the two tests of thought disorder; that the parents of autistic children showed closely comparable thought disorder scores to the parents of normals; and that there was no consistent association between thought disorder and anxiety. It appeared that the differences in findings between studies was probably in part a function of the effects of either treatment approaches or the test situation, in part an artefact of biases stemming from very high noncooperation rates in some of the studies, and possibly in part a function of different investigators looking at somewhat different groups of psychotic children. Whatever the explanation, it seems most *un*likely that any intrinsic thought disorder (as assessed by the tests used) is truly more common among parents of autistic children.

As already noted, Bender (Bender & Grugett, 1956) and Goldfarb (Meyers & Goldfarb, 1962) are the two main investigators to have found high rates of schizophrenia in the parents of psychotic children. (Both refer to "schizophrenic" rather than autistic children). Bender and Grugett (1956) reported that 22 out of 30 "schizophrenic" children had a schizophrenic parent compared with a figure of four out of 30 for children with nonschizophrenic psychiatric disorders. No criteria are given for the diagnosis of schizophrenia. Meyers and Goldfarb (1962) reported schizophrenia in 44% of the mothers of "organic" and 21% of the mothers of "nonorganic" schizophrenic children. The comparable figures for fathers were 15% and 7%. Again, no diagnostic criteria were given (although it is said that the term schizophrenia was used to include "schizoid personalities"—only two of the 84 parents studied had been hospitalized).

However, the problem of parental diagnosis was dealt with systematically in a very recent study by Goldfarb and his collaborators (1976). The Spitzer-Endicott interview schedule was utilized by researchers who did not know the index children; the data were then used to generate a computer diagnosis. Eighteen percent of the 56 parents of psychotic children showed schizophrenia compared with 8% of 130 adults in a general population control group. The diagnosis was largely based on social isolation in the absence of affective disturbance. Strikingly the groups did *not* differ appreciably on inappropriate behavior, retardation—lack of emotion, or suspicion—persecution—hallucinations. These negative findings and the high rate of schizophrenia in the general population control group strongly suggest that the diagnosis of schizophrenia refers to conditions (probably personality disorders) which would *not* be diagnosed schizophrenic in Europe (see Cooper *et al.*, 1972). It should be added that this recent study (Goldfarb *et al.*, 1972) did *not* confirm the diagnostic difference reported earlier (Meyers & Goldfarb, 1962) between fathers and mothers or that between the parents of "organic" and "nonorganic" types of childhood schizophrenia.

Almost all other studies have come up with entirely different findings. Thus, Creak and Ini (1960) found schizophrenia in only two out of 120 parents; Kanner (1954) found no cases among 200; and Lotter (1967) found one case of psychotic disorder among 60 parents. Rutter and Lockyer (1967) found that 11 parents of 63 psychotic children had received psychiatric treatment compared with 13 parents of 63 children with nonpsychotic mental disorders and in only one case—a control—was the diagnosis schizophrenia. Herzberg (1976) noted mental illness in four parents of 75 psychotic children and five parents of 44 controls. Kolvin *et al.* (1971b) found one case of schizo-affective disorder among the 92 parents of autistic children but six cases of definite schizophrenia among the 64 parents of children with late onset psychosis (difference statistically significant).

It seems clear that the frequency of schizophrenia is no greater among the parents of autistic children and that there is probably a *marked* increase among the parents of children whose psychosis begins in later childhood or early adolescence. The Bender and Goldfarb findings suggest a raised rate of chronic mental disorders (involving abnormalities of personality but not acute psychotic manifestations) among the parents of a heterogeneous group of "schizophrenic" children. To what extent these overlap with autism remains uncertain.

PARENTAL IQ AND SOCIAL CLASS

Kanner (1949, 1954) emphasized the high social status and outstanding academic accomplishments of the parents of autistic children and noted that he had never seen an autistic child who came of unintelligent parents. Subsequent investigations in both the United States and the United Kingdom have confirmed that parents of autistic children tend to be of somewhat above-average intelligence and somewhat superior social status compared with both the general population and also the parents of other child psychiatric patients (e.g., Alanen *et al.*, 1964; Allen *et al.*, 1971; Bender & Grugett, 1956; Cox *et al.*, 1975; Davids, 1958; Florsheim & Peterfreund, 1974; Lotter, 1967; Lowe, 1966; Kolvin *et al.*, 1971b; Rutter & Lockyer, 1967; Treffert, 1970). However, the same studies also show that the difference is only relative—autistic children come from all social classes and from parents of all levels of intelligence even though proportionately more autistic children come from intellectually and socially advantaged sections of the population. The only major study with findings different from these is that by Ritvo *et al.* (1971). The description of diagnostic criteria (Ornitz &Ritvo, 1968) suggests that their group probably includes severely retarded children whose main clinical features were stereotypies and with abnormal responses to sensory stimuli and who would not necessarily be regarded as autistic by other workers.

Two possible artefacts need to be considered in relation to these findings. First, the relative social class/IQ superiority could be due to a bias in referral policies. This bias can be eliminated only through the use of total population statistics. Such data confirm the clinic findings (Lotter, 1967; Treffert, 1970). Secondly, the supposed intellectual superiority could be a consequence of biased clinical perceptions. Systematic psychometric evaluation is required to rule out that possibility. However, IQ testing of the parents has confirmed the clinical findings. Thus, Lotter (1967) found that a third of the parents had scores in the top 5% on the Matrices and a fifth were in the top 5% on the Mill Hill Vocabulary Scale. Allen *et al.* (1971) reported a mean verbal IQ on the WAIS of 109 for mothers and 116 for fathers of autistic children compared with 104 and 101, respectively, for the mothers and fathers of brain-damaged retarded children. Other studies have reported mean IQs of 122 (Davids, 1958), 110–119 (Alanen *et al.*, 1964), and 108–118 (Florsheim & Peterfreund, 1974) for the parents of autistic children. It may be concluded that the evidence clearly indicates a tendency to higher IQ and superior social status among the parents but the difference is relative and of only moderate size.

DEVIANT PARENT-CHILD INTERACTION

Observations and accounts of the interaction between autistic children and their parents, usually in the course of psychotherapeutic treatment, have led to an incredible variety of hypotheses on deviant family interaction. Postulated pathogenic variables include too much stimulation; too little stimulation; inadequate structuring of the environment; excessive structuring and intrusion; lack of family roles and identities; lack of shared family pleasure; inadequate or abnormal communication patterns; and lack of suitable reinforcement (see Quay & Werry, 1972; Behrens & Goldfarb, 1958; Tinbergen & Tinbergen, 1972; Ward, 1970).

Several of the leading theories on deviant parent-child interaction have already been discussed when considering early stress and parental deviance. Thus, Bettelheim (1967) has emphasized parental rejection; the Putnam group (Rank, 1949, 1955; Putnam, 1955) lack of maternal affection; and Goldfarb (1967) parental perplexity. According to Goldfarb's views the family pattern of perplexity is associated with poor maternal communication to the autistic child, generally impaired family interaction, a lack of structure or control, and poor sensitivity to the child's signals. Bowen *et al.* (1959) have also noted a lack of dominance, indecisiveness, immaturity, and parenting insufficiency in the families of schizophrenic children. Inadequate structuring of parent-child interactions is also thought to be important by Fraknoi and Ruttenberg (1971).

Ferster (1961, 1966) has suggested a somewhat similar pattern from a behaviorist viewpoint. He has argued that autism consists of an impoverished repertoire of normal behaviors which are impaired because parents have failed to reinforce or respond to the child's initial social advances.

Tinbergen and Tinbergen (1972), on the other hand, have proposed the opposite—namely that autistic children withdraw because parents are too intrusive and give too much response to the child's social approaches.

Although all the theorists suggest that the abnormal patterns of parent-child interaction preceded and led to the autism, nevertheless all the theories derive from observations of such interactions *after* the child is already autistic. Thus, it should be possible to test the hypotheses by direct observation of the interactions between parents and their autistic child.

Donnelly (1960) did this through the use of the Fels Parent Behavior Rating Scales. She found that parental behavior toward the psychotic child differed significantly from that toward normal children in the same family with respect to warmth and control but not amount of contact or degree of infantilization. Parents tended to be less approving, more rejecting, more restrictive and stricter in discipline with their psychotic child. No control or comparison group was used.

Byassee and Murrell (1975) compared six families with autistic children, six families with normal children, and six families with emotionally disturbed children using the Ferreira and Winter Unrevealed Differences Task. The Unrevealed Differences Task yields four main measures of family interaction: spontaneous agreement, choice fulfillment, decision-making time, and index of normality. No differences were found between families with autistic children and those with normal children. On the other hand, families with emotionally disturbed children differed from the families of both normal and autistic children in showing less spontaneous agreement between the parents.

Lennard *et al.* (1965) found that families of autistic children asked more questions than families of normal children. It was concluded that these families were more "intrusive" and more discouraging of self-motivated behavior. The same data, however, could be used to support the conclusion that these families were simply trying to "draw out" their children to help them use the little speech they possessed.

King (1975) used case-note ratings to derive measures of "double-bind" parent-child relationships. Such relationships or attitudes were found much more frequently among the families of 12 autistic children than among the control families with a nonautistic child patient. However, the findings are of dubious validity in that the ratings were applied to unsystematic and potentially biased clinical judgments (e.g., M "put him through the paces") rather than to standardized observations of parent-child interaction. Other studies have shown the considerable difficulties in assessing double-bind phenomena (see Hirsch & Leff, 1975).

The most extensive evidence of deviant patterns of interaction between parents and their psychotic children is provided by the studies of Goldfarb and his colleagues. They have utilized several different approaches. First, family interaction scales (Behrens *et al.*, 1969) have been used to make ratings on home observations with "schizophrenic" and normal children (Behrens & Goldfarb, 1958; Goldfarb, 1961a; Meyers & Goldfarb, 1961). Parents of "nonorganic" schizophrenic children have been found to have difficulty structuring their children's environment and tend to show perplexity, inconsistency, vacillation, vagueness, uncertainty, absence of roles, detachment, and passivity. The family adequacy scores of "organic" schizophrenic children were broadly comparable to those found with normal children.

Second, a speech pathologist rated language samples from mothers of schizophrenic and of normal children (Goldfarb *et al.*, 1966a). It was found that the mothers of schizophrenics were inferior to mothers of normals in both their general speech adequacy and in their communications of meaning and mood. It was concluded that the mothers of schizophrenic children provided poor language models.

Third, the verbal encounter between the schizophrenic child and his mother was investigated in a meeting on a "surprise visit" to the hospital (Goldfarb *et al.*, 1966b). Measures of clarity of communication were obtained on some 2 minutes of tape recording. Compared with mothers of children hospitalized for orthopedic conditions, the mothers of "nonorganic" schizophrenic children showed more clarity errors (Goldfarb et al., 1972). It seemed that the mothers of autistic children failed to stimulate interest in communication, failed to maintain continuous conversation, failed to reinforce normal speech, used illogical expressions, failed to pick up the child's verbal cues, and generally failed to cope with the child's own verbal differences. Fourth, mothers were asked to describe certain objects to their children in order to aid them in a discrimination task. Mothers of schizophrenics were less able to assist their children than mothers of normals (Goldfarb et al., 1973).

These studies appear to provide an impressive documentation of deviant patterns of interaction in the families of schizophrenic children. However, numerous criticisms have been made of the investigations (Klein & Pollack, 1966; Howlin et al., 1973a; Cox et al., 1975; Baker et al., 1976) on the grounds of poor definition of the groups of children studied, failure to match groups on relevant variables, lack of suitable sampling techniques in the choice of controls, unsuitable circumstances for eliciting language samples, excessively short samples of language, and loose definitions of language categories.

Many of the studies suggesting deviant patterns of interaction in the families of psychotic children rely on comparisons with normal children. That this may be misleading is shown by an ingenious study by Gardner (1977) who made a detailed systematic study of mother-child interaction with autistic and with normal children. In order to examine the effects of the autistic child on interacting patterns he got both sets of mothers to interact with normal *and* with autistic children. In order to check on possible familiarity effects he also got the parents of normal children to interact with *someone else's* normal child and the parents of autistic children to interact with someone else's autistic child. He found that mothers spoke more to autistic children but used shorter utterances, more questions, more commands, and more verbal rewards and punishments. It was clear that the characteristics of the autistic children elicited different kinds of parental behavior. There were also some familiarity effects (e.g., mothers spoke less to their own children). After taking these child and familiarity effects into account (by analysis of variance) there were still some differences between the mothers of autistic and the mothers of normal children, but the findings do not allow any simple conclusion as to which interaction patterns was more deviant. Thus, the mothers of normal children elicited more cooperative behavior from both normal and autistic children, they spoke more, asked more questions, and made more commands. Mothers of normal children were more likely to achieve a mutual facing position with autistic children but mothers of autistic children were more likely to do so with normal children.

Because of the important child effects, in our own studies we compared the family interaction patterns of autistic children with those of "dysphasic" children (Cantwell *et al.*, 1977a,b). The two groups were behaviorally quite different (Bartak *et al.*, 1975) but they shared the common handicap of specific retardation in receptive language. To avoid the problem of diagnostic heterogeneity the group of autistic children all met the specific diagnostic criteria outlined by Rutter (1971). In order to allow direct comparisons with Goldfarb's work, identical measures of interaction were employed but the language interaction scores were based on a much longer language sample obtained in the more natural environment of the child's own home. However, because of the subjective nature of some of Goldfarb's ratings we also added various other assessments, utilizing a range of measures of demonstrated reliability which gave information pertinent to the main psychogenic hypotheses. Both systematic interviews and direct observations in the home were employed.

The interview data included a detailed minute-by-minute description of the child's "standard day," using the technique developed by Douglas and his colleagues (Douglas *et al.*, 1968; Lawson & Ingleby, 1974). Interactions with the child were classed as to whether they are "concentrated" (i.e., full attention to the child), "continuous" (i.e., considerable, but not full, attention), "available" (largely supervisory attention), or "available-not used" (largely supervisory, but child does not take advantage of the availability). Weighted general measures of attention were computed by combining these measures for each of the four types of interaction (Cantwell *et al.*, 1977a).

The two groups did not differ significantly on any of the various degrees of intensity of interaction. Autistic and "dysphasic" children experienced almost exactly the same amount (and proportions) of concentrated, continuous, and available interactions with other people. Moreover, the figures for autistic and dysphasic children were also similar when interactions with fathers and with mothers were considered separately. The only sizable difference between the groups concerned interactions with sibs and with peers. Dysphasic boys had much greater interaction with other children—almost certainly a consequence of autistic children's very serious difficulties in peer relationships.

Parent-child interaction was also assessed on the basis of parental interview data concerning the frequency of various specified activities which the child might consider positive or pleasurable (Brown & Rutter, 1966). The mothers of autistic children spent more time reading to them than did mothers of dysphasic children but otherwise there was little difference between the groups in mother-child interaction. On the other hand, the fathers of autistic children spent nearly twice as much time with their children. Much of this time was spent in roughhousing or other forms of boisterous play. In short, none of the interview measures provided any support for the hypothesis that autistic children suffered from a lack of parental attention. The findings could be interpreted as showiɾg greater intrusion on the part of parents of autistic children but equally it could be argued that the autistic child's difficulty in initiating interaction required greater involvement by parents. Whether this was beneficial or harmful can be better judged from actual observations of detailed sequences of parent-child interaction in the home.

This was carried out and evaluated in three very different ways. First, after a 4–6-hour period in the home including a family meal in which family interactions were carefully observed, the relevant portions of the Ittleson Center Scales (Behrens *et al.*, 1969) used by Goldfarb were filled out. Analysis of the results revealed no differences between the autistic and dysphasic groups with regard to: spontaneity of interactions, decisiveness, mode of relating to the child, imposition of routine, anticipation of the child's need, and meeting the demands of the child. The only significant difference between the two groups was that the fathers of the autistic children were more consistent in emotional relatedness than the fathers of the dysphasic children (Cantwell *et al.*, 1977a).

Second, detailed time-sampled observations were made of mother-child interaction over a 90-minute period in the home, using measures developed by Hemsley *et al.* (1977). The character and affective tone of activities were recorded every 10 seconds, thus building up a very detailed portrait of mother-child interaction. There were no differences between the groups in the porportions of time spent actively involved with the child, the amount and type of communication with the child, the frequency of physical contact, or the extent to which interactions were positive or negative in affective tone.

Third, a systematic and detailed analysis was made of audiotape recordings of the mother's communications with the child at home, using methods developed by Howlin *et al.* (1973a). This analysis (Cantwell *et al.*, 1977b) had four separate parts: (a) a functional analysis of language usage, (b) a linguistic analysis of the mother's utterances, (c) measurement of the "clarity" of expression of meaning, using the categories of Goldfarb *et al.* (1966b), and (d) assessments of the tone of voice used in each communication.

The first analysis showed that the same general patterns of language use were employed by both groups of mothers. Both groups made most use of questions, directions, and statements and least use of mitigated echoes, reductions, and directed mimicries. The only significant difference between the groups was that the mothers of autistic children made significantly more affectionate remarks to their children. The second analysis showed that the language of the two groups of mothers was equally complex, equally grammatical, and equally detailed. The third analysis showed no marked difference between the groups on clarity errors (but the mothers of autistic children tended to make fewer errors). The fourth analysis showed that in both groups most speech was neutral in tone but the mothers of autistic children used significantly more positive tone and significantly less neutral tone. In both groups approving comments were usually positive in tone and critical tone of voice was almost restricted to disapproving or correcting remarks. However, the two groups did differ significantly on the tone of voice used with disapproving comments; the mothers of autistic children were less likely to use negative tone and more likely to use neutral tone.

This last item was the only one which might be taken to reflect maladaptive functioning in that it could be said to make it more difficult for children to get the message. However, this was just one rather isolated significant difference which applied to only 3% of remarks. Overall, it was clear that the maternal model of speech was generally similar in the case of autistic and "dysphasic" children. The findings provide *no* support for the suggestion that autism is due in whole or in part to deviant patterns of parent-child communication.

OVERVIEW

The first conclusion from this review of the literature is that although psychogenic *hypotheses* on the causation of autism are diverse and contradictory, the *empirical* studies (with the one important exception of Goldfarb's studies) are generally in remarkably good agreement. This is particularly striking as the investigations span 20 years and two continents. Also they include measures ranging from questionnaires through systematic interview assessments to detailed observations in the home.

Thus, it is clear that autistic children tend to come from homes which are slightly favored in terms of social status and the intelligence of parents. Equally it is clear that they do *not* tend to come from broken homes. Although there are a few slightly contradictory findings it is also evident that most autistic children do *not* suffer early family stresses and the parental attitudes and child-rearing practices they encounter are mostly unexceptional.

All studies are agreed that the parents of autistic children show normal scores on tests of extroversion and they show no particular tendency to any form of neurotic trait (including obsessionality). Questionnaire and interview measures are also in agreement in showing normal empathy and sociability. With the exception of Bender and Goldfarb investigators agree in finding that the rate of schizophrenia is *not* higher in the families of autistic children.

There are some contradictions in the findings on parental thought disorder but the balance of evidence suggests that thought disorder is not particularly frequent in the parents of autistic children. Furthermore, there are good grounds for supposing that the differences between studies are a function of methodological issues and situational effects (see Schopler & Loftin, 1969b; Lennox *et al.*, 1977).

Finally, there is general agreement that the patterns of interaction in families of autistic children are much the same as those in families of children with other developmental handicaps.

The one major exception to this general picture of negative findings is provided by the series of studies undertaken by Goldfarb and his colleagues. Unlike almost all other studies they have consistently shown abnormal patterns of parenting, communication, and family interaction. It is important to consider the possible reasons for this disparity in results.

Several possibilities can be ruled out immediately. It is not just a question of different measures of family functioning. This is clear both because our own studies (Cantwell *et al.*, 1977a,b) have used the same measures as the Goldfarb group (Meyers & Goldfarb, 1961; Goldfarb *et al.*, 1966a,b) and yet still have come up with different findings, and because findings contrary to Goldfarb's have come from investigations using measures as varied as they could be. Thus, our own studies have used standardized questionnaires, a variety of specially designed interview measures and both structured and unstructured observations in the home (Cox *et al.*, 1975; Cantwell *et al.*, 1977a,b). The Great Ormond Street group used clinical observations (Creak & Ini, 1960) and questionnaires (Pitfield & Oppenheim, 1964). The Indiana group used systematic interviews (DeMyer *et al.*, 1972) and questionnaires (see Chapter 17 this volume), and so did the Newcastle group (Kolvin *et al.*, 1971b; Gonzales *et al.*, 1977). All these studies have come up with generally comparable findings in spite of a variety of measures used with different subject samples in different research centers.

Also, it is not a function of secular trends in ways of looking at autistic children and their families. Studies undertaken nearly 20 years ago (e.g., Klebanoff, 1959; Pitfield & Oppenheim, 1964) produced results closely comparable with those from the most recent investigations (e.g., DeMyer *et al.*, 1972; Gonzales *et al.*, 1977). Again, it is not a question of United States-United Kingdom differences as most studies from both sides of the Atlantic have produced closely comparable findings.

What, then, does explain the differences in findings? There is evidence in favor of three main factors: (1) choice of control group, (2) diagnosis of child, and (3) methodological bias.

Control Groups

Most of the studies which have found "abnormalities" in parent-child interaction have compared autistic children with either normal children or physically ill children. But this is an invalid comparison for assessing *parental* effects vis-à-vis parent-child interaction as it entirely ignores *child* effects (Bell, 1968, 1971). Handicapped children, because they are handicapped, are liable to elicit parental behaviors which are different from those elicited by normal children. It should be noted that this is not just a theoretical possibility as it has been found *in actual practice*. Thus, there is considerable evidence that the quantity and quality of adults' speech is affected by the verbal skills of the children with whom they interact (Siegel, 1963; Siegel & Harkins, 1963; Spradlin & Rosenberg, 1964; Moerk, 1974; Snow, 1972; Fraser & Roberts, 1975). Also, Gardner (1977) found that parents of normal children behaved differently toward autistic children from the way they behaved toward nonautistic children. Or again, Klebanoff (1959) found that parents of schizophrenics and parents of brain-injured children differed in child-rearing attitudes from parents of normal children, but did not differ between themselves. Studies of families with a congenitally handicapped child have also shown that this is associated with different parental behavior (Cummings *et al.*, 1966). Furthermore, experimental manipulation of children's behavior has been shown to alter the ways parents interact with them (Osofsky & O'Connell, 1972).

Accordingly, it is clear that if studies of parent-child interaction are to assess differences in parental behavior which are not simply the consequences of having a handicapped child, comparisons need to be made with families in which there is a child without autism but with some other form of developmental/behavioral handicap. Alternatively, a cross-over design such as that employed by Gardner (1977) is needed. It is likely that this difference in control groups between Goldfarb's studies (normal children or children with orthopedic problems) and our own (dysphasic children) provides part of the explanation for the differences in findings.

Diagnosis

It is striking that almost all of the studies showing some form of family dysfunction have been concerned with "schizophrenic" rather than with autistic children. This applies, for example, to all of Goldfarb's studies as well as those by Bender (Bender & Grugett, 1956), Meyer and Karon (1967), Rice *et al.* (1966), and Singer and Wynne (1963). This is an important factor as Kolvin *et al.* (1971b) have shown that the family backgrounds of autistic and schizophrenic children are very different. There is also much other evidence to indicate that the two conditions are different (Rutter, 1972).

Of course, Goldfarb's concept of "schizophrenia" is a broad one—he equates his diagnostic criteria (Goldfarb, 1974) with Creak's (1961) "nine points." This approach would be expected to include many autistic as well as nonautistic children. However, it should be noted that he finds verbal skills somewhat superior to nonverbal skills in both his "organic" and "nonorganic" schizophrenic groups (Goldfarb, 1967). This is different from the usual findings with autistic children (e.g., Lockyer & Rutter, 1970; Bartak *et al.*, 1975) and suggests that the children he has studied include many who would not fit the criteria for autism employed by other workers. It should be appreciated that, insofar as a diagnostic difference accounts for the fact that Goldfarb's findings differ from those of most other workers, this implies that family pathology may be associated with schizophrenia even if it is not with autism. On the other hand, using less rigorous measures, Bender (Bender & Grugett, 1956) unlike Goldfarb did not find family inadequacy to be characteristic of schizophrenic children.

Methodological Bias

Quite apart from control group and diagnostic effects, there are several important methodological biases which may have distorted findings. First, there is the question of how representative are the samples studied. Cox *et al.* (1977) have shown that people who do not cooperate with studies tend to be systematically different from those who do. This finding casts considerable doubt on those studies which have relied on volunteer control groups (e.g., Meyer & Karon, 1967; Ogdon *et al.*, 1968). Second, biases may be introduced through situational effects. Thus, Goldfarb has used very short (2-minute) maternal speech samples obtained in the artificial situation of a "surprise visit" to hospital. We have found that about 30 minutes is required to provide an adequate language sample (Howlin *et al.*, 1973a). We have also found that parent-child interaction in the clinic is often very different from that in the home (Howlin *et al.*, 1973b). In addition, interaction may well be altered through involvement in treatment programs (see Schopler & Loftin, 1969b), as well as by the child being placed in hospital. For all these reasons, we relied on lengthy samples of family interaction in the child's own home with autistic children who remained in the community (Cantwell *et al.*, 1977a,b). It is likely that this resulted in our obtaining a more valid picture of family functioning than that available to Goldfarb and his colleagues. Third, biases may arise through "criterion contamination." Thus, when Goldfarb *et al.* (1976) used independent assessments of parental psychiatric state there was no longer any difference between their "organic" and "nonorganic" groups. The earlier finding of a difference between these subgroups (Meyers & Goldfarb, 1961) was presumably an artefact. Of course, this source of bias is a major problem when observers of family interaction cannot possibly be unaware of which child is autistic. However, it does argue for the use of methods of measurement which are as objective as possible.

In this overview we have not referred to the various impressionistic reports based on therapeutic experience with patient groups. Not only are the observations likely to be biased as a result of therapeutic involvement but also no valid conclusions can be drawn without reference to some kind of appropriate comparison group. However, we have paid attention to systematic studies utilizing all kinds of measures and based on all kinds of theoretical assumptions.

It is not possible to be sure how far each of these three areas of bias has led to the differences in findings between Goldfarb's group and our own (as well as several others). However, it seems highly likely that between them they do account for the differences. If they do, it may be concluded confidently that there is no satisfactory evidence that deviant patterns of family functioning lead to infantile autism. The very few findings that suggest abnormal patterns are probably an artefact of parental responses to a handicapped child or methodological inaccuracies in measurement. Alternatively, they apply to groups of psychotic children who are not autistic.

Modes of Family Influence

In our introduction we noted that family influences might operate in several different ways. It remains to consider how the evidence we have re viewed bears on these. First, it was suggested that there might be both organically and psychogenically determined cases of autism. Goldfarb's studies have pointed to this possibility. However, his own most recent study based on independent assessments (Goldfarb *et al.*, 1976) failed to confirm the differentiation. Other investigators who have sought to relate family variables to the presence or absence of neurological signs have almost always found that the two were unrelated (e.g., Anthony, 1958; Gittelman & Birch, 1967; see also Chapter 17). The one exception (Gonzales *et al.*, 1977) found that the parents of autistic children with evidence of cerebral dysfunction tended to be more (rather than less) deviant in personality than the parents of autistic children without evidence of brain damage. Furthermore, studies such as our own (Cox *et al.*, 1975;

Cantwell *et al.*, 1977a,b), which have focused exclusively on higher IQ autistic children without evidence of neurological dysfunction, have shown essentially normal family functioning just as have those which have involved retarded autistic children (e.g, the Newcastle and Indiana groups). One cannot prove a negative but the evidence to date provides no good support for the differentiation into "organic" and "nonorganic" varieties of autism.

Second, it could be that the development of autism requires some psychogenic stress in addition to a biological deficit. However, as there is no good evidence that any psychogenic stress is consistently associated with autism, this hypothesis also lacks support. It might be argued that high social class and high parental IQ constitute such a psychosocial influence (although scarcely a stress). However, although the tendency for autistic children to come from such families is well documented it is also true that autistic children may be born to parents of all levels of intelligence in all social strata. None of the theories so far proposed to explain the social class association (e.g., Rimland, 1964; Moore & Sheik, 1971) appear at all plausible and all lack empirical support. The link between social class and autism does not appear to fit with a purely biological causation of autism, either, and it remains an awkward finding which still demands an explanation.

The third suggestion that the autistic child's biological deficit renders him vulnerable to ordinary family influences which are not hazardous to other children remains a possibility. So far there is no very satisfactory evidence for or against the hypothesis. In order to investigate this matter detailed studies of the social interactions of very young autistic children are needed.

The fourth possibility is that psychosocial influences are important only in the course of autism after it has developed. This review has not considered the evidence that psychosocial influences effect the later development of autistic children but this issue is considered in Chapter 20, with the conclusion that they do.

In summary, on all counts the evidence gives no support to the notion that certain familial factors are necessary causes for the syndrome of infantile autism to develop. It may be that in some children a basic cognitive defect is both a necessary and sufficient cause. In others this basic defect may be a necessary, but not sufficient cause and some other additional factors may contribute to the etiology. However, a careful review of the literature indicates a lack of hard evidence to support the idea that this necessary contributing factor is some type of familial environmental factor.

ACKNOWLEDGMENTS

This paper was supported in part by the following grants: NIMH Special Research Fellowship 1F03MH 52205-01, MH 08467-13, MCH 927, Scottish Rite Foundation.

REFERENCES

Alanen, Y. Some thoughts on schizophrenia and ego development in the light of family investigations. *Archives of General Psychiatry*, 1960, *3*, 650–656.

Alanen, Y., Arajarvi, T., & Viitamaki, R. Psychoses in childhood. *Acta Psychiatrica Scandinavica Supplement*, 1964, *174*, 5–30.

Allen, J., DeMyer, M. Norton, J., Pontius, W., & Yang, G. Intellectuality in parents of psychotic, subnormal and normal children. *Journal of Autism and Childhood Schizophrenia*, 1971, *1*. 311–326.

Anthony, J. An experimental approach to the psychopathology of childhood: Autism. *British Journal of Medical Psychology*, 1958, *31*, 211–225.

Baker, L., Cantwell, D., Rutter, M., & Bartak, L. Language and autism. In E. Ritvo (Ed.), *Autism: Diagnosis, current research and management*. New York: Spectrum Publications, 1976.

Bartak, L., Rutter, M., & Cox, A. A comparative study of infantile autism and specific developmental receptive language disorder: I. The children. *British Journal of Psychiatry*, 1975, *126*, 127–145.

Behrens, M. C., & Goldfarb, W. A study of patterns of interaction of families of schizophrenic children in residential treatment. *American Journal of Orthopsychiatry*, 1958, *28*, 300–312.

Behrens, M., Meyers, D., Goldfarb, W., Goldfarb, N., & Fieldsteel, N. The Henry Ittleson Center family interaction scales. *Genetic Psychology Monographs*, 1969, *80*, 203–295.

Bell, R. A reinterpretation of the direction of effects in studies of socialization. *Psychological Review*, 1968, *75*, 81–95.

Bell, R. Stimulus control of parent or caretaker behavior by offspring. *Developmental Psychology*, 1971, *4*, 63–72.

Bender, L., & Grugett, A. E. A study of certain epidemiological problems in a group of children with childhood schizophrenia. *American Journal of Orthopsychiatry*, 1956, *26*, 131–145.

Bene, E. A Rorschach investigation into the mothers of autistic children. *British Journal of Medical Psychology*, 1958, *38*, 226–227.

Bettelheim, B. *The empty fortress–infantile autism and the birth of the self*. New York: The Free Press, Collier-Macmillan, 1967.

Block, J. Parents of schizophrenic, neurotic, asthmatic, and congenitally ill children: A comparative study. *Archives of General Psychiatry*, 1969, *20*, 659–674.

Boatman, M. J., & Szurek, S. A clinical study of childhood schizophrenia. In D. Jackson (Ed.), *The etiology of schizophrenia*. New York: Basic Books, 1960.

Bowen, M., Dysinger, R., & Basamania, B. The role of the father in families with a schizophrenic patient. *American Journal of Psychiatry*, 1959, *115*, 1017–1020.

Brown, G., & Rutter, M. The measurement of family activities and relationships: A methodological study. *Human Relations*, 1966, *19*, 241–263.

Byassee, J., & Murrell, S. Interaction patterns in families of autistic, disturbed, and normal children. *American Journal of Orthopsychiatry*, 1975, *45*, 473–478.

Cantwell, D. P., Baker, L., & Rutter, M. Families of autistic and dysphasic children: II. Mothers' speech to the children. Submitted for publication, 1977. (a)

Cantwell, D. P., Baker, L., & Rutter, M. Families of autistic and dysphasic children. I. Family life and interaction patterns. Submitted for publication, 1977. (b)

Cooper, B. Parents of schizophrenic children compared with parents of non-psychotic emotionally disturbed and well children: A discriminant function analysis. Temple University, *Dissertation Abstracts*, 1964.

Cooper, J. E., Kendell, R. E., Burland, B. J., Sharpe, L., Copeland, J. R. M., & Simon, R. *Psychiatric diagnosis in New York and London*. Institute of Psychiatry/Maudsley Monographs No. 20. London: Oxford University Press, 1972.

Cox, A., Rutter, M., Newman, S., & Bartak, L. A comparative study of infantile autism and specific developmental receptive language disorder: II. Parental characteristics. *British Journal of Psychiatry*, 1975, *126*, 146–159.

Cox, A., Rutter, M., Yule, B., & Quinton, D. Bias resulting from missing information: Some epidemiological findings. *British Journal of Preventive and Social Medicine*, 1977, *31*, 131–136.

Creak, M. (Chairman). Schizophrenia syndrome in childhood: Progress report of a working party. *Cerebral Palsy Bulletin*, 1961, *3*, 501–504.

Creak, M., & Ini, S. Families of psychotic children. *Journal of Child Psychology and Psychiatry*, 1960, *1*, 156–175.

Cummings, S. J., Bayley, H. C., & Rie, H. E. Effects of the child's deficiency on the mother: A study of mothers of mentally retarded, chronically ill and neurotic children. *Americal Journal of Orthopsychiatry*, 1966, *36*, 595–608.

Davids, A. Intelligence in childhood schizophrenics, other emotionally disturbed children, and their mothers. *Journal of Consulting Psychology*, 1958, *22*, 159–163.

DeMyer, M., Pontius, W., Norton, J., Barton, S., Allen, J., & Steele, R. Parental practices and innate activity in autistic and brain-damaged infants. *Journal of Autism and Childhood Schizophrenia*, 1972, *2*, 49–66.

Despert, J. Schizophrenia in children. *Psychiatric Quarterly*, 1938, *12*, 366–371.

Despert, J. Some considerations relating to the genesis of autistic behavior in children. *American Journal of Orthopsychiatry*, 1951, *21*, 335–350.

Donnelly, E. M. The quantitative analysis of parent behaviour toward psychotic children and their siblings. *Genetic Psychology Monographs*, 1960, *62*, 331–376.

Douglas, J. W. B. Early hospital admissions and later disturbances of behaviour and learning. *Developmental Medicine and Child Neurology*, 1975, *17*, 456–480.

Douglas, J., Lawson, A., Cooper, J., & Cooper, E. Family interaction and the activities of young children. *Journal of Child Psychology and Psychiatry*, 1968, *19*, 157–171.

Eisenberg, L. The fathers of autistic children. *American Journal of Orthopsychiatry*, 1957, *27*, 715–724.

Ferster, C. B. Positive reinforcement and behavioral deficits of autistic children. *Child Development*, 1961, *32*, 437–456.

Ferster, C. B. The repertoire of the autistic child in relation to principles of reinforcement. In L. Gottschalk & A. H. Auerback (Eds.), *Methods of research in psychotherapy*. New York: Appleton-Century-Crofts, 1966.

Fish, B. Biologic antecedents of psychosis in children. In D. X. Freedman (Ed.), *Biology of the major psychoses*. New York: Raven Press, 1975.

Florsheim, J., & Peterfreund, O. The intelligence of parents of psychotic children. *Journal of Autism and Childhood Schizophrenia*, 1974, *4*, 61–70.

Fraknoi, J., & Ruttenberg, B. Formulation of the dynamic economic factors underlying infantile autism. *Journal of the American Academy of Child Psychiatry*, 1971, *10*, 713–738.

Fraser, C., & Roberts, N. Mother's speech to children of four different ages. *Journal of Psycholinguistic Research*, 1975, *4*, 9–16.

Gardner, J. Three aspects of childhood autism: Mother-child interactions, autonomic responsivity, and cognitive functioning. Ph.D. Thesis, University of Leicester, 1977.

Gittelman, M., & Birch, H. Childhood schizophrenia: Intellect, neurologic status, perinatal risk, prognosis, and family pathology. *Archives of General Psychiatry*, 1967, *17*, 16–25.

Goldfarb, W. *Childhood schizophrenia*. Cambridge, Mass.: Harvard University Press, 1961. (a)

Goldfarb, W. The mutual impact of mother and child in childhood schizophrenia. *American Journal of Orthopsychiatry*, 1961, *31*, 738–747. (b)

Goldfarb, W Factors in the development of schizophrenic children: An approach to subclassification. In J. Romano (Ed.), *The origins of schizophrenia*. Amsterdam: Excerpta Medica Foundation, 1967. Pp. 70–91.

Goldfarb, W. *Growth and change of schizophrenic children: A longitudinal study*. New York: Wiley, 1974.

Goldfarb, W., Goldfarb, N., & Scholl, M. The speech of mothers of schizohprenic children. *American Journal of Psychiatry*, 1966, *122*, 1220–1227. (a)

Goldfarb, W., Levy, D., & Meyers, D. The verbal encounter between the schizophrenic child and his mother. In G. Goldman & D. Shapiro (Eds.), *Developments in psychoanalysis at Columbia University*. New York: Hafner, 1966. (b)

Goldfarb, W., Levy, D., & Meyers, D. The mother speaks to her schizophrenic child: Language in childhood schizophrenia. *Psychiatry*, 1972, *35*, 217–226.

Goldfarb, W., Spitzer, R. L., & Endicott, J. A study of psychopathology of parents of psychotic children by structured interview. *Journal of Autism and Childhood Schizophrenia*, 1976, *6*, 327–338.

Goldfarb, W., Yudkovitz, E., & Goldfarb, N. Verbal symbols to designate objects: An experimental study of communication in mothers of schizophrenic children. *Journal of Autism and Childhood Schizophrenia*, 1973, *3*, 281–298.

Gonzales, S., Kolvin, I., Garside, R. F., & Leitch, I. M. Characteristics of parents of handicapped children. I. Preliminary findings. Submitted for publication, 1977.

Hemsley, R., Cantwell, D., Howlin, P., & Rutter, M. The adult-child interaction schedule. In preparation, 1977.

Herzberg, B. The families of autistic children. In M. Coleman (Ed.), *The autistic syndromes*. Amsterdam: North-Holland, 1976.

Hirsch, S. E., & Leff, J. P. *Abnormalities in parents of schizophrenics*. London: Oxford University Press, 1975.

Holroyd, J., & McArthur, D. Mental retardation and stress on the parents: A contrast between Down's syndrome and childhood autism. *American Journal of Mental Deficiency*, 1976, *80*, 431–436.

Howlin, P., Cantwell, D., Marchant, R., Berger, M., & Rutter, M. Analyzing mothers' speech to young autistic children: A methodological study. *Journal of Abnormal Child Psychology*, 1973, *1*, 317–339. (a)

Howlin, P., Marchant, R., Rutter, M., Berger, M., Hersov, L., & Yule, W. A home-based approach to the treatment of autistic children. *Journal of Autism and Childhood Schizophrenia*, 1973, *3*, 308–336. (b)

Kanner, L. Autistic disturbances of affective contact. *Nervous Child,* 1943, *2,* 217–250.
Kanner, L. Problems of nosology and psychodynamics of early infantile autism. *American Journal of Orthopsychiatry*, 1949, *19*, 412–426.
Kanner, L. To what extent is early infantile autism determined by constitutional inadequacies? *Proceedings of the Association for Research in Nervous and Mental Diseases*, 1954, *33*, 378–385.
Kasanin, J., Knight, E., & Sage, P. The parent-child relationship in schizophrenia. I. Overprotection-rejection. *Journal of Nervous and Mental Disease*, 1934, *79*, 249–263
Kaufman, I., Frank, T., Heims, L., Herrick, J., Rusir, D., & Willer, L. Treatment implications of a new classification of parents of schizophrenic children. *American Journal of Psychiatry*, 1960, *116*, 920–924.
Kaufman, I., Frank, T., Heims, L., Herrick, J., & Willer, L. Parents of schizophrenic children: Workshop, 1958: Four types of defence in mothers and fathers of schizophrenic children. *American Journal of Orthopsychiatry*, 1959, *29*, 460–472.
King, P. D. Early infantile autism: Relation to schizophrenia. *Journal of the American Academy of Child Psychiatry*, 1975, *14*, 666–682.
Klebanoff, L. Parental attitudes of mothers of schizophrenic, brain-injured and retarded and normal children. *American Journal of Orthopsychiatry*, 1959, *29*, 445–454.
Klein, D., & Pollack, M. Schizophrenic children and maternal speech facility. *American Journal of Psychiatry*, 1966, *123*, 232.
Kolvin, I., Garside, R., & Kidd, J. Studies in the childhood psychoses. IV. Parental personality and attitude and childhood psychoses. *British Journal of Psychiatry*, 1971, *118*, 403–406. (a)
Kolvin, I., Ounsted, C., Richardson, L., & Garside, R. Studies in the childhood psychoses. III. The family and social background in childhood psychoses. *British Journal of Psychiatry*, 1971, *118*, 396–402. (b)
Lawson, A., & Ingleby, J. D. Daily routines of pre-school children: Effects of age, birth order, sex and social class, and developmental correlates. *Psychological Medicine*, 1974, *4*, 339–415.
Lennard, H. L., Beaulieu, M. R., & Embrey, M. G. Interaction in families with a schizophrenic child. *Archives of General Psychiatry*, 1965, *12*, 166–183.
Lennox, C. Conceptual reasoning of parents of autistic children: A function of anxiety? B.Sc. Thesis, University of Toronto. Substudy 73.3, Clarke Institute Psychiatric Child and Adolescent Service, 1973.
Lennox, C., Callias, M., & Rutter, M. Cognitive characteristics of parents of autistic children. *Journal of Autism and Childhood Schizophrenia,* 1977, *7*, 243–261.
Lockyer, L., & Rutter, M. A five-to-fifteen-year follow-up study of infantile psychosis: IV. Patterns of cognitive ability. *British Journal of Social and Clinical Psychology*, 1970, *9*, 152–163.
Lotter, V. Epidemiology of autistic conditions in young children. II. Some characteristics of the parents and children. *Social Psychiatry*, 1967, *1*, 163–173.
Lowe, L. Families of children with early childhood schizophrenia. *Archives of General Psychiatry*, 1966, *14*, 26–30.
Meyer, R. G., & Karon, B. P. The schizophrenogenic mother concept and the T.A.T. *Psychiatry*, 1967, *30*, 173–179.
Meyers, D., & Goldfarb, W. Childhood schizophrenia: Studies of perplexity in mothers of schizophrenic children. *American Journal of Orthopsychiatry*, 1961, *31*, 551–564.
Meyers, D., & Goldfarb, W. Psychiatric appraisal of parents and siblings of schizophrenic children. *American Journal of Psychiatry*, 1962, *118*, 902–915.
Moerk, E. Changes in verbal child-mother interactions with increasing language skills of the child. *Journal of Psycholinguistic Research*, 1974, *3*, 101–116.

Moore, D. J., & Sheik, D. A. Toward a theory of early infantile autism. *Psychological Review*, 1971, *78*, 451–456.

Netley, C., Lockyer, L., & Greenbaum, G. Parental characteristics in relation to diagnosis and neurological status in childhood psychosis. *British Journal of Psychiatry*, 1975, *127*, 440–444.

Ogdon. D. P., Buss, C. L., Thomas, E. R., & Lordi, W. Parents of autistic children. *American Journal of Orthopsychiatry*, 1968, *38*, 653–658.

O'Gorman, G. *The nature of childhood autism* (2nd ed.). London: Butterworth, 1970.

Ornitz, E. M., & Ritvo, E. R. Perceptual inconstancy in early infantile autism. *Archives of General Psychiatry*, 1968, *18*, 76–98.

Osofsky, J. D., & O'Connell, E. J. Parent-child interaction: Daughters' effects upon mothers' and fathers' behaviors. *Development Psychology*, 1972, *7*, 157–168.

Pitfield, M., & Oppenheim, A. Child rearing attitudes of mothers of psychotic children. *Journal of Child Psychology and Psychiatry*, 1964, *5*, 51–57.

Putnam, M. C. Some observations on psychoses in early childhood. In G. Caplan (Ed.), *Emotional Problems of Early Childhood*. New York: Basic Books, 1955.

Quay, H., & Werry, J. (Eds.), *Psychopathological disorders of childhood*. New York: Wiley, 1972.

Quinton, D., & Rutter, M. Early hospital admissions and later disturbances of behaviour: An attempted replication of Douglas' findings. *Developmental Medicine and Child Neurology*, 1976, *18*, 447–459.

Rank, B. Adaptation of the psychoanalytic technique for the treatment of young children with atypical development. *American Journal of Orthopsychiatry*, 1949, *19*, 130–139.

Rank, B. Intensive study and treatment of preschool children who show marked personality deviations or "atypical development" and their parents. In G. Caplan (Ed.), *Emotional problems of early childhood*. New York: Basic Books, 1955.

Reiser, D. Psychosis in infancy and early childhood as manifested by children with atypical development. *New England Journal of Medicine*, 1963, *269*, 798–884.

Rice, C., Kepecs, J. C., & Yahalom, I. Differences in communicative impact between mothers of psychotic and non-psychotic children. *American Journal of Orthopsychiatry*, 1966, *36*, 529–543.

Rimland, B. *Infantile autism*. New York: Appleton-Century-Crofts, 1964.

Ritvo, E., Cantwell, D., Johnson, E., Clements, M., Benbrook, F., Slagle, S., Kelly, P., & Ritz, M. Social class factors in autism. *Journal of Autism and Childhood Schizophrenia*, 1971, *1*, 297–310.

Ruttenberg, B. A psychoanalytic understanding of infantile autism and its treatment. In D. Churchill, D. Alpern, & M. DeMyer (Eds.), *Infantile autism: Proceedings of the Indiana University Colloquium*. Springfield, Ill.: Charles C Thomas, 1971.

Rutter, M. Concepts of autism: A review of research. *Journal of Child Psychology and Psychiatry*, 1968, *9*, 1–25.

Rutter, M. The description and classification of infantile autism. In D. Churchill, G. Alpern, & M. DeMyer (Eds.), *Infantile autism: Proceedings of the Indiana University Colloquium*. Springfield, Ill.: Charles C Thomas, 1971.

Rutter, M. Childhood schizophrenia reconsidered. *Journal of Autism and Childhood Schizophrenia*, 1972, *2*, 315–337.

Rutter, M., & Brown, G. The reliability and validity of measures of family life and relationships in families containing a psychiatric patient. *Social Psychiatry*, 1966, *1*, 38–53.

Rutter, M., & Lockyer, L. A five-to-fifteen year follow-up study of infantile psychosis: I. Description of sample. *British Journal of Psychiatry*, 1967, *113*, 1169–1182.

Sarvis, M. A., & Garcia, B. Etiological variables in autism. *Sourcebook in Abnormal Psychology*. Boston: Houghton Mifflin, 1967.

Schopler, E., & Loftin, J. Thinking disorders in parents of young psychotic children. *Journal of Abnormal Psychology*, 1969, *74*, 281–287. (a)

Schopler, E., & Loftin, J. Thought disorders in parents of psychotic children. *Archives of General Psychiatry*, 1969, *20*, 174–181. (b)

Siegel, G. M. Adult verbal behavior with retarded children labeled as "high" or "low" in verbal ability. *American Journal of Mental Deficiency*, 1963, *68*, 417–424.

Siegel, G. M., & Harkins, J. P. Verbal behavior of adults in two conditions with institutionalized retarded children. *Journal of Speech and Hearing Disorders Monographs*, Supplement, 1963, 10.

Singer, M., & Wynne, L. Differentiating characteristics of parents of childhood neurotics and young adult schizophrenics. *American Journal of Psychiatry*, 1963, *120*, 234–243.

Snow, C. E. Mothers' speech to children learning language. *Child Development*, 1972, *43*, 549–565.

Spradlin, J. E., & Rosenberg, S. Complexity of adult verbal behavior in a dyadic situation with retarded children. *Journal of Abnormal Psychology*, 1964, *58*, 694–698.

Szurek, S. Psychotic episodes and psychotic maldevelopment. *American Journal of Orthopsychiatry*, 1956, *26*, 519–543.

Szurek, S., & Berlin, I. (Eds.), *Clinical studies in childhood psychoses*. New York: Brunner/Mazel, 1973.

Tinbergen, E. A., & Tinbergen, N. *Early childhood autism: An ethological approach.* Berlin: Paul Parey, 1972.

Tizard, B., & Rees, J. A comparison of the effects of adoption, restoration to the natural mother, and continued institutionalization on the cognitive development of four-year-old children. In A. M. Clarke and A. D. B. Clarke (Eds.), *Early experience: Myth and evidence*. London: Open Books, 1976.

Treffert, D. Epidemiology of infantile autism. *Archives of General Psychiatry*, 1970, *22*, 431–438.

Tustin, F. *Autism and childhood psychosis*. London: Hogarth Press, 1972.

Ward, A. J. Early infantile autism: Diagnosis, etiology and treatment. *Psychological Bulletin*, 1970, *73*, 350–362.

Williams, S., & Harper, J. A study of aetiological factors at critical periods of development in autistic children. *Australian and New Zealand Journal of Psychiatry*, 1973, *7*, 163–168.

Wing, J. Review of Bettelheim: "The Empty Fortress." *British Journal of Psychiatry*, 1968, *114*, 788–791.

Wolkind, S. The components of "affectionless psychopathy" in institutionalized children. *Journal of Child Psychology and Psychiatry*, 1974, *15*, 215–220.

19

Limits of Methodological Differences in Family Studies

ERIC SCHOPLER

Kanner's (1943) lucid clinical description of the autistic syndrome enabled clinicians from many different regions to identify similar children. Although such children were identified with increasing regularity, the origins of their strange behavior continued to be mysterious and puzzling. Among the first and most extensive efforts to clarify these unknown causes have been the research studies seeking links between the peculiar characteristics of the autistic child and his parents and their family life. A large proportion of these publications have claimed direct and indirect evidence that these parents show psychopathology of personality, communication styles, child-rearing proclivities, and family interactions. These pathological factors were widely considered as primary explanations for the child's autistic development. Yet another group of studies well illustrated by McAdoo and DeMyer (Chapter 17) and comprehensively reviewed by Cantwell *et al.* (Chapter 18) do not find evidence for pathogenic traits in parents of autistic children. These two opposing views on the parents of autistic children have been supported by large numbers of publications. (Over 100 were cited in Cantwell *et al.*)

The major explanation offered in the preceding two chapters for these contradictory findings on the same research issue revolves around methodological flaws and differences. This is best demonstrated in the parental characteristics reported from the Goldfarb studies (Goldfarb, 1961; Meyers & Goldfarb, 1962). He divided his group of children into

ERIC SCHOPLER · Department of Psychiatry, Division of Health Affairs, University of North Carolina School of Medicine, Chapel Hill, North Carolina.

organic and nonorganic subgroups and suggested that parents of the "organic" children were similar to parents of normal children, while the parents of the nonorganic group showed the deviant characteristics of personality and communication that many have associated with the child's autistic development. McAdoo and DeMyer (Chapter 17) have reported a carefully detailed replication of the Goldfarb studies. They differentiated their autistic children by the same criteria for organicity used by Goldfarb (1961), but the parental characteristics they found went in the opposite direction of those he reported. Unlike Meyers and Goldfarb (1962), McAdoo and DeMyer did not find a high incidence of psychosis in mothers of nonorganic children, nor did they find a significant number of such mothers in need of psychiatric help.

The striking differences between the findings of the Goldfarb studies and the McAdoo and DeMyer replication may be explained by some of the improvements in research design and methodology used in the replication. First, where the Goldfarb studies were confined to control groups of normal and physically handicapped subjects, McAdoo and DeMyer used more directly relevant controls. They included a group of adult outpatients, subjects who sought psychiatric help for themselves. They also included a control group of child-guidance parents whose children had milder forms of disturbance than did the nonorganic autistic subjects. Second, where the Goldfarb studies used clinical ratings and other measures developed for the particular study, McAdoo and DeMyer used the MMPI, a personality inventory which has been standardized more extensively than any other psychometric instrument. Third, unlike the McAdoo and DeMyer study the diagnostic grouping used in the Goldfarb research does not appear to conform to the broad criteria of autism used by most investigators. Goldfarb's selection bias is toward the higher level, verbal subjects. Thus it would appear that any of these methodological differences alone could account for the sharp differences in results between the Meyers and Goldfarb (1962) findings and the McAdoo and DeMyer replication (Chapter 17).

It is not likely that inconsistent use of diagnostic grouping alone accounts for the contradictory findings on parental characteristics. The early Goldfarb study (1961) used childhood schizophrenia synonymously with autism, a distinction now increasingly recognized and accepted (Rutter, Chapter 1). If the condition of the Goldfarb subjects was more similar to childhood schizophrenia than to autism, then his findings can be compared with the research on parents of schizophrenics. However, even in this group of studies the weight of empirical evidence goes against the theory linking parental psychopathology to the cause of the child's disorder.

Hirsch and Leff (1975) have compiled a scholarly review of almost

200 studies dealing with family factors and the transmission of schizophrenia. Through careful comparison of studies and analysis of methodological shortcomings, they demonstrate how the findings and interpretations of many past studies do not hold up. Their assessment of the evidence includes the following conclusions:

1. When mothers of schizophrenics are found to be overprotective, this is probably a response to an abnormal child rather than the reverse.

2. The characterization of mothers of schizophrenics as cold and rejecting is not sustained by the evidence.

3. While disharmony among parents of schizophrenics has been found there is no evidence linking parental deprivation and schizophrenia, and there is no evidence for unusual peculiarities in parental personalities.

4. There is little evidence that parents of schizophrenics communicate in peculiar ways and considerable evidence that preschizophrenic children affect their parents adversely.

The Hirsch and Leff monograph offers a most comprehensive review of the research evidence we have available for understanding characteristics of families with schizophrenics. It offers a useful supplement to the thoughtful, analytic review provided by Cantwell *et al.* (Chapter 18). There may be some overlap in the studies covered by the two reviews. Cantwell *et al.* include more recent research and focus on studies of autistic children and their parents, a group characterized by a younger age range and different diagnostic characteristics than the bulk of subjects in the Hirsch and Leff review. In spite of differences in the studies included, there are striking similarities between the conclusions of these two different reviews. Both reviews find no significant empirical evidence for parental pathology as a causal explanation for the child's autistic or schizophrenic development. Both reviews find that parental communication patterns are not remarkable as had been reported in so many research studies.

The questions arise: Why do these two reviews confirm each other's conclusions? Why do they both fail to substantiate the large body of psychogenic literature covered in their surveys? Methodological shortcomings are emphasized in both reviews to account for the reversal. However, improvement in methodology cannot be the whole story. After all, methodological problems are not eliminated from the studies concurring in finding no evidence for the psychogenic formulation.

For example, somewhat different diagnostic criteria were used in the McAdoo and the Cantwell studies. McAdoo included three symptoms: social withdrawal, language disorder, and nonfunctional object use. Cantwell, on the other hand, also used failure to relate and faulty language development, but instead of nonfunctional object use, he added early onset of before 30 months, and ritualistic and compulsive phenomena.

There are also significant differences in the selection of control groups, instruments used, and so on.

Because these methodological differences do not produce striking differences in the characteristics of the parents studied, it is questionable that methodological differences alone can account for the many opposite findings reported in the past. After all, sound research design and methodology only assure us of an empirically based, rational process by which certain questions can be answered. We must remember that the answer is only a probability statement about the relationship between the hypothesis and the results, under specifically defined circumstances.

I am suggesting that the research methods by which a question is answered may have little bearing on the process by which the investigator selects his research question. This is particularly true when the research has direct consequences for intervention. For example, some years ago there was much excitement and research in the treatment of schizo-phrenia with psychosurgery. The Nobel Prize was even awarded to the first perpetrator of this crude intervention, now almost totally disre-garded. We have had surges in drug studies for controlling hyperactivity, operant conditioning, and megavitamins, to name only a few. Most of these interventions were based on valid and significant research issues to begin with. However, the application of these interventions was carried out well beyond the scope of the research data. The treatment implica-tions were pushed beyond the limits of reasonable research design. I am referring to these phenomena as fads to the extent that they were applied beyond the intent or data of the related research.

Many other research-related fads have come and gone without adequate reason or explanation. The many publications linking autism and schizophrenia to parental pathology may have, at least partly, been the product of such a fad during a particular time period. From a so-ciology of knowledge perspective it can be observed that following the period of World War II, psychoanalytic theory became increasingly popularized, especially in the United States. It was no longer confined to the training of psychoanalysts, but permeated other professions such as the practice of general psychiatry, social work, nursing, and special edu-cation. It even influenced child-rearing practices, especially of upper middle-class parents. Since much of psychoanalytic theory was used for explaining both normal and deviant personality development as a function of early family experience, it was an automatic extension to explain both autistic and schizophrenic behavior in terms of family pathology.

If this fad hypothesis is true we may expect that the studies looking for connections between parental pathology and psychosis in offspring occurred about the same time psychoanalytic theory was popularized and overused. This time period should include the several decades following

World War II, but exclude the last decade when the empirical basis for research in autism made an increasing contribution.

Fortunately the data for testing this fad hypothesis are available. The excellent review in Cantwell *et al.* (Chapter 18) offers a most comprehensive survey of the studies published on this topic. Fifty-eight publications are cited in which parental pathology was implicated with childhood psychosis or autism. Thirty-nine or 67% of these reports were published before 1965. Forty-two publications were cited in which parents were found to be without significant pathology, but showing the effects of having a deviant child. Only six of these studies were made before 1965, and 36 or 86% were published after 1965. When these numbers were compared statistically, the X^2 analysis was $X^2 = 27.60, p < .001$.

This distribution clearly supports the hypothesis that a large number of publications concerned with parental pathology, published during a relatively brief time period, were at least partly the product of popularized psychoanalytic theories and beliefs. Some may argue that the conclusions drawn from studies finding no causal connection between parental behavior or attitudes and the child's abnormalities are the product of a fad, since the studies supporting these conclusions are also circumscribed by a particular time period. Such an assertion would be based on a failure to recognize the difference between theoretical and clinically related assertions on the one hand, and clinical and research data on the other. Cantwell *et al.* trace carefully and clearly the empirical and scientific bases for their conclusions. This is not to say that the conclusions they draw from the reviewed data could not tomorrow be extended to the horizonless scope of a fad. However, for the time being our understanding of family factors is anchored in meaningful data. Cantwell *et al.* have supplied us with a vital supplement to the knowledge summarized by Hirsch and Leff (1975).

Perhaps we have witnessed the wrap-up of a tortured research trail. More likely the issue of family factors is linked to the classification of the disorder. As new etiological factors are discovered, new specific subgroups of autistic children may be named. These subgroups may also be linked with family factors we have not yet thought of today.

REFERENCES

Goldfarb, W. *Childhood schizophrenia.* Cambridge, Mass.: Harvard University Press, 1961.

Hirsch, S. R., & Leff, J. P. *Abnormalities in parents of schizophrenics.* London: Oxford University Press, Ely House, 1975.

Kanner, L. Autistic disturbances of affective contact. *The Nervous Child*, 1943, *3*, 217–250.

Meyers, D. I., & Goldfarb, W. Psychiatric appraisal of parents and siblings of schizophrenic children. *American Journal of Psychiatry*, 1962, *118*, 902–915.

20

Psychotherapeutic Work with Parents of Psychotic Children

IRVING N. BERLIN

There has been a biphasic pattern to psychotherapeutic work with parents of children in treatment for psychiatric disorders. The first occurred after World War I. Under the influence of the pioneers in American child psychiatry, children and parents were worked with. There followed a period prior to and following World War II which was influenced by psychoanalysis when the child alone was treated. Since the late 1940s, there has again been an emphasis on therapeutic work with parents as well as children (Witmer, 1940; Krug *et al.*, 1964).

A BRIEF HISTORY

When Leo Kanner, in the 1930s, delineated the characteristics of autistic children, he described their parents as "refrigerator parents" (Kanner, 1943). He studied children primarily of parents in various professions, who, on evaluation, appeared very cold, withdrawn, isolated, and had difficulty in forming relationships with adults and with children. While he believed that autism was probably primarily of organic etiology, he believed that the environmental factors might be important in the symptomatic manifestations. However, he did not describe treatment processes for the child. Healy and Bronner (1952) provided data from

IRVING N. BERLIN · Department of Psychiatry, School of Medicine, University of California, Davis, California.

their work with delinquents which emphasized the importance of environmental influences on behavior. In the late 1930s and in the late 1940s at the Philadelphia Child Guidance Clinic, Allen (1942) was very much concerned with altering the experience of the child with adults, especially the parent, through new relationships. This corrective experience was intended to help the child feel differently about himself. While concentration was primarily on work with the child by child psychiatrists, there was some work with parents by social workers. The primary purpose was to help the parents understand the treatment process and not impede it. As Dawley (1939) described, despite every effort at focusing on the child, a good many parents engaged themselves in psychotherapeutic work because of their own needs which were not met by the focus on the child's problems. In the 1940s significant work was done by Szurek *et al.* (1942) with parents and children which helped influence current intensive work with parents. Since parents are critical factors in the child's environment, Szurek *et al.* sought to test the hypothesis that parents' neurotic conflicts, which contribute to the developmental disturbances of the child, could be altered by simultaneous treatment of child and parent. Initial efforts centered around psychotherapeutic work with parents and children with neurotic and antisocial problems.

PSYCHOTHERAPY OF PSYCHOTIC CHILDREN

Szurek and coworkers in the late 1940s began to test the previously described hypothesis in psychotherapy with psychotic and autistic children and their parents (Szurek *et al.*, 1942). The influx of psychoanalysts into the United States during Hitler's regime in the 1930s and during the Second World War shifted the priorities of the child-guidance clinics which were working with parents to a focus on the child alone. In many centers, in Boston with Putnam (1948) and Rank (1949) and in New York with Mahler (1952) and various other psychoanalysts, individual work with psychotic children was emphasized. However, the conflictful relationship of psychotic children with parents needed to be understood, especially the symbiosis that occurred between some parents and children (Szurek & Berlin, 1973).

Goldfarb, in his work at the Ittleson Center, began to delineate several kinds of psychotic children—those who have primarily an organic disorder and those whom he considered to be primarily of psychological etiology (Goldfarb, 1961). He described the fact that parents of these two kinds of children seemed to be rather different. The parents of the children with primarily organic disorders seemed to be very much less disturbed and their disturbances appeared to be reactive to the severe disturbance in the child and its impact upon the family environment. Whereas, with the parents of those children who had no signs of organicity, it was felt that the parents were much more psychologically disturbed and much more involved with the children in their psychotic disturbance.

DEVELOPMENTAL ISSUES IN AUTISM

Let me turn now, somewhat more specifically, to the developmental issues which may be related to autism. First, let me briefly review the studies of Krech *et al.* (1966) at the University of California at Berkeley which reveal that in a variety of animals, including primates, stimulus deprivation, especially environmental deprivation, of interaction with the mothering figure or stimulation from agemates leads to biochemical changes in the neurons with eventual decrease in the size and the number of brain cells. Thus, certainly in animals, environmental factors can alter brain biochemistry and finally neurological structure early in the life of primates. There appears to be little reversibility in neural structures and if extensive enough, as in the work of Cravioto *et al.* (1966) on malnutrition in infants, such deprivation may result in permanent impairment.

In quite another area, Lipsett and coworkers (Lipsett, 1967) at Brown describe the great amount of learning possible in the first few days of life, indicating the great potential for reactivity of infants in the neonatal period. Lipsett (1967) reveals how infants, through conditioning, can learn to rock their cribs to turn on lights or sound shows. The work of Bowlby (1969) and many others on attachment and bonding points to critical periods in the first weeks and months of life during which failures in bonding may lead to behaviors in infants which appear autistic, i.e., isolated, noninteractive behavior between the dyad of mother and child.

More to our concern with autism is the work of Sander *et al.* (1970). Some of it was carried out in England with neonates. All newborns in a large maternity hospital born within 24 hours were divided into two cohort groups of about ten babies each. All had normal Apgars and no neurologic abnormalities on examination at birth.

The first cohort was put on a strict bottle-feeding and changing schedule and no other handling allowed. The second cohort was fed, changed, and handled responsively by nurses chosen for their sensitivity to infants.

In seven days, the first cohort of strictly handled infants became very irritable, colicky, and there was no estabishment of regular eating or sleeping patterns, despite the strict schedule. By ten days, these infants looked, as noted on films, very much like brain-damaged babies, very irritable, unstable, colicky, etc. On neurologic examination, they showed early neurologic soft signs. The second cohort had begun to stabilize in seven days in their sleep and feeding rhythms; they were not irritable or colicky. At ten days, reversal of treatment with each cohort occurred with reversal of behavior.

The work is now being replicated in several laboratories. Implications are that early adverse experiences may produce serious behavioral changes and possible biochemical and, perhaps, then structural changes in infants' brains. The behaviors of these infants are similar to retrospective description by mothers of the very early infancy of some autistic children. Other mothers describe the very good, placid, not cuddly, passive infant who apparently seeks little interpersonal interaction.

IMPLICATIONS FROM EARLY INTERVENTION

Early intervention programs with high-risk infants of young unwed poverty mothers or with premature infants have been widely described. The experiences, as are noted in a wide variety of early stimulation programs, reported in a recent U.S. NIMH publication (Segal *et al.*, 1975), indicate both seriously disturbed infants and mothers can change greatly with help.

THERAPEUTIC EFFECT ON PARENTS

There can be no question of the therapeutic effect on mothers and fathers who learn to appropriately stimulate their infants. They can observe their infant's modified behavior in response to their efforts. The implications of the many early intervention studies for infantile autism I submit are major ones.

The identification shortly after birth of the maladaptive mother-infant pair may be important. Evidence of the mother's inability to accept or show pleasure in relation to the infant or the infant's apathy and lack of responsiveness to the mother or fussiness and overactive behavior which make it difficult for the mother to feel related to the infant are indications for help with parenting. Research findings (Bowlby, 1969) reveal how important these early interactions are for bonding to occur. His theory, as amplified by Szurek (1973), is that failures in bonding may be an etiologic factor in either autism or infantile psychosis. Altering the bonding (parenting) process presumes to reverse some elements of the autistic process.

The antero-prospective investigations with premature infants and with infants at risk due to mother-child alienation on the effect of early stimulation vs. no intervention may produce new data about the relevance of early parent-infant relationship to both infantile autism and infantile psychoses. After parents feel much more competent when their intervention does alter the infant's behavior and subsequent development, their own personalities may be altered also.

We now have some tools for early infant assessment like those of Bayley (1969) and Brazelton (1973). We also have, at least in the United States, access to babies and parents in early intervention studies and in controlled environments. Thus, we can now begin prospective controlled studies. It is now possible to accumulate another important set of data from the earliest period of life to be added to the beautiful experimental data presented in other chapters in this book. I would encourage more investigators to take seriously the experimental data already available to them on the earliest days and months of life and to use these data as a beginning of our understanding of biological and interpersonal factors which could add to our understanding of the early infantile phenomena relevant for early infantile autism.

The issues of early identification studies and early intervention and the help to parents are extremely important. No matter what the actual causes of infantile autism are, the child lives within the family environment. In an effort to help the child overcome developmental problems of whatever etiology, we need to work with the family unit. Since behavior disorders develop in the context of the family, this requires that along with the child the parents must be therapeutically involved.

ISSUES OF PARENTAL BLAME AS OBSTACLES TO COLLABORATIVE WORK WITH PARENTS

Most parents blame themselves for the problems of their psychotic children. In large part, this may be due to the emphasis on parental fault in the professional and popular literature. Distortions arise from the attitudes of some professionals. This is especially true when educators, child psychiatrists, and especially the milieu team feel terribly frustrated in their work with very severely disturbed children who do not progress very rapidly. Because they feel inept, their own professional competence is threatened. They, therefore, need to blame someone for their failure. The parent becomes a very convenient object of blame. The way in which the developmental history of the child is taken often gives parents the sense that they are believed to be directly at fault for the child's difficulties. In some instances, they discern a feeling that they have, with malice aforethought, somehow deliberately contributed to the difficulty of their child. There is no such evidence.

Where psychological issues seem to be predominant, it is usually the intercurrent events of the parents' lives and their own past experiences with their own nonnurturant parents which make it extremely difficult for them to nurture a small infant or child (Szurek & Berlin, 1956). Most children with very severe ego deficits as a result of neurointegrative disturbances or severe deprivation are extremely difficult for parents to cuddle, to feed, and to nurture. Their continued, constant restlessness, their constant crying, and the inability of the parents to actually satisfy the child make it difficult for the parent to feel rewarded by his or her attention to the child.

SOME INDICATIONS FOR THERAPEUTIC WORK WITH PARENTS

There are, however, other instances in which the self-blame of the parent is clearly the result of a previous set of experiences and a lifetime of masochistic behavior, beginning with their family or origin and evidenced with friends and their spouses. In these circumstances, development of the parents' sense of inadequacy and lack of capacity to nurture is part of the picture of incapacity and inability to feel effective in any sphere of living. It is very clear in the developmental history of some children that intercurrent life events, such as the death of a parent's parent or the death of a spouse as well as crisis situations may affect the caretaking parent. The occurrence of severe depression, especially of the mother, may deprive the infant of the required opportunities for nurturance and attachment. The infant then develops massive withdrawal and behavior which may be subsequently described as autistic.

COUNSELING PARENTS

Parents with troubles out of their own childhood or from very recent traumatic experience which interfere with parenting often do well with individual psychotherapeutic work. It has been our experience that as they deal with their conflictful feelings and are in less turmoil, they relate better to their children and spouse. When we have helped these parents through intensive psychotherapeutic work, the improvements noted in their autistic or, in some instances, psychotic children have been marked. In several instances, the changes in both behavior and development in interpersonal, cognitive, or social areas have been impressive (Berlin, 1973a).

Many parents of psychotic children, especially autistic children, primarily require help in understanding their child's problems, their origins, and the neurobiological issues which may be important factors. They also require understanding of how the environment contributes to and aggravates the behavioral manifestations of the neurobiological disorder, so that an attempt can be made to correct the child's very difficult behavior through planned interventions at home.

The process of counseling with the parents is often helped by observations of their own child with a variety of child-caring professionals. Home visits by trained personnel to help parents with an especially difficult behavior of the child, often around eating or going to bed and sleeping, are frequently very helpful. The parents' modeling after effective child-caring professionals and their learning how to interact effectively with their child are often major factors in reducing the child's behavioral disturbances.

Parents also may require some help in understanding how the child's behavior disturbances have affected the family unit and created certain problems in living for the entire family. Thus, both parents can profit from some brief family therapy aimed at helping them interact differently in the family situation with the psychotic child and to facilitate more adaptive behavior of other children. This relieves their feeling of being stuck not only with the very sick child but with the very difficult family situation that results from having an autistic or psychotic child. Parents can be helped to learn that being attentive to the needs of the other children in the family helps the autistic child experience a more normal place for himself in the family.

PERSONAL PSYCHOTHERAPY FOR PARENTS OF DISTURBED CHILDREN

Some parents require personal psychotherapy. In this instance, the parents' severe neurotic behavior interferes with any efforts to alter their behaviors with their child. Their severely sadomasochistic behavior and their inability to assert themselves interfere with their relationship to the child patient and their relationships with other members of their family. This greatly affects the behavior of the disturbed child. In these instances, we usually encounter severe marital problems and severe neurotic handicaps in ordinary living.

VIGNETTE

Regie J. came to us as a three-and-a-half-year-old child referred to the clinic by a psychologist who noted his frozen affect, his ritualistic play with blocks, his lack of speech, and his inability to relate to adults and to children. His history revealed relatively normal physical development until about age two. With the birth of a younger sibling and the simultaneous severe depression of mother around the death of her own mother, there came a series of housekeepers who focused their attention on the very cute baby girl and left Regie to himself. He said a few words by age two, but it was noted in the early pediatric records that he was extremely shy and had difficulty in relating both to his parents and to other adults. Subsequent developmental history indicated that Mrs. J. had suffered a moderate depression during and for about six months after her pregnancy with Regie when her husband was on overseas military duty. Mrs. J., who felt totally inadequate in most of her dealings with other people, was forced to take care of her child alone without the support of her husband, who had previously provided most of the direction for her living.

Shortly after the birth of his sibling when he was two, Regie stopped saying "Mommy," "bye-bye," and "go." He became extremely violent and aggressive toward his sibling, his mother, and other children and became destructive of furniture. He was, therefore, locked in his room for many hours during each day because neither mother nor the housekeeper could handle him.

Regie's father was an extremely effective businessman and a self-made man. He was contemptuous of his wife, whose life he managed. When we first saw Regie's mother, she could not drive a car, was unable to manage any household affairs, and felt totally helpless and dependent upon her husband. She was very frightened of Regie. The psychotherapeutic work with the parents was aimed at resolving some of their own problems which interfered with their closeness and capacity to interact and support each other in various aspects of their living together.

During the course of the two and a half years of treatment with Regie and his parents, Mrs. J. was able to resolve many of her conflicts about her own sense of inadequacy resulting from her experiences in her own upbringing. She was, as the youngest daughter, dealt with as a rather helpless baby by the rest of her family. She was considered pretty but not effective. As Mrs. J., through treatment, asserted herself and took on more of the family duties, learned to drive a car, and actually became involved in business affairs with her husband, their marital relationship changed. Mr. J., who had always, as an elder son, taken care of both his widowed mother and his younger siblings, was also able to feel less threatened by his wife's increased adequacy and capacity. As he could let himself be taken care of and babied, his severe psychosomatic symptoms, primarily a bleeding peptic ulcer and tension headaches with some episodes of hypertension, were greatly relieved. Simultaneously, the parents were better able to handle their child. They learned from observations of the milieu staff working with their child how they could deal more effectively with him at home. During the treatment's latter phases, Regie was able to develop speech that was near normal for his age. He began to relate relatively well to a few adults and tentatively to a number of children. His violent destructiveness ceased with the first year of his hospital treatment as he became much more playful and involved in learning experiences. We have described Regie in greater detail in our volume on child psychosis (Berlin, 1973b).

DISCUSSION

Parents, as they become more aware of the variety of etiological factors involved in psychosis of childhood, feel less blame and can involve themselves more fully in a treatment program which is aimed at reducing the developmental disorder of the child, as well as the disturbed behavior.

One of the most important factors in the therapist's capacity to help the parents is his increased awareness of the extraordinary problems that face families living with a psychotic child. Nothing short of living with such a child in an inpatient setting and taking care of him produces an awareness of the enormous wear and tear both physically and psychologically that occurs and its reciprocal effect on the parent-child relationship. In a program which, as in the case of Regie, involves the parents in learning to work with their child in the milieu setting, the parents feel much more adequate in dealing with their child.

In some instances psychotherapeutic work with the parents aimed at reducing the conflicts which interfere with the parents' capacity to relate with their child and spouse may alter the autistic behavior of the child.

It is evident that no matter what the etiological factors, a collaborative effort with parents is an essential part of undertaking the treatment of the autistic child.

REFERENCES

Allen, H. *Psychotherapy with children*. New York: Norton, 1942.

Bayley, N. *Manual for the Bayley Scales of Infant Development*. New York: Psychological Corporation, 1969.

Berlin, I. N. Intrapersonal, interpersonal and impersonal factors in the genesis of childhood schizophrenia. In S. A. Szurek & I. N. Berlin (Eds.), *Clinical studies in childhood psychoses*. New York: Brunner/Mazel, Inc., 1973. Pp. 551–593. (a)

Berlin, I. N. Simultaneous psychotherapy with a psychotic child and both parents. The reversal of mutism. In S. A. Szurek & In. N. Berlin (Eds.), *Clinical studies in childhood psychoses*. New York: Brunner/Mazel, Inc., 1973. Pp. 626–653. (b)

Bowlby, J. Attachment and loss. In *Attachment* (Vol. I). New York: Basic Books, 1969.

Brazelton, T. B. *Neonatal Behavioral Assessment Scale*. Philadelphia: J. B. Lippincott Co., 1973.

Cravioto, J., DeLicardie, E. R., & Birch, H. G. Nutrition, growth and neurointegrative development: An experimental and ecologic study. Supplement to *Pediatrics*, 1966, *38* (2, part II).

Dawley, A. Interrelated movements of parent and child in therapy with children. *American Journal of Orthopsychiatry*, 1939, 9, 748–754.

Goldfarb, W. *Childhood schizophrenia*. Cambridge, Mass.: Harvard University Press, 1961.

Healy, W., & Bronner, A. F. *New light on delinquency and its treatment*. New Haven: Yale University Press, 1952.

Kanner, L. Autistic disturbances of affective contact. *Nervous Child*, 1943, 2, 217–250.

Krech, D., Rosenzweig, M. R., & Bennett, E. L. Environmental impoverishment, social isolation and changes in brain chemistry and anatomy. *Physiology and Behavior*, 1966, 1, 99–104.

Krug, O., Gardner, G. E., Hirschberg, J. C., Lourie, R. S., Rexford, E. N., & Robinson, J. F. (Eds.), *Career training in child psychiatry*. Report of the Conference on Training in Child Psychiatry, Washington, D.C., January, 1963. Washington, D.C.: American Psychiatric Association, 1964. Pp. 108–109.

Lipsett, L. P. Learning in the human infant. In H. W. Stevenson, E. H. Hess, & H. L. Rheingold (Eds.), *Early behavior: Comparative and developmental approaches*. New York: Wiley, 1967. Pp. 225–247.

Mahler, M. On childhood psychosis and schizophrenia: Autistic and symbiotic infantile psychosis. In R. S. Eissler, A. Freud, H. Hartmann, & K. Kris (Eds.), *Psychoanalytic study of the child* (Vol. 7). New York: International University Press, 1952. Pp. 286–305.

Putnam, M. Case study of an atypical two and a half year old. *American Journal of Orthopsychiatry*, 1948, 18, 1–30.

Rank, B. Adaptation of the psychoanalytic techniques for the treatment of young children with atypical development. *American Journal of Orthopsychiatry*, 1949, 19, 130–139.

Sander, L. W., Stechler, G., Burns, P., & Julia, H. Early mother-infant interaction and 24-hour patterns of activity and sleep. *Journal of the American Academy of Child Psychiatry*, 1970, 9, 103–123.

Segal, J., Boomer, D., & Bouthilet, L. (Eds.). Research in the service of mental health, Report of the Research Task Force of the National Institute of Mental Health, DHEW Pub. No. (ADM 75-236), 1975.

Szurek, S. A. Attachment and psychotic detachment. In S. A. Szurek & I. N. Berlin (Eds.), *Clinical studies in childhood psychoses*. New York: Brunner/Mazel, Inc., 1973. Pp. 191–277.

Szurek, S. A., & Berlin, I. N. Elements of psychotherapeutics with the schizophrenic child and his parents. *Psychiatry*, 1956, 19, 1–9

Szurek, S. A., & Berlin, I. N. (Eds.). *Clinical studies in childhood psychoses*. New York: Brunner/Mazel, Inc., 1973.

Szurek, S. A., Johnson, A., & Falstein, E. Collaborative psychiatric therapy of parent-child problems. *American Journal of Orthopsychiatry*, 1942, 12, 511–516.

Witmer, H. L. *Psychiatric clinics for children*. New York: Commonwealth Fund, 1940.

21

Play, Symbols, and the Development of Language

AUSTIN M. DesLAURIERS

Some time ago, Kanner (1971) suggested that scientists and professionals concerned with the care and treatment of the autistic child would do well to focus less on the quality or novelty of their therapeutic "approaches" and concentrate more on giving a clear account of their understanding of this behavioral syndrome of early infancy. The implications of Kanner's suggestion would seem cogently simple: (a) there should be a rational (and optimally a scientific) base to the therapeutic "approach" to the autistic child's disabilities; and (b) the therapeutic rationale should fit in comprehension and in extension the totality of the syndrome of early infantile autism.

Most of the current therapeutic "approaches" used in dealing with autistic children do not meet these stringent requirements. In part, this results from the confusion that has developed in the understanding of the nature of this syndrome, the temptation always existing in the scientist and professional to reduce a complex situation not easily amenable to quick solutions to a more simple and more familiar one. In part also the loose therapeutic rationale in many "approaches" stems from parental, social, or educational pressures placed on therapists to do something about the autistic child's problems, whether these problems are understood or not.

AUSTIN M. DesLAURIERS · Center for Autistic and Schizophrenic Children, The Devereux Foundation, Devon, Pennsylvania.

REVIEW OF TREATMENT APPROACHES

On the assumption that there may exist an organic (biologically, neurologically, or biochemically envisaged) substratum to the behavioral manifestations of infantile autism, many current therapeutic approaches have relied on biochemicals to deal with these manifestations. In this respect, Coleman's (Chapter 12) review of biochemical studies of autistic infants appears to have raised more questions than answers. The questionable claims of megavitamin therapy (Hawkins & Pauling, 1971; Rimland, 1971) have little rationale to support them.

Bettelheim's (1967) "approach" was consistent with the "autistic anlage" of a child whose mother not only did not want this child but entertained death wishes toward it. The question here remained whether the children Bettelheim talked about were autistic in the Kannerian sense or whether they were schizophrenic, angry, confused, and disillusioned about life. Parents of autistic infants felt hurt and unjustly victimized by Bettelheim's indictment; they rejected him. His psychoanalytic speculations may have had some plausibility when applied to children who, having developed normally for the first three or four years of early infancy, showed from then on insidious evidence of regressive, delusional, and psychotic withdrawal from human contacts; the symptoms of early infantile autism appear much earlier in life and reflect developmental arrest with affective isolation and detachment rather than regression and withdrawal. Bettelheim's therapeutic "approach" has much to commend it, since it rests, within the psychoanalytic frame of reference, on a rational basis; but it seems to be addressed to the wrong population of children. One does not attack an "empty fortress"; one leaves it alone or enters it for historical or archeological purposes.

There is no pretense at therapeutic rationale in the behavior modification "approach" to the problems of the autistic infant. Here the socially unacceptable behaviors of the autistic child become the only currency to handle, regardless of the origin, the meaning or the complexity of the symptoms. In the United States, this "approach" has become extremely fashionable on two counts. It has, first, all the external accouterment of a scientifically designed method of treatment, with behavioral analysis, careful measurement of behaviors to be maintained or to be extinguished, implementation of specific reinforcement or avoidance devices, and graphic representation of the behavioral changes occurring under defined conditions; in the second place, the method is simple, untrammeled as it is with any attempt at understanding the nature of infantile autism. Considering the state of existing understanding of the nature of infantile autism, how could anyone ask for anything more? If pigeons, dogs, and monkeys can be taught to behave under rigidly controlled human conditions, why not the autistic child? The practical merits of the behavioristic "approach" to the social problems of the autistic child would seem incontrovertible, especially if one listens, without a critical ear, to its glamorous claims of success. Lovaas (Chapter 25) who has most ardently championed this approach admits to its tragic limitations and suggests that the long-range value of the approach may depend either on a complete reengineering of the human environment or on demanding that the parents of autistic children, like the souls in Dante's Inferno, give up all personal wishes and desires, and devote their entire life to teaching social conformity to their autistic child!

In England, the "approach" to the behavioral difficulties presented by the autistic child has been mostly of a formal educational type, with a good deal of common sense and genuine human concern added (Wing, 1976). Nourished by and derived from research studies of an epidemiological (Lotter, 1974) and experimental (Hermelin & O'Connor, 1970) nature, this approach emphasized what was viewed as a central cognitive defect in the autistic child's development (Rutter, 1974); it therefore led to educational procedures which in many ways paralleled those utilized with mentally defectives or with children suffering from a so-called learning disability. In a special way, stress was placed on the language deficit and also on cognitive impairments, but no specific techniques of remediation were derived from this plausible view. Furthermore, no attempt was made to clarify the processes involved in language receptivity as distinct from those of language expression, as if the same cognitive functions were at work, in a similar way, in both. Thus, in terms of Kanner's hopes, there seems to have been in England a real attempt made at clarifying and understanding some aspects of the autistic child's handicaps in personality development, but the educational "approach" reported to deal with these handicaps appears to have only limited relationship to the findings of the research studies presented.

THE DesLAURIERS-CARLSON HYPOTHESIS

A comprehensive effort to meet the implications of Kanner's comments, both in the understanding of the complete syndrome of early infantile autism and in deriving, on the basis of such understanding, a rationale to an appropriate therapeutic-educational "approach" to the child's problems, has been reported in the research work of DesLauriers and Carlson (1969). This research focuses on the totality of the autistic child's patterns of behavioral difficulties and relates the developmental arrest and learning problems reflected in the child's behaviors to a central functional impairment in the decoding of the affective components of sensory experiences.

On a neurophysiological level an imbalance is hypothesized between two arousal systems, the ascending reticular formation whose functional ascendancy dominates in the emission of responses to sensory stimulations (System I), and the midbrain limbic system which is central in the processing of the affective, pleasurable, or painful qualities of such stimulations (System II). In the light of this hypothesis, the behavioral syndrome of early infantile autism (affective isolation, preservation of sameness, ritualistic stereotypies, learning difficulties, and language development delays) can be seen as consequent to a high threshold of sensory–affective receptivity in the child, resulting in a low level of arousal in the midbrain limbic system (Arousal System II). This deficiency or impairment in the affective processing (decoding) of sensory experiences leaves the child unaffected by such experiences, and therefore these experiences have no meaning to the child, who finds no gratification in responding to them. Furthermore, these affectless and meaningless experiences having no gratifying value never become consolidated through short-term or long-term memory fixations. Therefore no real learning takes place and the responses to sensory stimulations are endlessly repeated in a stereotyped and inappropriate fashion. This neurophysiological model is elaborated in DesLauriers and Carlson (1969).

THERAPLAY

The test of this hypothesis required a methodological design which would address the central core of the autistic infant's problems, the defect in affective experience (low arousal of System II), by overcoming, in the child, the high threshold of affective receptivity. The test had to be for the autistic child a total human experience in which all sensory-stimulating contacts with the environment would be loaded with a *high affective impact* of such intensity duration that it would have the effect of overcoming the high-threshold barrier to affective responsiveness. We found that the most affectively impactful pattern of sensory-stimulating experiences for the child could be found in a play situation, where excitement, novelty, surprise, and fun had the total arousal value needed to bring about a meaningful affective response from the autistic child. We called our method—our therapeutic-educational "approach" and experiment —Theraplay.

There are many advantages to using fun and play in arousing and awakening the autistic child. Play is first of all a central element in the development of any child; it is a natural "reinforcement" of a child's activities (Bruner, 1976). Secondly, play allows for constant experimentation, discovery, and learning, in a pleasurable and nonthreatening way.

Thirdly, play permits many close, physical, and social contacts with people of a rewarding and growth-producing value. Finally, play makes unusual, unfamiliar, and fearful situations less frightening because of its open-ended, free, and inconsequential quality (Pulaski, 1976).

Theraplay and Play Therapy

Theraplay should not be understood to be Play Therapy (see Axline, 1976). In Play Therapy the therapist sets up for the child a play situation—or allows the child to set up his own play or game—and then observes the child at play or encourages directly the child's elaboration of his game, deriving from these observations and participations an understanding of the child's problems and helping him work through these problems.

Theraplay, in contrast, emphasizes the central factor of high affective impact in all stimulating contacts or communication with the autistic child; this is done through play, games, and fun. In this relaxed atmosphere, no special effort is made at teaching the child anything; there is no specific special education methods or tools utilized. What the child learns first is that it is good to be human with another human being, that it is fun to be a member of the human race, and that grown-up human beings are worth having around to make life pleasant. The parents are urged to observe these playful interactions between the child and his therapist and they are asked to try this fun stuff, in their own ways and in their own style, at home. It helps family life. Because the playful atmosphere is maintained in all stimulating contacts at home or in the therapy session, whatever the child learns in one or the other situation is easily and naturally generalized from one place to the other.

Theraplay and Language Development

To illustrate the therapeutic-educational effectiveness of Theraplay in working with autistic children, an attempt will be made here to demonstrate its impact on the development of language in the autistic child. To this end, the view will be offered that language development depends on the child's capacity to acquire an understanding of some level of symbolic representation and that this capacity in the child is closely related to his learning how to play.

THE CASE OF CHRISTOPHER

The case of Christopher will establish the proper context and direction of this argument.

Christopher was nearly four years old when he was first brought to me. He was a slender healthy-looking fellow with light brown hair and blue eyes.

Diagnosis

He presented the central characteristics of early infantile autism as described by Kanner (1943) and reported by Rutter (Chapter 1). The diagnosis had been established by a number of child psychiatrists, pediatricians, and psychologists during the preceding years when his parents had relentlessly been seeking help for him. Christopher was an affectively isolated child, indifferent to people, lacking any communicative speech, prone to distress and tantrums whenever changes were introduced in his routines, and endlessly absorbing himself in rolling, juggling, and bouncing balls which he had an uncanny knack of discovering or which he made himself with paper, socks, or any other pliable and soft materials. Neurologically, no findings of any importance had been determined to account for his behaviors, and his medical history was reported negative.

Treatment

Theraplay was consistently used with Christopher during the years of his treatment with me. Christopher—like any other autistic child—caught on rather quickly to the approach. When he came to his first therapy session he brought with him a red ball and ignored completely his parents and the therapist as he isolated himself in the corner of the room, spinning, rolling, juggling his little red ball on the floor or between his hands. I sat beside him on the floor and at short intervals would reach out, take the ball, and hold it for a few seconds, and immediately give it back to Christopher before he had time to become upset. After 30 minutes of such intrusion, Christopher appeared reassured that I was not going to take his toy from him. The next move was for me to keep the ball a little longer so that Christopher, slightly puzzled and surprised, would have to turn to me or reach out for the ball; then, I would immediately give it to him, either by putting it in his hand or rolling it to him. By the end of the session, Christopher knew that I was there and that he and his ball were safe.

In the early subsequent sessions the time lapse between my taking Christopher's ball and my giving it back to him increased steadily; the variety of playful maneuvers intended to stimulate interaction between Christopher and myself enlarged also considerably. Eventually Chris was off the floor, out of his corner, and laughingly running after me in the room to retrieve his ball. At that point, a slight shift in strategy was introduced: instead of taking the ball away from the child I would wait. Christopher would then become concerned that the game was over and he would himself give me the ball to run away with. Obviously Christopher had found it more fun to play with me and the ball than to isolate himself with it.

We were now friends and playmates. Whatever I wanted to do in the room, Christopher was ready to do it with me (the happy expression "He was game to do it" would be most appropriate here!). Playing with the ball became less and less central to our interactions, but it was never completely removed from our games. For instance, if we both sat on the floor to work on putting a puzzle together, Chris had to have the ball in front of him, between his legs. Then the ball game still remained available, but he played less and less with the ball. He played at stacking up plastic doughnuts on a spindle and at looking at me through each doughnut before placing it on the spindle. He played at jumping off the desk into my arms or at balancing himself on his stomach on the sole of my shoes. He had fun bouncing up and down on the couch and landing on my back on the floor. He learned to discriminate colors and shapes while putting puzzles together and laughing out loud if I happened, on purpose, to give him a wrong piece of the puzzle. During these sessions, the ball sat on the file cabinet, available, but for the most part untouched.

Thus Christopher played and learned. He became a happy child, looking forward to his visits with me twice each week. After six months, we were definitely good friends. Though no effort was made at teaching him anything specific, it was amazing how quickly he became interested in learning things, like playing hide-and-go-seek behind the desk, jumping on my back when I pretended to be a horse, and demanding more funny sounds when I growled like a lion or barked like a dog. Childish games, all, but reassuring and exciting to Christopher. One day, when he was rocking on the couch, I began singing the nursery rhyme: "Sing a song of six pence a pocket full of rye" to the exact pace and rhythm of his rocking. Christopher looked up, smiled in response, and continued his rocking. Little by little I increased the tempo of the song and Christopher increased his rocking. Then, I would slow down, and he would slow down too. Christopher demanded that this be repeated over and over and, later on, whenever we tired of more exhausting activities, we would always go back to our singing and rocking game to rest.

It was also on the couch that Christopher said his first word. We were putting together a Playschool puzzle of a monkey eating a banana. Christopher had no trouble putting the puzzle together; but the real game included singing the monkey song, sharing the banana with the monkey, and generally dramatizing the entire adventure of the monkey. The song, "Monkey eats the banana," was a real hit with Chris; he would work fast while I sang. Then I'd pretend to take a piece of the banana, put it to my mouth, and eat it; Chris was encouraged to do the same. After that came my giving a piece of make-believe banana to Chris, and his placing another piece in my mouth. We'd break down the puzzle and re-do it many times, with any number of funny variations on the singing, the sharing, the eating. And it was when I hesitated once in giving him his piece of banana, that Christopher laughingly shouted, "Banana!" This was definitely an instance of "organic vocabulary," as his teacher used to call it. The game didn't stop, yet whenever I acted as if I had forgotten my part, Christopher was right there to remind me of it by shouting, "Banana!"

Outcome

Christopher's vocabulary increased rapidly in subsequent sessions: one word sentences, then two words together, then short sentences. His parents at home found a wide variety of opportunities to enlarge his vocabulary, always in the context of pleasant situations. His articulation was not perfect and, frequently, in trying to speak, only the first and last words of his sentence came out understandable. To correct this defect, I learned that if I sang the sentence to him, he could repeat it much more intelligibly.

Today Christopher attends a special education classroom in a public school. He has no difficulty learning there, but the fun situation is considerably reduced. What he enjoys most are the songs the children sing. Otherwise he remains his independent self, having little to do with the other children. In contrast, he plays well at home with his brother and sister, both younger than he is, and he teases his mother and father with a considerable amount of mischief.

THE CASE OF TONY

The case of Christopher is not unique. A further illustration of the effectiveness of Theraplay in language development can be seen in the case of Tony.

Tony, a handsome looking youngster with a wistful smile and the physique of a promising all-American athlete, joined us at the Center for Autistic and Schizophrenic Children in January 1975, at the age of eight years, 9 months. The Center offers Theraplay as the therapeutic-educational method in working with 18 young children diagnosed as autistic or schizophrenic.

Diagnosis

Tony had never been in a residential center. His history was reported to be one of emotional deprivation, his parents having divorced when he was born. His entire upbringing was turned over to a governess. Nothing appeared physically wrong with Tony during infancy and early childhood. He passed all the developmental milestones with flying colors, except for language and speech. Because of his social withdrawal, his affective independence, his ritualistic absorptions in mechanical gadgets and his inability to use verbal communicative speech, he was repeatedly diagnosed as autistic. By the age of five he had acquired a few words; these were affective or as Kanner named them, metaphorical words, associated with anxious or emotional events in his life (like his "litany" recitation of: "Daddy sick, Mommy sick, Mochie (dog) sick, Sebastian (dog) sick"). His receptive language was quite adequate and his self-help skills were very acceptable. Tony demonstrated no interest or motivation in learning in school, but his intellectual endowment was assessed to be in the low normal range. We had some doubts as to Tony being an infantile autistic child, because so many traumatic events had surrounded his early infancy and because we had limited convincing evidence of the symptomatic aspects of his infant behaviors. It would have been difficult, on the other hand, to see him as schizophrenic; his development was certainly atypical, with strong autistic features.

Treatment

At the Center, Tony was placed in a small group of four children, with two counselor-therapists whose functions included being mother, teacher, and recreational companion in the various aspects of his daily program. In addition, Tony was seen twice per week in Speech Therapy and also in Individual Psychotherapy. The philosophy of Theraplay dominated all of his interactions at the Center.

For many months following his admission to the Center, Tony remained socially isolated, refusing to participate in any group activities and absorbing himself in riding his bicycle in the playground outside or toying with some mechanical gadget in the classroom. He never talked though he would repeat in a somewhat echolalic way some words. When he became upset he would repeat inappropriately the "litany" of metaphorical sentences which he had been reported to say at home ("Daddy sick, Mommy sick, etc."). With his counselors he progressively learned to play teasing games, which involved mostly his running away from them, wanting to be chased, or wanting to be coaxed into some playful social activity. He became very attached to his speech therapist and found pleasure in trying to articulate words for her. In individual therapy, he enjoyed mostly going on excursions which always ended up at an ice cream parlor where he quickly learned to say to the waitress what sort of ice cream dish he wanted.

Outcome

Tony's first indications of becoming socially aware of his peers took the form of aggressive intrusions upon them: He would sneak up to a child, push him hard, pull his hair, or hit him. It became clear that his target was always a child whom Tony had appeared to tolerate most; Tony, in his bullying, aggressive ways, was trying to be friendly! This awkward type of friendship developed rapidly when the object of his attacks began retaliating. We used the situation for make-believe war games. All manner of toy vehicles, boxes, and other materials were recruited in expanding the game, which included military rules and discipline. Tony drove his "tanks" making all the appropriate sounds. His "enemy" tried to out-fox his strategies. In learning how to pretend, Tony discovered he didn't need to hit or pull hair to achieve his victories.

With the discovery of make-believe in his play, Tony found himself also using many more words to communicate his plans, needs, and wishes. Tony was now acting like a three-year-old and his short sentences were articulated in a much freer way. We further encouraged this development by having Tony on frequent visits to the homes of some of his counselors, where he could play with small "normal" children. These experiences enlarged Tony's sociability and communicative speech. His speech with adults remained reticent, but he had learned, by playing with children, a new world of fun and make-believe.

DISCUSSION OF TREATMENT

These cases should not be viewed as exceptions. They illustrate most dramatically some of the central issues with which this presentation is concerned. Let me briefly review these issues. The first question touches on the very nature of the condition of early infantile autism. Regardless of etiological factors, early infantile autism is very much a human condition, a condition which affects the basic psychological requirements of human personality growth and development. The quality and level of human interaction basic to normal development is not available to the autistic infant. This child neither seeks such human contacts nor does he normally respond to them.

Secondly, the developmental arrest which characterizes the autistic condition prevents the acquisition and learning of uniquely human skills associated with reflective consciousness, symbol formation, language, and communication.

Finally, the deficits in these basic human functions appear to be the consequence of an affective rather than cognitive dysfunctioning.

The preceding understanding of infantile autism underlies the rationale of Theraplay utilized with Christopher and Tony. With this treatment approach Tony learned to trust the adults in his life, and eventually enlarge his playful contacts to include his peers with whom he felt comfortable in communicating. Christopher moved from a condition of isolation to a stable pattern of response involving human interaction, communication, and interpersonal relationship. This movement was neither forced nor contrived. It happened because the appropriate requirements for such a natural process to take place were established and maintained.

Conceptual Implications

In this understanding of autism, play is considered central to child development and to those symbolic functions which serve as a basis for language development. Most observers of human behavior agree on the striking importance of play in the early development of the child (i.e., Murphy, 1972; Erickson, 1972; White, 1975). If you observe an on-going play situation, with its fun qualities, its rules and regulations, its orderly and meaningful sequences, you will quickly recognize how much of a concatenation of signals, signs, behavioral substitutions, and symbols it involves. Even the earliest game a mother plays with her baby contains all these important components: In the peek-a-boo game which has a universally appealing quality to the baby, the mother "pretends" to disappear and to reappear so that the baby, surprised, puzzled, reassured, and delighted, very soon learns that it is possible to be there and not be there at the same time, that the signal for a disappearing act is to cover your eyes or your face, and that the sign that everything remains safe and secure is in the mother's predictable, anticipated, and expected reappearance (Bruner, 1974–1975; Bruner & Sherwood, 1976; Sylva *et al.*, 1976).

Lovaas (1968) himself found play to be one of the most effective reinforcers in the shaping and conditioning of human behavior, but was at a loss to establish the scientific basis for such an effect. Anyone who attempts to work with a young autistic child discovers very soon that the child does not know how to play: The sterile repetition of his rituals and obsessive self-stimulating behaviors contributes little to his enjoyment of life or to his developmental growth as a functioning human being. Undoubtedly such behaviors offer the autistic child a certain degree of security within the limited boundaries of his emerging self, but play offers a much more dynamic form of security. In Theraplay the child learns that one thing can be substituted for another, that one behavior can signal another, that one part of the game can be represented by another, that surprises, novelty, and rearrangements are never totally devastating to the orderly patterns and sequences of fun behaviors. These behavioral substitutions, these novel and surprising rearrangements, these signals and symbolic representations need to be safely and securely assimilated before becoming spontaneously expressed in the child's behavior.

Finally, without a symbolic function the child cannot develop that "inner language" (Lenneberg, 1967) required for meaningful, communicative speech. The entire world of dreams and fantasies (Meltzer, 1975) as well as of abstract ideations (Piaget, 1952, 1962) is dependent upon it. But the representational images of the world which need to be transformed into communicative speech reflect a special case of symbol formation directly related to the concrete, sensori-affective experience of the child. The child's first words are always "organic" in this sense, but their symbolic representational value far from being semantically precise and definite, extends to the totality of the sensori-affective experience it is intended to stand for. When Christopher, no different at that moment than any normal child, exclaimed "Banana!" he was not talking about the fruit but about the entire game in which he was engaged with his therapist friend. It is understandable then, that the word itself, "banana," pregnant with the sum total of those sensori-affective behavioral experiences which helped shape it and form it, acquired at that moment the stimulating power of reinitiating and triggering off the entire pleasurable situation. Kanner himself (1946) in a paper seldom referred to, pointed this out when he discussed the metaphorical language of the autistic child.

The child's first word, born out of the excitement and pleasure of a game with a protective adult friend, possesses from the beginning a socially communicative intentionality: It is meant to be spoken because it is part of the game. It therefore includes the rules of the game and its appropriateness is defined from the context of the game. It is not surprising, therefore, that, as Langer (1942) has theorized, its "basic grammar" is similar to that of a song, with its sound and rhythm, with its organic tempo, and its evocative feeling. The first communicative word a baby says is frequently the word of a song, a word which summarizes the total pleasurable experience in which he is involved when it is sung to him. Christopher's language development also required the sound and rhythm of music on which to attach words that said nothing special, because they said everything that meant anything to him.

Similarly, Tony's progress in language pointed to the importance of its affective components for its early acquisition, even though its use remained for a long time purely metaphorical. With the advent of fun and make-believe in his life, his discovery that he could represent his views and ideas through signs and symbols which substituted for so many new experiences of importance in his life as a child, Tony found it easier to speak. Even though he never sang, his new vocabulary had all the lyrical promise of an emerging song.

The adventures of both Christopher and Tony from isolation to communication, from games to symbols and language, are the total experience of Theraplay.

CONCLUSION

The rationale of Theraplay was derived from a conceptual understanding of the total syndrome of infantile autism. Central to this understanding is the affective deficit and impairment which accounts for the child's social and emotional isolation and interferes seriously with the learning of those skills essential to the child's development as a human being. This deficit and impairment of the affective functions can be said to be the most important characteristic of all autistic infants. By emphasizing its presence, there is less risk, for the researcher, to seek answers to the autistic condition in other functional areas that are not similarly impaired in all autistic children. Theraplay was designed to remediate the affective functional deficit and by the same token test the plausibility of our understanding of the autistic syndrome. As a methodological design its broad parameters encompass the totality of the autistic syndrome while remaining focused on the central issue of the child's affective response to sensory stimulating experiences. It is in that light that Theraplay can be considered an attempt in the direction of meeting Kanner's hope that progress in the treatment of early infantile autism will hinge eventually on establishing a focus of understanding of this syndrome and deriving from it a rational approach to its challenge.

REFERENCES

Axline, M. Play therapy procedures and results. In C. E. Schaefer (Ed.), *Therapeutic use of child's play*. New York: Jason Aronson, 1976.

Bettelheim, B. *The empty fortress*. New York: Free Press, 1967.

Bruner, J. S. From communication to language: A psychological prospective. *Cognition*, 1974–1975, *3*, 255–287.

Bruner, J. S. Nature and uses of immaturity. In J. S. Bruner, A. Jolly, & K. Sylva (Eds.), *Play–Its role in development and evolution*. New York: Basic Books, 1976.

Bruner, J. S., & Sherwood, V. Peekaboo and the learning of rule structures. In J. S. Bruner, A. Jolly, & K. Sylva (Eds.), *Play—Its role in development and evolution*. New York: Basic Books, 1976.

DesLauriers, A. M., & Carlson, C. F. *Your child is asleep: Early infantile autism*. Homewood, Ill. The Dorsey Press, 1969.

Erikson, E. H. Play and actuality. In M. Piers (Ed.), *Play and development*. New York: Norton, 1972.

Hawkins, D. R., & Pauling, L. C. (Eds.), *Orthomolecular psychiatry*. San Francisco: W. H. Freeman, 1971.

Hermelin, B., & O'Connor, N. *Psychological experiments with children*. London: Pergamon, 1970.

Kanner, L. Autistic disturbances of affective contact. *Nervous Child*, 1943, *2*, 217–250.

Kanner, L. Irrelevant and metaphorical language in early infantile autism. *American Journal of Psychiatry*, 1946, *103*, 242–246.

Kanner, L. Approaches: Retrospect and prospect. *Journal of Autism and Childhood Schizophrenia*, 1971, *1*, 453–459.

Langer, S. K. *Philosophy in a new key*. New York: Harvard, 1942.

Lenneberg, E. H. *The biological foundations of language*. New York: Wiley, 1967.

Lotter, V. Social adjustment and placement of autistic children in Middlesex: A follow-up study. *Journal of Autism and Childhood Schizophrenia*, 1974, *4*, 11–33.

Lovaas, O. I. Some studies on the treatment of childhood schizophrenia. In J. M. Schlein (Ed.), *Research in psychotherapy* (Vol. 3). Washington, D.C.: American Psychological Association, 1968.

Meltzer, D. Mutism in autism, schizophrenia and manic-depressive states: The correlation of clinical psychoapthology and linguistics. In D. Meltzer (Ed.), *Explorations in autism*. Aberdeen: Aberdeen University Press, 1975.

Murphy, L. B. Infant's play and cognitive development. In M. Piers (Eds.), *Play and development*. New York: Norton, 1972.

Piaget, J. *The origin of intelligence*. New York: Norton, 1952.

Piaget, J. *Play, dreams and imitation in childhood*. New York: Norton, 1962.

Pulaski, M. A. Play symbolism in cognitive development. In C. C. Schaefer (Ed.), *Therapeutic use of child's play*. New York: Jason Aronson, 1976.

Rimland, B. High dosage levels of certain vitamins in the treatment of children with severe mental disorders. In D. R. Hawkins & L. C. Pauling (Eds.), *Orthomolecular psychiatry*. San Francisco: W. H. Freeman, 1971.

Rutter, M. The development of infantile autism. *Psychological Medicine*, 1974, *4*, 147–163.

Sylva, K., Bruner, J. S., & Genova, P. The role of play in problem solving of children 3–5 yrs. old. In J. S. Bruner, A. Jolly, & K. Sylva (Eds.), *Play—Its role in development and evolution*. New York: Basic Books, 1976.

White, B. l. *The first three years of life*. Englewood Cliffs, New Jersey: Prentice-Hall, 1975.

Wing, L. *Early childhood autism: Clinical, educational and social aspects* (2nd ed.). Oxford: Pergamon Press, 1976.

22

Etiology and Treatment: Cause and Cure

MICHAEL RUTTER

Both Berlin and DesLauriers in the preceding chapters have urged the importance of planning treatment on the basis of what is known about the nature of autism and about its etiology. Berlin (Chapter 20) draws attention to the growing evidence on how early life experiences can shape later development and on how stresses and deprivation in infancy can cause persisting and serious disorders. He suggests, as does Szurek (1973), that a failure in initial bonding may be one cause of autism. It is argued that this bonding failure may often stem from parental difficulties or inadequacies and, therefore, that psychotherapy with the parents constitutes a crucial part of treatment in many cases. In his early writings (e.g., Szurek & Berlin, 1956) he stated that his therapeutic approach to autism was based on the hypothesis that the disorder was "entirely psychogenic." Berlin now recognizes that there may be constitutional components (Berlin & Szurek, 1973) but still argues (Berlin, 1973) that "a prime etiologic element is the interpersonal one" and that "the child's reactions usually result from the conflicted amalgam of the parents' feelings and behavior toward him."

DesLauriers (Chapter 21), in contrast, relates the autistic child's developmental arrest and learning problems to a "central functional impairment in the decoding of the affective components of sensory experiences," which in turn is thought to be a consequence of imbalance be-

MICHAEL RUTTER · Department of Child and Adolescent Psychiatry, Institute of Psychiatry, De Crespigny Park, Denmark Hill, London SE5 8AF, England.

tween arousal systems in the ascending reticular formation. Treatment consists of sensory stimulation with a high affective impact designed to overcome the sensory threshold barrier. It is considered that fun and play provide the best medium for such stimulating experiences. DesLauriers suggests that one of the outstanding merits of his therapeutic approach is the fact that it is derived from a "conceptual understanding of the total syndrome of infantile autism" and he is critical of both American behavior modification and English educational therapies for failing to link etiological concepts and therapeutic approaches.

LINKS BETWEEN CAUSE AND CURE

The attempts to develop treatment methods which are logically related to knowledge on etiology are certainly right. An effective "cure" for autism is only likely to be possible if it provides a remedy for the basic causes of the condition. This is true whether the cause is physical or psychogenic. Thus, an effective treatment for phenylketonuria required the identification of the essential biochemical fault. Similarly, the prevention of mental retardation in institution-reared children was based on an understanding of the importance of meaningful experiences in aiding intellectual growth (see Rutter, 1972). The alleviation of children's distress during hospital admission was dependent on a recognition of the importance of the young child's affective ties with his parents (Bowlby, 1969, 1973). In the same way, the planning of optimal alternative care when a child has to be separated from his parents relies on an appreciation of his emotional and social needs (see Robertson & Robertson, 1971). Our treatment methods for autism would be more effective if they could be designed to put right whatever it is went wrong in the child's development.

Nevertheless, we need to be aware of how rare it is in medicine to have true "cures" or even restoration of normal functioning in any of the chronically handicapping disorders. Furthermore, the effective treatments are often discovered or developed *without* knowledge of why they work. Thus, digitalis was used effectively as a treatment for congestive cardiac failure before there was any understanding of its pharmacological properties. Hypnotism worked in spite of the theory of "animal magnetism" being quite wrong. Or again, electroconvulsive therapy relieves some cases of severe depression in spite of its original rationale being mistaken and in spite of our continuing ignorance of why it is effective when it is (or indeed why it is not when it isn't). A variety of behavioral treatments (desensitization, flooding, etc.) have been conclusively shown to be effective in relieving specific phobias although the theoretical basis of several of them appears contradictory (see Rutter, 1975). The pad and bell treatment of nocturnal enuresis is much the most effective of all the available treatments but there is continuing controversy on how it works and it is highly unlikely that its mode of action has anything to do with the initial reason for the bedwetting.

EFFICACY OF TREATMENT METHODS

It follows from these considerations that the demonstration that a treatment is effective does not thereby mean that its theoretical basis was also correct. Conversely, because a theory is right it does not necessarily mean that the treatment methods which derive from it will be successful. However prestigious the theoretical pedigree the only way to determine if a treatment works is to measure its effectiveness in comparison with other therapeutic interventions. On this criterion both Berlin's psychotherapeutic work with parents and DesLauriers' "Theraplay" fall down. In neither case have there been systematic evaluations. Of course, that does not mean that they are useless but it does mean that the onus is on them to undertake therapeutic studies. Autistic children have been treated at the Langley Porter's Children's Service for over a quarter of a century (Berlin & Szurek, 1973) and although there has been a limited follow-up (Etemad & Szurek, 1973), there have been no trials of treatment. Clearly, they are needed.

Arousal and Decoding

In the meanwhile, we need to turn to the suggested etiological hypotheses. DesLauriers' argument has three parts: (1) There is a postulated abnormality in the reticular formation associated with disturbances in arousal, (2) this results in impaired responsiveness and impaired processing of sensory experiences, and (3) that play is the best way of overcoming this barrier. The empirical evidence on the first part is inconclusive. EEG studies as a measure of arousal are contradictory (Hutt *et al.*, 1965; Kolvin *et al.*, 1971; Hermelin & O'Connor, 1968; Creak & Pampiglione, 1969), as are investigations of variations in stereotypies according to environmental complexity (Hutt & Hutt, 1965; Ornitz *et al.*, 1970). Arousal is a complex concept and different measures of arousal do not always agree with one another but so far there is little indication that autistic children are generally over- or underaroused. Whether or not there is impaired functioning of the reticular system is equally difficult to say. Some of Ornitz' work (see Chapter 8; also Ornitz, 1973) is consistent with this suggestion but, as Yule (Chapter 10) points out, many of the necessary controls have still to be made.

On the other hand, there is no doubt that DesLauriers' second postulate is correct. As shown by a variety of well-controlled studies (see Hermelin & O'Connor, 1970), autistic children are indeed seriously impaired in their processing of incoming sensory information. What remains to be determined is the exact nature of this deficit and its anatomical and physiological origin.

There is very little direct evidence on the third step in the argument concerning the benefits of play. However, this suggestion is discussed more fully later in the chapter.

ENVIRONMENTAL INFLUENCES ON DEVELOPMENT

In contrast, Berlin (Chapter 20) bases his arguments on what is known about environmental influences on development. Unfortunately, his account is somewhat misleading in a few key respects. He argues that early adverse experiences during critical periods may produce biochemical and perhaps structural changes in infants' brains which are associated with autistic-like behavior. It is necessary to consider the different elements in this argument separately before discussing it as a whole.

Effects of Early Life Experiences

First, as has been shown repeatedly in both human and animal studies (Hinde, 1970; Rutter, 1972, 1977), early life experiences can have long-lasting effects on behavior. Moreover, these effects are particularly seen in the development of language, intelligence, and social relationships—features which are central to the abnormalities involved in the syndrome of autism. However, there the similarities end. Berlin, like Szurek (1973), emphasizes the importance of failures in bonding. These are most readily observed in children reared in institutions with a large number of ever-changing caretakers. Such children tend not to develop deep personalized relationships and they may have widespread difficulties in socialization (Tizard & Rees, 1975; Tizard & Hodges, 1977). However, quite unlike autistic children, they tend to be clinging and overfriendly, they are not intellectually retarded, and they develop normal language.

Of course, an environment which lacks adequate active experiences and conversational interchange can also lead to impaired intelligence and delayed language (see Rutter, 1972). On the other hand, autism is characterized by *deviant* language more than by delayed language (see Chapter 6) and that is *not* a characteristic of children reared in depriving environments. Moreover, there is abundant evidence that if the child's environment improves so does his intelligence and language (Clarke & Clarke, 1976). Complete or almost complete (and fairly rapid) social and intellectual recovery has been shown even for children very severely deprived up to age six years (Davis, 1947; Koluchova, 1972, 1976). There is no equivalent of changes of that degree or that speed in autism.

Critical Periods

Berlin points to the evidence that even very young infants are remarkably responsive to subtle environmental changes. Certainly this is so, as shown by studies of both perception (Bower, 1974) and of mother-child relationships (Bowlby, 1969, 1973). On the other hand, the suggestion that the early weeks or months constitute a critical period for development lacks supporting evidence. So far as first bonding is concerned, it may well be that there is a sensitive period for its development but if so it certainly extends up to at least age two years and probably well above that (Rutter, 1977). The critical period for language is said to extend up to age 12 years (Lenneberg, 1967) and there is no well-defined critical period for intelligence although the early years may be especially important (Rutter, 1977). So far as cognition and socialization are concerned (the two aspects of development most affected in autism), if there are critical periods they must extend over several years.

Environmental Influences on Brain Development

A further link in Berlin's chain of argument concerns the evidence that environmental influences can alter the structure and chemistry of the developing brain. Although at first sight a perhaps startling suggestion, the general notion that experiences help shape physical growth has good support. This is most obvious in terms of the rapid muscular (and later bony) atrophy which follows disuse or the lasting hypertrophy which is a consequence of frequent violent exercise. It is also well demonstrated in terms of the way visual experiences are necessary for the normal development of the retina and visual cortex (Riesen, 1965). The effects of visual privation are long lasting. However, with most functions the striking feature of brain development is its plasticity and modifiability so that, for example, unilateral lesions in early life have very little long-term effect on language development (Rutter, *et al.*, 1970).

Berlin also refers to the work of Rosenzweig and his associates which has shown that various forms of environmental deprivation or enrichment affect brain chemistry and structure (Rosenzweig, 1976). Other studies have confirmed that the nervous system is indeed modified by experience

(Horn *et al.*, 1973). However, contrary to Berlin's claims, the work has *not* been undertaken with primates. Of course, the effect may apply also to higher species including man but this has yet to be shown. Moreover, Rosenzweig (1976) found that the brain deficits caused by general isolation were *readily reversible*.

Most weight is laid on Sander's study (Sander *et al.*, 1970) which is said to show the development of "early neurologic soft signs" in human babies who have had only very restricted handling. In fact the report he cites shows no such thing. What it does show is effects on patterns of crying, sleeping, and motility. These are important results which rightly warrant attention but there is no evidence of brain damage. We have much yet to learn about environmental influences on brain development and it may be that the effects are greater than current investigations suggest (most marked changes are results of quite gross and artificial environmental distortions). On the other hand, what requires explanation in the development of autistic children is the frequent emergence of epileptic seizures in adolescence. There is *nothing* in the research to date which indicates that that is likely to be due to early life experiences.

Family Stress and Autism

So far, the arguments have been entirely by analogy and by drawing parallels. Although the published studies all indicate that early stresses and deprivation tend to lead to behaviors quite different from the syndrome of autism, clearly it is possible that other kinds of adverse experiences could result in autistic behavior. What evidence, then, is there that autistic children have experienced abnormal patterns of parenting? The findings are reviewed by Cantwell *et al.* (Chapter 18) who conclude that the data run counter to this suggestion. Berlin refers to Goldfarb's (1961) hypothesis that there are both organic and psychogenic varieties of autism and his finding that parental pathology is most evident in the latter group. However, Goldfarb's own systematic study (Goldfarb *et al.*, 1976) *failed* to confirm his earlier subjective observation (Meyers & Goldfarb, 1962) that schizophrenia was more common in the parents of his nonorganic group. McAdoo (see Chapter 17) has also failed to find parental differences between autistic children who do and those who do not have evidence of organic pathology. In short, the available data provide no convincing support for the notion that family stresses cause autism.

TREATMENT STRATEGIES

Where does this leave considerations of treatment strategies? Clearly the etiological arguments of both Berlin and DesLauriers provide a most unsatisfactory basis for planning therapeutic intervention. Do other views provide any better basis? That remains to be seen but certainly there have been attempts. DesLauriers (Chapter 21) has criticized English workers for failing to differentiate the processes involved in language receptivity from those involved in language expression. In fact this is not so. Hermelin and O'Connor (1970; Hermelin, 1976), Rutter (1974; see also Chapter 6), and Ricks and Wing (1976) have all gone to some lengths to emphasize that the autistic child's problems lie in symbolic processing and *not* just in speech production. DesLauriers went on to suggest that no attempt had been made to derive treatment methods from the view that autism was associated with cognitive impairments. This also is not so. Rutter and Sussenwein (1971) tried to do just that.

However, both Berlin and DesLauriers are surely right in urging that treatment methods must be designed to foster more normal development and not just to remove deviant behaviors by human engineering. This was very much the burden of Rutter and Sussenwein's (1971) paper and, as with Berlin, use was made of Bowlby's (1969) work on attachment to suggest ways in which social development might be facilitated. It was recognized that parents faced many difficulties because of the autistic child's unresponsiveness. They need help in picking up the child's cues and in insuring that there can be regular and intensive parent-child interaction which is meaningful and pleasurable to the child and enjoyed by the parents. There must be an emphasis on language but "because the autistic child's defect lies first in prelinguistic skills and in language comprehension rather than the production of words, the first goal is to aid the development of these skills" (Rutter & Sussenwein, 1971). Again, social play seems to be the most appropriate means to do this. Language is first and foremost a communication skill which develops in a social context. As stressed by Schopler and Reichler (1971), most emphasis was placed on working with parents to help them in their interactions with their children. Rutter and Sussenwein (1971) also urged the need for skilled and experienced parental counselling to help deal with the family stresses and conflicts associated with bringing up a handicapped child. There is evidence now that this general approach is effective in helping autistic children and their families (see Chapter 33). As Lansing and Schopler clearly show (Chapter 29), their own very similar approach has much to offer also in the field of education.

CONCLUSION

Exception may be taken to several of the specific arguments put forward by Berlin (Chapter 20) and by DesLauriers (Chapter 21). On the other hand, there seems to be general agreement now with their overall view that whatever the treatment rationale, the therapeutic endeavors should involve working closely with parents and should utilize social play to aid the autistic child's language and social development. The debate goes on, however, as to exactly how this should be done.

REFERENCES

Berlin, I. N. Intrapersonal, interpersonal, and impersonal factors in the genesis of childhood schizophrenia. In S. A. Szurek & I. N. Berlin (Eds.), *Clinical studies in childhood psychoses.* New York: Brunner/Mazel; London: Butterworth, 1973. Pp. 551–593.

Berlin, I. N., & Szurek, S. A. Psychoses of childhood, retrospect and prospect. In S. A. Szurek & I. N. Berlin (Eds.), *Clinical studies in childhood psychoses.* New York: Brunner/Mazel; London: Butterworths, 1973. Pp.3–9.

Bower, T. G. R. *Development in infancy.* San Francisco: W. H. Freeman, 1974.

Bowlby, J. *Attachment and loss, Vol. 1: Attachment.* London: Hogarth Press, 1969.

Bowlby, J. *Attachment and loss, Vol. 2: Separation, anxiety and anger.* London: Hogarth Press, 1973.

Clarke, A. M., & Clarke, A. D. B. *Early experience: Myth and evidence.* London: Open Books, 1976.

Creak, M., & Pampiglione, G. Clinical and EEG studies on a group of 35 psychotic children. *Developmental Medicine and Child Neurology,* 1969, *11,* 218–227.

Davis, K. Final note on a case of extreme isolation. *American Journal of Sociology,* 1947, *52,* 432–437.

Etemad, J. G., & Szurek, S. A. A modified follow-up study of a group of psychotic children. In S. A. Szurek and I. N. Berlin (Eds.), *Clinical studies in childhood psychoses.* New York: Brunner/Mazel; London: Butterworth, 1973. Pp. 303–347.

Goldfarb, W. *Childhood schizophrenia.* Cambridge, Mass.: Harvard University Press, 1961.

Goldfarb, W., Spitzer, R. L., & Endicott, J. A study of psychopathology of parents of psychotic children by structured interview. *Journal of Autism and Childhood Schizophrenia,* 1976, *6,* 327–338.

Hermelin, B. Coding and the sense modalities. In L. Wing (Ed.), *Early childhood autism: Clinical, educational and social aspects,* (2nd ed.). Oxford: Pergamon Press, 1976. Pp. 135–168.

Hermelin, B., & O'Connor, N. Measures of the occipital alpha rhythm in normal, subnormal and autistic children. *British Journal of Psychiatry,* 1968, *114,* 603–610.

Hermelin, B., & O'Connor, N. *Psychological experiments with autistic children.* Oxford: Pergamon Press, 1970.

Hinde, R. A. *Animal Behaviour* (2nd ed.). New York: McGraw-Hill, 1970.

Horn, G., Rose, S. P. R., & Bateson, P. P. G. Experience and plasticity in the central nervous system. *Science,* 1973, *181,* 506–514.

Hutt, C., & Hutt, S. J. Effects of environmental complexity upon stereotyped behaviours in children. *Animal Behaviour,* 1965, *13,* 1–4.

Hutt, S. J., Hutt, C., Lee, D., & Ounsted, C. A behavioural and electroencephalographic study of autistic children. *Journal of Psychiatric Research,* 1965, *3,* 181–198.

Koluchova, J. Severe deprivation in twins: A case study. *Journal of Child Psychology and Psychiatry,* 1972, *13,* 107–114.

Koluchova, J. The further development of twins after severe and prolonged deprivation: A second report. *Journal of Child Psychology and Psychiatry,* 1976, *17,* 181–188.

Kolvin, I., Ounsted, C., & Roth, M. Studies in the childhood psychoses: V. Cerebral dysfunction and childhood psychoses. *British Journal of Psychiatry*, 1971, *118*, 407–414.

Lenneberg, E. H. *Biological foundations of language.* New York: Wiley, 1967.

Meyers, D. I., & Goldfarb, W. Psychiatric appraisal of parents and siblings of schizophrenic children. *American Journal of Psychiatry*, 1962, *118*, 902–915.

Ornitz, E. M. Childhood autism—a review of the clinical and experimental literature. *California Medicine*, 1973, *118*, 23–47.

Ornitz, E. M., Brown, M. B., Sorosky, A. D., Ritvo, E. R., & Dietrich, L. Environmental modifications of autistic behaviour. *Archives of General Psychiatry*, 1970, *22*, 560–565.

Ricks, D. M., & Wing, L. Language, communication and the use of symbols. In L. Wing (Ed.), *Early childhood autism: Clinical, educational and social aspects* (2nd ed.). Oxford: Pergamon Press, 1976. Pp. 93–134.

Riesen, A. Effects of early deprivation of photic stimulation. In S. F. Osler & R. E. Cooke (Eds.), *The Biosocial basis of mental retardation.* Baltimore: Johns Hopkins Press, 1965.

Robertson, J. & Robertson, J. Young children in brief separations: A fresh look. *Psychoanalytic Study of the Child*, 1971, *26*, 264–315.

Rosenzweig, M. R. Effects of environment on brain and behaviour in animals. In E. Schopler, & R. J. Reichler, (Eds.), *Psychopathology and child development: Research and treatment.* New York and London: Plenum Press, 1976. Pp. 35–50.

Rutter, M. *Maternal Deprivation Reassessed.* Harmondsworth: Penguin, 1972.

Rutter, M. The development of infantile autism. *Psychological Medicine*, 1974, *4*, 147–163.

Rutter, M. *Helping troubled children.*Harmondsworth: Penguin, 1975.

Rutter, M. Maternal deprivation 1972–1977: New findings, new concepts, new approaches. Submitted for publication, 1977.

Rutter, M., Graham, P., & Yule, W. *A neuropsychiatric study in childhood.* Clinics in Developmental Medicine Nos. 35/36. London: SIMP/Heinemann, 1970.

Rutter, M., & Sussenwein, F. A developmental and behavioral approach to the treatment of pre–school autistic children. *Journal of Autism and Childhood Schizophrenia*, 1971, *1*, 376–397.

Sander, L. W., Stechler, G., Burns, P., & Julia, H. Early mother–infant interaction and 24–hour patterns of activity and sleep. *Journal of the American Academy of Child Psychiatry*, 1970, *9*, 103–123.

Schopler, E., & Reichler, R. J. Developmental therapy by parents with their own autistic child. In M. Rutter (Ed.), *Infantile autism: Concepts, characteristics and treatment.* London: Churchill-Livingstone, 1971. Pp. 206–227.

Szurek, S. A. Attachment and psychotic development. In S. A. Szurek & I. N. Berlin (Eds.), *Clinical studies in childhood psychoses.* New York: Brunner/Mazel; London: Butterworth, 1973. Pp. 191–277.

Szurek, S. A., & Berlin, I. N. Elements of psychotherapeutics with the schizophrenic child and his parents. *Psychiatry*, 1956, *19*, 1–9.

Tizard, B., & Hodges, J. The effect of early institutional rearing on the behaviour problems and affectional relationships of eight year old children. *Journal of Child Psychology and Psychiatry*, 1977, in press.

Tizard, B., & Rees, J. The effect of early institutional rearing on the behaviour problems and affectional relationships of four year old children. *Journal of Child Psychology and Psychiatry*, 1975, *16*, 61–73.

23

Pharmacotherapy

MAGDA CAMPBELL

Psychoactive agents have been in use for the treatment of psychotic children since the 1930s. In those early years only hypnotics and anticonvulsants were available and they were administered chiefly for the management of excitement and agitation. The discovery of chlorpromazine (Delay & Deniker, 1952) and other neuroleptics resulted in a revolution in the treatment of psychoses of adults. It was hoped that these drugs would arrest or decrease the psychotic process in the child and that after the reduction of anxiety, hyperactivity, and/or aggressiveness he would become more amenable to special education and other treatment modalities. These hopes were not always fulfilled. Drug therapy has come under attack because of improper use, especially with institutionalized retarded children. This abuse has included high dosage over prolonged periods of time, multiple drug usage—children receiving from two to eight drugs at the same time—no drug holidays, little or no evidence of drug monitoring, and usage of sedative types of phenothiazines which probably suppress cognitive learning functions (Lipman, 1970).

In schizophrenia of adults there is scientific evidence concerning the efficacy of antipsychotic drugs as compared to other treatment modalities and their interaction. The superiority of neuroleptics over psychosocial treatments has been established (GAP, 1975). The acquisition of this knowledge was possible because of relatively good agreement on diagnostic criteria, use of sophisticated methodology in drug studies, and availability of a large number of homogeneous patient samples in collaborative studies. It has been shown that pharmacotherapy not only

MAGDA CAMPBELL · New York University Medical Center, New York, New York.

decreases symptoms but also modifies the course of schizophrenia (WHO 1967).

In psychotic children such effect of drug treatment has not been demonstrated even though drugs are being widely used in their management. There is a lack of critical evaluation of these agents and of systematic investigation of their role in the treatment of this patient population. Well-designed and controlled studies in homogeneous and sizable samples of children hardly exist. There is a great need to explore drug treatment and other treatments (psychoanalytic play therapy, psychotherapy, behavior therapy, etc.) independently and conjointly with drugs.

When drug treatment is terminated, symptoms frequently reappear. Irwin (1974) pointed out that the only lasting effect of a drug is indirect: a result of modified interaction of the individual with his environment. It is conceivable that a therapeutic drug can diminish certain symptoms, make a child more amenable to conditioning procedures, and hasten behavioral improvements. Our current controlled study of haloperidol and behavior therapy represents the first systematic, planned investigation of drug and conditioning in a sizable sample (40) of preschool-age psychotic children (Campbell *et al.*, 1978a). Even though many young psychotic children's overall functioning is on a retarded level, the effect of drugs on cognition has not been investigated.

Anterospective studies, comparing the natural history of illness with pharmacologic or any other intervention are nonexistent. Nevertheless, statements have been made that treatment or a type of treatment does not affect the outcome (Gajzago & Prior, 1974; Eisenberg, 1956). In addition, diagnosis is still an unresolved or, at best, a difficult problem, although from De Sanctis (1925), Bender (1947), Kanner (1943), and Creak (1964) to Rutter (1966; this volume, Chapter 1) efforts were made to spell out diagnostic criteria (for review see Goldfarb, 1970; Rutter, 1967; Wing, 1966).

Psychotic disorders of childhood and adolescence were divided into three broad groups on the basis of age of onset (Eisenberg, 1966; Rutter, 1967, 1972). This classification is useful in terms of phenomenology, course of illness, and prognosis. It may have merits so far as the patient's response to drug treatment is concerned since the age of onset may be correlated with the degree of central nervous system involvement (Fish, 1971, 1975). The earlier the onset of illness, the more globally the individual child may be affected. This, in turn, may shape the nature and severity of the disturbed behavior, and degree of mental retardation (Bender & Faretra, 1972; Pasamanick & Knobloch, 1961; Kolvin, 1971; Rutter & Lockyer, 1967; Torrey *et al.*, 1975).

Though evidence is admittedly inconclusive, it is possible that the degree of central nervous system involvement may influence the individual patient's drug response. Satterfield and associates (Satterfield *et al.*, 1974) found a marked correlation between the degree of evidence of brain dysfunction obtained from electroencephalographic (EEG) abnormalities, skin conduction levels and neurologic findings, and the clinical response to methylphenidate treatment in hyperactive boys. Baseline low arousal was associated with good drug response.

In small samples of preschool-age psychotic children we, too, have found differential drug response. In a crossover study of lithium and chlorpromazine, the child's baseline EEG and EEG response appeared to be a predictor of his behavioral response to the drug (Campbell *et al.*, 1972b).

Fish (1960) and this author (for review see Campbell, 1975, 1976a) have found that the young psychotic child is frequently sedated by small doses of neuroleptics, such as chlorpromazine. It is conceivable that the excessive sedation is a result of the degree of cerebral dysfunction. On the other hand, a small number of these low-IQ young psychotics showed behavioral improvement on imipramine (Campbell *et al.*, 1971a), lithium (Campbell *et al.*, 1972b), levodopa (Campbell *et al.*, 1976), and T_3 (Campbell *et al.*, 1978d), drugs which are rarely therapeutic in schizophrenia of adults. These findings need further clinical investigation.

It is hoped that delineation of subgroups, using various parameters, including demographic information, will eventually enable us to predict which individual child within a descriptive diagnostic category will respond to drug treatment or even to a specific drug.

Until the diagnostic entities of childhood are more precisely defined by historical, demographic, behavioral, and biologic criteria, children's psychopharmacology will remain on an empirical basis. Thus, we feel strongly that there is a great need to collect demographic information in psychoses of childhood since these disorders are heterogeneous not only behaviorally, but most probably also biologically and etiologically (Chess, 1971; Kanner, 1969). The Children's Personal Data Inventory (CPDI) of the NIMH-ECDEU Pediatric Packet (Addendum, *Psychopharmacology Bulletin*, Special Issue, 1973) is a comprehensive questionnaire, but it lacks a detailed family history for mental illness and the child's pre- and perinatal history. The Children's Psychopharmacology Unit at New York University developed a demographic template for these additional items (Campbell, 1976b, 1977a), a total of 153. The use of this template should facilitate communication between various research centers.

METHODOLOGICAL ISSUES

In drug research it is important to clearly define the population under study, including variables such as IQ, age, sex, social class, severity of illness, family history of mental illness, and pre- and perinatal history (complications of pregnancy and birth) which may contribute to the illness. Most drug studies do not report these. It is also important to assess baseline, target behavior. In our research unit at New York University the following rating scales are used for preschool psychotic children: Children's Behavior Inventory (CBI, Burdock & Hardesty, 1967), Global Clinical Impressions (CGI), the Conners Parent-Teacher Questionnaire (PTQ, Addendum, *Psychopharmacology Bulletin,* Special Issue, 1973), the first 28 items of the Children's Psychiatric Rating Scale (CPRS, Addendum, *Psychopharmacology Bulletin*, Special Issue, 1973), and an additional four items (short attention span, social initiation, language comprehension, and initiation of speech) from Fish's Scale (1968), as well as a timed behavioral rating scale (Cohen *et al.*, 1978).

Before starting drug treatment, a minimum of a two-week placebo period is necessary, not only to "wash out" previously used drugs, but also to evaluate the relationship of some symptoms to the patient's family or other environment, and to screen out those who are placebo responders.

To reduce bias, patients are randomly assigned to drug or placebo (and/or other treatment) and the trial is conducted under double-blind conditions. Even with randomization the groups might not be alike: only large samples give balanced groups. It is desirable to stratify the patient sample for age, sex, IQ, or chronicity (Lipman, 1968) and to randomize thereafter. The differences in groups on baseline may still need adjustment (Nash, 1960).

A proper experimental design is essential; it should aim for differentiation of drug effects from other treatments and environmental manipulations (e.g., milieu treatment, special education, parental counseling): All of these could obscure or confound drug effects or interact with drugs, or they could be more powerful than drugs.

Chassan (1960) recommended a single subject design: an intensive evaluation of each subject. Since individual patients often showed different responses, objections have been raised to apply group statistics in this design. Others proposed combining the intensive evaluation of each subject with an extensive evaluation across patients (Turner *et al.*, 1975).

Qualification and quantification of pharmacologically induced changes with behavioral and objective instrumental measurements (performance changes) described above, should be done with an adequate number of, preferably trained, independent observers.

Behavioral as well as other drug-dependent variables should be assessed on baseline and at fixed points not too long after drug ingestion (Minde & Weiss, 1970).

Rating conditions or ecological observations are important in the process of evaluating treatment effects (Gleser, 1968). The psychotic child may be quite different in a structured or semistructured interview (Campbell *et al.*, 1978a; Fish *et al.*, 1968) than in the more natural conditions at home, or in the classroom. The description of rating conditions is conspicuously absent in the literature.

The child's behavior should be rated by independent observers in his natural environment, using videotapes when possible. We have found the timed behavioral rating scale in combination with the CGI, NGI, PTQ, and CPRS to be useful instruments for rating behavior changes.

Drug administration in research should be time-limited to help differentiate drug effects from maturational changes. More detailed discussions of methodological issues can be found in Eisenberg and Conners (1971), Sprague and Werry (1971), Campbell *et al.* (1977), and Bradford Hill (1971).

REVIEW OF DRUG STUDIES

Psychopharmacology of autistic and schizophrenic children and adolescents is still in a primitive state. Drug studies with large samples of diagnostically homogeneous populations, controlled for age, IQ, and other pertinent variables, are almost nonexistent.

There is no evidence based on well-controlled, double-blind studies, in large samples of children, that pharmacotherapy is more effective than administration of placebo or any other treatment. Furthermore, it was not investigated whether drug administration is more effective when combined with some other treatment modality (for review see Campbell, in press).*

THERAPEUTIC EFFECTS
Neuroleptics

Phenothiazines

Of all psychoactive agents, chlorpromazine and thioridazine, two sedative types of phenothiazines, are most widely used with psychotic and retarded children. However, well-designed and controlled studies are sparse. Fish (1960), in a diagnostically heterogeneous population of children, found chlorpromazine (dose range 1 to 4 mg/kg/d, average dose 2 mg/kg/d) to be useful with schizophrenic children, though the hypoactive and apathetic patients tended to be depressed. In a double-blind study (Fish & Shapiro, 1965; Korein *et al.*, 1971) hospitalized children, 6 to 12

*Since this chapter was written, one such study has been carried out (Campbell *et al.*, 1978a).

years of age, after a two-week placebo washout, were treated with either chlorpromazine, diphenhydramine, or placebo. Sixteen of the 45 children were diagnosed as schizophrenic; the remaining were moderately to mildly distrubed with a variety of diagnoses. Eighty percent of the more severely disturbed children improved on chlorpromazine (100–200 mg/d, 2–9 mg/kg/d), while none of these improved on placebo. Diphenhydramine (200–800 mg/d, 3–26 mg/kg/d) was less effective than chlorpromazine and only minimally superior to placebo. The differences were statistically significant. In a double-blind crossover study of ten preschool-age children, seven of whom were autistic, changes from baseline to treatment with chlorpromazine (9 to 45 mg/d, mean = 17.3 mg/d) were not statistically significant (Campbell et al., 1972b). On daily doses as low as 9 to 15 mg excessive sedation, motor excitation, irritability, insomnia, worsening of psychosis, and catatonic-like states were observed.

Because the prepuberty psychotic child was frequently sedated even by small doses of chlorpromazine and this interfered with the child's functioning, Fish and associates explored trifluoperazine, a potent phenothiazine which was found to be less sedative in adults. In a double-blind, controlled study involving 22 retarded, autistic children, two to six years of age, after a three-week placebo washout, the treatment group received trifluoperazine up to maximum individual tolerance (0.11-0.69 mg/kg/d), while the control group, matched for severity of illness and language impairment, received increasing doses of amphetamine. Amphetamine was substituted by placebo when behavioral untoward effects were noted (Fish et al., 1966). Trifluoperazine produced statistically significant changes only in the most retarded children. Therapeutic changes included increased alertness, social responsiveness, motor initiation and language production. Assessments were done on a symptom severity scale developed by Fish (1968).

Trifluoperazine was also explored in 16 schizophrenic school-age children, in a controlled, double-blind study (Wolpert et al., 1967). In doses of 13 to 20 mg/d, it was found to be as effective as thiothixene in reducing withdrawal and stereotypies, and improving appetite. Assessments were based on clinical global impressions, nurses' rating scale, Bender Gestalt, and Human Figure drawings.

Fluphenazine was explored in two double-blind studies and compared to haloperidol. The assignments were randomized and the dosages individualized. In a study of 30 children, Engelhardt et al. (1973) found that 93% of the children given fluphenazine (mean dose 10.4 mg/d) improved at a statistically significant level as rated on a 19-item symptom rating scale developed by the authors. The improvements were in the areas of self-awareness, constructive play, compulsive acts, and self-mutilation. Faretra *et al.* (1970) explored these drugs in 60 inpatients, ages 5 to 12 years, 52 of whom were schizophrenics. Fluphenazine in doses of 0.75 to 3.75 mg/d caused overall improvement in more than 50% of the children, mainly in decreasing anxiety. Neither of these studies found differences in the efficacy of the two drugs.

Butyrophenones

There are two reports on the effectiveness of trifluperidol (Fish *et al.*, 1969a; LeVann, 1968). However, this drug is no longer used in the United States because of the high rate of extrapyramidal side effects and narrow therapeutic index.

Haloperidol was found to be effective in reducing withdrawal and other symptoms in doses of 0.75 to 3.75 mg/d (Faretra *et al.*, 1970), 0.5 to 4 mg/d (Campbell *et al.*, 1978a), up to 11.9 (mean dose 10.4) mg/d (Engelhardt *et al.*, 1973).

Thioxanthenes

There is only one controlled double-blind study of thiothixene (Wolpert *et al.*, 1967). However, in this clinical trial the methods of assessment were not entirely satisfactory, as noted above. On the basis of clinical judgments, thiothixene did not appear to be different from trifluoperazine. Sixteen children were randomly assigned to thiothixene or trifluoperazine. Four improved on the former, and three on the latter drug. In another study a single-blind design was used (Waizer *et al.*, 1972). The doses utilized range from 1 to 30 mg/d, in children 3 to 15 years of age. On the basis of their findings it seemed that thiothixene had a wide therapeutic margin. The reduction of psychotic symptoms was not accompanied by sedation.

Dihydroindolones

Molindone is a neuroleptic with some characteristics of antidepressants: It has activating, stimulating properties. There is only one published study available with children. In this small sample of preschool-age autistics, molindone was effective in doses of 1 to 2.5 mg/d. It reduced anergy and hypoactivity, and increased verbal production and speech (Campbell *et al.*, 1971b). However, this was a pilot study and cannot be considered conclusive without further exploration.

Tricyclic Dibenzoxazepines

The effects of loxapine succinate, a new antipsychotic agent, were evaluated with schizophrenic children, 13 to 18 years of age. They presented acute psychosis or exacerbation of chronic schizophrenia (Pool *et al.*, 1976). The study was done in double-blind fashion, where the 75 subjects were randomly assigned to loxapine (average dose 87.5 mg/d), haloperidol (average dose 9.8 mg/d), or placebo. Both drugs were significantly superior to placebo in reducing psychotic symptoms. Differences between the two drugs were not significant.

Diphenylbutylpiperidines

Pimozide was explored in an open and single-blind fashion, in a pilot study involving children 9 to 14 years of age (Pangalila-Ratulangi, 1973). Eight of the 10 children were diagnosed as schizophrenic. In doses of 1 to 2 mg/d, pimozide was effective in improving affective contact and social behavior, as evidenced by ratings on various instruments.

Miscellaneous Drugs

Methysergide and LSD

Both were explored in psychotic children. The results of these poorly controlled studies suggested that these drugs were not sufficiently therapeutic to warrant use with these children (Bender *et al.*, 1966; Fish *et al.*, 1969b; Rolo *et al.*, 1965; Simmons *et al.*, 1966, 1972).

Lithium Carbonate

This drug was administered in a placebo-controlled, double-blind study, in a sample of 18 children (ages 8 to 22 years). Eleven of them were diagnosed as infantile autistics. Each subject was assigned randomly to one of the two treatment conditions, in a crossover design (Gram & Rafaelsen, 1972). Lithium was effective in reducing hyperactivity, agressiveness, and stereotypies, among other symptoms. In a small sample of preschool-age children ($N = 10$), lithium was not more effective than chlorpromazine, and the changes from baseline to termination were not statistically significant (Campbell *et al.*, 1972b). However, in one child with a lifelong history of self-mutilation, lithium produced dramatic improvement. It is noteworthy that this patient's autoagressiveness was a most treatment-resistant symptom, that had failed to respond to a variety of drugs.

Levodopa

While one placebo controlled study involving four autistic children failed to show therapeutic effects of levodopa (Ritvo *et al.*, 1971), another report indicated that this drug merited further exploration, particularly in anergic, hypoactive autistic children, with a history of insult to CNS (Campbell *et al.*, 1976).

Tricyclic Antidepressants

This class of drugs had a mixture of therapeutic and untoward effects in this population (Kurtis, 1966; Campbell *et al.*, 1971a). However, well-controlled studies were not done.

Amphetamines

Both dextroamphetamine (Campbell *et al.*, 1972a) and levoamphetamine (Campbell *et al.*, 1976b) cause worsening of behavior in this population.

Triiodothyronine (T3)

Two studies, with inadequate samples and poor controls, indicated that T_3 had positive behavioral effects in euthyroid autistic children (Campbell *et al.*, 1973; Sherwin *et al.*, 1958). In a double-blind, placebo-controlled study of 30 autistic children, 2 to 7 years of age, the results with T_3 were less impressive (Campbell *et al.*, 1978d).

Megavitamins

Though in wide use (Rimland, 1973), therapeutic effectiveness of these drugs has not yet been demonstrated in a double-blind, controlled study (Greenbaum, 1970).

CLINICAL IMPRESSIONS

In our experience drugs can be a valuable though temporary treatment modality in the total treatment of many psychotic children. However, some investigators have reported that psychoactive agents are helpful for schizophrenic children, but not for those with infantile psychosis (Ornitz & Ritvo, 1976).

Neuroleptics are most effective in diminishing psychomotor excitement. This is a direct and usually predictable effect on a target function of behavior. The more enduring effects develop slowly, as a result of the modified interaction between the individual and his environment (Irwin, 1974). The hyperactive child with short attention span, when calmed down by a drug, may be able to focus his attention on a task and thus acquire some reading and writing skills. When agitated, assaultive, or self-mutilating symptoms are eliminated or reduced with an effective psychoactive medication, the patient may develop better social adaptation and improved learning. Drugs themselves do not create learning or intelligence, nor do they necessarily alter parental attitude. However, they can make the patient more amenable to environmental treatments or manipulations. In our clinical experience, though this has not been systematically investigated, the patient seems to respond more to drug therapy in the early stages of illness rather than after other, frequently inappropriate, therapies have failed.

Clearly, in the formative years of the individual, drugs alone never suffice. The choice of other treatments (environmental manipulations, remedial education, individual psychotherapy, group therapy, parental counseling), as well as hospitalization, will depend on contributing factors, associated handicaps of the individual patient, and the family. Many patients, even after the cessation of symptoms such as hyperkinesis, agitation, hallucinations—no longer in need of drug therapy—will require continuation of other treatments and follow-up.

The more severely disturbed and impaired low-IQ patients require potent psychoactive agents such as the neuroleptics. They respond less favorably or even fail to respond to treatment with milder drugs, such as diphenhydramine (Fish, 1960; Fish & Shapiro, 1964, 1965). Fish (1960) suggested that in high-IQ schizophrenics, diphenhydramine be tried as a first step in drug treatment because of its safety and the ease with which it can be regulated. The less-impaired psychotic children may show clinical improvement even on placebo (Fish & Shapiro, 1964, 1965) or milieu treatment (Fish et al., 1966).

Clinical experience has shown that the prepubertal, particularly the preschool-age psychotic child is often excessively sedated by the aliphatic type of phenothiazines such as chlorpromazine, on doses which diminish some of the symptoms (Campbell et al., 1972b, 1972c; Fish, 1960, 1970). The piperazine type of phenothiazine, trifluoperazine, with its stimulating actions, proved to be somewhat better in that respect (Fish et al., 1966). The youngest age group of schizophrenic patients respond to drugs in similar fashion as the adult chronic schizophrenics who are anergic, apathetic, and withdrawn (Fish, 1970). It is conceivable that the excessive sedation of the child at very low doses of chlorpromazine, for example, may be a result of the degree of cerebral dysfunction.

These observations led to a series of drug trials involving psychoactive agents with stimulating properties (Campbell et al., 1970, 1971b, 1972a, c, 1973; Fish et al., 1969a), which were safe and effective in chronic schizophrenics and may be suitable for some of these children.

There is some indication that a psychoactive drug capable of producing behavioral improvement in schizophrenics, including adolescents, may also alter positively certain biochemical and physiological parameters (Brambilla & Penati, 1971; Brambilla et al., 1974).

A number of pilot studies have been carried out to determine whether biochemical abnormalities, including that of serotonin metabolism, exist in psychotic, autistic children (Schain & Freedman, 1961; Boullin *et al.*, 1970; Campbell *et al.*, 1974, 1975; Cohen *et al.*, 1974; for review see Coleman, 1973, and Ornitz & Ritvo, 1976). An attempt to correct possible biochemical aberrations with drug treatment which will result in clinical improvement has failed (Ritvo *et al.*, 1971) and there was no correlation between blood serotonin changes and behavioral response to levodopa in a small sample of psychotic children (Campbell *et al.*, 1976).

For adolescents who show the clinical picture of acute schizophrenia, the purpose of treatment is to restore normal functioning. For the preschool child with onset of psychosis in infancy, one would also wish to promote development. Emerging or nonexistent functions, such as speech and adaptive skills, have to be fostered. Whereas in adults and adolescents with acute schizophrenia, diminution in reactivity via neuroleptics is a desirable effect (Himwich, 1960), the same is considered an untoward effect in the young child who is apathetic, anergic, and lacking any motor initiative.

Thus, drugs such as thiothixene, haloperidol, and molindone seem to be more valuable than the sedative type of aliphatic or piperidine phenothiazines.

UNTOWARD EFFECTS

All these pharmacological agents may cause immediate and long-term untoward effects. Some of these may still be unknown. As always in medicine, the possible untoward effects of treatment should be weighed against the untoward effects of untreated illness (Campbell *et al.*, 1978b; Eisenberg, 1956; Kanner *et al.*, 1972; Rutter *et al.*, 1967).

The immediate untoward effect of neuroleptics is behavioral toxicity; for review DiMascio (1970) and DiMascio *et al.* (1970a) are recommended. Cutaneous disorders, hepatic damage, agranulocytosis, and extrapyramidal effects are infrequently seen in children. (These are elaborated in Campbell, in press-a; DiMascio *et al.*, 1970b; Shader, 1970; Shader & DiMascio, 1970, and will not be discussed here.) Phenothiazines generally lower seizure threshold, and chlorpromazine tends to increase seizures in patients with prior history of convulsive disorder (Tarjan *et al.*, 1957).

Untoward effects due to excess drug, including the parkinsonian symptoms, can be eliminated by decreasing the dose. Routine administration of antiparkinson agents is not recommended since they decrease the plasma levels of neuroleptics and their clinical efficacy (Rivera-Calimlim *et al.*, 1976; Chan *et al.*, 1973; for review see Ayd, 1975). Administration of diphenhydramine (25–50 mg orally or intramuscularly) results in relief of acute dystonic reactions.

Information concerning *long-term untoward effects* is limited. Since psychoactive agents affect the neurotransmitters which control the secretion of hypothalamic neurohormones, caution should be exercised in their administration to prepubertal children and adolescents. They may influence growth, central nervous, endocrine, and reproductive systems. Chlorpromazine decreases growth hormone secretion in adults (Sherman *et al.*, 1971); there are no reports on children. Currently we are investigating the effects of maintenance haloperidol on growth hormone secretion in young psychotic children. There is some evidence that abnormal growth patterns are seen in psychotic children who never received drug treatment (Campbell & Hollander, in preparation; Campbell *et al.*, 1978c; Dutton, 1964; Simon & Gillies, 1964). Menstrual irregularities, amenorrhea, galactorrhea, aspermia, and particularly marked weight gain have been noted on maintenance with neuroleptics. Prolonged administration of phenothiazines may affect IQ (McAndrew *et al.*, 1972).

Reports are available concerning a neurologic syndrome in children resembling tardive dyskinesia in adults (Schiele *et al.*, 1973). McAndrew *et al.* (1972) found that of 125 hospitalized patients, age 8 to 15 years, ten developed involuntary movements of the upper extremities with akathisia after the abrupt withdrawal of phenothiazines. In addition, six patients showed facial tics. These neurological effects were first observed three to ten days after drug withdrawal. They ceased within three to six months. Comparison of these ten patients with those who remained asymptomatic showed that the median duration of drug intake was 32 months in the symptomatic group versus four months in the asymptomatic group. The daily termination dose was 400 mgm of chlorpromazine equivalents, in accordance with the standard dose conversion table (Hollister, 1970) in the symptomatic and only 99 mgm in the asymptomatic group; the median gram intake was 403 in the first and 8.7 in the latter group.

Polizos *et al.* (1973) found that 14 out of 34 outpatient childhood schizophrenics showed similar symptoms (involuntary movements, primarily in the extremities, trunk and head, associated with ataxia) after withdrawal of neuroleptics, which included haloperidol and thiothixene. Both abrupt total withdrawal and gradual, graded withdrawal with weekly reduction of dose by 25% gave the same results. The relationship of this apparently reversible syndrome to persistent, tardive dyskinesia in adults has not been determined (Engelhardt, 1974).

In an attempt to understand the mechanism of these involuntary movements, Winsberg *et al.* (1976) studied cerebrospinal fluid in six psychotic children after probenecid administration. Those who developed withdrawal emergent symptoms had decreased levels of homovanillic acid (HVA) and probably of 5-hydroxyindoleacetic (5-HIAA) as compared to those without dyskinesia.

CONCLUDING REMARKS

As the review of literature shows, psychopharmacology of autistic and schizophrenic children is still in a primitive state. Improved research design and methodology is urgently needed.

There is a paucity of controlled drug studies in homogeneous populations with detailed diagnostic work-up. Studies comparing drug to other treatments are also lacking. Whereas drug administration may result in long-term and not sufficiently known untoward effects, other treatment modalities have shortcomings too (lack of availability and/or great amount of time invested in each patient without yielding dramatic results).

In adult schizophrenia one of the aims of neuroleptics is to correct abnormal mentation. In childhood psychosis on the other hand, an effective drug should decrease psychotic symptoms and correct delays and deficits in development. Such a drug is not available at the present time. On the contrary, the sedative type of neuroleptics are thought to affect cognitive functions adversely.

Biochemical and other biological anomalies found in children with early psychosis suggest that there is a biological basis underlying the abnormal development of behavior. We suggest that with improved research methodology, subgroups within the etiologically heterogeneous population of young autistics may be formed. These should enable us to predict whether a certain child can benefit from a specific drug.

In our present state of knowledge, the use of available therapeutic psychoactive agents cannot be considered a long-term treatment modality, but rather as a temporary though often essential adjunct in the total treatment. Drugs should not be used if the risk and toxicity outweigh the possible therapeutic gains.

ACKNOWLEDGMENTS

This research was supported in part by Public Health Service Grant MH-04665 from the National Institute of Mental Health. Part of this paper was discussed in Psychopharmacologic Treatment of Psychoses in Childhood and Adolescence, in *Psychopharmacology in childhood and adolescence* (J. M. Wiener, Ed.) (Campbell, 1977b). The author wishes to express her thanks to Miss Lynn Wickham for her assistance with the bibliography.

REFERENCES

Addendum—Children's ECDEU Battery. *Psychopharmacology Bulletin*, Special Issue, Pharmacotherapy of Children, 1973, 196–239.

Ayd, F. J. Treatment resistant patients: A moral, legal and therapeutic challenge. In. F. J. Ayd (Ed.), *Rational psychopharmacotherapy and the right to treatment*. Baltimore: Ayd Medical Communications, Ltd., 1975. Pp. 37–61.

Bender, L. Childhood schizophrenia: Clinical Study of 100 schizophrenic children. *American Journal of Orthopsychiatry*, 1947, *17*, 40–56.

Bender, L., Cobrinik, L., Faretra, G., & Sankar, D. V. S. The treatment of childhood schizophrenia with LSD and UML. In M. Rinkel (Ed.), *Biological treatment of Mental Illness*. New York: L. C. Page & Co., 1966. Pp. 463–491.

Bender, L., & Faretra, G. The relationship between childhood schizophrenia and adult schizophrenia. In A. R. Kaplan (Ed.), *Genetic factors in "schizophrenia"*. Springfield, Ill.: Charles C. Thomas, 1972. Pp. 28–64.

Boullin, D. J., Coleman, M., & O'Brien, R. A. Abnormalities in platelet 5-hydroxytryptamine efflux in patients with infantile autism. *Nature*, 1970, *226*, 371–373.

Bradford Hill, A. *Principles of medical statistics*. New York: Oxford University Press, 1971.

Brambilla, F., Guerrini, A., Riggi, F., & Ricciardi, F. Psychoendocrine investigation in schizophrenia. *Diseases of the Nervous System*, 1974, *35*, 362–367.

Brambilla, F., & Penati, G. Hormones and behavior in schizophrenia. In D. H. Ford (Ed.), *Influence of hormones on the nervous system*. Basel: Karger, 1971. Pp. 482–492.

Burdock, E. I., & Hardesty, A. S. Contrasting behavior patterns of mentally retarded and emotionally disturbed children. In J. Zubin & G. A. Jervis (Eds.), *Psychopathology of mental development*. New York: Grune & Stratton, 1967. Pp. 370–386.

Campbell, M. Pharmacotherapy in early infantile autism. *Biological Psychiatry*, 1975, *10*, 399–423.

Campbell, M. Biological interventions in psychoses of childhood. In E. Schopler and R. J. Reichler (Eds.), *Psychopathology and child development: Research and treatment*. New York: Plenum Press, 1976. Pp. 243–270. (a)

Campbell, M. Children's personal data inventory (additional items). Early clinical drug evaluation unit program (ECDEU). *Intercom*, 1976, *5*, 12–21. (b)

Campbell, M. Demographic parameters of disturbed children: A template to the CPDI and CSH, its rationale and significance. *Psychopharmacology Bulletin*, 1977, *13*, 30–33. (a)

Campbell, M. Psychopharmacologic treatment of psychoses in childhood and adolescence. In J. M. Wiener (Ed.), *Psychopharmacology in childhood and adolescence*. New York: Basic Books, Inc., 1977. Pp. 101–118. (b)

Campbell, M. Psychopharmacology for children and adolescents. In J. D. Noshpitz (Ed.), *Basic handbook of child psychiatry*. New York: Basic Books, Inc., in press.

Campbell, M., Anderson, L.T., Meier, M., Cohen, I.L., Small, A.M., Samit, C., & Sachar, E.J. A comparison of haloperidol and behavior therapy and their interaction in autistic children. *Journal of the American Academy of Child Psychiatry*, 1978, *17*, 640–655. (a).

Campbell, M., Fish, B., David, R., Shapiro, T., Collins, P., & Koh, C. Response to triiodothyronine and dextroamphetamine: A study of preschool schizophrenic children. *Journal of Autism and Childhood Schizophrenia*, 1972, *2*, 343–358. (a)

Campbell, M., Fish, B., David, R., Shapiro, T., Collins, P., & Koh, C. Liothyronine treatment in psychotic and nonpsychotic children under 6 years. *Archives of General Psychiatry*, 1973, *29*, 602–608.

Campbell, M., Fish, B., Korein, J., Shapiro, T., Collins, P., & Koh, C. Lithiumchlorpromazine: A controlled crossover study in hyperactive severely disturbed young children. *Journal of Autism and Childhood Schizophrenia*, 1972, *2*, 234–263. (b)

Campbell, M., Fish, B., Shapiro, T., & Floyd, A., Jr. Thiothixene in young disturbed children. A pilot study. *Archives of General Psychiatry*, 1970, *23*, 70–72.

Campbell, M., Fish, B., Shapiro, T., & Floyd, A., Jr. Imipramine in preschool autistic and schizophrenic children. *Journal of Autism and Childhood Schizophrenia*, 1971, *1*, 267–282. (a)

Campbell, M., Fish, B., Shapiro, T., & Floyd, A., Jr. Study of molindone in disturbed preschool children. *Current Therapeutic Research*, 1971, *13*, 28–33. (b)

Campbell, M., Fish, B., Shapiro, T., & Floyd, A., Jr. Acute responses of schizophrenic children to a sedative and "stimulating" neuroleptic: A pharmacologic yardstick. *Current Therapeutic Research*, 1972, *14*, 759–766. (c)

Campbell, M., Friedman, E., DeVito, E., Greenspan, L., & Collins, P. J. Blood serotonin in psychotic and brain damaged children. *Journal of Autism and Childhood Schizophrenia*, 1974, *4*, 33–41.

Campbell, M., Friedman, E., Green, W. H., Collins, P. J., Small, A. M., & Breuer, H. Blood serotonin in schizophrenic children. A preliminary study. *International Pharmacopsychiatry*, 1975, *10*, 213–221.

Campbell, M., Geller, B., & Cohen, I.L. Current status of drug research and treatment with autistic children. *Journal of Pediatric Psychology*, 1977, *2*, 153–161.

Campbell, M., Hardesty, A.S., Breuer, H., Jr., & Polevoy, N. Childhood psychosis in perspective: A follow-up of 10 children. *Journal of the American Academy of Child Psychiatry*, 1978, *17*, 14–28. (b)

Campbell, M., & Hollander, C.S. Findings suggestive of hypothalamic involvement in young psychotic children. In preparation.

Campbell, M., Petti, T., David, R., Genieser, N.B., Green, W.H., & Cohen, I.L. Some physical parameters of young autistic children. Paper presented at the 25th Annual Meeting of the American Academy of Child Psychiatry, San Diego, Oct. 25–29, 1978. (c)

Campbell, M., Small, A.M., Collins, P.J., Friedman, E., David, R., & Genieser, N.B. Levodopa and levoamphetamine: A crossover study in schizophrenic children. *Current Therapeutic Research*, 1976, *18*, 70–86.

Campbell, M., Small, A.M., Hollander, C.S., Korein, J., Cohen, I.L., Kalmijn, M., & Ferris, S. A controlled crossover study of triiodothyronine in autistic children. *Journal of Autism and Childhood Schizophrenia*, 1978, *8*, 371–381. (d)

Chan, T. L., Sakalis, G., & Gershon, S. Some aspects of chlorpromazine metabolism in humans. *Clinical Pharmacology and Therapeutics*, 1973, *14*, 133.

Chassan, J. B. Statistical inference and the single case in clinical design. *Psychiatry*, 1960, *23*, 173–184.

Chess, S. Autism in children with congenital rubella. *Journal of Autism and Childhood Schizophrenia*, 1971, *1*, 33–47.

Cohen, D. J., Schaywitz, B. A., Johnson, W. T., & Bowers, M. Biogenic amines in autistic and atypical children. *Archives of General Psychiatry*, 1974, *31*, 845–853.

Cohen, I.L., Anderson, L.T., & Campbell, M. Measurement of drug effects in autistic children. *Psychopharmacology Bulletin*, 1978, *14*, 68–70 (vol. 4).

Coleman, M. Serotonin and central nervous system syndromes of childhood. A review. *Journal of Autism and Childhood Schizophrenia*, 1973, *3*, 27–35.

Creak, M. Schizophrenic syndrome in childhood: Further progress report of a working party. *Developmental Medicine and Child Neurology*, 1964, *4*, 530–535.

Delay, J., & Deniker, P. 38 cas de psychoses traitées par la cure prolongée et continue de 4560 RP-Leme Congrès des Alien et Neurologique de Langue Française, Luxembourg, 21–27 Juillet, 1952, p.503.

De Sanctis, S. La neuropsychiatria infantile. *Infanzia Anormale*, 1925, *18*, 633–661.

DiMascio, A. Behavioral toxicity. In A. DiMascio & R. I. Shader (Eds.), *Clinical handbook of psychopharmacology*. New York: Science House, 1970, Pp. 185–193.

DiMascio, A., Shader, R. I., & Giller, D. R. Behavioral toxicity. Part III: Perceptual-cognitive functions and Part IV: Emotional (mood) states. In R. I. Shader & A. Di-Mascio (Eds.), *Psychotropic drug side effects*. Baltimore: The Williams & Wilkins Company, 1970. Pp. 132–141. (a)

DiMascio, A., Soltys, J. J., & Shader, R. I. Psychotropic drug side effects in children. In R. I. Shader and A. DiMascio (Eds.), *Psychotropic drug side effects*. Baltimore: The Williams & Wilkins Company, 1970. Pp. 235–260. (b)

Dutton, G. The growth pattern of psychotic boys. *British Journal of Psychiatry*, 1964, *110*, 101–103.

Eisenberg, L. The autistic child in adolescence. *American Journal of Psychiatry*, 1956, *112*, 607–612.

Eisenberg, L. The classification of the psychotic disorders in childhood. In L. P. Eron (Ed.), *The classification of behavior disorders*. Chicago: Aldine, 1966. Pp. 87–114.

Eisenberg, L., & Conners, C. K. Psychopharmacology in childhood. In N. B. Talbot, J. Kagan, & L. Eisenberg (Eds.), *Behavioral science in pediatric medicine*. Philadelphia: W. B. Saunders Co., 1971. Pp. 397–423.

Engelhardt, D. M. CNS consequences of psychotropic drug withdrawal in autistic children: A follow-up report. Paper presented at the Annual ECDEU Meeting, NIMH, Key Biscayne, Florida, May 23–25, 1974.

Engelhardt, D. M., Polizos, P., Waizer, J., & Hoffman, S. P. A double-blind comparison of fluphenazine and haloperidol. *Journal of Autism and Childhood Schizophrenia*, 1973, *3*, 128–137.

Faretra, G., Dooher, L., & Dowling, J. Comparison of haloperidol and fluphenazine in disturbed children. *American Journal of Psychiatry*, 1970, *126*, 1670–1673.

Fish, B. Drug therapy in child psychiatry: Pharmacological aspects. *Comprehensive Psychiatry*, 1960, *1*, 212–227.

Fish, B. Methodology in child psychopharmacology. In D. H. Efron, J. O. Cole, J. Levine, & J. R. Wittenborn (Eds.), *Psychopharmacology, review of progress, 1956–1967*. Washington, D. C.: U. S. Government Printing Office, 1968. (U. S. Public Health Service Publication No.1836.) Pp. 989–1001.

Fish, B. Psychopharmacologic response of chronic schizophrenic adults and predictors of responses in young schizophrenic children. *Psychopharmacology Bulletin*, 1970, *6*, 12–15.

Fish, B. Contributions of developmental research to a theory of schizophrenia. In J. Hellmuth (Ed.), *Exceptional infant, Vol. 2. Studies in abnormalities*. New York: Brunner/Mazel Inc., 1971. Pp. 473–482.

Fish, B. Biological antecedents of psychosis in children. In D. X. Freedman (Ed.), *The biology of the major psychoses: A comparative analysis*. New York: Raven Press, 1975. Pp. 49–83.

Fish, B., Campbell, M., Shapiro, T., & Floyd, A., Jr. Comparison of trifluperidol, trif-luoperazine and chlorpromazine in preschool schizophrenic children: The value of less sedative antipsychotic agents. *Current Therapeutic Research*, 1969, *11*, 589–595. (a)

Fish, B., Campbell, M., Shapiro, T., & Floyd, A., Jr. Schizophrenic children treated with methysergide (sansert). *Diseases of the Nervous System*, 1969, *30*, 534–540. (b)

Fish, B., & Shapiro, T. A descriptive typology of children's psychiatric disorders II: A behavioral classification. Psychiatric Research Report, 18, American Psychiatric Association, 1964, Pp. 75–86.

Fish, B., & Shapiro, T. A typology of children's psychiatric disorders I: Its application to a controlled evaluation of treatment. *Journal of the American Academy of Child Psychiatry*, 1965, *4*, 32–52.

Fish, B., Shapiro, T., & Campbell, M. Long-term prognosis and the response of schizo-phrenic children to drug therapy: A controlled study of trifluoperazine. *American Journal of Psychiatry*, 1966, *123*, 32–39.

Fish, B., Shapiro, T., Campbell, M., & Wile, R. A classification of schizophrenic children under five years. *American Journal of Psychiatry*, 1968, *124*, 1415–1423.

Gajzago, C., & Prior, M. Two cases of "recovery" in Kanner Syndrome. *Archives of General Psychiatry*, 1974, *31*, 264–268.

Gleser, G. C. Psychometric contributions to the assessment of patients. In D. H. Efron, J. O. Cole, J. Levine, & J. R. Wittenborn (Eds.), *Psychopharmacology, review of progress, 1956–1967.* Washington, D.C.: (U.S. Public Health Service, U.S. Government Printing Office, Publication No. 1836.) 1968. Pp. 1029–1037.

Goldfarb, W. Childhood psychosis. In P. H. Mussen (Ed.), *Carmichael's manual of child psychology*, (3rd ed., Vol. 2). New York: John Wiley and Sons, 1970. Pp. 765–830.

Gram, L. F., & Rafaelsen, O. J. Lithium treatment of psychotic children. A controlled clinical trial. In A. L. Annell (Ed.), *Depressive states in childhood and adolescence.* Stockholm: Almquist & Wiksell, 1972. Pp. 488–490.

Greenbaum, G. H. An evaluation of niacinamide in the treatment of childhood schizophrenia. *American Journal of Psychiatry*, 1970, *127*, 129–132.

Group for the Advancement of Psychiatry (GAP). Pharmacotherapy and psychotherapy: Paradoxes, problems and progress. Vol. IX, Report No. 93, 1975.

Himwich, H. E. Biochemical and neurophysiological action of psychotropic drugs. In L. Uhr & J. G. Miller (Eds.), *Drugs and behavior.* New York: John Wiley & Sons, 1960. Pp. 41–48.

Hollister, L. E. Choice of antipsychotic drugs. *American Journal of Psychiatry*, 1970, *127*, 186–190.

Irwin, S. How to prescribe psychoactive drugs. *Bulletin of the Menninger Clinic*, 1974, *38*, 1–13.

Kanner, L. Autistic disturbances of affective contact. *Nervous Child*, 1943, *2*, 217–250.

Kanner, L. The children haven't read those books. Reflections on differential diagnosis. *Acta Paedopsychiatrica*, 1969, *36*, 2–11.

Kanner, L., Rodriguez, A., & Ashenden, B. How far can autistic children go in matters of social adaptation? *Journal of Autism and Childhood Schizophrenia*, 1972, *2*, 9–33.

Kolvin, I. Psychoses in childhood—a comparative study. In M. Rutter (Ed.), *Infantile autism: Concepts, characteristics, and treatment.* Edinburgh: Churchill Livingstone, 1971. Pp. 7–26.

Korein, J., Fish, B., Shapiro, T., Gerner, E. W., & Levidow, L. EEG and behavioral effects of drug therapy in children. Chlorpromazine and diphenhydramine. *Archives of General Psychiatry*, 1971, *24*, 552–563.

Kurtis, L. B. Clinical study of the response to nortriptyline on autistic children. *International Journal of Neuropsychiatry*, 1966, *2*, 298–301.

LeVann, L. J. A new butyrophenone: Trifluperidol. A psychiatric evaluation in a pediatric setting. *Canadian Psychiatric Association Journal*, 1968, *13*, 271–273.

Lipman, R. S. Methodology of drug studies in children. Paper presented at the American Orthopsychiatric Meeting, Chicago, Illinois, March 1968.

Lipman, R. S. The use of psychopharmacological agents in residential facilities for the retarded. In F. Menolascino (Ed.), *Psychiatric approaches to mental retardation.* New York: Basic Books, Inc., 1970. Pp. 387–398.

McAndrew, J. B., Case, Q., & Treffert, D. Effects of prolonged phenothiazine intake on psychotic and other hospitalized children. *Journal of Autism and Childhood Schizophrenia*, 1972, *2*, 75–91.

Minde, K. K., & Weiss, G. C. The assessment of drug effects in children as compared to adults. *Journal of American Academy of Child Psychiatry*, 1970, *9*, 124–133.

Nash, H. The design and conduct of experiments on the psychological effects of drugs. In L. Uhr & J. G. Miller (Eds.), *Drugs and behavior.* New York: John Wiley & Sons, 1960.

Ornitz, E. M., & Ritvo, E. R. The syndrome of autism: A critical review. *American Journal of Psychiatry*, 1976, *133*, 609–621.

Pangalila-Ratulangi, E. A. Pilot evaluation of Orap® (Pimozide, R6238) in child psychiatry. *Psychiatria, Neurologia, Neurochirurgia*, 1973, *76*, 17–27.

Pasamanick, B., & Knobloch, H. Epidemiologic studies on the complications of pregnancy and the birth process. In G. Caplan (Ed.), *Prevention of mental disorders in children*. New York: Basic Books, Inc., 1961. Pp. 74–94.

Polizos, P., Engelhardt, D. M., Hoffman, S. P., & Waizer, J. Neurological consequences of psychotropic drug withdrawal in schizophrenic children. *Journal of Autism and Childhood Schizophrenia*, 1973, *3*, 247–253.

Pool, D., Bloom, W., Mielke, D. H., Roniger, J. J., & Gallant, D. M. A controlled evaluation of loxitane in seventy-five adolescent schizophrenic patients. *Current Therapeutic Research*, 1976, *19*, 99–104.

Rimland, B. High dosage levels of certain vitamins in the treatment of children with severe mental disorders. In D. Hawkins & L. Pauling (Eds.), *Orthomolecular psychiatry*. San Francisco: W. H. Freeman & Company, 1973. Pp. 513–539.

Ritvo, E. R., Yuwiler, A., Geller, E., Kales, A., Rashkis, S., Schicor, A., Plotkin, S., Axelrod, R., & Howard, C. Effects of L-dopa in autism. *Journal of Autism and Childhood Schizophrenia*, 1971, *1*, 190–205.

Rivera-Calimlim, L., Nasrallah, H., Strauss, J., & Lasagna, L. Clinical response and plasma levels: Effect of dose, dosage schedules, and drug interactions on plasma chlorpromazine levels. *American Journal of Psychiatry*, 1976, *133*, 646–652.

Rolo, A., Krinsky, L., Abramson, H., & Goldfarb, L. Preliminary method study of LSD with children. *International Journal of Neuropsychiatry*, 1965, *1*, 552–555.

Rutter, M. Behavioural and cognitive characteristics of a series of psychotic children. In J. K. Wing (Ed.), *Early childhood autism*. Oxford: Pergamon Press, 1966. Pp. 51–81.

Rutter, M. Psychotic disorders in early childhood. In A. J. Coppen & A. Walk (Eds.), *Recent developments in schizophrenia: A symposium*. London: R.M.P.H., 1967. Pp. 133–158.

Rutter, M. Childhood schizophrenia reconsidered. *Journal of Autism and Childhood Schizophrenia*, 1972, *2*, 315–337.

Rutter, M., Greenfeld, D., & Lockyer, L. A five to fifteen year follow-up study of infantile psychosis. II. Social and behavioural outcome. *British Journal of Psychiatry*, 1967, *113*, 1183–1199.

Rutter, M., & Lockyer, L. A five to fifteen year follow-up study of infantile psychosis. I. Description of sample. *British Journal of Psychiatry*, 1967, *113*, 1168–1182.

Satterfield, J. H., Cantwell, D. P., & Satterfield, B. T. Pathophysiology of the hyperactive child syndrome. *Archives of General Psychiatry*, 1974, *31*, 839–844.

Schain, R. J., & Freedman, D. X. Studies of 5-hydroxyindole metabolism in autistic and other mentally retarded children. *Journal of Pediatrics*, 1961, *58*, 315–320.

Schiele, B. C., Gallant, D., Simpson, G., Gardner, E. A., & Cole, J. O. Tardive dyskinesia. *American Journal of Orthopsychiatry*, 1973, *43*, 506, 888.

Shader, R. I. Endocrine, metabolic, and genitourinary effects of psychotropic drugs. In A. DiMascio & R. I. Shader (Eds.), *Clinical handbook of psychopharmacology*. New York: Science House, 1970. Pp. 205–212.

Shader, R. I., & DiMascio, A. (Eds.). *Psychotropic drug side effects*. Baltimore: The Williams & Wilkins Company, 1970.

Sherman, L., Kim, S., Benjamin, F., & Kolodny, H. D. Effect of chlorpromazine on serum growth-hormone in man. *New England Journal of Medicine*, 1971, *284*, 72–74.

Sherwin, A. C., Flach, F. F., & Stokes, P. E. Treatment of psychoses in early childhood with triiodothyronine. *American Journal of Psychiatry*, 1958, *115*, 166–167.

Simmons, J. Q., III, Benor, D., & Daniel, D. The variable effects of LSD-25 on the behavior of a heterogeneous group of childhood schizophrenics. *Behavioral Neuropsychiatry*, 1972, *4*, 10–16, 24.

Simmons, J. Q., III, Leiken, S. J., Lovaas, O. I., Schaeffer, B., & Perloff, B. Modification of autistic behavior with LSD-25. *American Journal of Psychiatry*, 1966,. *122*, 1201–1211.

Simon, G. B., & Gillies, S. M. Physical characteristics of a group of psychotic children. *British Journal of Psychiatry*, 1964, *110*, 104–107.

Sprague, R. L., & Werry, J. S. Methodology of psychopharmacological studies with the retarded. In N. R. Ellis (Ed.), *International review of research in mental retardation.* New York: Academic Press, 1971. Pp. 147–219.

Tarjan, G., Lowery, V. E., & Wright, S. W. Use of chlorpromazine in two hundred seventy-eight mentally deficient patients. *American Medical Association Journal of Diseases of Children*, 1957, *94*, 294–300.

Torrey, E. F., Hersh, S. P., & McCabe, K. D. Early childhood psychosis and bleeding during pregnancy. *Journal of Autism and Childhood Schizophrenia*, 1975, *5*, 287–297.

Turner, D. A., Purchatzke, G., Gift, T., Farmer, C., & Uhlenhuth, E. H. Intensive design in evaluating anxiolytic agents. In F. G. McMahon (Ed.), *Principals and techniques of human research and therapeutics*. J. Levine, B. C. Schiele, W. J. R. Taylor (Eds.), Vol. VIII, *Psychopharmacological agents*. Mt. Kisco, New York: Futura Publishing Company, 1975.

Waizer, J., Polizos, P., Hoffman, S. P., Engelhardt, D. M., & Margolis, R. A. A single-blind evaluation of thiothixene with outpatient schizophrenic children. *Journal of Autism and Childhood Schizophrenia*, 1972, *2*, 378–386.

Wing, J. K. Diagnosis, epidemiology, aetiology. In J. K. Wing (Ed.), *Early childhood autism*. Oxford: Pergamon Press, 1966. Pp. 3–49.

Winsberg, B. G., Hurwic, M. J., Perel, J. M., Sverd, J., Castells, S., & Yepes, L. Neurochemistry of withdrawal emergent symptoms in children. Paper presented at the Annual ECDEU Meeting, NIMH, Key Biscayne, Florida, May 20–22, 1976.

Wolpert, A., Hagamen, M. B., & Merlis, S. A comparative study of thiothixene and trifluoperazine in childhood schizophrenia. *Current Therapeutic Research*, 1967, *9*, 482–485.

World Health Organization (WHO) Scientific group on psychopharmacology: *Research in psychopharmacology*. World Health Organization Technical Report Series No. 371. Geneva: World Health Organization, 1967.

24

Therapy with Autistic Children

THEODORE SHAPIRO

If any therapy is to be investigated scientifically one must dispassionately, without romanticism acknowledge that: (1) autism is not a static disorder; (2) we know something about the factors related to outcome (Rutter, 1974) which include initial I.Q., degree of language impairment, total symptom score, and experience of schooling; (3) we adhere to the usual rules of scientific investigation which include statistical comparisons and attempt to decrease the bias of selection. Finally, we must include in all our studies some consideration of the natural history of autism so that when we claim significant change, we are not doing less than or equal to that which would be expected when the child is reared with the humanistic supports that are available. We have on hand Eisenberg's (1957) careful literature review indicating that if communicative speech is developed by age five, there is a 50% chance for some social recovery, and that within the entire group of psychotic children improvement rates of better than 25% must be recorded in order to do better than natural outcome alone. Our own studies (Shapiro *et al.*, 1974; Shapiro, 1976) on the language development of young psychotic children point to certain parameters which might be used as early prognosticators such as the persistence of echoic speech—the development of out-of-context bizarre utterance, as well as the percentage of communicativeness after brief (three months) observation in nursery milieu.

In order to approach the issue of therapy with autistic children the

THEODORE SHAPIRO · Department of Psychiatry, Cornell University Medical College; Department of Child and Adolescent Psychiatry, Payne Whitney Clinic, New York, New York 10021.

distinctions between *organism* and *person,* and *structure* and *function* must be made: The person-organism distinction is by far the easiest! In most people's mind an organism is pictured as some interactional system referring to a substrate of mutually separably functional units all contributing to larger coherent homeostatic units. This may include such notions as the interplay of the cytoarchitecture of an organ or an organelle with an organ or an organ within a larger biological system. One problem in dealing with the level, organism, concerns the hope of the Helmholzian tradition, i.e., that all behavior and functions would be reduced to structure (biochemical or otherwise). However, reductionism does not have to be the outcome of such investigation. The *person,* on the other hand, is a creature created by the Humanist tradition. He or she is looked at as the outcome of a long line of evolution, currently finding its terminus in homosapiens. Observation leads us to the notion that persons live in contexts, social and personal. In that sense, society can also be looked at as an organism and the distinction between person and organism is threatened. Ontogenetically, the growing organism lives in an expanding psychological as well as social milieu ranging from immediate caretakers, usually within a family grouping and then spreading out to peers and communities. In order to somehow do away with the biological and social discontinuities, animal behaviorists (Schneirla & Rosenblatt, 1961) talk about the biosocial and psychosocial forces that dictate later behavior of individuals. By momentarily belaboring these distinctions we will be aided to bring order to the broad span of therapeutic interventions that have been used to treat autistic children.

THE BIOLOGICAL ORGANISM AND PHARMACOTHERAPY

Campbell, my colleague for many years, has presented data (Chapter 23) which might best be considered at a biological organismic level. She cautions us that though drugs effect *some* substrate, we are not yet in the position to know *what* substrate and with what specificity. However, we are in a position to make a number of empirical judgments which have "cash value": Drug treatment results in autistic children (as in the case of psychotropic drug treatment for other psychiatric conditions) should be judged on the basis, not of the knowledge of end organ biochemical site of action, but on the basis of behavioral criteria. Thus, while Campbell works on the organismic level, she must also of necessity deal with the person, his/her behavior, and general adaptation. Indeed, she states that the most appropriate use of drugs for young psychotic children might be to set development in motion so that other techniques at the level of education could be utilized to increase potentials for learning, or to use drugs to diminish those behaviors which serve to interrupt learning. I take this as a fortunate and seminal notion, because it considers the idea that while we may find a biochemical deficit one day in psychotic children, the finding may not necessarily provide us with an outcome where an intact person emerges who is able to function in the world. While Campbell calls for larger statistical studies in addition to her collaborative studies with other methods, I would offer a caveat: As I am sure she is aware, single case studies also have much to offer at this stage of our knowledge. Since we both agree that autism is a developmental disorder, we ought to be most curious about how one can distinguish developmental processes from behaviors we consider indications of treatment progress. What we do not know about these children is how the timetable of maturation is fixed, i.e., how can we deduce and influence the "functional" from the behavior agreed on? If we consider Bender's (1947) and Fish's (1961) concepts of dysmaturation and deduce a process which disrupts the temporal flow of maturation in childhood psychotics, we may be able to postulate the possibility that there are piloting and guiding biological clocks that can be influenced from within and without. The question we then must ask in relation to psychopharmacological agents is, do they work by *trigger processes* that are instrumental in forwarding maturation or are they *continuously necessary to sustain* essential biological mechanisms as exogenous thyroxin might be in hypothyroidism? Such questions can only be answered once we get a closer look at individual children who may indeed not even share a common biochemical defect and see how the developmental process is affected by drugs.

Campbell's increased attention to distinguishing those features that are vital for matching samples such as demographic, behavioral, etc., may be important at a certain stage of scientific development, but I am sure she has kept in mind the model of syphilis versus the model of fever. In the former, the same substrate interaction between a host and microorganism may show effects in different organ systems at different times in the life cycle while in the latter the symptom may be a general biological indicator of a great variety of processes. In other words, we sometimes must question whether or not autism is a single disorder at the organismic level just because common features in behavior are selected to diagnose the syndrome. Coleman (1974) has certainly shown us autistic-appearing children who have lesions in uric acid metabolism as well as other varieties of biochemical lesions, while Chess (1971) has pointed to the fact that autistic syndromes may accrue following congenital rubella. We do not know what the future will bring with respect to the elaboration of the interplay between substrate and behavior, but more germane to the issue of therapy we may have to accept varied treatments tailored to specific lesions rather than one "medicine" for all.

In addition to pharmacotherapy there have been still other treatments which address the substrate. During a time when convulsive therapies were being empirically tried for many ills in psychiatry, Bender (1971) utilized this treatment for acutely regressed young psychotic children. While she claimed this therapy aided in the reorganization of the children's adaptive behaviors, it was never widely used and has been largely discontinued.

NEURODEVELOPMENTALLY DIRECTED THERAPIES

Bender (1947) and her followers have also been responsible for a view of autism that has directed therapeutic intervention. She looked at autism in a continuum with childhood schizophrenia suggesting that the behavioral difficulties of these children were secondary to a large array of neuroperceptual lags, dysmaturation, and lack of integration. Fish's (1971) conclusion that autism is a disorder of "timing and integration of neurological maturation" echoes Bender's (1947) citation that one should not look for the answer to autism and childhood schizophrenia in the cytoarchitecture of the brain. This formulation led some therapists to perceptual training techniques designed to strengthen integration (see Fish, 1975, 1976). Similar approaches have been advocated and described by Frostig (1972). Ayres and Heskett (1972) describe progress of a seven-year-old girl via "careful use of tactile and vestibular stimulation" yielding gains in I.Q. and adaptive behavior. Similar recommendations have been made regarding autism and training techniques for learning disabilities. In opposition to simple "perceptual training" Mann and Goodman (1976) criticized the approach on the grounds that perception is an abstraction while "skills for living" should be the real task-oriented educational aims. Indeed, reports of transfer from perceptual training programs to adaptive skills without direct training in these skills are few.

EDUCATIONAL-MILIEU THERAPIES

A more tailored general educational approach has been described by Fenichel (1974), former director of the League School in New York. His use of the day school rather than a hospital and educational rather than therapeutic approaches helped autistics to be brought into a more usual normalizing milieu. They lived at home, went to school, etc. He believed, however, as did the physician, that these children suffered from a central nervous system disorder with cognitive deficits requiring special educational techniques based on interdisciplinary psychoeducational assessment that covered some of the same areas suggested by Bender (1947) and others (Fish, 1971, 1975).

Fenichel's attempts to "detoxify" the atmosphere by making therapy education helped also to educate mothers and fathers to new roles in aiding their children. This general aim was also served in the formulation of the Community Services approach to developmental therapy begun in North Carolina by Schopler and Reichler (1971), which was legislatively established as TEACCH (Treatment and Education of Autistic and Related Communications Handicapped Children) and is described in Chapter 29. This approach grew out of the realization of the need for special educational and therapeutic skills of a variety of professionals to be coordinated for the care and education of autistic children. The educational facilities in this system remain within the public school system. The diagnostic process is coordinated and prescriptions carried out by therapeutic workers, under careful supervision, in a "total push" program that supports parents and engages them as adjunct teacher-therapists as well. Similar efforts to engage State agency responsibility for education of autistic children have been a central focus of many parent groups that lobby for their children.

PLAY AND INDIVIDUAL THERAPIES

It is not of passing interest that such programs sometimes evolve out of a countermeasure against therapies that seemed to blame mothers for the deviant behavior of their children. Theories of autistogenesis and interactional etiology sometimes supported the ready guilt of parents struggling to find cause for their children's unusual behavior. Schopler and Loftin's (1969) work was instrumental in demonstrating that mothers' difficulties in thinking are specifically aroused around test performances and are primed by concern with their ill child. Similarly, Goldfarb's designation (1974) "parental perplexity" seemed related to parents' difficulties in understanding and dealing with their autistic schizophrenic children. Thus, while therapy proceeds best when guilt is lessened, mothers and fathers of autistic children do seem to respond differentially toward their autistic children than their normally developing siblings. This difference in response is increasingly recognized as parental response to a difficult child (see Chapters 17 and 18).

While Bettelheim (1967) met the matter of parental influence with parentectomies and prolonged separations during therapy, Mahler and Furer (1972) adapted their psychoanalytically modeled therapy in their later work to include the mothers in what they call "action research." Bettelheim's claim of good results in treating autistics is clearly related to his tough selection criteria. Though many of his children did not speak at time of admission to the Orthogenic School they had regressed from speaking and were not as severely impaired. Fish (1975) has repeatedly emphasized that the more severely disordered autistic child can be recognized earlier than the less severe. Greater severity increases the probability of direct genetic involvement and of disabling symptoms. Such early cases become fixed in older age groups and more resistant to any treatment. Indeed, Wenar and Ruttenberg (1976) used before and after treatment scores using their Behavior Rating Instrument for Autistic Children (BRIAC) on 46 children undergoing a wide range of therapies. Sensitive attentive staff whether behaviorally or analytically oriented were better than simple custodial care. This latter fact certainly should cause us to take notice that while most agree that autism has a biological basis we must not ignore the person emerging despite these deficits. Thus, humanistic, person-oriented therapeutic approaches may, at times, err in excessive reliance on environmental influence, but they need not ignore the contribution of biology to the disorder or even of the influence of biological factors on cognition and symbolization.

Man's most distinguishing feature is thought by some to be his capacity to symbolize. A symbol usually applies to a relationship between a referent and its reference. In easy jargon it pertains to a thing or idea and the word that stands for it. In the highest symbolic form such as abstraction the relationship between the referent and its reference is generally arbitrary. While we can gain some agreement as to what a symbol is, play as a symbolic vehicle becomes a more difficult problem. Millar (1968) outlines many past and existing theories. No one would quibble that at one stage of ontogenetic development play becomes a vehicle for symbolic expression with the possibility that it even subsumes multiple functions (Waelder, 1936) central to adaptation. Indeed, with increasing structuralization of mind play undergoes a number of ontogenetic alterations for cognitive psychologists (Piaget, 1962; Stern, 1974). DesLauriers presents a number of syllogisms: The autistic child has no symbolic function and therefore he has no language. He has no language and therefore has no communicative speech. He does not speak because he has nothing to say, because he lacks the development of symbolic function, and furthermore—and this is probably the most important link in the chain—he has no symbolic function because he has not learned to play, thus his designation "Theraplay." This view of the child in his essentially human role as symbolizer *in statu nascendi* is the fact that tears more hearts of parents and therapists alike and indeed, increases our interest as investigators.

We see little people full of apparent integrity both physically and in their isolation. Their behavioral stance has led some to feel that they keep a deep inner secret (Bettelheim, 1967). I would further suggest that the fascination of many professionals with autistic children is the belief that there is such a secret little person somewhere within the screened isolation especially insofar as they often show a number of developmental high points which startles the examiner out of the notion that they are simply retarded. However, while our human sympathies permit the most humane interventions we must not ignore propositions which may have more scientific value.

Again, I will make a distinction—language is not speech—language is a formal organization or code which has rules and structure that can be studied apart from any particular vehicle for its expression such as speech. In fact, DeSausure (1959) suggested that speech is a particular performance of a language presented in a vocal-auditory vehicle. In the human condition, the auditory-vocal channel is primary in developing linguistic skills (Lenneberg, 1967).

Children learning language become remarkably proficient with the code which is then used as an important social device so quickly that many believe that the achievement of language competence does not follow simple rules of learning. Rather it has been suggested by Chomsky (1965), Lenneberg (1967), and even by some behaviorists that in the S-R relationship there must be an Sm (S meaning) which may be looked at as an intervening variable corresponding to an organismic factor to account for rapid language learning. At the highest points in development we can deal with the territory by use of a mental map we call language even in the absence of immediately perceived things, i.e., a stored representational reality is a required postulate. Werner and Kaplan (1963) provide us with the notion that as this distancing occurs, it parallels the separation-individuation (Mahler, 1968) between mother and infant. This proposition is useful because it states in modest form a naturally occurring early developmental interaction. However, it does not suggest that one side of the interaction is more critical than the other. The postulation of a biologically based maturation of language led to the experiment of Frederick II of the Holy Roman Empire. He wanted to know which language was first, Hebrew or Greek, so he subjected two babies to wet nurses, who were instructed not to prattle to the children or sing to them and see which language they developed. The children died because of some variety, I imagine, of marasmic syndrome.

It may not be possible to identify experimentally the priority of either side in the mother-infant interaction. In a similar fashion, unlike the implication of DesLauriers (Chapter 21), play may not be prior to speech. Rather, play is probably concomitant with aspects of symbolic function, and speech like play is a particular vehicle by which symbolic function is expressed. They may enhance each other developmentally, but it is more likely that there is a common substructure which permits the development of both. There is very clear experimental evidence by Sinclair-de-Zwart (1960) that linguistic learning does not enhance subsequent cognitive competence.

LEARNING THERAPIES

While DesLauriers attends to the naturalness of play, the Behaviorists make careful scrutiny of its contingencies in order to arrive at more adaptive solutions for autistic children whose behavior seems so deviant. They are most interested in the stimulus circumstances that the proximal environment unwittingly sets up that in turn enhance nonadaptive behavior. They pay little attention in their therapeutic approaches to the organism except as a given, and attend more to the stimuli that support unwanted or undeveloped behaviors. This analysis leads to therapeutic programs that follow the general laws of learning usually using operant conditioning rather than classical conditioning techniques. Lovaas *et al.* (1974) describe Ferster's early demonstration that reinforcement schedules are feasible to train autistics and Lovaas, himself, suggests that therapy ought to address itself to three classes of behavior: self destructive, self stimulative, and the teaching of appropriate but undeveloped behaviors such as language.

The bulk of the behavioral modification work has been directed toward operant conditioning but aversive techniques have also been administered. A recent review (Lichstein & Schreibman, 1976) of 12 studies suggest that there is a 25 : 5 ratio of positive-to-negative results. Aversive treatment of this sort meets with the same moralizing disapproval as ECT did in the earlier history of treatment. It is considered as inhumane by many despite the positive results claimed by those who use it. As in the case of convulsive therapy the medical dictum of *primum non nocere* should prevail and the positive results should clearly outweigh the possible harm if such techniques are to be included in our armamentarium.

Indeed, behavioral approaches have also come under some general criticism as being mechanical, inhumane, and designed to control behavior rather than direct and modulate it. However, the results in diminishing self-destructive behavior in low-I.Q. autistics are significant. The most interesting work concerns operant conditioning of verbal behavior. Hewett (1965) first and then Lovaas *et al.* (1966, 1974) have produced the most extensive summaries to date of the results of conditioning. Lovaas claims good stimulus generalization effecting nonverbal behaviors as well and some durability. However, the drop-off when therapy is discontinued is significant. While Lovaas claims good results no child so treated "approaches normality." Similar reports by Freeman *et al.* (1975), suggest that positive reinforcement of a child with rapid echolalia could change his behavior to appropriate speech. While these changes seem to include an increased repertoire of lexical items and phrases they do not seem to increase linguistic complexity which may be a feature of cognitive and grammatical maturation and would be most significant in developing the human use of language.

The positive effect of reinforcement was recognized earlier than Lovaas' work, *but* some took special note that humans so frequently met with aversion that Goodwin and Goodwin (1969) used a machine type-writer approach to help these children in their language learning. Colby (1973; Colby & Kraemer, 1975) reported that 13 of his 17 nonspeaking autistic subjects showed linguistic improvement when using a computer-based, keyboard-controlled audio-visual display apparatus. The approach described enabled the discovery itself to be self-reinforcing without the need for mediation by food reward which they felt was away from their aim. Hargrave and Swisher (1975) describe a similar treatment mode using a Bell & Howell language master showing that taped voice was as instrumental as live voice to stimulate progress.

As the latter data suggest, interest in speech and language acquisition have been of central importance to the therapists of the autistic child. Spurred on by recent success in teaching subhuman primates American Sign Lanuage, Fulwiler and Fouts (1976) trained a five-year-old autistic for 20 hours to acquire signs that then led to an increase in vocal speech and generalization to other situations with better social interaction. These results are interesting in view of an earlier paper by DeMyer *et al.* (1972) suggesting that body imitation is deficient in autistics and their best performance is with objects where the solution to the problem is implied by the structure. Again, the same issue prevails with sign language as with imitative vocal reinforcement schedules—i.e., does increased lexicon signal or trigger better language capacity? Thus far the gains have not been demonstrated to be better than natural course would require.

CONCLUSION

We have gone through a gamut of therapeutic and investigative possibilities because they exist in such an array and profusion in the literature of autism. The variety is testimony to the fact that there is no *adequate* therapy in the current state of the art which will withstand controlled investigation. Anyone who purports to be a therapist without presenting his investigative instruments and his outcome, is doing what any humanistically oriented person would do for another person, but if they present their results in scientific form for the public to see and to judge by more precise standards they may add to our knowledge. The ultimate aim of rational therapy based on rational understanding of etiology is dependent upon the continuing observations of people working at every level of organization from organism to person, from behavior to function. Until our data are firmer many therapies will continue to exist. Moreover, individual parents will reach longingly to a variety of aids and some will even gain considerably from a great variety of professionals. Also, significantly, education and therapy will merge and be indistinguishable or blurred in some hands as we approach organism, learning, and psychic trust from many directions. Indeed, in closing, we may say that with autistic children therapy at its best is education and education at its best is therapy.

REFERENCES

Ayres, J. A., & Heskett, W. M. Sensory integrative dysfunction in a young schizophrenic girl. *Journal of Autism and Childhood Schizophrenia*, 1972, *2*, 174–181.

Bender, L. Childhood schizophrenia: A classical study of 100 schizophrenic children. *American Journal of Orthopsychiatry*, 1947, *17*, 40–56.

Bender, L. Remission rate and long term results of convulsive therapy of schizophrenic children. Paper read at the Society of Biological Psychiatry, Los Angeles, California, 1971.

Bettelheim, B. *The empty fortress*. New York: Free Press, 1967.

Chess, S. Autism in children with congenital rubella. *Journal of Autism and Childhood Schizophrenia*, 1971, *1*, 33–47.

Chomsky, N. *Aspects of a theory of syntax*. Cambridge, Mass.: M.I.T. Press, 1965.

Colby, K. M. The rationale for computer-based treatment of language difficulties in nonspeaking autistic children. *Journal of Autism and Childhood Schizophrenia*, 1973, *3*, 254–260.

Colby, K. M., & Kraemer, H. C. An object measurement of nonspeaking children's performance with a computer-controlled program for the stimulation of language behavior. *Journal of Autism and Childhood Schizophrenia*, 1975, *5*, 139–146.

Coleman, M. A crossover study of allopurinol administration to a schizophrenic child. *Journal of Autism and Childhood Schizophrenia*, 1974, *4*, 231–240.

DeMyer, M. D., Alpern, G. D., Barton, S., DeMyer, W. E., Churchill, D. W., Hingtgen, J. N., Bryson, C. Q., Pontius, W., & Kimberlin, C. Imitation in autistic, early schizophrenic, and nonpsychotic subnormal children. *Journal of Autism and Childhood Schizophrenia*, 1972, *2*, 264–287.

DeSausure, F. *Course in general linguistics*. New York: Philosophic Library, 1959.

Eisenberg, L. The course of childhood schizophrenia. *American Medical Association, Archives of Neurology and Psychiatry*, 1957, *78*, 69–83.

Fenichel, C. Special education as the basic therapeutic tool in treatment of severely disturbed children. *Journal of Autism and Childhood Schizophrenia*, 1974, *4*, 177–186.

Fish, B. The study of motor development in infancy and its relationship to psychological functioning. *American Journal of Psychiatry*, 1961, *117*, 1113–1118.

Fish, B. Contributions of developmental research to a theory of schizophrenia. In J. Hellmuth (Ed.), *Exceptional infant* (Vol. 2). New York: Brunner/Mazel, 1971. Pp. 473–482.

Fish, B. Biologic antecedents of psychosis in children. In D. X. Freedman (Ed.), *Biology of the major psychoses*. New York: Raven Press, 1975. Pp. 49–80.

Fish, B. An approach to prevention in infants at risk for schizophrenia: Developmental deviations from birth to 10 years. *Journal of the American Academy of Child Psychiatry*, 1976, *15*, 62–82.

Freeman, B. J., Ritvo, E., & Miller, R. An operant procedure to teach an echolalic autistic child to answer questions appropriately. *Journal of Autism and Childhood Schizophrenia*, 1975, *5*, 169–176.

Frostig, M. Visual perception, integrative functions and academic learning. *Journal of Learning Disabilities*, 1972, *5*, 1–15.

Fulwiler, R. L., & Fouts, R. S. Acquisition of American sign language by a nonspeaking autistic child. *Journal of Autism and Childhood Schizophrenia*, 1976, *6*, 43–51.

Goldfarb, W. *Growth and change of schizophrenic children: A longitudinal study*. New York: John Wiley and Sons, 1974.

Goodwin, M. S., & Goodwin, T. C. In a dark mirror. *Mental Hygiene*, 1969, *53*, 550–563.

Hargrave, E., & Swisher, L. Modifying the verbal expression of a child with autistic behavior. *Journal of Autism and Childhood Schizophrenia*, 1975, *5*, 147–154.

Hewett, F. M. Teaching speech on an autistic child through operant conditioning. *American Journal of Orthopsychiatry*, 1965, *35*, 927–936.

Lenneberg, E. H. *Biological foundations of language*. New York: John Wiley and Sons, 1967.

Lichstein, K. L., & Schreibman, L. Employing electric shock with autistic children: A review of the side effects. *Journal of Autism and Childhood Schizophrenia*, 1976, *6*, 163–173.

Lovaas, O. I., Berberich, J. P., Perloff, B. F., & Schaeffer, B. Acquisition of imitative speech in schizophrenic children. *Science*, 1966, *151*, 705–707.

Lovaas, O. I., Schreibman, L., & Koegel, R. L. A behavior modification approach to the treatment of autistic children. *Journal of Autism and Childhood Schiophrenia*, 1974, *4*, 111–129.

Mahler, M. *On human symbiosis and the vicissitudes of individuation*. New York: International Universities Press, 1968.

Mahler, M. S., & Furer, M. Child psychosis: A theoretical statement and its implications. *Journal of Autism and Childhood Schizophrenia*, 1972, *2*, 213–218.

Mann, L., & Goodman, L. Perceptual training: A critical retrospect. In E. Schopler & R. J. Reichler (Eds.), *Psychopathology and child development: Research and treatment*. New York: Plenum, 1976. Pp. 271–288.

Millar, S. *The psychology of play*. Baltimore: Penguin Books, 1968.

Piaget, J. *Play, dreams and imitation in childhood*. New York: Norton Library, 1962.

Rutter, M. The development of infantile autism. *Psychological Medicine*, 1974, *4*, 174–263.

Schneirla, T. C., & Rosenblatt, J. S. Behavioral organization and genesis of the social bond in insects and mammals. *American Journal of Orthopsychiatry*, 1961, *31*, 223–253.

Schopler, E., & Loftin, J. Thought disorders in parents of psychotic children: A function of test anxiety. *Archives of General Psychiatry*, 1969, *20*, 174–181.

Schopler, E., & Reichler, R. J. Developmental therapy by parents with their own autistic child. In M. Rutter (Ed.), *Infantile autism: Concepts, characteristics and treatment*. London: Churchill Livingstone, 1971.

Shapiro, T. Language behavior as a prognostic indicator in schizophrenic children under 42 months. In E. Rexford, L. Sander, & T. Shapiro (Eds.), *Infant psychiatry*. New Haven: Yale University Press, 1976.

Shapiro, T., Chiarandini, I., & Fish, B. Thirty psychotic children. *Archives of General Psychiatry*, 1974, *30*, 819–825.

Sinclair-de-Zwart, H. Developmental psycholinguistics. In E. Elkind and J. H. Flavell

(Eds.), *Studies in cognitive development: Essays in honor of Jean Piaget.* New York: International Universities Press, 1960.

Stern, D. The goal and structure of mother. *Infant Play*, 1974, *13*, 402–421.

Waelder, R. The principle of multiple function. *Psychoanalytic Quarterly*, 1936, *5*, 45–62.

Wenar, C., & Ruttenberg, B. A. The use of BRIAC for evaluating therapeutic effectiveness. *Journal of Autism and Childhood Schizophrenia*, 1976, *6*, 175–191.

Werner, H., & Kaplan, B. *Symbol formation.* New York: J. Wiley and Sons, 1963.

25

Parents as Therapists

O. IVAR LOVAAS

In general, the results one obtains by treating children with behavior modification are encouraging and lead one to become optimistic about this form of treatment for such children. However, there are certain problems or weaknesses in this approach which should be identified, so as to improve the treatment model. In describing these problems we draw heavily upon our own data, but this is mostly for purposes of illustration, because other investigators report similar problems across a rather wide range of children. Let us describe the problems we have encountered.

In 1973 we published a study which evaluated the effects we had achieved by using behavior therapy with autistic children (Lovaas *et al.*, 1973). We obtained objective measurements of each child's behaviors before, during, and after treatment. We closely examined the records of 20 such children. At intake, most of the children were severely undeveloped and all showed a very poor prognosis; in fact, we had selected them on that basis. Three main problems arose. (1) The treatment gains were often situation specific: that is, there was often limited generalization from the treatment environment to the outside. (2) Follow-up measures taken from one to four years after treatment showed large differences depending upon the posttreatment environment. Children who were discharged to their (trained) parents retained their gains or improved. (3) Most often, the treatment proceeded slowly, required a massive effort which one could only achieve with a large, enthusiastic and well-trained staff.

It is possible to entertain a variety of guesses as to why we observed these limitations in the treatment effects, and we speculated on some of

---

O. IVAR LOVAAS · University of California, Los Angeles, California.

these. Essentially, we saw the problem as our basic ignorance of motivation. In behavioral terms this comes down to a question of what are effective and normal reinforcing stimuli ("feedback stimuli" such as praise and disapproval). So far as we can determine, autistic children are largely indifferent to these stimuli. We would do well with autistic children if we could help them acquire the value of those (conditioned and normal) reinforcing stimuli. However, we do not know how to build normal reinforcers, neither do we know exactly what are, or are not, reinforcing stimuli for normal children.

In the absence of knowing more about normal reinforcers, we had to use the "artificial" reinforcers. That is, we knew that we could motivate the children by use of food and other exaggerated feedback. The use of such artificial reinforcers necessitates that the child be explicitly taught, even the most elementary behaviors, in a large variety of environments by a large variety of people.

The primary disadvantage in the use of artificial motivational stimuli is that the behaviors one builds using those stimuli become restricted to those environments in which these stimuli are available. Both the learning process and the maintenance of what is learned, therefore, had to be *controlled*, or somehow monitored.

Consistent with these speculations, our follow-up data showed clearly that children whose parents were trained to carry out behavior therapy continued to improve after we discharged them, while children who were sent to institutions or foster homes with untrained parents regressed. These data are shown in more detail in Fig. 1. These data concerned 13 children on whom we have particularly thorough follow-up data. The children may be divided into two groups: those who were discharged to a state hospital and those who remained with their parents. The figure shows percent occurrence of the various behaviors on the ordinate before (B) and after (A) 12 to 14 months of treatment, and the latest follow-up measures (F) taken one to four years after termination of our treatment. "I" denotes children discharged to state institutions after our treatment, while "P" denotes data on children discharged to parents who had been taught to treat the children. As can be seen, the children who were discharged to state hospitals lost what they had gained in treatment with us; their psychotic behaviors increased in frequency (self-stimulation and echolalia). They appear to have lost all they had gained of social nonverbal behavior, and they lost much of what they had gained in appropriate verbal and play behaviors. On the other hand, the children who stayed with their trained parents maintained their gains or improved further. A brief reinstatement of behavior therapy, which we tried on two of the children who were institutionalized, temporarily reestablished some of the therapeutic gains they had made while in our program.

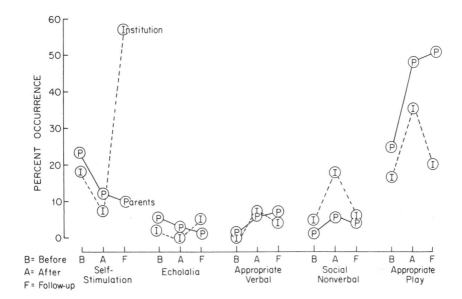

Fig. 1. Comparison of behaviors of autistic children before and after treatment, and at follow-up. I = children discharged to institutions; P = children discharged to trained parents.

The results of this follow-up research led us to deemphasize institutional treatment. Even though we could show that the children improved with institutional treatment, the improvement showed limited transfer and did not maintain itself on the outside. More and more we became involved in parent-training because the parents, as the child's therapists, could overcome these problems. That is, they could restructure the child's total environment and provide him with continuous treatment, which protected against the situational effects and the reversibility.

PRACTICAL PROCEDURES

How extensively one needs to train a child's parents is probably a function of how disturbed or retarded the child is. Since autistic children have extensive behavioral retardation, their parents probably need extensive training to be of measurable help. In the parent-training program we have available, the training proceeds as follows.

Once a child is accepted for treatment, the parents are informed about the time requirements for treatment, namely, that one of them (usually the mother, since for financial reasons the father most often works) will be expected to work with the child for most of the day for at least one year. If both parents work, this necessitates one of them quitting work, postponing having more children, and the like. If the parents understand and agree to that, they sign a legal contract with us which states in general terms what they can expect from us, and what we will expect from them. For example, the family is allowed to miss only two clinic visits per month. They have to learn to take reliable data within the first month of treatment, to shape a behavior on their own during the second month of treatment, and so on. We explain that failure to meet these criteria may lead to termination of the project. On the other hand, they can expect us to train them in these procedures, to spend at least ten hours a week in their home helping to treat their child, to help them find suitable schools for their child, to train and assist the child's teacher, to provide additional evaluations, etc. They are told that they should feel free to withdraw their child from treatment at any time, and if they do we will help them find alternate clinic placements if they so desire. I emphasize this contract because from the very beginning we attempt to establish a working relationship between colleagues, rather than a doctor-patient relationship. The reason for avoiding the latter kind of relationship may become clear as the program is described.

Once this "work-contract" is understood and signed, treatment begins which allows the parents to learn from us in an apprenticeship fashion. That is, the parents are required to be present at all supervisory clinic sessions and to be present at at least 50% of the sessions at home. The parents are treated like equal members on the treatment staff, which means that no treatment is introduced without their knowledge and consent. They receive all feedback, be it compliments or criticism, just as other staff members do. The sessions are totally work- and child-oriented. Parents with personal emotional problems are referred elsewhere for individual treatment, unless these problems are minimal or acute. In general, we seem correct in assuming that if the parent can help his child in a substantial way, then this is the best form of treatment for the parent's anxiety about the child, obsessive indecisions about how to handle him, and so on.

Typically, the first treatment session begins with the child and the treatment team (which consists of the clinic supervisor, parent, a graduate student, and three or four undergraduate student volunteers) together in one room. One "expert" member (that is, a person with treatment experience) may then demonstrate how to build the first response, which may be for the child to sit in a chair for 5 to 10 seconds, when asked to do so. Usually this involves the therapist sitting in one chair and the child sitting in a chair close by, facing him. As soon as the senior therapist makes some progress on this task (usually 15 to 30 minutes into the first session), the parent is asked to "take over" and repeat what she has observed the senior therapist doing. The parent's first session as a therapist may last 5 minutes, and we will heavily reward her for what we judge to be correct, and point out her mistakes. A new member of the team then takes the parent's place, and is trained in the same way as the parent (similar rewards and criticisms). The parent may then take over again, so as to become more skilled and comfortable on this first task. It is probably important to try to make some progress on the child's tantrums this first visit, since they typically increase with the new demands such as the child having to sit, even for a short time. At least we try to reach some agreement on an initial treatment plan for tantrums. In order to further encourage equality, we discourage "virtuoso" performances by experienced therapists, since they may make the parent feel inadequate ("I must be a horrible mother, because my child behaves so well with you"), or cause her to shrug responsibility ("Since you do so well, why don't you do the whole job?").

We make an all-out effort, even during the first visit, to heavily reinforce the parents for their correct behaviors, and we arrange the situations for this to occur, so that we set an optimistic and happy tone for the time we spend together. This is also realistic since we can set the task at such a level that the child does show some improvement. Sitting in a chair for 5 to 10 seconds is not a major task.

Whatever we succeed in teaching during this first session is immediately transferred to the home. That is, that very afternoon, the mother will practice having her child sit, periodically during the day, for short periods in the kitchen, in the living room, etc. We have ample data which show that the child *and* the mother will not transfer what they learn in the clinic to the home without explicit training in doing so. Incidentally, this problem is not limited to autistic children and their parents, but occurs in other clinical populations as well. Therefore, one or more of the therapists will work with the parent in the home, even on the first day. This is very important.

Initially we build all new behaviors in the clinic, where the parent comes to visit two or three times a week during the first month, and the initial parent training consists primarily of transferring clinic-acquired behaviors to the home. The first behavior may be sitting in a chair for up to a minute at a time, the second behavior may be visually fixating on the therapist's face ("look at me," for 2 seconds or so initially), followed by *nonverbal imitation* training, such as raising arms in imitation of the therapists, followed by imitating the therapist touching his shoe, then clapping hands, touching table, and then on to increasingly elaborate imitations as we have described in earlier publications. Shortly thereafter, we begin training on *receptive language* ("sit in chair," "look at me," "raise your hands," "touch your shoe," "clap your hands," etc.). We usually begin with the same behaviors the child has learned in nonverbal imitation, fading out the visual cues and substituting verbal ones. About one month into training we are ready to begin *verbal imitation* training (teaching the child to imitate the adult verbally, starting simply with sounds like "a" and "m"), and so on, to *expressive language.*

Note that as such a program is easy and gradual for the child, it is likewise easy and gradual for the parent; something guaranteed for both to succeed in. After a few months in the program, as the parent is catching on to the technique, the situation typically develops where the senior therapist is reinforcing the parent, who then is reinforcing the child. We now serve more as *consultants* to the parent, who is assuming the main treatment responsibility. At this point, the parents are well on the way to build new and complex behaviors in the home, without first starting them in the clinic. At this time we may drop to once-a-week visits to the clinic. However, we maintain the extensive home visits, 10 to 20 hours of student help per week, because there is an enormous amount of work to be done with the child. Furthermore, since we have been "in the business" for a long time and have become experts in behavior therapy, we know more than the parent about certain problems and, therefore, remain as the expert advisors. The parents usually know more about the child than we do and are in a position to provide essential information about what he already knows, what motivates him, etc. One can save months of work by having the parents demonstrate such information.

The teaching situation relies most heavily on the parents learning about therapy by working in an apprenticeship fashion with senior therapists. As the parents become skillful in concrete techniques we gradually introduce didactic material. This includes teaching the technical terminology, processes such as prompts, S^Ds, prompt fading, extinction, etc. We introduce the parents to this area by referring them to Patterson and Gullion's book, *Living with Children* (1968), and our own manual for teaching children, *Teaching Slow Children Fast* (Lovaas *et al.*, 1977). The creative use of behavior modification necessitates an understanding of the abstract terms involved as subsequent research will show. On the other hand, mere exposure to the abstract terms, as can be gained by reading basic texts on behavior modification appears inadequate without exposure to the actual practice.

REVIEW OF RESEARCH

The research literature on teaching parents is very extensive, even though this field is less than 10 years old. The extensive use of parents as therapists probably reflects a number of needs, such as financial ones, which limit the availability of professional people. But it may also reflect the growing awareness that if children are a product of their environment, that environment may have to be changed. In such an enlightened environment everyone is a therapist, and that includes parents. It is beyond the scope of this paper to provide an extensive review of the research literature on parent training; excellent reviews have already been provided by Berkowitz and Graziano (1972), Johnson and Katz (1973), and Baker and Heifetz (1976). Recently, the Banff (Canada) Conference on Behavior Therapy was devoted to the role of parents as therapists, and two excellent volumes on parent training emerged from that conference (Mash *et al.*, 1976a, b).

The techniques in training parents can be roughly divided into three areas, depending on the procedures employed: (1) Those who use essentially an instructional model, as when parents are instructed individually or in a group, or read specific treatment manuals, or read about basic learning theory and learning-theory-derived treatment procedures in texts. An example of such a procedure would be to have parents read texts such as Patterson and Gullion (1968) or Becker (1971). These are basic and popular texts on principles in social learning theory written for parents. Such reading may then be supplemented by classroom discussions of various concepts. (2) Those who use modeling procedures, whereby the parents observe and are encouraged to model a senior therapist working with their child. Wolf *et al.* (1964) reported the first attempt to have the parents of an autistic child observe behavior therapy. They began their observation and training in the clinic, and then gradually faded the parents in as the child's primary therapists, as the child was moved from the institution to his home. Obviously we rely most heavily on this procedure in our work. (3) The use of videotape procedures, whereby the parents' interaction with their child is videotaped, and then played back to the parents and a senior therapist-teacher for discussion. Bernal's (1969) programs have relied heavily on such videotape procedures. Most therapists probably use combinations of these techniques, as in the work of Patterson *et al.* (1968), where the parents initially watched the treatment in the clinic, then were taught to record the child's behavior. They were then taught to model the therapist's interactions, first in the clinic, then gradually moving to the home. They also instructed parents in social-learning theory.

How well these procedures will hold up and compare in effectiveness will probably depend on a number of variables, such as the parents' motivational and educational level, the extent of their child's problem, and so on. The largest single category of parent training has centered on the reduction of maladaptive behaviors in relatively normal children (such as oppositional behavior, enuresis, antisocial behavior, etc.). Perhaps relatively short-term parent-training programs can be helpful in such instances. However, in the case of autism, a comprehensive approach is essential, stressing both the reduction of maladaptive behavior and the teaching of complex behavior, such as language. Only a comprehensive training program would seem to work in such cases.

In one of the few attempts to evaluate the relative efficiency of multiple training procedures (modeling, group discussion, instruction on basic concepts, etc.), Johnson and Brown (1969) found that modeling by the therapist was the most effective in changing parent behavior. In an important study working with parents of autistic children, Koegel *et al.* (1977) evaluated various parent-training procedures (reading texts, viewing videotapes, having correct teaching procedures modeled, etc.). They closely monitored how well the parents were doing in teaching the autistic children on unfamiliar (not specifically trained, but "testing") tasks. They evaluated the parents on their use of prompts, clarity of instructions, use of rewards, etc. They found that it was essential for the parents to know the abstract and theoretical principles basic to behavior modification (in addition to modeling) in order for the parents to teach their children effectively on new (not previously instructed) tasks. That is, the parents' creative use of their training necessitated that they know the theoretical basis of the treatment procedures. There are a great number of other issues being researched in the area of parent training at this time, and the reader is encouraged to consult the reviews given earlier in this section for a more comprehensive account of this area.

TEACHING HOMES

Not all parents of autistic children can be taught to become effective teachers or therapists for their children. This is true in cases where the parents are divorced and the mother has to work for financial reasons. Sometimes parents cannot be trained because of their own personal psychological problems or a lack of interest in their child, or the child may have problems which are so severe (he is so self-destructive or so strong and assaultive) that he has to be cared for elsewhere.

To avoid institutionalization, which is the most common trap and frightening future for autistic children, we have begun exploring training professional parents to take care of the autistic child in family residences ("Teaching Homes") within the child's own community, near his parents, and schools and medical facilities that are already available. Four to six children per home constitute a heterogeneous group, so that the more regressed autistics can model after nonpsychotic peers. The children are cared for by "Teaching Parents," who are college graduates enrolled in Master's programs (Education, Social Welfare, or Psychology), and whom we explicitly train in behavior therapy techniques. At the same time we take explicit measures on the child's progress within those Teaching

Homes, and explicitly arrange for surveillance and input by the child's natural parents to protect the child's welfare and improve upon such a program. The efficient, short-term education of parents to become *professional* parents is probably the critical variable in assuring the success of this operation. Autistic children are difficult to manage, and without an efficient training procedure which can turn out a large number of (para) professionals to successfully care for them, they all too often end up in the large traditional and deadly (state) institutions. This program has been described in more detail elsewhere (Russo *et al.*, 1974).

UNIQUE ADVANTAGES

In the introduction to this paper we discussed that the primary reason for involving the parents of the autistic child in the treatment program as explicitly trained cotherapists concerned efforts to generalize the treatment gains to all of the child's environments, and to protect his gains from relapse once formal therapy is terminated. It was also pointed out that since autistic children have such a profound deficiency, nothing short of a major therapeutic-educational effort, necessitating one-to-one treatment on an all-day basis, could be considered sufficient. For all practical purposes such a plan must involve the child's parents.

It may be appropriate to point out two more considerations. Since behavior therapy is quite controversial, both in its explicit statement concerning *control* of children, and in its occasional use of negative reinforcers (particularly aversives), it is absolutely essential, for both moral and legal purposes, that the parents be intimately familiar with how their child is treated. Such a public involvement helps to protect a democratic and enlightened approach to the rights of individuals.

It is probably also helpful to point out that the autistic child presents only the extreme of child-rearing problems to his parents (and they are extreme), and that to some extent all parents are in the difficult position of the autistic child's parents: They lack knowledge in how to cope with the child's behavior. What we learn about helping the autistic child's parent may then some day be useful for the parent of the normal child. We do not possess knowledge and evaluations of child-raising procedures yet, but they may be crucial for the maintenance of our society.

REFERENCES

Baker, B. L., & Heifetz, L. J. The read project: Teaching manuals for parents of retarded children. In T. D. Tjossem (Ed.), *Early intervention with high risk infants and young children*. Baltimore: University Park Press, 1976.

Becker, W. C. *Parents are teachers: A child management program*. Champaign, Ill.: Research Press, 1971.

Berkowitz, B. P., & Graziano, A. M. Training parents as behavior therapists: A review. *Behavior Research and Therapy,* 1972, *10,* 297–317.

Bernal, M. E. Behavioral feedback and the modification of brat behaviors. *Journal of Nervous and Mental Disorders,* 1969, *148,* 375–383.

Johnson, C. A., & Katz, R. C. Using parents as change agents for their children: A review. *Journal of Child Psychology and Psychiatry,* 1973, *14.* 181–200.

Johnson, S. M., & Brown, R. A. Producing behavior change in parents of disturbed children. *Journal of Child Psychology and Psychiatry,* 1969, *10,* 107–121.

Koegel, R. L., Glahn, T. J., & Niemenson, G. *Generalization of behavioral techniques.* Manuscript submitted for publication, 1977.

Lovaas, O. I., Ackerman, A., Alexander, D., Firestone, P., Perkins, M., & Young, D. *Teaching slow children fast.*In preparation, 1977.

Lovaas, O. I., Koegel, R. L., Simmons, J. Q., & Long, J. Some generalization and follow-up measures on autistic children in behavior therapy. *Journal of Applied Behavior Analysis,* 1973, *6,* 131–166.

Mash, E., Hanley, L., & Hamerlynch, L. (Eds.). *Behavior modification approaches to parenting.* New York: Brunner-Mazel, 1976. (a)

Mash, E., Hanley, L., & Hamerlynch, L. (Eds.). *Behavior modification and families.* New York: Brunner-Mazel, 1976. (b)

Patterson, G. R., & Gullion, M. E. *Living with children: New methods for parents and teachers.* Champaign, Ill.: Research Press, 1968.

Patterson, G. R., Ray, R. S., & Shaw, D. A. Direct intervention in families of deviant children. *Oregon Research Institute Research Bulletin,* 1968, *8* (9).

Russo, D. C., Glahn, T. J., Miners, W., & Lovaas, O. I. *Use of teaching homes for the treatment of psychotic children.* Paper presented at the meeting of the American Psychological Association, New Orleans, September 1974.

Wolf, M. M., Risley, T., & Mees, H. Application of operant conditioning procedures to the behavior problems of an autistic child. *Behavior Research and Therapy,* 1964, *1,* 305–312.

26

Treating Autistic Children in a Family Context

R. HEMSLEY, P. HOWLIN, M. BERGER, L. HERSOV, D. HOLBROOK, M. RUTTER, and W. YULE

Infantile autism is a complex disorder which involves abnormalities of development and of behavior, both of which need to be taken into account in planning any type of treatment program. During the 1960s there was an increasing number of reports of the successful use of behavioral techniques in alleviating these many and various problems in individual children. Operant approaches were employed to aid the development of positive or normal features such as imitation, eye contact, social skills, speech usage and communication (Hingtgen *et al.*, 1965, 1967; Hingtgen & Trost, 1966; Lovaas, 1966, 1967; Lovaas *et al.*, 1965b, Metz 1965; McConnell, 1967; Halpern, 1970; Mathis, 1971; Risley & Wolf, 1967; Sloane *et al.*, 1968). Operant procedures were also used to eliminate a wide range of deviant behaviors including self-injury, temper tantrums, aggressive behavior, hyperactivity, encopresis, and negativism (Lovaas *et al.*, 1965a, b, 1966; Tate & Baroff, 1966; Graziano, 1970; Wetzel *et al.*, 1966; Wolf *et al.*, 1964, 1967; Maier, 1971; Marshall, 1966; Sulzbacher & Costello, 1970; Brown & Pace, 1969). It became clear that operant approaches provided a useful set of therapeutic interventions which could achieve fairly rapid results in terms of alteration of specific behaviors, if the child

R. HEMSLEY, P. HOWLIN, M. RUTTER, and W. YULE · Institute of Psychiatry, De Crespigny Park, Denmark Hill, London SE5 8AF, England. M. BERGER · Institute of Education, 20 Bedford Way, London WC1, England. L. HERSOV and D. HOLBROOK · The Maudsley Hospital, Denmark Hill, London SE5 8AF, England.

was treated daily by skilled professionals on a one-to-one basis in a well-controlled inpatient setting.

However, several clinical problems remained. First, it was soon evident to therapists, and later confirmed by systematic studies (Browning, 1971; Lovaas et al., 1973) that the gains were often lost after the child was discharged home. Behavioral changes in hospital did not adequately generalize to other situations and it seemed that parents would have to be actively involved in the therapeutic program if the benefits to the child were to be lasting. Even during the mid-1960s there had been several reports of the successful training of nonprofessionals in the use of behavioral techniques (Patterson & Brodsky, 1966; Wetzel et al., 1966) and there came to be an increasing emphasis on the use of parents as cotherapists (Berkowitz & Graziano, 1972; O'Dell, 1974). This approach was extensively developed by Schopler and Reichler (Schopler & Reichler, 1971a, b; Reichler & Schopler, 1976) and became a central feature of their therapeutic program. Independently, we had developed a broadly similar approach in which the focus was on working together with parents (Howlin et al., 1973b; Rutter & Sussenwein, 1971; Rutter, 1973; Berger et al., 1974). However, there was a difference in that we worked directly with parents at *home,* rather than at the clinic. We had found that autistic children often behaved very differently in the clinic from the way they were at home; also some behaviors by their nature could only be observed at home. Accordingly, we shifted completely to a home-based approach in which we trained parents in the home to deal with problems as and when they arose there.

Second, there was a need to broaden the therapeutic aims. Much of the early work was concerned with helping autistic children to produce words in the experimental setting (Yule & Berger, 1972). Less attention was paid either to the spontaneous use of spoken language for social communication or to the understanding of language. Moreover, there was a neglect of nonvocal means of communication (such as gesture). However, during the last decade there has been a shift of focus from words to language and from speech to social communication. Thus, many workers, including ourselves (Howlin et al., 1973b), have been concerned to develop language comprehension and the use of gesture as well as the production of words (Webster et al., 1973; Fulwiler & Fouts, 1976; Bonvillian & Nelson, 1976; Miller & Miller, 1973).

Third, there was a need to extend the range of therapeutic techniques. At first, most operant treatments largely used food as a reward for desired behaviors and either withdrawal of attention or physical punishment in order to extinguish deviant behaviors. However, it became apparent that food was neither a necessary nor a desirable reward for most autistic children. Rather, it was possible to employ more natural *social* rewards (praise, attention, opportunity to engage in a favorite activity, etc.) provided the setting and approach were appropriately structured and planned. In addition, it became clear that there were many advantages in combining operant techniques with other behavioral methods such as graded change to deal with abnormal attachment to objects (Marchant *et al.*, 1974); desensitization and graded exposure to treat phobias (Howlin *et al.*, 1973a); the "Bell and Pad" to control enuresis (Howlin *et al.*, 1973b) and social intrusion to increase social interaction and improve patterns of play (Rutter & Sussenwein, 1971; Howlin *et al.*, 1973b). Of course, too, the problems of an autistic child cannot be considered in isolation from his family. The stresses created by an autistic child are often considerable and we have found that casework or counseling techniques were a valuable adjunct to behavioral methods, particularly when family difficulties intrude on the therapeutic program (Rutter & Sussenwein, 1971; Howlin *et al.*, 1973b).

Fourth, together with other workers, we became aware of the need to adapt treatment to the developmental level of the individual child (Rutter & Sussenwein, 1971; Howlin *et al.*, 1973b). It was important to make optimum use of existing skills but not to set goals beyond the child's capabilities at that point. This demanded a full and careful developmental and behavioral evaluation of each child in order to plan the most effective form of therapeutic intervention.

It was in the light of these considerations that we set up a research project in 1970 to evaluate the efficacy of treating autistic children in a family context using behavioral methods and developmental principles. This chapter outlines some of the key findings which are available so far.

RESEARCH ISSUES

Before describing the study it is necessary to consider the central research issues which had to be investigated. Laboratory and inpatient studies had already shown that behavioral methods could bring about major changes in individual symptoms. However, we still had to determine whether similar results could be obtained in the home where there was less control over the environment and with much more limited professional time. That was our first task, which meant examining changes in individual behaviors in relation to treatment.

Second, we had to find out whether or not our treatment methods were *superior* to other approaches in bringing about immediate changes in behavior. In order to do that we compared our experimental group with a matched control group (A) in terms of changes in the children's behavior over a six-month period.

Third, it was necessary to determine whether or not we were effective in altering *parental* behavior. Previous studies which had utilized parents as cotherapists had all assessed outcome in terms of changes in the *children's* behavior. However, this is an inadequate evaluation for methods, such as ours, which have as their goal the improvement of parenting skills. Accordingly, we used the same matched control group (A) to examine changes in parental behavior over a six-month period, and compared the findings with those in the experimental group over the same time span.

Fourth, it was important to determine the *long-term* benefits of treatment. Most evaluations have considered only short-term improvement. However, with a chronically handicapping condition like autism, the real test is whether treatment makes any difference to the children's social development. So far as our own study was concerned, there was also the question whether or not the same gains might be achieved with a lesser investment of resources. Although the cost of a home-based program is very much less than inpatient treatment, nevertheless it is still much more expensive than sporadic outpatient attendance. Is that extra cost worthwhile? To answer that question we used a second individually matched control group (B) of children seen at our own clinic outside the home-based project. These had been dealt with on the basis of the same therapeutic principles but, because they lived a long way from the clinic or because they were referred before the project began, they had been seen only occasionally for advice.

METHODS

Treatment Program

The methods we use in treatment have been previously described in some detail (Rutter & Sussenwein, 1971; Howlin *et al.*, 1973b) and will be outlined here only briefly. Before treatment begins, the major areas of difficulty are identified for each child as well as the conditions under which such difficulties appear to arise. For this purpose, a combination of interviews, observations, and psychological testing is used. Initial treatment is directed to those behaviors which have given rise to most parental concern. A functional analysis of these behavior problems is undertaken with the parents to determine when, where, how often, and for what apparent reason the behaviors occur. Detailed schemata of treatment are then drawn up for each child with clearly defined goals together with the stages leading to the goals.

Reinforcement techniques (in which praise, encouragement, and rewards are used to increase normal behavior) are widely employed but do not constitute the only techniques. Physical punishment is very rarely advised but "time out" methods, such as removal to a bathroom or nonpreferred area of the house, may be used to control tantrums. Graded exposure *in vivo* is used to treat phobias and to overcome resistance to changes in the environment; and graded change to modify rituals and abnormal attachment to objects. Deliberate intrusion into the child's solitary activities, in ways found to be pleasurable to the child, is used to increase social interaction.

Parents are helped to be consistent in their style of interaction and handling. They set aside *short* periods each day to teach the child specific social and communication skills, making the sessions as pleasurable as possible to both child and parents. At first many parents expressed misgivings about the sessions as they felt that their child would object to their new role of teacher. In fact the children came to enjoy and to look forward eagerly to their daily session when they received their parents' undivided attention. However, parents are also encouraged to lead a normal life with adequate time for recreation with each other and with their nonautistic children.

Our major aim is to enable parents to deal with their own child efficiently and sensitively without our direct help. The general principles of behavior modification techniques are taught by explanation, discussion, the use of programmed learning texts (e.g., Patterson & Gullion, 1968) and modeling by the therapist. Parents are taught how to look at a problem and break it down into small steps, to be specific about targets and goals and to keep records on the child's behavior. Emphasis is placed on teaching parents to formulate their own functional analysis of their children's problems, to plan what needs to be done, and to monitor the success (or failure) of their actions.

Opportunities are provided to discuss feelings of guilt, confusion, or ambivalence and counseling for family problems is used when needed. Many parents need help to keep their child's problems in perspective and care must be taken to ensure that other children do not suffer unduly from the parents' overinvolvement with the autistic child. Practical advice may be given, for example, on how to find baby-sitters; practical help is also available (as in arranging holiday breaks or short-term care at times of family crisis); and close liaison is maintained with schools.

Two full-time clinical psychologists are engaged on the project and between them deal with 16 families living within a 60-mile radius of the center of London. Visits to the home vary in frequency from once or twice a week in the early stages of treatment to once a month or less when treatment is well established.

Samples

The experimental cases consisted of 16 boys who met the criteria for autism outlined by Rutter (1971); who had a nonverbal IQ of 60 or greater; who did not have an overt neurological disorder or sensory impairment; who were in the age range 2-11 years; who were not already receiving adequate treatment; and who lived within a range which made home-visiting feasible (the furthest away was 60 miles but ease of access as well as distance were taken into account). Fourteen of the boys were new referrals who were taken consecutively into the project. The other two cases were nonspeaking autistic boys who had already been attending the Maudsley Hospital as outpatients, but who had not been receiving intensive treatment.

Control group A consisted of another 14 boys who met the same diagnostic criteria; but the group was chosen on the basis that the families were *not* in treatment or were receiving some different form of treatment. This group was obtained with the help of the National Society for Autistic Children and was group-matched with the experimental cases on social class, age, IQ, and use of language at the time of initial assessment in the project (see Table 1); there was no restriction on place of residence. Originally it had been hoped to take a group of autistic children all receiving the same form of nonbehavioral treatment. A psychoanalytic center agreed to cooperate with us but there were so few new referrals of autistic children to the center that this plan had to be abandoned. Group A was seen on just two occasions six months apart using measures identical to those employed with the experimental group. During the six months we offered no help beyond that already being given by other agencies. But after the second assessment, if they wished it, families not already receiving treatment were provided with whatever help was required.

Table 1.   *Matching Cases with Control Group A*

Measures	Cases (n = 16)		Control Group A (n = 14)	
	Mean	*(S.D.)*	*Mean*	*(S.D.)*
Age at initial assessment	73.9 mo	(29.1)	76.8 mo	(28.4)
Nonverbal IQ at initial assessment	88.8	(17.9)	88.4	(24.8)
Social class 1 and 2	50.0%		57.0%	
Social class 3	12.5%		14.5%	
Social class 4–6	37.5%		28.5%	
Language: Mute	31.25%		35.7%	
Mainly words	25.0%		21.4%	
Phrase speech	43.75%		42.9%	

NOTE: No differences statistically significant.

Table 2. Matching Cases with Control Group B

Measures	Cases (n = 16)		Control Group B (n = 16)	
	Mean	*(S.D.)*	*Mean*	*(S.D.)*
Age at referral	59.5 mo	(18.5)	61.9 mo	(17.4)
Nonverbal IQ at referral	89.1	(17.6)	88.0	(17.8)
Meecham language score	25.0 mo	(21.2)	21.1 mo	(16.5)
Social class 1 and 2	50.0%		62.5%	
Social class 3	12.5%		18.75%	
Social class 4–6	37.5%		18.75%	

NOTE: No differences statistically significant.

Control group B (see Table 2) consisted of a further 16 boys, again meeting the same diagnostic criteria, and *individually* matched with the experimental cases on age, severity of behavioral handicap, nonverbal IQ, and language level at the time of referral. All these children had attended the same clinic as the experimental cases and while some had been seen only for initial diagnostic assessments others had been given advice on the same general principles as the experimental families. However, because of distance or date of referral they were seen only occasionally as outpatients in contrast to the intensive home-based treatment given to the experimental group. Because autism is a relatively uncommon condition it was not possible to match also for duration of follow-up and from the way the group had to be selected it necessarily happened that group B was followed longer than the experimental cases.

Measures on the Child

The Merrill-Palmer Scales (Stutsman, 1931) or the Wechsler Intelligence Scale for Children (Wechsler, 1949) were used to assess the children's nonverbal intelligence at the start of the project for all groups, and at follow-up for the experimental group and control group B. The Meecham Language Scale (1958) was used for the initial matching of cases and control group B. The Reynell Development Language Scales (1969) were also used at follow-up for the experimental group and control group B.

Specially devised measures of syntactical competence and functional language usage were obtained from audio-tape recording of the child's speech as used at home during the middle half hour of an hour and a half's period of time with his mother. The measures have been found to have good reliability and validity (Cantwell *et al.*, 1977). There were used at the start of the project and at six monthly intervals thereafter for the experimental cases; at the initial evaluation and at follow-up for control group A; and at follow-up only for control group B.

For the experimental group and control group A, the child's behavior at home was assessed by means of a standardized direct observation technique utilizing time sampling. Seventeen specified child behaviors (including speech, gesture, mannerisms, rituals, and play) were coded according to their occurrence during each one of a series of 10-second time samples which covered an hour and a half's period in which the child and his mother were together. Good interrater reliability was obtained for all measures (Hemsley et al., in preparation).

For the experimental group and control group A, a standardized interview with the mother, similar to that used in our other studies of autistic children (Bartak et al., 1975; Folstein & Rutter, 1977), was employed to obtain systematic data on the nature, severity, and frequency of abnormalities in language, social relationships, and behavior. It also gave information on early development and on family circumstances.

For the purpose of individually matching the cases and control group B, case-note data were used in a standard way. For both groups the same standard form of history taking, observation, and psychological testing had been employed.

Finally, individually tailored measures were devised and used to chart the progress of individual behaviors which were the target of individual treatment. The measures were applied at much more frequent intervals as appropriate for each behavior.

In addition to specific behavior counts and charts, parents also kept general daily diaries in which they noted down any changes in routine, any changes in the child's behavior, either progress or deterioration, how the child behaved at various stages in the day and on what areas they were working in the training sessions. Parents often used these diaries as a means of expressing irritation, feelings of depression and hopelessness, or other problems which they might not raise in face-to-face interviews, but which could be taken up on later visits.

Measures on the Family

The same standardized direct observation technique used to measure the child's behavior at home was also employed to assess the mother's behavior in interaction with her autistic child. Twenty-one specified parental behaviors (including physical contact, praise, play, instruction, etc.) were coded according to their occurrence in each 10-second time sample over the course of one and a half hour's observation at home. Again good interrater reliability was obtained for all measures (Hemsley et al., in preparation).

Audio-tape recordings of the mother's speech during these observational sessions with her child were used to provide a fine-grain analysis of the way she talked to her child. A 17-category system of demonstrated reliability (Howlin *et al.*, 1973a) was used to differentiate between various types of speech usage, such as questions, prompts, corrections, directions, and expression of disapproval.

The "Standard Day Interview" developed by Douglas and his colleagues (Douglas *et al.*, 1968; Lawson & Ingleby, 1974) was used to assess the type, quantity, and quality of parent-child interactions over a 24-hour period in the experimental group and control group A. A systematic interview is used to obtain the details of what the child was doing, with whom, for how long, and with what intensity of interaction over the whole of the 24 hours previous to the interview. The method has been found to provide measures of good reliability and validity. In addition, Brown and Rutter's (1966) interview measure of "positive interaction" was used to assess parent-child interaction over the span of a week in terms of activities such as reading, rough and tumble play, outings, etc. This measure, too, has been shown to have good reliability.

These measures were all taken at the same time intervals as the observations and interviews regarding the children—i.e., initially and at six-month intervals thereafter in the experimental group, and initially and six months later for control group A.

Other measures on parental coping skills and on parental attitudes to the child were also obtained but findings on these are not included in this chapter.

RESULTS

Examples of Changes in Individual Behaviors

Attentional Behavior

For some children it was necessary to start by building up the attentional behavior itself before work could begin on language or other skills. Figure 1 shows the time spent sitting and attending by an 11-year-old boy who had never had to conform in this way at home or at school until our intervention. His father initially held him in his seat for a very brief period which was greatly resisted. After the introduction of a kitchen timer to set explicit limits on the length of time he was required to stay in his seat the sessions were rapidly lengthened without causing any distress to the boy. After about six weeks he was attending for 30 minutes and this was felt to be a reasonable time to expect him to remain seated and sessions were not extended beyond that length.

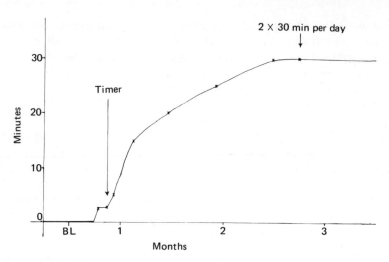

Fig. 1. Attentional behavior in an 11-year-old hyperactive boy.

Spoken Language

In these daily sessions much of the basic language training was carried out, in addition to the teaching of skills such as motor imitation, gestural training, doing up buttons and zippers, teaching simple reading and writing, working on imaginative play, teaching rudimentary social

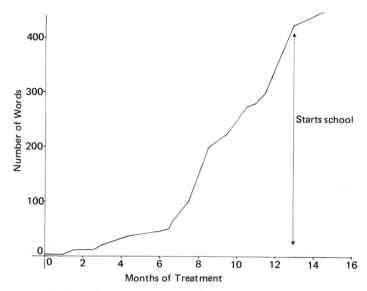

Fig. 2. Expressive language in a three-and-one-half-year-old.

skills for approaching other children, and building up cooperative behavior between the child and his parents generally. Figure 2 uses records kept by the mother to show the acquisition of spoken language in a three-and-one-half-year-old child.

Initially he was using only the words "Oh dear," which he would repeat quite meaninglessly. For the first five to six months training concentrated on building up his comprehension before working on his expressive language directly. Now, at the age of six years, his use and understanding of language are almost at an age-appropriate level.

Maladaptive Attachment to Objects

However, the problems shown by individual children varied widely, and many did not arise during the treatment sessions. Techniques had to be taught to the parents for dealing with each of these problems as they arose throughout the day. A technique of gradual alteration of the environment was introduced for children with maladaptive attachments to objects or rigid adherence to routines. In this approach, the object or routine was very gradually changed until the child voluntarily gave it up (Marchant *et al.*, 1974).

Figure 3 shows the reduction of an attachment to a cot blanket in a 4-year-old, which had persisted for three years. This blanket was carried everywhere and was so large that it interfered with all activities requiring the use of both hands. Acute distress followed its direct removal, which was the policy followed by his mother until our suggestion that she gradually reduce the blanket by cutting pieces off it each night. This proceeded

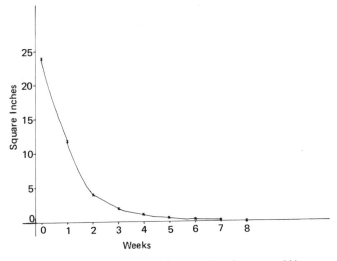

Fig. 3. Reduction of blanket attachment in a four-year-old boy.

rapidly until after about six weeks the "blanket" consisted of only a few threads knotted together. These were finally voluntarily discarded and new objects which were carried, such as plastic buses, were reduced in the same way. This procedure caused minimal distress to the child and enabled many other activities to be undertaken. The new attachments formed seemed far less intense and reduction of the objects proceeded more rapidly than with the blanket.

It is only possible in this chapter to give a few illustrations of the changes brought about in individual behaviors. However, it is clear from these examples that behavioral treatment applied by parents at home could produce changes comparable to those achieved by the more intensive application of similar techniques by professionals in the hospital or laboratory setting.

A variety of other behavioral techniques were also employed. Thus, systematic desensitization was taught in order to deal with phobias of baths, balloons, birds, etc. Graded exposure was also used for phobias. "Time out" was employed to treat temper tantrums, deliberate soiling, aggression, destructiveness, or running away from home (see Howlin et al., 1973b, for details). Training was also given in toileting procedures and in the use of the "Bell and Pad" for bedwetting.

These techniques had to be employed consistently throughout by the parents whenever the appropriate opportunity arose and required cooperation by everyone involved with the child such as other members of the family, teachers, school-car escorts, and baby-sitters. Parents were taught how to seize on commonplace activities to increase their child's communication skills or improve comprehension, and to expand on specific concepts taught in the sessions. Many parents became proficient in using routine situations to educate and to build up social skills in their child. However a few parents remained fairly rigid in their interpretation and use of the techniques, failing to generalize from sessions to normal situations unless the potential of such opportunities was made explicit to them.

The intensive and prolonged nature of the involvement of the therapist with the family in the home situation inevitably led to the consideration of family problems other than the behavior of the autistic child himself. Many parents expressed feelings of guilt and depression both in relation to their fears about the cause of the autism and the possibility of their having contributed to the development of the disorder. Conflicts between parents over handling difficulties, and feelings of hostility toward the handicapped child, had to be recognized and discussed. We were anxious not to increase the level of overall family stress by asking the parents to take additional responsibility for the child. However we found that our advice was usually the first practical help parents had been given for handling their child and generally they were very pleased to cooperate with any programs we suggested. Throughout the intervention we were aware that by focusing our attention on the handicapped child we might provoke jealousy in sibs and we encouraged parents to distribute their free time as evenly as possible so that these problems should not arise within the family.

Six-Month Changes in Child Behavior: Comparison and Control Group A

The findings on individual behaviors indicated that marked changes of the kind intended could and did occur during home-based treatment. However, it was next necessary to determine whether or not these changes could be due to what we had done and, in particular, whether our treatment methods were more effective than other approaches in bringing about changes in behavior. The first step in answering these questions was to compare the changes in children's behavior over a six-month period in the experimental group and in control group A (who received no treatment or a variety of other forms of treatment: speech therapy, psychotherapy, etc.).

Over the first six months of treatment the children in the experimental group almost doubled the amount they spoke (as indicated by the total number of utterances made during an audio-tape recorded session with the mother at home—see Table 3). These data derive from a detailed analysis of the audio-tape recordings of a half-hour sample of the child's speech at home (Howlin *et al.*, 1973a). More importantly, there was a significant increase in the proportion of socialized utterances and a significant decrease in the proportion of nonverbal utterances. There was also a reduction in the proportion of echolalic or "abnormal" (i.e., stereotyped, metaphorical, or thinking aloud) utterances, although this difference fell short of statistical significance. In other words, not only did the treated

Table 3.   Children's Language in Cases and Control Group A

	Cases		Controls (A)	
	Initial (mean)	6 months (mean)	Initial (mean)	6 months (mean)
Total number of utterances	105.7[a]	205.1[a]	144.7	184.7
% Socialized utterances	41.1[a]	64.1[a]	50.8	46.5
% Echolalic/"abnormal" utterances	21.2	14.0	13.2	17.5
% Nonverbal utterances	57.0[a]	38.0[a]	58.7	56.4

NOTE: Changes in functional speed calculated with respect to children with a score on that measure initially.
[a]Differences between initial and 6 months; $p < 0.05$ (t test).

Table 4. Communication in Cases and Control Group A

	Cases		Controls (A)	
	Initial (mean)	6 months (mean)	Initial (mean)	6 months (mean)
Communicative noises	5.8	8.9	8.1	16.2
Communicative words	22.6^a	58.6^a	34.1	27.0
Communicative phrases	63.9^a	87.3^a	126.3	125.0
Noncommunicative noises	41.7^a	20.4^a	47.3	46.6
Noncommunicative words	6.9	11.4	1.0	0.8
Noncommunicative phrases	18.1	6.5	4.3	3.3

NOTE: The figures in the body of the table refer to mean frequency of occurrence of items per one-and-one-half-hour observation period. Changes calculated with respect only to children with scores on that measure initially.
^{a}Differences between initial and 6 months; $p < 0.05$ (Wilcoxon matched pairs test).

autistic children come to speak more but also they came to make a greater use of speech for social communication. In sharp contrast, the control children failed to show changes on any of these measures.

The child's use of speech was also assessed by the time-sampled observations of the child's behavior over one and a half hours at home (Hemsley *et al.*, in preparation). These recorded the child's usage of various kinds of communicative and noncommunicative utterances. As is usually found with autistic children, in both the cases and the controls there were many noncommunicative utterances. However, whereas the control children showed no change in their pattern of usage over a six-month period, the experimental children greatly increased their use of communicative words and phrases and halved their output of noncommunicative noises over the first six months of treatment. It is possible that the experimental group's increase in communicative phrases could be an artefact of the base-line difference between cases and controls* on this variable (although it fell short of statistical significance). However this could not account for the increase in communicative words or the decrease in noncommunicative noises where there were no base-line differences. Taking the findings from Tables 3 and 4 together it is clear that the first six months of treatment resulted in substantial improvements in the children's socialized usage of spoken language, whereas no such improvement was evident in the control group. It should be added that both tables exclude the two experimental children who developed phrase

*This base-line difference does not reflect a lack of matching of the groups. All the change data are based on children who obtained a score on the items at the initial assessment. As it happened there were slightly more experimental children with communicative phrases although that subgroup was somewhat less advanced than comparable children with phrase speech in the control group.

Table 5. Children's Behavior in Cases and Control Group A

Behaviors	Cases		Controls	
	Initial (mean)	6 months (mean)	Initial (mean)	6 months (mean)
Not occupied	52.3	44.2	84.1	90.6
Oppositional	8.9	5.2	5.4	14.0
Cooperative, goal directed	99.2[b]	170.2[b]	87.8	98.4
Solitary, goal directed	46.8	33.8	46.3	27.1
Play with mother	25.9	14.9	28.8	13.2
Other (nondisruptive, noncooperative)	66.9	31.7	47.6	56.7
Total number of 10-second periods	300	300	300	300
Rituals[a]	58.4	26.5	40.8	52.1
Distress	6.6	7.5	6.7	8.4

NOTE: No differences between cases and controls on initial measures statistically significant.
[a] These last two items refer to actual frequencies whereas the other items all concern the number of 10-second periods (out of a total of 300) in which the specified behaviors occurred.
[b] Difference between initial and 6 months; $p < 0.01$ (t test).

speech during the course of treatment. No children in the control group developed phrase speech over the six-month period.

Table 5 summarizes some of the main findings on nonlinguistic behaviors. Again, the control group showed no significant changes. However, after six months of treatment, the experimental children showed only half as much ritualistic behavior (this difference fell short of the 5% level of significance) and nearly doubled the amount of time spent on cooperative goal-directed behaviors (such as doing puzzles or building things or looking at picture books with their mothers). It may be concluded that our treatment methods were successful in bringing about greater short-term changes in behavior than those found in the control group.

Six-Month Changes in Parental Behavior: Comparison with Control Group A

The next question is why had the children's behavior improved? The whole basis of the project was the use of parents as the main agents of change. Accordingly it was necessary to determine how far we had been successful in altering parental behavior in the desired direction. Our approach involved a specific focus on *how* the parents interacted with the child. The program did *not* require the parents to spend any more time with their autistic child, only to improve the way that time was used.

Table 6.   Positive Interaction in Cases and Control Group A

	Cases		Controls (A)	
	Initial (mean)	6 months (mean)	Initial (mean)	6 months (mean)
Mothers	18.3	24.2	19.1	20.0
Fathers	12.7	12.7	14.3	9.8

Tables 6 and 7 provide some data on the amount of parent-child interaction in the two groups. The cases and controls did not differ in their positive interaction scores (a measure of potentially pleasurable interactions such as rough and tumble play, games, joint outings, reading together, etc.), and there were no significant changes over time in either group (see Table 6). The standard day data (Table 7) indicates the number of minutes during the previous 24 hours that the parents spent with the autistic child. Again there were no changes over the six-month period in either group (the greater time spent with the child in the experimental group is largely an artefact of there being fewer children at nurseries or schools). Thus, as planned, the treatment program had meant *no* increase in the overall time parents spent with their child.

The next issue is whether we had succeeded in altering the way parents spent their time with their children. Table 8 summarizes the findings with respect to the one-and-a-half-hour observation period of mother-child interaction at home. We had sought to get the mothers to be more actively involved with their child during their time together. We succeeded in this. The mothers in the experimental group increased their active involvement with the child by some 50%; there was no change in the control group. The experimental mothers also greatly (and significantly) increased the number of remarks they made to the child during the one-and-a-half-hour period; again there was no change in the control group.

We also tried to get mothers to provide their children with better immediate feedback, praising the child for things done well or appropriately and correcting him for things done wrongly. Our success in this was shown by the finding that experimental group mothers more than

Table 7.   Parental Time with Child in Minutes (Standard Day Data)

	Cases		Controls(A)	
	Initial (mean)	6 months (mean)	Initial (mean)	6 months (mean)
Mother alone	360	361	225	286
Father alone	58	78	43	40
Both parents together	139	73	49	40

Table 8. Parental Behavior in Cases and Control Group A

Behavior	Cases		Controls	
	Initial (mean)	6 months (mean)	Initial (mean)	6 months (mean)
Active involvement with child	125.3	185.1[b]	116.6	111.6
Ignoring child	81.8	53.3	102.8	91.2
Other	92.9	61.6	80.6	97.2
Total number of 10-second periods	300	300	300	300
Total remarks[a]	170.1	276.1[c]	253.1	223.1
Neutral remarks	83.4	125.8[b]	137.8	138.2
Praise	15.1	32.5[b]	12.3	7.9
Corrections	5.7	17.8[b]	7.0	7.5
Directions	53.9	75.0	71.0	.1
Other speech	12.6	24.9	25.4	25.2
Gestures	4.1	11.4[b]	14.3	9.6
Directing physical contact	0.6	5.4[b]	7.1	4.2

NOTE: Difference between cases and controls on initial assessment for remarks and for directing physical contact significant at 0.01 level.
[a] All the categories in the lower half of the table concern actual frequencies and not number of time periods.
[b] Difference between initial and 6 months; $p < 0.01$ (t test).
[c] Difference between initial and 6 months; $p < 0.001$ (t test).

doubled their use of both praise and corrections while the control mothers showed no significant change. The experimental-group mothers also increased significantly in their use of gesture and physical prompts, styles of interactions we had encouraged in view of the children's difficulty in understanding spoken language.

Nevertheless, we also focused on the way the mothers spoke to their children. Our aim was to increase the amount of verbal interaction (as there is a tendency to reduce the amount said to a child who fails to respond), to get mothers to use speech to encourage communication by the child, and to use language at a degree of complexity appropriate for the child's level of understanding. Table 9 shows that the experimental-group mothers markedly increased the number of utterances* made during the half-hour audio-tape recorded period at home. Moreover, they significantly altered their pattern of language usage. There was a significant increase in the proportion of mothers' utterances which were directed to the child's use of language (i.e., questions, expansions, prompts, corrections, etc.—see Table 1 in Howlin *et al.*, 1973a) and a significant

---

*N. B. "Utterances," as used in Table 9, involve a much finer breakdown of what is said by mothers compared with the broader concept of "remarks" used in the one-and-a-half-hour observation period findings in Table 8.

Table 9. Mothers' Language to Child in Cases and Control Group A

	Cases		Controls	
Utterances	Initial (mean)	6 months (mean)	Initial (mean)	6 months (mean)
% Language-related utterances	26.0	45.8[b]	32.9	36.6
% Nonlanguage-related utterances	65.0	48.1[a]	56.4	49.7
% Interjections and incomprehensible remarks	9.4	6.1	10.7	13.6
Total number of utterances	253.6	407.6[c]	310.4	286.8

[a]Difference between initial and 6 months significant; $p < 0.05$.
[b]Difference between initial and 6 months significant; $p < 0.01$.
[c]Difference between initial and 6 months significant; $p < 0.001$.

decrease in utterances not so directed (i.e., directions, general statements, suggestions, etc.). Once more, there were no significant changes in the control group.

The findings clearly show that six months of intervention can bring about significant and worthwhile changes in the parental behavior. These changes were not found in the control group and therefore it may be concluded that the changes were due to our treatment program. It is more difficult to determine whether the modifications brought about in parental behavior were responsible for improvements in the child, although it seems probable that they were—at least in part. Group comparisons are mostly inappropriate for relating changes in parental behavior to alterations in the child as, to a considerable extent, parents needed to change in rather different ways. Obviously, the best way of dealing with a mute autistic child is not the same as the optimal response to an autistic child already using speech for communication. Similarly, the manner of coping with social withdrawal is not the same as that for managing aggressive outbursts. Nevertheless, some impression of the links between parent and child behaviors can be obtained by considering individual children.

Figures 4 and 5 show the changes over 18 months in the pattern of interaction shown by one mother-child pair on the home observation measures. This boy was of low intellectual level and was initially mute. He spent most of the first observation session playing ritualistically with chalk. He did not interact with his mother and she sat passively and silently watching him. After 18 months of treatment, the mother spent very little time passively watching but rather actively directed the child's activities, physically prompting him, using gestures herself, and correcting his mistakes. In parallel with these changes, the child's ritualistic behavior had ceased; he was involved in activities with his mother for most of the session and he was using some gestures and making word approximations.

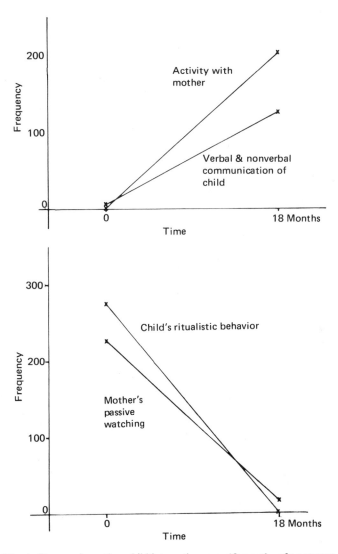

Fig. 4. Changes in mother-child interaction over 18 months of treatment.

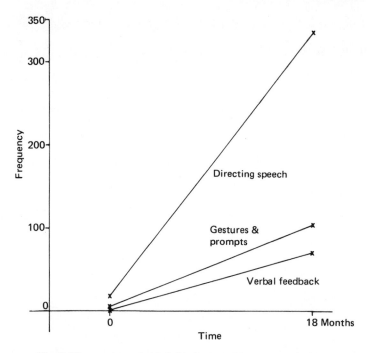

Fig. 5. Changes in mother's behavior over 18 months of treatment.

Figure 6 shows the interaction changes of another mother and child pair in which the child had acquired communicative phrase speech by the 18-month follow-up observation. The mother's total amount of speech increased over the 18 months but in addition by the end of treatment she was using gestures and providing verbal corrections and praise which had been totally absent initially. The changes in the child were comparable. On the first occasion he spent very little time involved with his mother but in the last session he spent most of the time actively interacting with her.

These findings all refer to mothers' behavior as observed at home during single one-and-a-half-hour periods of observations. However, our goal was not just to improve parents' responses to the children as they were, but also to increase their skills in coping with new problems and difficulties as they arose. This is more difficult to measure but we attempted to assess this by examining various aspects of parental coping skills in relation to some of the commoner problems shown by autistic children. The analysis of these findings is still incomplete and will be reported elsewhere.

Longer-Term Follow-Up: Comparison with Control Group B

All the measures used at the six-month follow-up were repeated again at 18 months for the cases (but not for control group A). The findings are consistent in showing that in all cases the improvements had been maintained but also that the gains in the second 12 months were much less than those in the first six months. In short, it was usually possible to bring

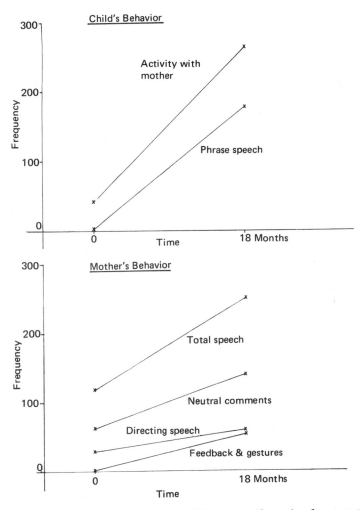

Fig. 6. Interaction changes of a mother-child pair over 18 months of treatment.

Table 10. 18-Month Follow-Up on Child Behaviors (Cases Only)

	Initial (mean)	6 month (mean)	18 month (mean)
Cooperative goal-directed behavior	99.4	170.2[a]	176.6[a]
Ritualistic behavior	58.4	26.5	22.8
Communicative phrases	63.9	87.3[b]	94.5[b]
Proportion of socialized utterances	41.1	64.1[b]	82.1[b]

NOTE: The figures in the body of the table refer to mean frequency of occurrence of items per one-and-one-half-hour observation period, apart from the figures for proportion of socialized utterances which refer to the mean percentage for that item.
[a] Differences from initial assessment significant at 1% level.
[b] Differences from initial assessment significant at 5% level.

about quite rapid short-term changes in the child's behavior and, equally, it was usually possible to maintain these improvements. On the other hand, the *pace* of change generally slowed quite markedly and few of the differences between six months and 18 months were statistically significant. Table 10 gives a few of the key findings which illustrate this pattern. The results for other measures of the child and for parental behavior are closely similar.

These findings clearly raise the question of how far the treatment influenced long-term outcome. To answer that question we compared the experimental group with control group B (an individually matched sample of autistic children seen only for assessment or treated at the same clinic in a much less intensive fashion with outpatient attendance only). The average length of time between first clinic attendance and follow-up was 33 months for the experimental group and 63 months for the controls.*

Table 11 summarizes the language data at follow-up for the two groups. Because the groups had been followed for different periods of time and because language skills would be expected to improve over time even without treatment, a regression analysis was used to partial out differences due to duration of follow-up (t values are based on differences between the regression lines for the two groups). On all measures of language competence the slope of the regression line for the experimental group was greater than that for the control group, although not significantly so. This means that the *rate* of improvement in language in the cases tended to be greater. In absolute terms, the experimental group had a marginally lower level of language on most (but not all) measures, but the difference was less than that expected on the basis of the age difference

*Control group A was followed for only six months as it was considered both impractical and unethical to ask control families to cooperate for longer without being offered the treatment program. For reasons already discussed, it was not possible to obtain a comparison group treated over the same period of time by some other treatment method.

Table 11. Longer Term Follow-Up (Language Measures)

	Cases		Controls (B)		Significance
	Mean (months)	(S.D.)	Mean (months)	(S.D.)	(t value)
Age at follow-up	93.9	(29.0)	125.6	(28.3)	3.12[a]
Nonverbal IQ at follow-up	85.4	(21.5)	83.8	(28.7)	0.18
Language tests					
Reynell Expressive Language	32.4	(21.2)	38.9	(27.6)	0.84
Reynell Language Comprehension	37.4	(18.7)	46.5	(21.3)	0.58
Spontaneous language sample (½ hour)					
Number of comprehensible utterances	176.8	(127.2)	142.7	(127.6)	1.15
Number of socialized communicative					
utterances	150.1	(112.8)	120.3	(118.3)	1.45
Number of correct phrases	38.4	(43.5)	44.5	(70.6)	0.95
Number of morphemes	79.3	(92.5)	103.3	(164.1)	0.87
Number of transformations	42.2	(67.3)	37.6	(83.9)	0.36
Parental interview data					
Language deviance/delay score	34.1	(33.0)	41.8	(33.4)	0.65

[a]$p < 0.005$.

between the groups. Thus, the experimental children had a Reynell expressive language score only six months below the controls in spite of a 32-month age difference. Moreover, on the measures which assessed functional language usage (rather than level of language complexity) the cases were actually superior in spite of being younger. Thus, the experimental children made more comprehensible utterances and more socialized, communicative utterances. None of these differences reach the 5% level of statistical significance because of the enormous spread on all measures within groups. Thus, the experimental group includes children with no comprehensible utterances but also others with fluent, socialized conversational speech. Analyses are still being undertaken to determine the reasons for these huge individual differences between autistic children all of whom have a nonverbal IQ of at least 60.

However, the fact of great within-group spread suggests that, as the groups were individually matched, a child-by-child comparison might be more informative than group means. Table 12 summarizes these findings. There was a general tendency for the cases to be superior to their matched controls. This difference was statistically significant at the 5% level for the number of grammatical transformations used and at the 10% level for number of comprehensible utterances and the proportion of socialized communicative utterances. There was no difference between the groups on either the Reynell language comprehension score or the language deviance/delay score based on parental interview data. It appeared that, on the whole, the treatment program had had more effect on the children's use of speech for communication than on their level of basic language competence.

The longer-term case-control differences with respect to abnor-

Table 12. Child-by-Child Comparisons on Long-Term Follow-Up
(After Partialing Out Effects of Time)

Measures	Cases superior to matched controls (B) N	Cases inferior to matched controls (B) N
Number of comprehensible utterances	11	5[a]
Number of socialized communicative utterances	11	5[a]
Number of correct phrases	10	6
Number of morphemes	10	6
Number of transformations	13	3[b]
Language deviance/delay score	8	8
Reynell Expressive Language	9	7
Reynell Language Comprehension	9	7

[a] $\chi^2 = 3.12; p < 0.10$.
[b] $\chi^2 = 6.12; p < 0.02$.

Table 13. Longer Term Follow-Up (Behavioral Measures)

Parental interview data	Cases		Controls (B)		Significance	
	Mean	(S.D.)	Mean	(S.D.)	t	(p value)
Social deviance	10.9	(4.7)	18.2	(6.7)	3.56	(<0.005)
Abnormal response to parents	1.8	(1.9)	4.8	(3.1)	3.33	(<0.005)
Social disruption	4.6	(2.9)	7.0	(3.5)	2.13	(<0.05)
Abnormal peer relationships	4.6	(1.2)	6.5	(1.9)	3.49	(<0.005)
Stereotyped play	13.5	(8.6)	20.8	(10.9)	2.09	(<0.05)
Obsessions and rituals	4.6	(4.0)	10.6	(5.4)	3.59	(<0.005)
Behavioral abnormalities	4.8	(3.9)	10.8	(7.6)	2.82	(<0.01)

malities in behavior were very much greater, as shown by Table 13. The parents were asked systematically about the frequency and severity of various social and behavioral abnormalities. These were grouped under seven main headings and in each case the scores on individual items were summed to provide a total score in which 0 reflected normality. The findings indicated that the cases showed less abnormality in all areas; they were more socially responsive both with their family and with other people; they showed more flexible and imaginative play; there were fewer tantrums and social disturbance; and there was less ritualistic behavior, abnormal attachments, and morbid preoccupations. While there was considerable individual variation within both groups, the spread was much less than that for the language measures.

DISCUSSION

Several important conclusions may be drawn from our study. We have shown that it is possible to apply systematic and standardized methods of evaluation to a flexible and comprehensive treatment program in the children's own homes. Obviously, when working with parents at home it is not possible (or desirable) to impose the same degree of rigid experimental control as in a laboratory or hospital setting. Nevertheless, we have shown that reliable quantitative assessments of both the child's behavior and parental functioning can be obtained at home. Such assessments have the major advantage of evaluating outcome with respect to the child's behavior in his natural environment. This is particularly important with autistic children in view of the well-established finding that they tend *not* to generalize changes from one setting to another.

The study design was novel in five major respects. First, as already noted, behavioral change was assessed at home and not at the clinic. Second, not only were changes examined through use of the children as their own controls, but also there was a systematic comparison with control groups treated in other ways. This is an essential part of any treatment evaluation because (a) autistic children tend to improve with increasing age, irrespective of active treatment; (b) there is huge individual variation in both the rate and extent of improvement (with and without treatment); and (c) it is necessary to determine not only that children benefit from treatment but also that they benefit more from one form of treatment than from another. Third, the control groups were well matched on variables (such as IQ and language impairment) known to be related to outcome. This has been lacking from previous group comparisons (cf. Wenar et al., 1967; Wenar & Ruttenberg, 1976) but is crucial because of the extensive evidence that children of low IQ and/or more severe language impairment have a much worse prognosis than other autistic children (Rutter, 1970). Fourth, the efficacy of treatment was evaluated in relation to a relatively long follow-up. While this has been undertaken in some previous investigation of behavioral methods (e.g., Lovaas et al., 1973), control groups have been lacking. Finally, unlike most previous studies utilizing parents as cotherapists, we have made direct assessments of our success in altering parental behavior.

As a result of these considerations, it is possible to have considerable confidence in the validity of the findings. The results show that our treatment program brought about quite dramatic changes during the first six months of intervention; that these changes were maintained and increased over the next year in spite of decreasing professional involvement; and that the long-term gains were both worthwhile and superior to those which followed other methods of treatment. In particular, we found that a home-based approach was superior to outpatient treatment based on similar principles. It may be concluded that the treatment program well proved its worth.* We were able to alter parents' ways of dealing with their autistic children and this was associated with marked benefits for the children.

It should be noted that these changes were obtained with comparatively little professional time spent with each family. The parents were working for about 30 minutes a day with their child on specific skills and the therapists were spending at the most two hours a week at the home at the beginning of treatment and as little as one hour per month in the later stages (although also they could be contacted by telephone). Parents

---

*Of course, it is not suggested that this is sufficient in itself. Specialized educational treatment is also crucial, as previously demonstrated (Bartak & Rutter, 1973; Rutter & Bartak, 1973; see also Chapters 28 and 29 in this volume).

were, of course, encouraged to be consistent in their handling of problem behavior throughout the day, and to use the principles taught for increasing specific skills when involved in other activities. But the actual amount of time spent on teaching each child was relatively short. We strongly disagree with Lovaas (see Chapter 25) that it is necessary for mothers to give up work and postpone having more offspring in order to spend most of the day with their autistic children. Indeed we think it positively undesirable that they be forced to do this in view of the needs of the nonautistic children and of the mothers' own needs for work and recreational outlets. There is no evidence that this family sacrifice brings about more improvement than that obtained in our program.

In planning our treatment approaches with each child, careful attention was paid to the child's developmental level with respect to nonverbal skills, language, and socialization. Intervention was made appropriate to the child's level in each of these areas so that, as far as possible, development followed the normal stages. This was especially the case with language programs (Yule & Berger, 1976) but it also applied to programs aimed at increasing imaginative play, cooperative activities with other people, and general self-help skills. Training in expressive use of language was never begun, for example, until a minimal level of comprehension had been established and simple representational play was taught before more abstract imaginative games.

Parental training in operant procedures was extended to include other behavioral techniques such as systematic desensitization for phobias, graded change of the environment for obsessional behavior, and deliberate intrusion into the child's ritualistic activities to foster social interaction and attachment. By starting on those problems most worrying to the family, we were usually able to motivate them to work on other areas we recognized as abnormal, and eventually we generally covered most aspects of each child's behavior. In addition to working on all problems shown by the autistic child we also had to be concerned with other problems within the family. Some of the sibs showed jealousy toward the autistic child, some had language or other problems of their own, and some suffered from the lack of attention consequent upon family preoccupation with the autistic child's disturbance. Often it was necessary to assist in finding suitable school placements and, where possible, implement parallel programs there. Sometimes we needed to arrange relief for parents at time of stress by finding appropriate baby-sitters or holiday placements. In addition there was the important part played by supportive counseling for the parents. Some had to be helped with feelings of guilt or depression and in other cases we needed to resolve conflict between parents over the child or deal with parental marital problems (which were present in about a quarter of our families). Our involvement frequently involved treatment of such family problems by casework or behavioral approaches and, in some cases, treatment of maternal depression or the resolution of a severe marital crisis had to take precedence for a while over the autistic child's problems.

The therapeutic program emphasized the need to improve the social skills and social involvement of the autistic child, both to increase his acceptability and adjustment within his natural environment and also to improve the quality of his social contacts with other people. Techniques for achieving this had to be acceptable to the parents as therapists and programs employing approaches they felt they could carry out were always worked out jointly with them. Parents were helped to adapt their own styles and methods of child handling to produce a more structured, consistent, and effective environment to foster optimal development in all spheres.

This usually required some degree of intrusion into the child's activities. The Tinbergens (1972, 1976) have argued that this approach should make the autism worse, whereas our results clearly indicate the reverse. Far from aggravating the autistic condition, the treatment program led to lasting and substantial benefits. It is necessary to consider why there is this apparent contradiction. The first point is that the Tinbergens' views have very little empirical support (Wing & Ricks, 1976) so the clash is between their *views* and our *empirical findings* (and also those from other studies) (see Chapters 21, 25, 26, and 29 in this volume). Nevertheless, Richer has attempted to test Tinbergen's views. His study (Richer & Richard, 1975) found that autistic children showed less withdrawal when an adult looked at them passively than when the adult smiled, looked away, or adopted a timid posture. It is evident both that the findings ran counter to Tinbergen's (and Richer's) hypothesis and also that none of their experimental conditions remotely approximated ours. The whole rationale of our intervention was to insure that we intruded in a way which proved pleasurable and rewarding to the child. In doing this, of course, we based our approach on how the child *actually responded* and not on any theoretical notions as to what *"ought"* to be rewarding. By these means it was readily possible to intrude and so foster conditions likely to encourage socialization and communication. Whereas inept and clumsy intrusion may indeed induce tension and withdrawal, skilled and sensitive intrusion leads to pleasurable social involvement. That was our objective.

Lastly, some comments are needed on variations in outcome. First, there is the difference between progress in language and reductions in behavioral abnormalitites. The longer-term control-group comparison showed that the therapeutic program had been much more successful in treating all manner of behavioral abnormalities than it was in fostering normal language development. This difference was not an artefact of different forms of measurement as it applies most strongly when *all* comparisons are made using parental interview data. In interpreting this finding, it is important to note that the variance on language measures was much greater than that on other measures. Some children moved rapidly to spontaneous use of grammatical sentences whereas others still had only a few words after most prolonged and intensive therapeutic intervention. The reasons for this marked individual variation are still being studied but it may well be that there are considerable constraints on development implicit in the severity of the children's basic cognitive handicap.

Of course, this does not mean that treatment is without effect on language development. To the contrary, there were marked benefits in the short-term, and modest, but still worthwhile, benefits in the longer term. It appears that although treatment does something to facilitate better competence in language, it is of most value in helping the children to make better social use of the language skills they possess. As the main purpose of language is social communication this is a most important gain.

Although the social and behavioral abnormalities proved to be more modifiable than the language deficit, nevertheless there was still considerable individual variation. At follow-up some of the children were inept in their peer relationships but otherwise were without serious behavioral disturbance. On the other hand, some remained a major management problem in spite of considerable improvement. The analysis of variables associated with this individual variation is still under way. However, as others (Wenar & Ruttenberg, 1976) have noted, the most handicapped children tend to be those for whom treatment did not begin until middle or later childhood. Whether they would have made better progress if home-based treatment had been available during the preschool years is uncertain. The matter requires further study. Nevertheless, the available findings indicate that early help with socialization and communication *might* avoid some of the later secondary behavioral handicaps. On the other hand, the same findings show that many autistic children continue to need help throughout childhood as new and different problems emerge with increasing age. On the basis of our evidence to date, home-based treatment involving parents, using behavioral and casework procedures as well as a developmental framework, will help meet the needs of autistic children and their families.

ACKNOWLEDGMENTS

We are indebted to the National Society for Autistic Children for assistance in obtaining control group A and to the D. H. S. S. whose grant made the study possible. Most of all, we are grateful to the families who, with patience and tolerance, agreed to the multiple measurements and assessments needed for this systematic evaluation of treatment.

REFERENCES

Bartak, L., & Rutter, M. Special educational treatment of autistic children: A comparative study. I. Design of study and characteristics of units. *Journal of Child Psychology and Psychiatry,* 1973, *14,* 161–179.

Bartak, L., Rutter, M., & Cox, A. A comparative study of infantile autism and developmental receptive language disorder: 1. The children. *British Journal of Psychiatry,* 1975, *126,* 127–145.

Berkowitz, B. P., & Graziano, A. M. Training parents as behavior therapists: A review. *Behavior Research and Therapy,* 1972, *10,* 297–317.

Berger, M., Howlin, P., Marchant, R., Hersov, L., Rutter, M., & Yule, W. Instructing parents in the use of behaviour modification techniques as part of a home-based approach to the treatment of autistic children. *Behaviour Modification Newsletter,* 1974, Issue 5, 15–27.

Bonvillian, J. D., & Nelson, K. E. Sign language acquisition in a mute autistic boy. *Journal of Speech and Hearing Disorders,* 1976, *41,* 339–347.

Brown, G. W., & Rutter, M. L. The measurement of family activities and relationships: A methodological study. *Human Relations,* 1966, *19,* 241–263.

Brown, R. A., & Pace, Z. S. Treatment of extreme negativism and autistic behavior in a 6-year old boy. *Exceptional Children,* 1969, *36,* 115–122.

Browning, R. M. Treatment effects of a total behavior modification program with five autistic children. *Behavior Research and Therapy,* 1971, *9,* 319–328.

Cantwell, D., Howlin, P., & Rutter, M. The analysis of language level and language function: A methodological study. *British Journal of Communication Disorders,* 1977, *12,* 119–135.

Douglas, J. W. B., Lawson, A., Cooper, J. E., & Cooper, E. Family interaction and the activities of young children. Method of assessment. *Journal of Child Psychology and Psychiatry,* 1968, *9,* 157–171.

Folstein, S., & Rutter, M. Infantile autism: A genetic study of 21 twin pairs. *Journal of Child Psychology and Psychiatry,* 1977, *18,* 297–321.

Fulwiler, R. L., & Fouts, R. S. Acquisition of American sign language by a noncommunicating autistic child. *Journal of Autism and Childhood Schizophrenia,* 1976, *6,* 43–52.

Graziano, A. M. A group treatment approach to multiple problem behaviors of autistic children. *Exceptional Children,* 1970, *36,* 765–770.

Halpern, W. I. The schooling of autistic children: Preliminary findings. *American Journal of Orthopsychiatry,* 1970, *40,* 665–671.

Hemsley, R., Cantwell, D., Howlin, P., & Rutter, M. The adult-child interaction schedule (in preparation).

Hingtgen, J. N., Coulter, S. K., & Churchill, D. W. Intensive reinforcement of imitative behavior in mute autistic children. *Archives of General Psychiatry,* 1967, *17,* 36–43.

Hingtgen, J. N., Sanders, B. J., & DeMyer, M. K. Shaping cooperative responses in early childhood schizophrenics. In L. Ullmann & L. Krasner (Eds.), *Case studies in behavior modification.* New York: Holt, Rinehart & Winston, 1965.

Hingtgen, J. N., & Trost, F. C. Shaping cooperative responses in early childhood schizophrenics. II. Reinforcement of mutual physical contact and vocal responses. In R. Ulrich, T. Stachnik, & J. Mabry (Eds.), *Control of human behavior.* Chicago: Scott, Foresman, 1966.

Howlin, P., Cantwell, D., Marchant, R., Berger, M., & Rutter, M. Analyzing mothers' speech to young autistic children: A methodological study. *Journal of Abnormal Child Psychology,* 1973, *1*, 317–339. (a)

Howlin, P., Marchant, R., Rutter, M., Berger, M., Hersov, L., & Yule, W. A home-based approach to the treatment of autistic children. *Journal of Autism and Childhood Schizophrenia,* 1973, *3*, 308–336. (b)

Lawson, A., & Ingleby, J. P. Daily routines of preschool children: Effects of age, birth order, sex, and social class, and developmental correlates. *Psychological Medicine,* 1974, *4*, 399–415.

Lovaas, O. I., Schaeffer, B., & Simmons, J. Q. Experimental studies in childhood schizophrenia: Building social behavior in autistic children by use of electric shock. *Journal of Pergamon Press,* 1966.

Lovaas, O. I. Behavior therapy approach to the treatment of childhood schizophrenics. In J. Hill (Ed.), *Minnesota symposia on child psychology* (Vol. 1). Minneapolis: University of Minnesota Press, 1967.

Lovaas, O. I., Berberich, J. P., Perloff, B. F., & Schaeffer, B. Acquisition of imitative speech in schizophrenic children. *Science,* 1966, *151*, 705–707.

Lovaas, O. I., Freitag, G., Gold, V. J., & Kassorla, I. C. Experimental studies in childhood schizophrenia. *Journal of Experimental Child Psychology,* 1965, *2*, 67. (a)

Lovaas, O. I., Koegel, R., Simmons, J. Q., & Long, J. S. Some generalization and follow-up measures on autistic children in behavior therapy. *Journal of Applied Behavior Analysis,* 1973, *6*, 131–165.

Lovaas, O. I., Schaeffer, B., & Simmons, J. O. Experimental studies in childhood schizophrenia: Building social behavior in autistic children by use of electric shock. *Journal of Experimental Research in Personality,* 1965, *1*, 99–109. (b)

Maier, K. An operant approach to the modification of hyperkinetic behavior: A preliminary report. Paper presented to the Annual Conference of the British Psychological Society, Exeter, April, 1971.

Marchant, R., Howlin, P., Yule, W., & Rutter, M. Graded change in the treatment of the behavior of autistic children. *Journal of Child Psychology and Psychiatry,* 1974, *15*, 221–227.

Marshall, G. R. Toilet training of an autistic eight-year old through conditioning therapy. A case report. *Behavior, Research and Therapy,* 1966, *4*, 242–245.

Mathis, M. I. Training of a "disturbed" boy using tne mother as therapist: A case study. *Behavior Therapy,* 1971, *2*, 233–239.

McConnell, O. L. Control of eye contact in an autistic child. *Journal of Child Psychology and Psychiatry,* 1967, *8*, 249–255.

Mecham, M. J. *Verbal Language Development Scale,* California: Western Psychological Services, 1958.

Metz, J. R. Conditioning generalized imitation in autistic children. *Journal of Experimental Child Psychology,* 1965, *2*, 389–399.

Miller, A., & Miller, E. E. Cognitive-developmental training with elevated boards and sign language. *Journal of Autism and Childhood Schizophrenia,* 1973, *3*, 65–85.

O'Dell, S. Training parents in behavior modification. *Psychological Bulletin,* 1974, *81*, 418–433.

Patterson, G. R., & Brodsky, G. A behavior modification program for a child with multiple problem behaviors. *Journal of Child Psychology and Psychiatry,* 1966, *7*, 277–295.

Patterson, G. R., & Gullion, M. L. *Living with children: New methods for parents and teachers.* Champaign, Ill.: Research Press, 1968.

Reynell, J. *Reynell Developmental Language Scales.* Windsor, Berks., England: N. F. E. R. Publishing Co., 1969.

Reichler, R. J., & Schopler, E. Developmental therapy: A program model for providing individualized services in the community. In E. Schopler & R. J. Reichler (Eds.), *Psychopathology and child development: Research and treatment.* New York: Plenum, 1976.

Richer, J., & Richards, B. Reacting to autistic children: The danger of trying too hard. *British Journal of Psychiatry,* 1975, *127,* 526–529.

Risley, T., & Wolf, M. Establishing functional speech in echolalic children. *Behavioral Research and Therapy,* 1967, *5,* 73–88.

Rutter, M. Autistic children: Infancy to adulthood. *Seminars in Psychiatry,* 1970, *2,* 435–450.

Rutter, M. The description and classification of infantile autism. In D. W. Churchill, G. D. Alpern, & M. K. DeMyer (Eds.), *Infantile autism: Proceedings of the Indiana University Colloquium.* Springfield, Ill.: Charles C Thomas, 1971.

Rutter, M. The assessment and treatment of preschool autistic children. *Early Child Development and Care,* 1973, *3,* 13–29.

Rutter, M., & Bartak, L. Special educational treatment of autistic children: A comparative study. II. Follow-up findings and implications for services. *Journal of Child Psychology and Psychiatry,* 1973, *14,* 241–270.

Rutter, M., & Sussenwein, F. A developmental and behavioral approach to the treatment of preschool autistic children. *Journal of Autism and Childhood Schizophrenia,* 1971, *1,* 376–397.

Schell, R. E., & Adams, W. P. Training parents of a young child with profound behaviour deficits to be teacher-therapists. *Journal of Special Education,* 1967–8, *2,* 439–454.

Schopler, E., & Reichler, R. J. Developmental therapy by parents with their own autistic child. In M. Rutter (Ed.), *Infantile autism: Concepts, characteristics and treatment.* Edinburgh: Churchill-Livingstone, 1971.(a)

Schopler, E., & Reichler, R. J. Parents as cotherapists in the treatment of psychotic children. *Journal of Autism and Childhood Schizophrenia,* 1971, *1,* 87–102. (b)

Sloane, H. N., Johnston, M., & Harris, F. R. Remedial procedures for teaching verbal behavior to speech deficient or defective young children. In H. N. Sloane & B. D. MacAulay (Eds.), *Operant procedures in remedial speech and language training.* New York: Houghton Mifflin, 1968.

Stutsman, R. *Mental measurement of preschool children.* Tarrytown-on-Hudson, New York: World Book Co., 1931.

Sulzbacher, S. I., & Costello, J. M. A behavioral strategy for language training of a child with autistic behaviors. *Journal of Speech and Hearing Disorders,* 1970, *35,* 256–276.

Tate, B. G., & Baroff, G. S. Aversive control of self-injurious behavior in a psychotic boy. *Behavioral Research and Therapy,* 1966, *4,* 281–287.

Tinbergen, E. A., & Tinbergen, N. Early childhood autism—an ethological approach. *Advances in Ethology,* 1972, *10.*

Tinbergen, E. A., & Tinbergen, N. The aetiology of childhood autism: A criticism of the Tinbergens' theory: A rejoinder. *Psychological Medicine,* 1976, *6,* 545–550.

Webster, C. D., McPherson, H., Sloman, L., Evans, M. A., & Kichar, E. Communicating with an autistic boy by gestures. *Journal of Autism and Childhood Schizophrenia,* 1973, *3,* 337–346.

Wechsler, D. Wechsler Intelligence Scale for Children. New York: Psychological Corporation, 1949.

Wenar, C., & Ruttenberg, B. A. The use of BRIAC for evaluating therapeutic effectiveness. *Journal of Autism and Childhood Schizophrenia,* 1976, *6,* 175–192.

Wenar, C., Ruttenberg, B. A., Dratman, M. L., & Wolf, E. G. Changing autistic behavior: The effectiveness of three milieus. *Archives of General Psychiatry,* 1967, *17,* 26–35.

Wetzel, R. J., Baker, J., Roney, M., & Martin, M. Outpatient treatment of autistic behavior. *Behavioral Research and Therapy,* 1966, *4,* 169–177.

Wing, L., & Ricks, D. M. The aetiology of childhood autism: A criticism of the Tinbergens' ethological theory. *Psychological Medicine,* 1976, *6,* 533–544.

Wolf, M., Risley, T., Johnston, M., Harris, F., & Allen, E. Application of operant conditioning procedures to the behavior problems of an autistic child: A follow-up and extension. *Behavioral Research and Therapy,* 1967, *5,* 103–111.

Wolf, M., Risley, T., & Mees, H. Application of operant conditioning procedures to the behavior problems of an autistic child. *Behavioral Research and Therapy,* 1964, *1,* 305–312.

Yule, W., & Berger, M. Behavior modification principles and speech delay. In M. Rutter & J. A. M. Martin (Eds.), *The child with delayed speech.* Clinics in Developmental Medicine, No. 43. London: SIMP/Heinemann, 1972. Pp. 204–219.

Yule, W., & Berger, M. Communication, language and behavior modification. In C. C. Kiernan & F. P. Woodford (Eds.), *Behavior modification with the severely retarded* (IRMMH Study Group 8). Amsterdam: Associated Scientific Publishers (Excerpta Medica), 1976. Pp. 35–65.

27

Changing Parental Involvement in Behavioral Treatment

ERIC SCHOPLER

Behavior modification techniques have had a growing effect on the treatment and management of autistic children during the past two decades. This trend has been accompanied by an immense literature and also several new journals in which to publish it. Most of these publications have attested to the efficacy of operant conditioning for changing many kinds of specific behavior from language (Lovaas, 1961) to self-injurious behavior (Tate & Baroff, 1966).

The involvement of parents with behavioral approaches to their own autistic children is even more recent (Schopler & Reichler, 1971; Howlin *et al.*, 1973). It is likely that the impact of behaviorism on autistic children has been greater through the application with their parents, than it has through the use of behavioral techniques with the children only. The preceding two chapters by Lovaas and Hemsley *et al.* review some of the most current issues and data on the parents' role in the treatment of autistic children. However, in order to fully appreciate the strides made in the treatment of autistic children and their families, I will briefly sketch out the historical roots of operant conditioning and also the parental role with severely disturbed children.

For centuries it has been recognized that human behavior could be shaped by responses in the natural and social environment. If such environmental responses had a pleasurable or desirable effect the behavior

ERIC SCHOPLER · Department of Psychiatry, Division of Health Affairs, University of North Carolina School of Medicine, Chapel Hill, North Carolina.

was likely to continue; if the response was painful or aversive the eliciting behavior was apt to be changed or discontinued. With the evolution of psychology as a social science during this century behavior theory was reformulated by Watson (1908) and gained impetus for behavioral engineering under the formulations of Skinner (1953). Early experimental research on the application of behaviorism was primarily carried on in the animal laboratory with pigeons and rodents. Translating this animal research to the application with autistic children was the pioneering task undertaken in the work of Lovaas *et al.* (1965) and Hingtgen *et al.* (1965). The attitudes of behaviorists toward parents during the 1960s could be characterized in two ways. Parents usually were either left out of behavioral studies because it was too difficult to work them into the research design, or they were regarded as the source of the conditioning history which provided the favorite explanation for the autistic behavior (Ferster, 1961).

In spite of the negligible role assigned to parents in the early application of operant conditioning the groundwork was formed for parental participation in the treatment enterprise. Prior to the mid 1960s (Rimland, 1964), autism was widely regarded as the child's primary social withdrawal from pathological family relationships. Changes in parental personality and attitudes were considered primary for improvement in the child. Parents were regarded and treated as patients, with the expectation of changing their personality and improving their psychopathology. Empirical evidence for this view of autistic children and their parents was lacking (Rutter, 1968). Instead it was based on theoretical views derived from psychoanalytic theory and administrative considerations (Schopler, 1971).

Two important shifts have been taking place in the treatment of autistic children and their parents during the past decade. One is the shift of parental role from cause of the disorder to cause of fostering optimum development. This trend is reflected in the contractual relationship between parent and therapist. The second is the evolution of behavioral techniques from an experimental research method to a therapeutic treatment approach. I will discuss several issues pertaining to these two trends, raised in the two preceding chapters, and from our experience with a statewide program for autistic children (Reichler & Schopler, 1976).

TREATMENT CONTRACT

From his behavioral orientation in Chapter 25, Lovaas refers to a contract between the therapist and the child's parents. The purpose of this contract is to change specific behaviors in the child, with explicit techniques to be learned and applied by the parent. The basis for therapist and parent working together and also for terminating their relationship is spelled out. In the traditional parent-therapist relationship, the presumption was that the parent misconstrued the essential elements of the parent-child relationship and the professional on the other hand understood it. A shift has taken place in this position. In the behavior orientation, the professional still lays claim to expert knowledge on changing behavior. He acknowledges that the parent may know more about his own child. This change in the contractual relationship between professional and parent has had a substantial effect in improving the coping abilities of both parents and their autistic children. Although there appears to be a noticeable difference between the traditional patient-doctor contract available to parents of autistic children and the therapeutic contract available within the behavioristic orientation, there are also some far-reaching differences among the treatment contracts possible in different behavioral programs. Some interesting comparisons can be made between the Lovaas experience (Chapter 25), the Hemsley home-treatment program (Chapter 26), and our statewide program in North Carolina. Lovaas formulates a "legal" contract requiring parents to make a substantial commitment to his treatment regime. This includes an attendance requirement, the expectation that one of two working parents give up work to devote himself or herself to the autistic child, and that they postpone having other children. Clearly the parents' ability to meet their commitments to their other children and to each other is in the balance.

Several important questions may be raised about such a rigorous contract. Are the demands made on the parents to change their style of life for the treatment of their autistic child based in clinical necessity, and the potential for cure, or are they part of the behaviorists' experimental research origin in which maximum control was required for subject and experimental conditions? In Lovaas' chapter no data were presented to indicate that his autistic subjects were cured. It is, therefore, instructive to consider some of the clinical data that informed the treatment contract used in the Lovaas program. From his clinical experience we learned that treatment outcome was highly variable and dependent on the posttreatment environment. Children who returned to their parents maintained their gains while children who went to institutions lost their gains. These observations led to the rigorous parent treatment contract. It could be that this contract was based on erroneous conclusions. Perhaps the critical outcome factor suggested by Lovaas is not just whether an autistic child goes with his parents or into an institution. Perhaps the critical factor is the degree of the child's impairment. The more severely impaired children may maintain the most limited gains and they would also be the strongest applicants for institutional care. If this is the case, perhaps parents should not be asked to alter their life style and to sacrifice educational opportunities of their other children to the limited benefit of the severely impaired child.

The Hemsley *et al.* chapter illustrates a different kind of parent contract also within the behavioral framework. Having recognized that families behave differently in the clinic than they do at home, they reasoned that a more accurate assessment of the child and his family may be obtained by seeing him at home in his natural environment. The assessment and observation of families with severely disturbed children in their own homes is not new. It has been reported by Erikson (1950) and Henry (1967). However, the concept of confining the treatment effort to the child's own home had not been reported before. It is an exciting clinical innovation which also has the effect of equalizing the treatment contract with parents. The burden of meeting for treatment sessions is transferred from the parents coming to the clinic to the therapist going to the home. Consistent with this attitude is the demand on time spent by parents with their autistic child. The Maudsley group does not expect parents to spend any more time with their autistic child as a result of their special training efforts. However, they do expect parents to make better use of the time they spend with their child. The emphasis in this home-treatment program is on language development and they report data showing that such qualitative improvement in the parent-child interaction occurs without the degree of parental sacrifice demanded by the Lovaas contract. The difference in the treatment contracts of these two behavioral programs no doubt represents a shift in research orientation, a movement from the rigorous controls of experimental research with laboratory animals to the clinical arena of the less controllable human family. It also represents a shift from viewing parents as sources of the wrong conditioning history who require intensive retraining to viewing parents as collaborators.

Our state program reported in Chapter 29 (see also Reichler & Schopler, 1976) uses a treatment contract more like the one in the Hemsley group than the Lovaas. However, in some ways we move even further than either of these programs in the direction of giving greater control to parents in defining the treatment contract. Our emphasis is on helping with the parent-defined adaptational problems. This does not mean that we do not share our assessment of the child's special developmental needs (Schopler & Reichler, 1976) and our treatment recommendations. It does mean that our treatment contract is defined individually by maximum parental participation. This contract may include additional parental effort spent with the child on home programs or less time when the child is transferred to one of our classrooms. These differences between the treatment contract in the Maudsley Program and the North Carolina Program also relate to the differences in program mandates. The Maudsley program has a clinical-research mandate focused on language problems, and the selection of autistic subjects with performance IQ of 60 or better. The North Carolina program has a service mandate of helping all autistic children in the state with any of their adaptational problems, including 76% of our children with performance IQ less than 60.

To summarize, behavioral approaches can now be used in treatment contracts which vary in the degree of control assumed by the experimental or clinical researcher. On one end of the continuum is a high degree of professional control over conditions of treatment and on the other end is a high degree of parental participation and control over the details of the treatment contract. This represents a shift from viewing the parental environment as the source of faulty conditioning procedures, requiring intensive training, to viewing parents as active collaborators who have a right and responsibility to shape the direction of their child's adaptation.

FROM LABORATORY RESEARCH TO ACTIVE HELP

I do not wish to suggest that the purposes of experimental research and therapeutic or educational help are incompatible. However, there are some important strains between the two kinds of enterprise, stresses which can be modified with different procedures. These differences include: (1) Experimental research is conducted on discrete questions which can be answered by available methods. Treatment involves complex adaptational questions requiring an answer without established solutions. (2) Experimental research occurs in a laboratory with maximum control over subjects and experimental conditions. Treatment occurs via patients or parents who are relatively free to identify their own problems and discomforts, and to seek the available treatment of their choice. (3) Experimental research requires homogeneous subjects selected to fit the research design. In such studies individual differences appear as statistical noise and investigator irritation. Treatment is administered to relatively self-selected populations whose idiosyncratic pains and problems are the focus of concern.

I review these incompatibilities or stresses in experimental research and treatment in order to identify some procedures by which they are being resolved within the behavioral approaches. A common criticism against the use of operant conditioning with autistic children has been the charge that laboratory techniques devised with animal subjects are used, and that the children are turned into robots. These critics confuse the historical origins of operant conditioning with the application of the technique with human beings. Some indications of this shift can be found in Chapters 25 and 26 and include the following.

Within the behavioral framework different approaches to parent interaction are used. In Lovaas' system three specific training techniques are emphasized. They include: (1) instructional techniques based on reading and theory; (2) modeling procedures in which the parent is expected to imitate and model himself after an expert therapist; and (3) videotape techniques in which the parent observes tapes of his own interaction with the child and has it criticized by a senior therapist. All three of these methods are based on experimentally demonstrated techniques. The Hemsley group's approach to parents is less experimental and more clinical. The child's major difficulties are identified and evaluated. Priority is given to those problems giving the most concern to the parents. A functional analysis of the problems and the conditions under which they arose is made with the parents. When the problems are complex they are broken down into smaller more manageable units. This approach maintains focus on the problems for which help is sought rather than on the techniques available. The problem-solving orientation described by the Hemsley group is also used in our Program. Clinically we have found that it is more useful to engage parents in a problem-solving dialogue than to offer them expert techniques. We formulate home programs in collaboration with parents. These teaching programs are modified individually on the basis of our ongoing dialogue.

Another manifestation of the same shift is Lovaas' suggestion that the appropriate use of behavior modification techniques requires parents to understand abstract and technical terminology. No doubt it can be a source of satisfaction for some parents to have mastered some technical jargon. However, it is also possible for the professional to translate behavior technology into the daily living routines of a family. The Hemsley group tends to follow this direction. They also discuss directly with parents their feelings of guilt, depression, confusion, whatever. This is also the approach followed in our Program. We do not usually have to pressure parents into using appropriate behavioral techniques, nor do we need to plan special negative reenforcers, like threat of termination, to move them toward appropriate behavior management. In the majority of cases the most effective reenforcer is the improvement in the child-parent interaction. This makes life easier and more pleasurable for parents, and most of them will continue their efforts for this.

Another significant shift from experimental technology to therapy can be seen in the shift from the emphasis on modifying specific behaviors to more complex behavior units. Change in the frequency of specific behaviors was the criterion of experimental outcome. In Chapter 25 Lovaas now discusses improvement in more complex behavior like appropriate play and social behavior. It is true that these are more difficult to count or rate. In fact the rating methodology was not reported. However, there is little doubt that such ratings can be made reliably and are reported in the Hemsley chapter.

More important, the experimental approach to behavior modification has no systematic way of selecting priorities in teaching or modifying behavior. This shortcoming could have a direct connection with the frequently expressed concern over the lack of generalization in the child's specific behavioral improvement. Lovaas speculated that such lack of generalization may be due to insufficient knowledge about the child's motivation. Perhaps even more important is the question of how the tasks and behaviors to be taught to the child are selected. This behavior selection process is often not evaluated systematically in most behavioral programs. The Hemsley group evaluates each child developmentally. Our own experience suggests that a developmental evaluation is necessary (Reichler & Schopler, 1976). Autistic children are characterized by irregularities in their learning skills. Tasks that are too difficult and tasks that are too easy for individual children must be identified separately from emerging skills. If this assessment is not taken seriously, the child's motivation is lost in items beyond or below his developmental level, and unrealistic goals and treatment contracts can develop.

While it is true that each of us can improve whatever skills we have with practice, it is an oversimplification to suggest that this improvement potential is unlimited and that we can avoid the assessment of these limits. This assessment of a child must play an important part in the treatment contract made with the parents, especially when they are regarded as the major agents for generalizing the child's newly learned adaptational skills.

The shift in behavioral approaches from experimental lab to the human service arena has also been reflected in the research methodology. Prior to the advent of behavior modification, virtually no systematic research was conducted on the treatment of autistic children. Most published reports were anecdotal and based more on theoretical views than clinical or objective data. Perhaps partly as a reaction to this trend, early operant conditioning was scrupulously rigorous. A specific behavior was selected to be modified, behavior frequency was counted, reenforcers specified, modified behavior counted and graphed. It is not surprising that this research model is questioned for studying treatment outcome (Azrin, 1977), and rarely used for reporting parent-child interaction.

A wider range of behavioral changes in both children and their parents are noted in the Hemsley chapter. Treated and untreated controls are used for demonstrating treatment effects. Although this design approaches more closely the clinical needs of autistic children and their families than does the experimental laboratory research of the early behaviorists, some inconsistencies between clinical and research requirements are in evidence. Because of research interest in language development and the effects of treatment on language, most of the outcome measures involved language. Moreover, the subjects included had a nonverbal IQ of 60 or higher. They no doubt had a better language potential than autistic children with nonverbal IQ of less than 60. The latter children were excluded as was appropriate to the language research focus. However, the incompatibilities of research requirements and clinical requirements persist. Confusion between clinical outcome and research outcome is likely, perhaps inevitable. For example, a mute child in a language study might have learned the word "milk." The acquisition of this single word and the mother's increased verbalization in teaching it, could count toward the child's successful response to the treatment. If at the same time the child's toilet training was completed, or if it regressed for that matter, neither could easily show up in the research findings.

Although it seems unlikely that all incompatibilities between research and clinical requirements can be resolved, I believe it is possible to move further in the direction suggested by the two chapters discussed above. In our state program no autistic child can be excluded, even when functioning in the severely retarded range. Autism is recognized as a handicap with varying severity. Treatment aims are formulated with priority for the difficulty parents experience with their child and also special education goals set jointly between staff and parents. Individual goals and methods by which they are to be met, and assessment of the extent to which these goals are met are recorded. Treatment outcome for both specific and general goals is evaluated through independent interviews with parents, therapist, and teaching staff. Assessment of outcome is validated by the extent to which they agree on their evaluation of success or failure. This consensual validation design has advantages over the ABAB design and also over the comparison with the external control group. It incorporates the recognition that family adaptation with an autistic child can have many different forms. What a given family decides they can and want to do for optimum adaptation in collaboration with their professional consultant may be closer to the truth than any external criteria.

SUMMARY

In summary, Chapters 25 and 26 show evidence of important shifts within the application of behaviorism to the treatment of autistic children and their parents. The differences among various data-based behavioral programs are less striking than are the differences between them and theory-based psychodynamic treatment. A shift seems to be in progress moving from the experimental rigors underlying operant conditioning to research efforts including complex family problems. In a similar vein there is a shift from excluding parents in treatment research to giving them a prominent role. Both factors could indicate more effective and humane treatment of this handicapped population.

REFERENCES

Azrin, N. H. A strategy for applied research: Learning based but outcome oriented. *American Psychologist*, 1977, *32*, 140–150.

Erikson, E. *Childhood and society*. New York: W. W. Norton Company, Inc., 1950.

Ferster, C. B. Positive reenforcement and behavioral deficits of autistic children. *Child Development*, 1961, *32*, 437–456.

Henry, J. My life with the families of psychotic children. In G. Handel (Ed.), *The psychosocial interior of the family*. Chicago: Aldine, 1967. Pp. 30–46.

Hingtgen, J. N., Sanders, B. J., & DeMyer, M. K. Shaping cooperative behavior in early childhood schizophrenics. In L. Ullman & L. Krasner (Eds.), *Case studies in behavior modification*. New York: Holt, Rinehart & Winston, 1965.

Howlin, P., Marchant, R., Rutter, M., Berger, M., Hersov, L., & Yule, W. A home based approach to the treatment of autistic children. *Journal of Autism and Childhood Schizophrenia*, 1973, *3*, 308–336.

Lovaas, O. I. Interaction between verbal and nonverbal behavior. *Child Development*, 1961, *32*, 329–336.

Lovaas, O. I., Freitag, G., Gold, V. J., & Kassorla, I. C. Experimental studies in childhood schizophrenia: Analysis of self-destructive behavior. *Journal of Experimental Child Psychology*, 1965, *2*, 67–84.

Reichler, R. J., & Schopler, E. Developmental therapy: A program model for providing individual services in the community. In E. Schopler & R. J. Reichler (Eds.), *Psychopathology and child development*. New York: Plenum Publishing Corporation, 1976. Pp. 347–371.

Rimland, B. *Infantile autism*. New York: Appleton-Century-Crofts, 1964.

Rutter, M. Concepts of autism: A review of research. *Journal of Child Psychology and Psychiatry*, 1968, *9*, 1–25.

Schopler, E. Parents of psychotic children as scapegoats. *Journal of Contemporary Psychotherapy*, 1971, *4*, 17–22.

Schopler, E., & Reichler, R. J. Parents as co-therapists in the treatment of psychotic children. *Journal of Autism and Childhood Schizophrenia*, 1971, *1*, 87–102.

Schopler, E., & Reichler, R. J. *Psychoeducational profile*. Chapel Hill, North Carolina: Child Development Products, 1976.

Skinner, B. F. *Science and human behavior*. New York: Macmillan, 1953.

Tate, B. G., & Baroff, G. S. Aversive control of self-injurious behavior in a psychotic boy. *Behavior Research and Therapy*, 1966, *4*, 281–287.

Watson, J. B. Recent literature on mammalian behavior. *Psychological Bulletin*, 1908, *5*, 195–205.

28

Educational Approaches

LAWRENCE BARTAK

The teacher, faced with the task of teaching autistic children, whether in ones and twos or greater numbers, will often find it difficult to know where to start. Very frequently, autistic children present a bewildering array of disturbed behaviors and disabilities and it seems impossible to get the child to learn anything. Often, it is very difficult to find out what the child knows and what he does not. In such circumstances, it might be expected that teachers would turn to the special educational literature in an attempt to develop systematic teaching programs. However, relatively little information has been easily available until recently. This chapter aims to review knowledge on different approaches to the teaching of autistic children. It will lean heavily upon the results of a study carried out by the author and his colleagues in the United Kingdom.

The results of follow-up studies have suggested that the provision of special schooling for autistic children is associated with scholastic and social benefits (Lockyer & Rutter, 1969; Rutter *et al.*, 1967; Russell, 1975). Recent follow-up studies (Goldfarb, 1974; Rees & Taylor, 1975) have yielded much detailed information about the development of autistic children over time but have devoted relatively little attention to treatment regimes. There have been no well-controlled evaluations of schooling and the study carried out by the author was an attempt to provide a direct, controlled investigation of the results of special educational treatment.

LAWRENCE BARTAK · Faculty of Education, Monash University, Clayton, Victoria, Australia 3168.

In spite of the pressing need and the availability of techniques for therapeutic studies (Robins & O'Neal, 1969; Tizard, 1966), there have been very few attempts to evaluate specific forms of treatment for autistic children (Goldfarb, 1969). There have been a few investigations of a variety of residential regimes (Davids *et al.*, 1968). Other studies have examined programmed or structured teaching (Fischer & Glanville, 1970; Graziano, 1970; Halpern, 1970), group therapy (Speers & Lansing, 1965), psychotherapy (Havelkova, 1968) and operant methods (Wolf *et al.*, 1967; Ney *et al.*, 1971; Lovaas *et al.*, 1973; Browning, 1974; Carr *et al.*, 1975; Romanczyk *et al.*, 1975). However, these vary in their rigor and control and most include inadequate information on the children studied (Yule & Berger, 1972). They also suffer from a number of defects characteristic of educational follow-up studies (Guskin & Spicker, 1968). There have been four studies which have attempted to provide a direct comparison between different methods of treatment. Ney (Ney, 1967; Ney *et al.*, 1971) compared operant conditioning and play therapy in a controlled crossover design and found that the former was more effective. However, the study employed inexperienced therapists and extended over a few months only with no longer term follow-up. Accordingly, the findings are of limited relevance with respect to educational progress.

Goldfarb *et al.* (1966) compared matched groups of psychotic children in day and residential treatment. However, it is probable that not all of the children were autistic and there were no direct measures of the treatment regimes in the two settings. Although the groups were appropriately matched, little evidence was provided of direct assessment of the respective treatments involved. This could well give rise to different conclusions (Johnson & Christensen, 1975).

Wenar *et al.* (1967) provided some evaluation of each of the three milieus they studied (custodial institution, state institution with an active therapeutic program, and a psychoanalytically oriented day care unit). However, the findings were limited by their failure to control for individual differences between the children.

Russell (1975) compared the progress of autistic children attending day training centers run by the state mental health authority for retarded children with that of autistic children in special units. The children were followed up over approximately 12 months. Groups were matched in terms of age and sex but there were some differences in symptoms presented and in fathers' socioeconomic status, so that the groups were not strictly comparable. However, the results are of interest and suggested that children had shown greater improvement in social and verbal skills over one year when they had been attending units specifically designed to serve autistic children.

THE MAUDSLEY STUDY

In the few years prior to the beginning of the study, there had been a rapid growth in the number of special schools and classes for autistic children in the United Kingdom. However, the units set up varied considerably in their orientation, structure, and in the type of relationships provided for the children. When the study began, there was little evidence upon which to judge the merits of these different approaches and accordingly it set out to make some assessment of what type of schooling was best geared to the needs of autistic children.

Variations between units chiefly concern five areas: (a) theoretical orientation; (b) emphasis on formal teaching; (c) degree of structure in the environment; (d) segregation from or mixing of autistic children with other children; (e) staff-child ratio.

The main theoretical difference lay between those who viewed autism as an emotional disorder due to a failure in the early development of adequate parent-child relationships and those who regarded autism as a developmental disorder involving specific handicaps in language and perception. With regard to teaching, the difference lay between those who regarded the prime aim as the fostering of a strong emotional bond between adult and child (teaching beginning only when this had been established) and those who considered that the main need was for the deliberate teaching of specific skills.

Similarly, some workers emphasized the need for a permissive flexible environment so that the autistic child could develop in his own way, whereas others believed that a firm external structure must be imposed just because the handicaps of the autistic child prevented him from developing internal controls in the normal way. In some units, autistic children were mixed with others in order to provide more normal peer models for adequate communication and social interaction. In others, it was thought that mixing served only to dilute the special care necessary for autistic children. Finally, there was a difference between those emphasizing the importance of a one-to-one staff-child ratio and those who believed that it was necessary to progress from a one-to-one relationship to individualized teaching in small group settings.

Thus, some workers (Bettelheim, 1967; Mahler & Furer, 1968) have emphasized the supposedly psychogenic and emotional nature of autism with treatment being of a regressive kind within the context of a dependent relationship (Alpert & Pfeiffer, 1964; Dundas, 1968; Kaufman *et al.*, 1957). The use of educational methods takes second place to the fostering of a strong emotional bond between therapist and child.

In sharp contrast, others have based treatment on the assumption that the child has a developmental disorder (Cordwell, 1968; Elgar, 1966; Wing, 1976; Taylor, 1976). Within a structured environment, individual learning programs are implemented to help the child minimize his disabilities.

Still others, while recognizing the importance of relationships of a rich and gratifying nature (Williams, 1968) suggest that they should be combined with skilled educational techniques. Furneaux (1966) has suggested that regression should be combined with skilled educational methods.

Study Design

The detailed structure of the study and its results are presented elsewhere (Bartak & Rutter, 1973; Rutter & Bartak, 1973) so that this account focuses on the general approach and the main findings.

The study was designed to assess the value to autistic children of special educational treatment and to assess the various effects of different approaches to such treatment. Answers were sought to five questions: How much formal educational progress can be made by autistic children? To what extent is this dependent on the type of schooling provided? To what extent is formal educational progress dependent on the nature and severity of the child's handicaps? To what extent does formal educational improvement correlate with social and behavioral progress? Does the pattern of social, behavioral, and educational progress vary according to the educational methods used?

In order to answer these questions, a search was made for well-established units that differed on the dimensions outlined above. Three units were selected. One was primarily a psychotherapeutic unit with little emphasis on teaching, one provided a structured and organized setting for the teaching of specific skills, and one used a more permissive classroom environment in which regressive techniques and an emphasis on relationships were combined with special educational methods. The units will be referred to here as Units A, B, and C.

The study involved three stages. First, the children were studied in 1967 using a combination of standardized tests, interview, and direct observation. This served to confirm diagnostic criteria and provide a baseline for later progress. Second, there was a systematic study of the characteristics of each unit using time-sampled observations in the unit. Interobserver reliability was assessed for all observational variables and found to be satisfactory (Bartak & Rutter, 1975). Third, the children were followed up in 1970-71, giving a three-and-one-half to four-year follow-up to assess educational, social, and behavioral progress.

The Units

Unit A employed regressive techniques with minimal attention to the teaching of specific skills. A child psychotherapist was in charge of a group of lay helpers who worked one-to-one with the ten children attending the unit at the outset of the study. Children attended for four days per week, spending much of their time in free play in the presence of their individual staff member.

Unit B was an educational unit with no standard structure. A permissive classroom environment was combined with the application of special educational methods. An attempt was made to match the regime to the needs of the individual. Regimes might vary from fairly structured to a free play setting. Some of the children at this unit were not autistic and at the outset of the study, there were 25 children with seven staff members, all of whom were teachers.

Unit C was a school for autistic children only and exemplified the approach in which perceptual, motor, and cognitive handicaps were seen as primary and around which education was to proceed. It was suggested that the child's environment should be structured, organized, and logical, rather than permissive. The main emphasis lay in teaching the child specific skills. As in Unit B, there were 25 children and seven staff members at the beginning of the study and staff were teachers.

Sample

Children for whom a diagnosis of autism had been made were considered for inclusion in the study and finally selected if they had a disorder beginning before 30 months of age in which all three key features of autism described by Rutter (1971) were present (failure of social development, delayed receptive language development, presence of ritualistic or quasi-obsessive features).

There were 50 children in the study, six boys and two girls at Unit A, ten boys and eight girls at Unit B, and eighteen boys and six girls at Unit C. The mean age at the outset was similar in all units (7 to 9 years) but there were more adolescents at Unit C and more young children at Units A and B. Mean IQ was somewhat higher at Unit C. Mean IQ on nonverbal tests was 48 at Unit A, 52 at Unit B, and 66 at Unit C.

Methods

Children were assessed on measures of intelligence, language, social behavior, and educational attainment. Intelligence was measured on the

Merrill-Palmer and WISC Performance Scales. Language was tested with the English Picture Vocabulary Test and samples of speech were obtained in standard settings. Social behavior was assessed by direct observation of behavior in the classroom or unit setting, at play, and by parental report of behavior at home. Systematic observations in the unit were used to measure task orientation in the classroom, presence of autistic symptoms in behavior, and types of play. Social behavior at home was assessed in terms of social responsiveness, degree of social handicap resulting from disruptive behavior, and degree of deviant behavior such as obsessions or tantrums and self-injury.

Educational attainment was based on assessment of reading and arithmetic skills. The Neale Analysis of Reading Ability was used, supplemented by scales in which the child obtained points for identification of single sounds and for ability to carry out elements in a graded series of numerical operations ranging from counting and identifying integers up to solving problems involving algebra and geometry.

The units were assessed in two ways. First, staff were questioned and observed regarding timetables, methods, equipment, and materials available. Second, direct observational methods involving time and event sampling were used to assess the nature of staff-child interactions. Styles of staff-child interaction as well as the contingent qualities of the staff's response to children's behavior (Becker *et al.*, 1967) were assessed. Staff behavior was characterized according to various categories of approval or disapproval, both verbal and nonverbal, as well as aspects of tone of voice in the case of verbal responses. Other categories included aspects of instruction such as questioning or demonstrating methods.

Results

Units varied in their timetabling, principally in the amount of formal teaching. This ranged from as little as 40 minutes per week at the outset in Unit A, through about four hours per week in Unit B to about nine hours of formal teaching per week in Unit C. Teaching methods were broadly similar in approach in all units; jigsaw puzzles and other equipment were used to foster size and shape discrimination, as well as flash cards to teach reading and language. Methods were generally tailored to the individual needs of each child.

However, there were a number of differences among the units in staff activity. In Units B and C instruction was frequent whereas in Unit A playing with the child was the most frequent activity. Acts of approval were twice as frequent in Unit A as in either of the other units, although acts of disapproval were about equal in frequency in all units. Tone of voice also differed between units. In Unit A warm tone was frequently evident; it was less common in Unit B and unusual in Unit C. Conversely, critical tone of voice was common at Unit C, less so at Unit B, and very rare at Unit A. In summary, staff of Unit A spent much of their time in warm, approving play with children, whereas instruction was more common at the other units. Units B and C differed in the manner of instruction, warmth and physical approval being more characteristic of Unit B.

Event sampling of staff-child behavior sequences showed further differences in response to deviant behavior. In both Units B and C deviant behavior was generally followed by disapproval but in Unit A deviant behavior was as likely to be approved as disapproved. Thus it seemed that children there might find it difficult to discriminate right and wrong since behavior of all kinds was liable to be followed by praise. Similar findings were obtained with respect to tone of voice. Critical tone was never observed in Unit A and staff response was similar whatever the child's behavior. In Unit C deviant behavior was generally followed by critical tone of voice while acceptable behavior was usually followed by neutral tone. In Unit B tone of voice was most clearly used as a differential response, acceptable behavior often being followed by warm tone with little use of critical tone and deviant behavior frequently being followed by critical tone with little use of warm tone.

These findings suggested that in Unit A warmly expressed approval tended to follow whatever the child did whereas in Unit C approval and disapproval were contingent on child behavior with warmth infrequently expressed through tone of voice. In Unit B, staff showed quite a lot of warmth and by the end of the study were using approval and disapproval in a highly discriminating way.

Follow-Up: Overall Progress

Behavior in the Unit

Children's behavior in the unit when directed to some activity was time-sampled (using a series of 15-second observations). In Unit A most of the children's time (71%) was spent unoccupied or engaged in irrelevant or stereotyped activities. In contrast, in Unit C children were engaged in the activity planned for them for 59% of the observed time. Unit B was intermediate in this respect. This result was paralleled by that obtained from observation of free play. In Unit C 17% of time was spent in parallel or cooperative play compared with 5% in Unit A and 2% in Unit B. While this difference is statistically significant, it is noteworthy that in all three units the children spent a very large proportion of their time unoccupied or engaged in solitary activities. This finding has been borne out by other workers (Black *et al.*, 1975).

These differences in the behavior of the children in the three units were striking and seemed to reflect the orientation of the units. In Unit C there was continuous concern to insure that the children remained on task and did not engage in maladaptive or disruptive behavior. In contrast, in both the other units it was thought to be therapeutic to allow the children some choice in how they spent their time, and it was considered undesirable to impose undue controls on their behavior. This appeared to result in less on-task behavior and less play with other children.

Behavior at Home

Behavior outside the unit was assessed on the basis of systematic reports from parents. These showed that the group as a whole had become less deviant and more socially responsive. In 1968, two-fifths of the group were markedly deviant and only one-fifth slightly deviant. By 1971 this ratio was reversed with less than one-fifth markedly deviant and over two-fifths reported as only slightly deviant. Similarly, in 1968 only 14% of the children showed good social responsiveness (as shown by eye-to-eye gaze, play, facial expression, etc.) but by 1971, 54% did so. However, unlike the findings for behavior at the unit, there were no differences between units on either deviance or social responsiveness as reported by parents. There was a tendency for less improvement in the children in Unit A but it fell short of statistical significance.

Behavior in the Test Situation

Similar measures were made by direct observation in the test situation. Over the whole sample, 61% of children increased in social responsiveness but there were no differences between units in the proportion of children showing increased social responsiveness.

Of the children displaying deviant behaviors in the test situation, just over half (58%) showed decreased deviance at follow-up compared with their score in 1968. However, again there were no significant differences between units in this respect.

These findings, together with those based on parental information suggest that the differences in the behavior of the children while in the units failed to generalize to other situations. This is in keeping with the view that the behavioral control of the children while in the unit is a fairly direct response to the specific milieu, rather than a reflection of some fundamental modification of personality or cognitive functioning in the children.

Speech

In order to increase comparability between units, attention here was restricted to children aged at least ten years at follow-up. At Unit C, 17 of 22 children were using phrase speech communicatively whereas at Units

A and B, only half the children were doing so (three of seven at Unit A; seven of 16 at Unit B). The rest were without speech or using single words. Similarly, of the nine children at Unit C who were initially without speech, four gained phrase speech during the course of the study. This happened with two of the nine children with no speech in Unit B and none of the three in Unit A.

Educational Attainment

As with speech, educational achievement was greater at Unit C than for children at either of the other units. Attainment in reading accuracy and comprehension and in arithmetic were examined. Attention was confined to children aged ten or more at follow-up.

Reading accuracy and comprehension were measured with the Neale Analysis of Reading Ability. Just over half the Unit C children had a reading age for accuracy of eight years or more (12 children) compared with only one of the 16 Unit B children and none of those from Unit A. Nearly half the children scored below the floor of the test and most of these were in Units A or B. The Neale tests ability to read prose. Some of the children who did not score could read isolated words on cards. This emphasizes a problem common to many autistic children. They can learn something in one context but have great difficulty in applying it to another. Other findings show that they have much difficulty in understanding what they learn. Much is learned by rote. This is shown by the findings for reading comprehension. Again, the Unit C children did better. However, it is equally important to note that of the whole group of 45 children only two had a reading age for comprehension of eight years or better. One was in Unit B and the other in Unit C. The test requires the child to answer questions about the prose passage and is harsh in view of the language disabilities present. Nevertheless, it shows that for children with reasonably adequate skills in mechanical reading, understanding of what is read lags well behind.

Similar results were obtained with arithmetical skills. Two-thirds of the Unit C children were able to use all four basic processes (addition, subtraction, division, multiplication) with double-figure numbers compared with two of 16 children at Unit B and one of the seven at Unit A. Of the 16 children who had all four basic skills, eight were able to apply them to problems which required the child to decide which processes to use. All eight children were at Unit C.

Child Characteristics and Educational Attainment

An examination was made to see how children's characteristics at the start of the study predicted progress over the follow-up period. In keeping with previous studies, IQ proved to be the best predictor of educational and social progress. IQ scores from nonverbal testing at the outset of the study showed correlations of +0.42 with reading accuracy, +0.33 with arithmetic score, +0.59 with on-task behavior in the unit, 0.32 with social responsiveness, +0.43 with deviant behavior and +0.48 with social handicap, all at follow-up. IQ scores obtained at the beginning were highly stable as well and correlated +0.91 with IQ at follow-up, three to four years later. These results show the relevance of IQ as a predictor of the extent of scholastic and social progress. IQ was also found to predict whether there would be any appreciable scholastic progress. Of the 31 children with IQ of 50 or more, 29 made some measurable improvement in reading accuracy. In sharp contrast, only two of the 19 children with IQ less than 50 did so. These results suggest that IQ not only predicts how much a child is likely to gain in reading skill but also whether he will learn at all. These findings need to be considered in any comparison of the three units since there were rather fewer children of really low IQ at Unit C. It was necessary to take IQ into account and multiple regression analysis was used for this purpose.

Follow-Up Regression Analysis Findings

Reading Accuracy

Calculations were based upon the 34 children who obtained a score on the Neale test for accuracy. A regression equation with age and IQ as predictors was computed and used to calculate predicted reading scores, prediction thus being based upon the performance of the group as a whole. Obtained and predicted reading scores were then compared for each child. In Unit A the mean reading age was one year four months below that expected on the basis of the children's age and IQ. For Unit B it was about nine months below expectation, whereas in Unit C children were reading on average about five months better than predcction. The difference between Unit C and each of the other units was statistically significant at the 5% level. These results show that even after taking into account the children's characteristics at the outset of the study, the average reading achievement at Unit C was significantly better than in either of the other units.

Arithmetic

A similar equation was calculated for arithmetic scores. Again Unit C children did better than expected whereas those in Units A and B did worse, with the mean achievement in Unit C being significantly better than that in either of the other units.

Classroom Behavior

For this comparison, all children with an IQ of 80 or more were excluded as the regression departed from linearity at this point. The children remaining were then subdivided into those with an IQ of 50 or more and those with an IQ of 49 or less. On-task behavior was measured in the way previously described. In this case, the regression equation was based on IQ alone. As before, discrepancies between actual and predicted scores were computed.

There were no differences between the units for those children with IQ of 50 or more. However, for the lowest IQ group, the children in Unit C did significantly better than expectation. For the higher IQ group, children in all units were on task for about two-thirds of observed time. In Units A and B the frequency of on-task behavior for the low-IQ children was about half that figure i.e., about one-third of the observed time. However, in Unit C, there was no difference in on-task behavior between the two IQ groups—both running about two-thirds of observed time.

These figures suggest that children of IQ 50 or more have reasonably high levels of on-task behavior in the unit, regardless of the type of unit. (The extent to which this results in learning is dependent on the type of unit as shown above, but that is a separate issue.) Children with IQ of less than 50, however, have generally low rates of on-task behavior; however, this appears to be modifiable by the structure of the child's environment, as shown by the Unit C findings.

Social Behavior at Home

Parental information was used to test whether this apparent advantage extended to the children's behavior outside the special unit. The previously described scales covering social handicap, deviant behavior, and social responsiveness were used. In each case, the scale had correlated significantly and positively with IQ and regression equations with IQ as predictor were computed. However, no significant differences were found among the three units on any of these scales. Although all children had made considerable progress in social behavior, there were no differences between units after taking IQ into account.

DISCUSSION

How much educational progress can be made by autistic children? Results suggest qualified optimism. In all three units, most of the children had been rejected by other schools as unmanageable. Records showed that at admission prior to the study, most of the children were mentally retarded, socially unresponsive, and markedly disturbed behaviorally. In spite of this, the children made considerable progress in a number of directions.

To what extent is progress dependent upon the type of unit? Unit C children did best in terms of formal scholastic achievement. However, it is not possible to establish causal links between unit charactersitics and results. Unit C had the most structure and organization and the greatest emphasis upon formal teaching. It appears that this emphasis led to greater attainments. Schopler *et al*. (1971) have shown that autistic children respond better to a structured setting and their findings support the conclusion here that autistic children need systematic teaching in an ordered environment.

It is important, however, to clarify what is meant by the notion of "structuring." A structured situation may be defined simply as one in which there is a task orientation and the adult determines what the child should be doing. This amounts to a structuring of the child's *responses*. A second aspect of structuring is that the environment surrounding the child should be limited, planned, and organized. This amounts to a structuring of *stimuli* impinging. Both aspects of structuring may be viewed as providing external organization upon the child and his immediate world in a situation where he is as yet unable to organize his own behavior and environment for himself (Bartak & Pickering, 1976). Spicker (1971) has emphasized the importance of structuring relevant activities and others have stressed the problem in deciding what is relevant (Winett & Winkler, 1972; O'Leary, 1972). A number of studies (Cunningham, 1968; Hermelin & O'Connor, 1970; Churchill, 1972; Bartak *et al*., 1975; Browning, 1974; Carr *et al*., 1975; Dalgleish, 1975; Simmons & Baltaxe, 1975) have indicated specific deficits in language skills and social competence that are characteristic of autism. It is therefore of importance that a structured setting be used by staff to provide all children with specific teaching with respect to language skills and social competence. In the case of language, adequate and detailed programs are becoming available now (Lee *et al*., 1975; Snyder *et al*., 1975). However, there are fewer options available for social skill training. This may involve systematic teaching of play (Romanczyk *et al*., 1975) or may first have to involve further research upon such constructs as self-control (Kurtz & Neisworth, 1976) or interest (Kirkland, 1976) which autistic children seem characteristically to lack without specific rote learning.

Social skills training may not work unless the child can explicitly make use of rote skills (Cole, 1975). What may be needed may be a social learning program which emphasizes social input skills. In other words, we may first have to teach the autistic child to understand what is going on in the social situations in which he is immersed before we can expect him to do other than learn social output skills parrot fashion, i.e., by rote. We are currently undertaking a project in which autistic children will be introduced to graded problems in the domain of person perception. It is hoped that the development of social comprehension will be followed by the development of social competence together with a lessening of behavioral disturbance. If this should prove to be so, it would parallel the suggestion that comprehension is accompanied by greater emotional responsiveness (Scanlan *et al.*, 1963). Further controlled studies are needed.

Other directions in which further work is urgently necessary are in preschool education (Rutter, 1973) and in the continuing "adult" education of the older autistic person. A start has been made (Elgar, 1975; Hughson & Brown, 1975). However, although there is evidence for the utility of special education both from the study described here as well as from other studies (Russell, 1975) in effecting cognitive, scholastic, and social skills to a limited degree, grave problems arise in the placement in employment of autistic adults (Jackson, 1975) and many further structured programs for adults which are job-oriented are needed (Hughson & Brown, 1975).

REFERENCES

Alpert, A., & Pfeiffer, E. The treatment of an autistic child. *Journal of the American Academy of Child Psychiatry*, 1964, *3*, 594–616.

Bartak, L., & Pickering, C. Aims and methods of teaching. In. M. P. Everard (Ed.), *Some approaches to teaching autistic children*. Oxford: Pergamon, 1976.

Bartak, L., & Rutter, M. Special educational treatment of autistic children: A comparative study. I. Design of study and characteristics of units. *Journal of Child Psychology and Psychiatry*, 1973, *14*, 161–179.

Bartak, L., & Rutter, M. The measurement of staff-child interaction in three units for autistic children. In J. Tizard, I. Sinclair, & R. V. G. Clarke (Eds.), *Varieties of residential experience*. London: Routledge and Kegan Paul, 1975.

Bartak, L., Rutter, M., & Cox, A. A comparative study of infantile autism and specific developmental receptive language disorder: I. The children. *British Journal of Psychiatry*, 1975, *126*, 127–145.

Becker, W. C., Madsen, C. H., Arnold, C. R., & Thomas, D. R. The contingent use of teacher attention and praise in reducing classroom behavior problems. *Journal of Special Education*, 1967, *1*, 287–307.

Bettelheim, B. *The empty fortress: Infantile autism and the birth of the self.* London: Collier-Macmillan, 1967.

Black, M., Freeman, B. J. & Montgomery, J. Systematic observation of play behavior in autistic children. *Journal of Autism and Childhood Schizophrenia*, 1975, *5*, 363–371.

Browning, E. R. The effectiveness of long and short verbal commands in inducing correct responses in three schizophrenic children. *Journal of Autism and Childhood Schizophrenia,* 174, *4*, 293–300.

Carr, E. G., Schreibman, L., & Lovaas, O. I. Control of echolalic speech in psychotic children. *Journal of Abnormal Child Psychology*, 1975, *3*, 331–351.

Churchill, D. W. The relation of infantile autism and early childhood schizophrenia to developmental language disorders of childhood. *Journal of Autism and Childhood Schizophrenia*, 1972, *2*, 182–197.

Cole, P. G. The efficacy of a social learning curriculum with borderline mentally retarded children. *Australian Journal of Mental Retardation*, 1975, *3*, 191–199.

Cordwell, A. Educational techniques likely to benefit autistic children. In *Autism: Cure tomorrow, care today*. Warradale, South Australia: Autistic Children's Association of South Australia, 1968.

Cunningham, M. A. A comparison of the language of psychotic and non-psychotic children who are mentally retarded. *Journal of Child Psychology and Psychiatry*, 1968, *9*, 229–244.

Dalgleish, B. Cognitive processing and linguistic reference in autistic children. *Journal of Autism and Childhood Schizoprhenia*, 1975, *5*, 353–361.

Davids, A., Ryan, R., & Salvatore, P. D. Effectiveness of residential treatment for psychotic and other disturbed children. *American Journal of Orthopsychiatry*, 1968, *38*, 469–475.

Dundas, M. H. The one to one relationship in the treatment of autistic children. *Acta Paedopsychiatrica*, 1968, *35*, 242–245.

Elgar, S. Teaching autistic children. In J. K. Wing (Ed.), *Early childhood autism: Clinical, educational and social aspects*. Oxford: Pergamon, 1966.

Elgar, S. First year at Somerset Court. *Special Education*, 1975, *2*, (2), 14–15.

Fischer, I., & Glanville, B. Programmed teaching of autistic children. *Archives of General Psychiatry*, 1970, *23*, 90–94.

Furneaux, B. The autistic child. *British Journal of Disorders of Communication*, 1966, *1*, 85–90.

Goldfarb, W. Therapeutic management of schizophrenic children. In J. G. Howells (Ed.), *Modern perspectives in international child psychiatry*. Edinburgh: Oliver and Boyd, 1969.

Goldfarb, W. *Growth and change of schizophrenic children: A longitudinal study*. New York: Wiley, 1974.

Goldfarb, W., Goldfarb, N., & Pollack, R. C. Treatment of childhood schizophrenia. *Archives of General Psychiatry*, 1966, *14*, 119–128.

Graziano, A. M. A group treatment approach to multiple problem behaviors of autistic children. *Exceptional Children*, 1970, *36*, 765–770.

Guskin, S. L., & Spicker, H. H. Educational research in mental retardation. In N. R. Ellis (Ed.), *International review of research in mental retardation, 3*. New York: Academic Press, 1968.

Halpern, W. I. The schooling of autistic children: Preliminary findings. *American Journal of Orthopsychiatry*, 1970, *40*, 665–671.

Havelkova, M. Follow-up study of 71 children diagnosed as psychotic in pre-school age. *American Journal of Orthopsychiatry*, 1968, *38*, 846–857.

Hermelin, B., & O'Connor, N. *Psychological experiments with autistic children*. Oxford: Pergamon, 1970.

Hughson, E. A., & Brown, R. I. A bus training programme for mentally retarded adults. *British Journal of Mental Subnormality*, 1975, *21*, 79–83.

Jackson, R. The effect of secondary handicaps in the employment adjustment of educable mentally handicapped adolescents in Scotland. *British Journal of Mental Subnormality*, 1975, *21*, 61–70.

Johnson, S. M., & Christensen, A. Multiple criteria follow-up of behavior modification with families. *Journal of Abnormal Child Psychology*, 1975, *3*, 135–154.

Kaufman, I., Rosenblum, E., Heims, L., & Willer, L. Childhood schizophrenia: Treatment of children and parents. *American Journal of Orthopsychiatry*, 1957, *27*, 683–690.

Kirkland, J. Interest: Phoenix in psychology. *Bulletin of the British Psychological Society*, 1976, *29*, 33–41.

Kurtz, P. D., & Neisworth, J. T. Self-control possibilities for exceptional children. *Exceptional Children*, 1976, *42*, 213–218.

Lee, L. L., Koenigsknecht, R. A., & Mulhern, S. *Interactive language development teaching: The clinical presentation of grammatical structure*. Evanston, Ill.: Northwestern University Press, 1975.

Lockyer, L., & Rutter, M. A five to fifteen year follow-up study of infantile psychosis—III. Psychological aspects. *British Journal of Psychiatry*, 1969, *115*, 865–882.

Lovaas, O. I., Koegel, R., Simmons, J. Q., & Long, J. S. Some generalization and follow-up measures on autistic children in behavior therapy. *Journal of Applied Behavior Analysis*, 1973, *6*, 131–166.

Mahler, M. S., & Furer, M. *On human symbiosis and the vicissitudes of individuation. Vol. 1: Infantile psychoses*. New York: International Universities Press, 1968.

Ney, P. Operant conditioning of schizophrenic children. *Canadian Psychiatric Association Journal*, 1967, *12*, 9–15.

Ney, P. G., Palvesky, A. E., & Markely, J. Relative effectiveness of operant conditioning and play therapy in childhood schizophrenia. *Journal of Autism and Childhood Schizophrenia*, 1971, *1*, 337–349.

O'Leary, K. D. Behavior modification in the classroom: A rejoinder to Winett and Winkler. *Journal of Applied Behavior Analysis*, 1972, *5*, 505–511.

Rees, S. C., & Taylor, A. Prognostic antecedents and outcome in a follow-up study of children with a diagnosis of childhood psychosis. *Journal of Autism and Childhood Schizophrenia*, 1975, *5*, 309–322.

Robins, L. N., & O'Neal, P. L. The strategy of follow-up studies, with special reference to children. In J. G. Howells (Ed.), *Modern perspectives in international child psychiatry*. Edinburgh: Oliver and Boyd, 1969.

Romanczyk, R. G., Diamant, C., Goren, E. R., Trunell, G., & Harris, S. L. Increasing isolate and social play in several disturbed children: Intervention and post-intervention effectiveness. *Journal of Autism and Childhood Schizophrenia*, 1975, *5*, 57–70.

Russell, S. *The development and training of autistic children in separate training centres and in centres for retarded children*. Special Publication No. 6. Victoria: Mental Health Authority, 1975.

Rutter, M. The description and classification of infantile autism. In D. Churchill, G. Alpern, & M. DeMyer (Eds.), *Infantile autism: Proceedings of the Indiana University Colloquium*. Springfield, Ill.: Charles C Thomas, 1971.

Rutter, M. The assessment and treatment of pre-school autistic children. *Early Child Development and Care*, 1973, *3*, 13–29.

Rutter, M., & Bartak, L. Special educational treatment of autistic children: A comparative study. II. Follow-up findings and implications for services. *Journal of Child Psychology and Psychiatry*, 1973, *14*, 241–270.

Rutter, M., Greenfeld, D., & Lockyer, L. A five to fifteen year follow-up study of infantile psychosis. II. Social and behavioural outcome. *British Journal of Psychiatry*, 1967, *113*, 1183–1199.

Scanlan, J. P., Leberfeld, D. T., & Freibrun, R. Language training in the treatment of the autistic child functioning on a retarded level. *Mental Retardation*, 1963, *1*, 305–310.

Schopler, E., Brehm, S., Kinsbourne, M., & Reichler, R. J. Effect of treatment structure on development in autistic children. *Archives of General Psychiatry*, 1971, *24*, 415–421.

Simmons, J. Q., & Baltaxe, C. Language patterns of adolescent autistics. *Journal of Autism and Childhood Schizophrenia*, 1975, *5*, 333–351.

Snyder, L. K., Lovitt, T. C., & Smith, J. O. Language training for the severely retarded: Five years of behaviour analysis research. *Exceptional Children*, 1975, *42*, 7–15.

Speers, R. W., & Lansing, C. *Group therapy in childhood psychosis*. London: Oxford University Press, 1965.

Spicker, H. H. Intellectual development through early childhood education. *Exceptional Children*, 1971, *37*, 629–640.

Taylor, J. E. An approach to teaching cognitive skills underlying language development. In L. Wing (Ed.), *Early childhood autism* (2nd ed.). Oxford: Pergamon, 1976.

Tizard, J. The experimental approach to the treatment and upbringing of handicapped children. *Developmental Medicine and Child Neurology*, 1966, *8*, 310–321.

Wenar, C., Ruttenberg, B., Dratman, M., & Wolf, E. Changing autistic behaviour. The effectiveness of three milieus. *Archives of General Psychiatry*, 1967, *17*, 26–34.

Williams, S. The educational programme at North Ryde. In *Autism: cure tomorrow, care today*. Warradale, South Australia: Autistic Children's Association of South Australia, 1968.

Winett, R. A., & Winkler, R. C. Current behavior modification in the classroom: Be still, be quiet, be docile. *Journal of Applied Behavior Analysis*, 1972, *5*, 499–504.

Wing, L. The principles of remedial education for autistic children. In L. Wing (Ed.), *Early childhood autism*. (2nd ed). Oxford: Pergamon, 1976.

Wolf, M., Risley, T., Johnston, M., Harris, F., & Allen, E. Application of operant conditioning procedures to the behaviour problems of an autistic child: A follow-up and extension. *Behavior Research and Therapy*, 1967, *5*, 103–111.

Yule,W., & Berger, M. Behaviour modification principles and speech delay. In M. Rutter & J. A. M. Martin (Ed.), *The child with delayed speech*. Clinics in Developmental Medicine No. 43. London: SIMP/Heinemann, 1972. Pp. 204–219.

29

Individualized Education:
A Public School Model

MARGARET D. LANSING and ERIC SCHOPLER

Only recently have teachers been given a prominent role in the task of working with autistic children. Many special education techniques have been developed for these children, sometimes with competition for priority in the educational marketplace. However, the individual variations in behavioral symptoms, intellectual impairment, and learning styles found under the label of autism prevent the exclusive suitability of any one technique for every autistic child. Although the need for individualized curricula for autistic children is widely recognized, the practice is not always implemented. The North Carolina state program for autistic children (Division TEACCH) is cited as a model for developing administrative structures to facilitate individualization and assessment procedures for defining individual curricula within a developmental structure. Special education techniques are adapted for individualized use, and the involvement of parents promotes transfer of learning from the classroom to the home.

HISTORY OF EDUCATIONAL INTERVENTION

Although autistic children were not described and so labeled until the 1940s (Kanner, 1943), we can go back nearly 200 years to find the record-

MARGARET D. LANSING and ERIC SCHOPLER · Division TEACCH, University of North Carolina, Chapel Hill, North Carolina.

ings of educational techniques still in use today. In 1795, Itard (1962) took
on the education of a boy found in the wilderness, showing behavior we
now consider autistic. From this child Itard learned special education
principles still useful today. They include the importance of daily routines
and teaching of self-help skills. He used the written word as an aid in
developing the boy's language skills. He had to improve the boy's receptive understanding before he could teach expressive labeling, and he developed methods of sensory training.

When Kanner published the label we now use, the cause of these
behaviors was presumed to be primarily environmental, and specifically
caused by early parental attitudes and responses. It was therefore logical
to seek a "cure" by remediating these early experiences. The educational
approach took a back seat until relatively recently when research and
clinical experience questioned these early formulations and pointed once
again to education as the treatment of choice (Rutter, 1968; Rimland,
1964; Wing, 1976; Hewett, 1965; Schopler & Reichler, 1971a). As the
emphasis shifted from parental pathology (Schopler, 1971) to cognitive
deficits of the children, parents became politically active in demanding
and developing educational services for their children. The National
Societies for Autistic Children have been a powerful group in this movement. In the United States this has resulted in recent legislation mandat-
ing the right to education, requiring school systems across the country to
develop public school facilities for autistic children. Unfortunately, the
educators have not been anticipating this development and they find a
number of unanswered questions standing in their way. What is an autistic
child? Is he to be taught as retarded, brain-damaged, learning disabled, or
emotionally disturbed? Each of these categories has trained teachers with
their own professional terminology and their own curricula and classroom
techniques. Yet we find many children labeled autistic who fit only partially in some groups and completely in none.

EDUCATIONAL DEFINITION OF AUTISM

The definition of autism for educators has been a problem since the
term was first introduced. A behavioral definition such as Creak's (1964)
or Rutter's (Rutter *et al.*, 1969) offers useful criteria for identifying children, but is less useful for planning an individual program, since variations in types and degrees of deviancy are enormous. When the purpose
for labeling a child is to find an appropriate classroom or educational
program, we find that some "autistic" behaviors are more significant than
others.

To make an appropriate classroom placement it is less helpful to know the precise catalog of the child's autistic characteristics than it is to know their educational implications. How much individual attention does the child require to function in the classroom? At what developmental level are his language and other cognitive skills? Which of his autistic behaviors are incompatible with his learning or disrupt the other children? This kind of information is needed to decide the degree and kind of classroom structure the child requires (Schopler *et al.*, 1971).

In order to design the child's optimum educational curriculum, the educator also needs to know what skills, if developed, will enable a child to better function—regardless of the environment he is in. Through what special educational methods can he best acquire these new skills and information? Knowledge of his intellectual level and developmental milestones provides useful guidelines for long-range expectations.

These children show different primary deficits which are not in themselves unique to autism. Some with special education significance have been demonstrated in experimental research. Short-term memory deficits were found by Bryson (1972) and Pribram (1970) in both visual and auditory modalities. Variations in receptor preference have also been reported (Schopler, 1966; Hermelin & O'Connor, 1970). Difficulties in cross-modal integration and the corresponding problems with associative thinking and encoding abilities have been demonstrated by Hermelin and O'Connor (1970). Disorders of eye movements, scanning, and visual avoidance behaviors brought focus to the variations in patterns of visual attention seen in autistic children (Hermelin & O'Connor, 1970). Delayed development of sequential memory, temporal concepts, and causality reasoning has been pointed out by many investigators (Frith & Hermelin, 1969; Taylor, 1976; Dalgleish, 1975; Hermelin, 1976). The inability to imitate, to initiate and respond, and the corresponding dependence on the adult (Hartung, 1970; DeMyer, 1975; Wing, 1976) place unusual demands on the teacher. It may be necessary, for example, to actually manipulate the child's hands to initiate activity (Oppenheim, 1974). While this list contains some of the learning deficits found in autistic children it is by no means an exhaustive inventory.

SPECIAL EDUCATION TECHNIQUES

A number of useful educational techniques have been developed or rediscovered recently. Typically they were reported effective for an individual or limited numbers of autistic children. However, in the United States where the schools are under increasing legislative pressure to educate all children, these techniques are often overused beyond their demonstrated effectiveness. They are not necessarily oversold by their primary investigators, but rather by educational systems desperate for systematic answers. Often useful techniques are converted into disappointing fads. Even when applied in a technically correct fashion, all special education methods may be at risk to be ineffective when their goals are overgeneralized or when used with children they do not fit.

For example, detailed lesson plans of perceptual training to improve cognitive reading and writing skills have been published (Frostig & Horne, 1964; Dubenoff, 1968). The long-term results have been disappointingly negative. Although these programs have been useful for training certain perceptual skills, Mann and Goodman's (1976) evaluation shows that they are not effective in training perception as a general concept. Our own clinical experience has been that commercial materials do supply some useful exercises for improving some eye-hand coordination skills. However, they do not usually solve underlying perceptual problems shown by these children.

Ayers (1973) has developed a variety of perceptual-motor exercises organized along a normal sequence of maturational development of perceptual-motor skills. Autistic children do not necessarily follow this developmental sequence. We see children who have developed the conceptual skills necessary for reading and writing, for example, but who are deficient in earlier stages of perceptual-motor development; children, for example, who can read slogans and spell words from memory who cannot draw a simple geometric shape. The specific exercises which utilize, for example, swinging baskets and obstacle-course equipment are stimulating and pleasurable for some autistic children. Their educational value to the individual child must be determined on the basis of that child's profile of skills and deficits.

For the many autistic children who do develop verbal expression, detailed language programs are available (Gray & Ryan, 1973; Kent, 1974; Dunn & Smith, 1966; Engelman & Osborn, 1970) to provide sequential lesson plans. More recently Creedon (1975) has presented a program fostering "simultaneous communication" for the nonverbal child. Intensive training in signing accompanies the conceptual and motor curriculum. Signing provides an important new channel of communication and can help to improve the child's adaptation in other respects, including behavior control, attention, and peer relations. However, to use signs the child needs the visual skills for observing and remembering sign patterns, and the initiative to reproduce them when needed. He needs appropriate receptive skills to understand simple hand gestures. The extent to which autistic children have these skills is quite variable. For children with only minimal expressive language signing can facilitate learning speech as a manual prompt to the spoken word. These prompts should be faded out when they are no longer needed. The right timing to fade manual cues can only be accomplished by close observation of each child's progress. Language development, especially with the more handicapped child, requires coordination between school and home. The use of the same objects, signs, and names in both places is required to give the child the necessary repetition and consistency. Only individualized understanding of his language problem and appropriate curriculum goals can facilitate optimum language acquisition.

Operant conditioning with autistic children is reviewed by Lovaas in Chapter 25 and has been found extensively useful to many. Behavior modification techniques are published from procedures developed in a tightly controlled behavioral laboratory. Too frequently these specific techniques are accepted as scientifically established and are used under any circumstance when a child shows that particular behavior. There are, however, important differences between modifying behaviors in a controlled laboratory and in the relatively free circumstances of the family environment. To forget this difference is, of course, as unscientific as is the mechanical application of conditioning procedures from the laboratory to the home environment. This error is frequently made by, though not confined to, novices in the application of behavioral theory. No doubt it contributes to the poor generalization of improved behavior acknowledged by Lovaas *et al.* (1973) and others. Specific operant (conditioning) procedures must be adapted to the individual needs of a child and to the practicalities of his home and school environment.

EDUCATIONAL GOALS

Not only behavior management but also the goals of special education can be achieved only by adapting techniques to the individual child's learning patterns within the unique circumstances and life style of his own family. There are probably few educators who would today disagree with the importance of such individualization. In fact, these children and their idiosyncratic learning needs make it almost impossible to do otherwise. However, there are strong forces acting against the special educational mandate to individualize curricula. These include: (1) the pressures of many children and small number of teachers or classes; (2) lack of adequate training for teachers in how to develop individualized curricula; (3) administrative pressures to use one teaching system to the exclusion of others; and (4) inadequate communication with parents to facilitate transfer from classroom to home and back. These pressures have been avoided or reduced organizationally (Reichler & Schopler, 1976) in our North Carolina state program for autistic children (Division TEACCH).

INDIVIDUALIZATION IN THE DEVELOPMENTAL
THERAPY MODEL

In our North Carolina state program for autistic children we have developed several different organizational structures to maximize individualization and carry-over to the child's daily life. It is a statewide program, consisting of five centers and twenty-two classrooms. The centers are used for counseling and teaching individual children and their families. Each region also includes from four to six classrooms located within the public schools but under the center's direction. Classrooms handle from six to eight children with a teacher and teacher aide. They are as near to the children's home community as possible. This decentralization enables us at a reasonable cost to keep the relatively small teacher-child ratio necessary for meaningful individualized work.

One of the most important features of our program is the direct involvement of parents as cotherapists in the center (Schopler & Reichler, 1971a). A curriculum for home teaching is worked out on an individual basis, taking into consideration the child's developmental level and the parents' priorities and resources. Curricula and teaching techniques are developed jointly by therapists and parents, written out in a Home Program, and demonstrated in the clinic weekly sessions. This same process is being extended into the classroom although the curriculum emphasis has expanded to include group teaching and school routines. The direct involvement of parents at the center and in the classrooms provides carry-over between the child's learning experiences at school and at home. Parent involvement also adds a significant contribution toward the adaptation of the individual child to his own particular family circumstances.

Therapists from the center consult with classroom teachers or parents as requested. These therapists are not hired from any one professional specialty. They are hired for their qualifications as generalists, and their in-service training is directed toward this end. Therapists, teachers, and aides who meet the certification requirements of their local school systems are also selected through a special screening procedure. The final part of an applicant's screening involves the following steps. The applicant is asked to spend 15 minutes with one of our autistic children, with the instruction to formulate some initial suggestions regarding the child's special educational needs and how to meet them. These suggestions are to be written up in a brief report with supporting evidence only from the 15-minute observation. The applicant also conducts a brief interview with one of our parents. Here the task is to get an idea of what kinds of problems the parent is most concerned with, and to formulate suggestions for how to help with these problems. The applicant writes up a brief report on this parent interview, documenting his/her suggestions with the interview material. This is a most effective method for seeing quite rapidly whether an applicant can adapt his/her past experience and training to the individual needs of the child and parent.

No individualized curriculum can be developed without careful assessment of each child's unique learning patterns and deficits.

ASSESSMENT

Our diagnostic evaluation is roughly a six-hour procedure including the following components: (1) administering our Psychoeducational Profile (PEP); (2) assessment on standard intelligence tests; (3) parent interviews; and (4) direct observation of parent-child interactions.

The Psychoeducational Profile (PEP) presents materials typically used with younger children such as soap bubbles, clay, beads, books, etc. These tasks are grouped into six areas of function: imitation, perception, motor behavior, eye-hand integration, and expressive and receptive language skills.

The scoring is structured for flexible administration. If the child successfully completes the task, the response is scored pass. If he cannot or will not respond it is scored fail. If he responds in a peculiar fashion, but demonstrates some sense of what the task is about, the item is scored as emerging. It is the profile of these emerging items which makes up the basis for his individualized home program and classroom curriculum. This test provides a rough screening across the child's different mental functions. More detailed and specialized tests are used if indicated.

Standardized psychological tests appropriate for psychotic children are also used (Schopler & Reichler, 1971b). However, the main purpose of these tests is to obtain a measure of the child's intelligence quotient, rather than for prescriptive teaching. Educators are aware that these scores are often misunderstood or misinterpreted. In spite of these faults, research has shown IQ scores to be the best available long-range predictors of treatment outcome (Gittelman & Birch, 1967; Wing, 1976). This is important to both teachers and parents. First it provides a helpful timetable for planning expectations for progress. A child who has progressed at half the normal speed cannot be expected to suddenly speed through a normal one-year curriculum. Secondly these intelligence measures serve as a reminder for accepting the child's underlying handicap. Both parents teaching at home and teachers in the classroom have a tendency to blame themselves when their child does not progress as fast as they would like. They search for new techniques; they increase the length and frequency of lessons; they suffer from feelings of guilt and incompetence which increase their irritation with the child. The resulting bombardment of new activities, new routines, increased pressure and demands is both confusing and frustrating. It often leads to a sharp increase of psychotic behavior in the child. When we have some measure of the child's developmental limitations or impairment, teachers and parents are free to work out an individualized curriculum within the limits of his appropriate developmental expectations.

The parent interview and our direct observations of the parent-child interaction provide essential information for integrating on an individual basis the child's learning environment at home and at school. We learn about the size of the family, their life style, and resources for support. Many of our parents carry out daily structured teaching activities with their child at home. However, for a low-income working mother with five other young children to care for, a ten-minute training session may be optimum.

Special effort is made to coordinate home and school activities to maximize generalization of learning. A child who is told at home to "cut off the light" or "put it in the trash" is confused at school by the directions to "turn off" or "put in the wastebasket." The individualization that we plan for in developing content and techniques for teaching must also include the vocabulary and routines of the child's home life. The information obtained during the initial parent interview sets a pattern for an ongoing dialogue between therapist and parent at the clinic, and teacher and parent in the classroom. The parent-child interaction gives us a picture of communication and interaction patterns that are habitual in the family.

PLANNING THE INDIVIDUALIZED CURRICULUM FOR SCHOOL AND HOME

To facilitate the needed individualization, we found it necessary to develop the special organizational structure and diagnostic procedures discussed above. The diagnostic procedure enables us to limit and define the curriculum within the range of the child's developmental level. For example, an autistic child who is also severely retarded is directed toward self-help tasks rather than normal school participation. Within the child's developmental range the appropriate curriculum is derived from the teacher's observations and learning trials with the individual child. When the teacher has had years of experience with this group of children, experimental trials with tasks can be shortened, but this step cannot be eliminated. Prepackaged programs are often useful for a number of children, but they do not fit equally the vocabulary, interests, abilities, and handicaps of every autistic child in the group. Our program is planned on principles that pertain to the child's developmental level and his unique learning patterns. These are determined during the initial evaluation and an extended diagnostic treatment period of 6 to 8 weeks. We use materials and activities at the emerging level of ability that the child will find familiar and useful in daily life. The tasks are presented in a fashion that utilizes the child's particular strengths (often through manipulation of concrete materials) and that provides practice and training of deficit areas (often auditory memory and sequential organization).

We begin with a series of specific activities called a Home Program. These teaching activities are written out in detail and then used with the child by both his parents at home and his teacher in the classroom. Parents and teacher/therapists observe each other, and together make adjustments in the techniques of teaching and in the expectations that are appropriate for the child. As this process continues, parent and teacher share observations, ideas, the experimentation with each other, and thus develop an individual curriculum plan tailored to both the home and school. This mutual experience of selecting meaningful goals, task analysis, flexible experimentation, observing and recording progress provides a basis for continued integration between home and school experiences. This process has facilitated parent/school agreement in determining classroom placement and also has anticipated recent federal legislation requiring individualized educational programming (IEP) for handicapped children.

CASE ILLUSTRATION OF TASK ANALYSIS

From the educational assessment and the parent interview, tasks are chosen that are both within the child's ability and of importance to his daily life. For example, one six-year-old girl had learned to discriminate shape and size in puzzle pieces. She had a beginning vocabulary for labeling her immediate needs. However, she did not initiate self-help skills and was not dressing herself. At school she had no spontaneous interest in her tasks and extreme hypoactivity was her usual response. The teacher and parent mutually decided to teach "big and little" and *buttoning*. Shared observation from home and school pooled the information that Susan was very interested in her mother's jewelry, that she watched her mother sew, and that at school she learned best through daily repetition of structured routines.

Susan was quickly frustrated and rejected tasks that were clearly difficult. Buttoning her clothes was too difficult due to poor fine motor skills, poor visual attention, and lack of coordination between both hands.

The buttoning task was broken down into steps: beginning with pushing large poker chips through a cardboard slot, then to large buttons, smaller buttons, and finally to buttons attached to cloth. Both parent and teacher worked simultaneously and observed Susan's rate of increased skill, keeping the tasks at a level of possible success. Big and little were taught first in sorting tasks using jewelry, serving utensils, and food as well as school materials. The objects were labeled by size, and once descriptive labels were understood, Susan was required to use these in daily routines—asking for a "big cookie," "a big swing," "the little earring," etc. Susan was inconsistent in her performance from day to day. As her mother described it, "Some days she seems to get out on the wrong side of the bed." Both parent and teacher observed that on her irritable un-cooperative days they got better cooperation if they immediately simplified the tasks, reducing the number or using bigger buttons and giving increased help with her "big," "little" verbal responses. The process of task analysis continued from day to day.

Flexible experimentation and shared observations between home and school made it possible to expand her training from specific structured tasks to many daily routines. New ideas for individualization occurred to both parent and teacher based on their own observations of Susan's behavior. The teacher felt Susan could be expected to carry her tray in the school cafeteria, and this would be an important goal. She shared this decision with the parents. Susan's mother began at home to require her to carry dishes to the table, and to help carry groceries from the car. She observed that Susan watched her own hands when holding food she liked or a plate with a favorite food on it. By allowing Susan to have some initial accidents at home, like spilling the food from the tilted dish, this mother quickly taught increased visual attention and greater caution, and this generalized to the school cafeteria. At school the teacher noticed Susan was particularly interested in one boy and watched him as he dressed for outdoor play. Susan's father began asking Susan to help him button his coat in the morning. This proved successful. Flexible experimentation continued as new goals were chosen.

THE RECORDING OF A CHILD'S RESPONSES
AND PROGRESS

The recording of a child's responses and progress is a part of the home program process from the beginning and behavior management techniques, an immediate necessity as teaching begins, are selected on an individual basis that considers the child's level of awareness, the parents' priorities and general style, and the feasibility in the home and school environment.

Parents record the child's spontaneous language, his achievements in self-help skills, and responsibility for daily routines. They share this information with the teacher in order to increase the consistency of expectations and demands between home and classroom. Behavioral goals, the choice of rewards and punishments are also discussed. Parents are asked to chart the child's responses to these techniques. Some parents may find this bookkeeping distasteful. It may be unnecessary for the development of a consistent and rational approach with the child.

INCORPORATING NEW APPROACHES INTO THE INDIVIDUAL PROGRAM

The contributions of educational research and experimental studies can have a practical application in individual programs, both in the classroom and in home activities. A few examples of ideas gleaned from published results of new educational approaches that we have adapted to individual programs will illustrate this. The programmed language acquisition (Dunn & Smith, 1966; Kent, 1974; Gray & Ryan, 1973; Engelman & Osborn, 1970) materials provide us with guidelines for planning the development of language structures. A modified communications board (Ratusnik & Ratusnik, 1974) placed in the family kitchen has been useful to one family whose child could read but could not process spoken phrases. Another child who could not read used an apron containing pictures representing his predicted needs (spoon, cup, toothbrush, etc.). Activities developed by Ayers (1973) to stimulate sensory integration have been adapted into home program activities using tactile stimulation with brushes, powder, and lotion. We find that fathers enjoy activities that train gravity and postural responses such as jumping, being swung and flipped, and simple gymnastic exercises. Hopscotch, jump rope, balance beams, obstacle courses, sack and potato races can be adapted to the backyard situation and can often include siblings as well. Dancing, rhythm bands, and singing games (Euper, 1968) have been used in some families and all classrooms. The lyrics of these singing games can be adjusted to incorporate the current language goals of the children. The sequential progression from body actions, to distal senses, to language outlined by Miller and Miller (1973) was used with one child to teach up, down, in, and under. He was moved in space; he moved himself as he saw the directing gestures; he heard the word, made the gesture, and finally understood the word in a variety of contexts. Communication through signing (Creedon, 1975) has been adapted into programming for some children who did not have any expressive language, or who found that this visual and kinesthetic cue aided word retrieval. The development of initiative in discovering rules outlined by Taylor (1976) has been useful to our teachers and parents in incorporating this step into their sorting and sequential arranging activities. The understanding of why this is such a vital step encourages teachers to maintain the patience needed for giving the child time to initiate and draw conclusions. Montessori routines emphasizing orderly procedures have been helpful at home and in the classroom in fostering independent initiative and responsibility (Elgar, 1966) in self-help skills and household routines.

SUMMARY

The education of autistic children, both in the home and in structured classes predates the descriptive label. Educators have increasingly assumed the responsibility for treatment of these handicapped children while they remain at home. Many educational techniques are being developed. These techniques are frequently recommended and used on a systematic basis for all autistic children. Because of characteristic variability in learning abilities and handicaps, individualization of curriculum is necessary. Administrative structures allowing individualization in a state program were presented. Diagnostic procedures were described for modifying and combining specific techniques on an individualized basis. Parents' involvement has been the critical factor for promoting carry-over from classroom to home.

ACKNOWLEDGMENTS

The directors of Division TEACCH have made a significant contribution to this chapter and the program on which it is based. We gratefully acknowledge the helpful suggestions of Dr. Louis P. Semrau, Dr. Bernard B. Harris, Dr. Jerry L. Sloan, and Dr. Lee M. Marcus.

REFERENCES

Ayers, A. J. *Sensory integration and learning disorders.* Los Angeles: Western Psychological Services, 1973.

Bryson, C. Short term memory and cross-modal information processing in autistic children. *Journal of Learning Disabilities,* 1972, *5,* 25–35.

Creak, M. Schizophrenic syndrome in childhood: Further progress of a working party. *Developmental Medicine and Child Neurology,* 1964, *6,* 530–535.

Creedon, M. Appropriate behavior through communication: A new program in simultaneous language for nonverbal children. *Dysfunctioning Child Center Publication.* Chicago: Michael Reese Medical Center, 1975.

Dalgleish, B. Cognitive processing and linguistic reference in autistic children. *Journal of Autism and Childhood Schizophrenia,* 1975, *5,* 353–363.

DeMyer, M. K. The nature of neuropsychological disability in autistic children. *Journal of Autism and Childhood Schizophrenia,* 1975, *5,* 109–128.

Dubenoff, B. *Dubenoff school program.* Boston: Teaching Resources, 1968.

Dunn, L. M., & Smith, J. O. *Peabody language development kits.* Minneapolis: American Guidance Service, 1966.

Elgar, S. Teaching autistic children. In J. Wing (Ed.), *Early childhood autism.* New York: Pergamon Press, 1966. Pp. 205–237.

Engelman, E., & Osborn, J. *Distar language program.* Chicago: Science Research Associates, Inc., 1970.

Euper, J. A. Early infantile autism. In E. T. Gaston (Ed.), *Music in therapy.* New York: Macmillan, 1968. Pp. 181–191.

Frith, U., & Hermelin, B. The role of visual and motor cues for normal, subnormal, and autistic children. *Journal of Child Psychology and Psychiatry,* 1969, *10,* 153–163.

Frostig, M., & Horne, D. *The Frostig program for the development of visual perception.* Chicago: Follett Publishing Co., 1964.

Gittelman, M., & Birch, H. G. Childhood schizophrenia: Intellect, neurologic status, perinatal risk, prognosis, and family pathology. *Archives of General Psychiatry,* 1967, *17,* 16–25.

Gray, B., & Ryan, B. *A language program for the non-language child.* Champaign, Ill.: Research Press, 1973.

Hartung, R. O. A review of procedures to increase verbal imitation skills in autistic children. *Journal of Speech and Hearing Disorders,* 1970, *35,* 203–217.

Hermelin, B. Coding and the sense modalities. In L. Wing (Ed.), *Early childhood autism* (2nd ed.). Oxford: Pergamon Press, 1976.

Hermelin, B., & O'Connor, N. *Psychological experiments with autistic children.* Oxford: Pergamon Press, 1970.

Hewett, F. M. Teaching speech to an autistic child through operant conditioning. *American Journal of Orthopsychiatry,* 1965, *35,* 927–936.

Itard, J. G. *The wild boy of Aveyron.* New York: Appleton-Century-Crofts, 1962.

Kanner, L. Autistic disturbances of affective contact. *Nervous Child,* 1943, *2,* 217–250.

Kent, L. R. *Language acquisition program for the severely retarded.* Champaign, Ill.: Research Press, 1974.

Lovaas, O. I., Koegel, R., Simmons, J. Q., & Stevens, J. Some generalization and follow-up measures on autistic children in behavior therapy. *Journal of Applied Behavior Analysis,* 1973, *6,* 131–166.

Mann, L., & Goodman, L. Perceptual training: A critical retrospect. In E. Schopler & R. J. Reichler (Eds.), *Psychopathology and child development.* New York: Plenum, 1976.

Miller, A., & Miller, E. Cognitive developmental training with elevated boards and sign language. *Journal of Autism and Childhood Schizophrenia,* 1973, *3,* 65–85.

Oppenheim, R. *Effective teaching methods for autistic children.* Springfield, Ill.: Charles C Thomas, 1974.

Pribram, K. Autism: A deficiency in context-dependent processes? *Proceedings of the National Society for Autistic Children.* Rockville, Maryland: Public Health Service, U. S. Department of HEW, 1970.

Ratusnik, C. M., & Ratusnik, D. L. A comprehensive communication approach for a ten-year-old nonverbal autistic child. *American Journal of Orthopsychiatry,* 1974, *44,* 393–403.

Reichler, R. J., & Schopler, E. Developmental therapy: A program model for providing individual services in the community. In E. Schopler & R. J. Reichler (Eds.), *Psychopathology and child development.* New York: Plenum, 1976.

Rimland, B. *Infantile autism.* New York: Appleton-Century-Crofts, 1964.

Rutter, M. Concepts of autism: A review of research. *Journal of Child Psychology and Psychiatry*, 1968, *9*, 1–25.

Rutter, M., Lebovici, S., Eisenberg, L., Sneznevskij, A. V., Sadoun, R., Brooke, E., & Lin, T-Y. A tri-axial classification of mental disorder in childhood. *Journal of Child Psychology and Psychiatry*, 1969, *10*, 41–61.

Schopler, E. Visual versus tactual receptor preference in normal and schizophrenic children. *Journal of Abnormal Psychology*, 1966, *71*, 108–114.

Schopler, E. Parents of psychotic children as scapegoats. *Journal of Contemporary Psychotherapy*, 1971, *4*, 17–22.

Schopler, E., Brehm, S., Kinsbourne, M., & Reichler, R. J. Effect of treatment structure on development in autistic children. *Archives of General Psychiatry*, 1971, *24*, 415–421.

Schopler, E., & Reichler, R. J. Parents as cotherapists in the treatment of psychotic children. *Journal of Autism and Childhood Schizophrenia*, 1971, *1*, 87–102. (a)

Schopler, E., & Reichler, R. J. Problems in the developmental assessment of psychotic children. *Excerpta Medica International Congress Series No. 274*, 1971, 1307–1311. (b)

Taylor, J. An approach to teaching cognitive skills underlying language development. In L. Wing (Ed.), *Early childhood autism* (2nd ed.). Oxford: Pergamon Press, 1976. Pp. 205–220.

Wing, L. Diagnosis, clinical description and prognosis. In L. Wing (Ed.), *Early childhood autism* (2nd ed.). Oxford: Pergamon Press, 1976. Pp. 15–48.

30

Educational Aims and Methods

MARIA CALLIAS

The field of education generally is not renowned for its sound scientific basis. Although teachers, clinicians, and researchers have used extensive experience to form considered opinions and theories about the education of autistic children, empirical data are largely lacking (e.g., Rutter, 1970; Wing, 1976; Bartak & Pickering, 1976; Elgar, 1976; Gallagher & Wiegerink, 1976). The study of Bartak (see Chapter 28) constitutes an important exception.

Its strength lies in its direct examination of (a) what actually went on in schools holding different educational philosophies, and (b) the relationships between these educational practices and long-term changes in the children. The study is not without methodological problems. The types of unit are represented by only one school each; the number of children within each school is variable and small; the three groups of children were not exactly comparable in terms of age or ability. The classroom observations and assessment of children were not done blind. Bartak is aware of the problems and has taken appropriate action wherever possible. For example, the possibility that teacher-child interaction, and curriculum planning, could change over the three to four years of the study was checked by repeating the observations toward the end of the study. And indeed, some small changes were found for one of the units. A more problematic issue was that of taking into account differences in age and intelligence, the latter having been found consistently to be the most important variable associated with later status (Lockyer & Rutter, 1969).

MARIA CALLIAS · Department of Psychology, Institute of Psychiatry, London, England.

The regression analysis taking age and ability into account confirmed that scholastic progress was associated with the nature of schooling. All the schools used essentially the same type of educational material, but differed markedly in the amount of time spent on educational pursuits, as well as in the degree of structure imposed.

The study comes up with some other interesting findings which are less easy to explain. For example, there was very little social play in any of the units, the children at Unit C showing the most. It is not possible from this study to know whether this was associated with individual differences in the children such as age, IQ or "personality," and/or active fostering of play by the staff.

Drawing attention to some of the problems of this study does not detract from its valuable and important contributions. First, it clearly demonstrates that autistic children learn specifically what teachers actively set out to teach. Second, the behavioral improvements noted at school (in all units) were situation-specific, with very little spontaneous generalization to home or other situations. These are important findings which have implications for changes in educational strategies. Further, the study highlights problems which need to be studied further. For example: How can the amount of unoccupied time spent by the children be reduced? What can be done about teaching children to play with each other? At a more general level, one of the most important contributions of this study is that it shows that controversial aspects of education can be studied sensitively. Moreover, the results of such studies can provide an empirical basis for decisions about change in the aims, content, or methods of teaching. It should not be necessary to leave these to the mercies of unsubstantiated opinion, or, at best, educated guesses.

Lansing and Schopler (see Chapter 29; also Reichler & Schopler, 1976) have opened up somewhat different educational avenues. By starting to provide a comprehensive educational service from scratch in 1972, TEACCH has been able to build into the organization of educational services ways of avoiding the problems encountered in maintaining treatment/training gains once children have left intensive treatment settings (e.g., Lovaas *et al.*, 1973) or in effecting generalization to home (Bartak, see Chapter 28; Rutter & Bartak, 1973). The most impressive features of the program are the planned collaboration between home and school, the careful assessment of the child leading to individually planned educational goals, the collaboration of psychologists and teachers which allows for new developments to be incorporated into teaching rapidly and the use of a developmental framework for selecting educational goals.

These innovations offer considerable promise on what might be achieved in the education of autistic children. It is particularly important that the TEACCH program aims to provide remedies for some of the limitations in the units studied by Bartak. It will be vital to determine the extent to which collaboration between home and school is possible, how this is affected by the selection of children and families, and the reasons for breakdowns when they occur. The TEACCH classes are based in ordinary schools. How far does this provide opportunities for integration and what snags arise? As with all major innovations and ventures, the next challenge is evaluation. This is crucial both at the molecular level of what works best with individual children and also at the molar level of how TEACCH compares with other approaches in terms of its achievements, limitations, and cost benefits. Systematic monitoring and comparative studies are required if we are to assess the real value of what TEACCH has to offer.

EDUCATIONAL AIMS

There seems little argument about the general aims of education for autistic children (Rutter, 1970; Wing, 1976). Briefly, these are (1) to prevent or reduce secondary behavior handicaps, (2) to find approaches which circumvent the primary handicaps, and (3) to find ways of helping children develop functions which are involved in the primary handicaps. "The purpose of education is to help the handicapped person derive as much satisfaction and enjoyment from life as possible" (Wing, 1976, p. 197). It is the translation of these broad aims into the nitty-gritty of curriculum content and teaching methods that is still controversial.

CURRICULUM CONTENT

Lansing and Schopler (Chapter 29) state that, in their project, individual curricula are drawn up within a developmental framework to select tasks that are within the child's abilities and are of importance to his daily life. Bartak (Chapter 28) has shown that it is only when the "Three Rs" are actively taught that children master these skills. Setting specific goals and taking into account the child's current status in each area of development is crucial (Bartak & Pickering, 1976).

The question of whether or not the normal developmental sequence necessarily provides the best basis for deciding what to teach is a controversial issue currently evident in the important area of language training (Ruder & Smith, 1974). In general, the question of whether it is best to concentrate teaching effort on skills which are emerging or developmentally just ahead of the child's present level of functioning, or to disregard the sequence of normal development in selecting goals, is open to empirical investigation. At present, our best guide to deciding what to teach rests on a careful analysis of the child's present competencies, the guidance of (but not slavish adherence to) a developmental framework, and our social values or judgment of what is important to the child.

The main content areas which are acknowledged to be important and which have received some successful attention both in and outside school have been self-help skills, language, reading and number concepts, and various problem behaviors (e.g., Rutter & Bartak, 1973; Howlin et al., 1973; Hemsley & Howlin, 1976; Lovaas et al., 1976; see also Chapters 28 and 29).

The content of language and communication programs remains more controversial than that of most other areas. Differences of opinion occur particularly over the group of children who have reasonable comprehension but fail to develop speech in spite of skilled teaching. As Lansing and Schopler have shown, the use of systematic sign language and other innovations can be valuable in facilitating communication. Yet many teachers are reluctant to introduce alternate forms of communication for fear that these will supplant or inhibit the development of spoken speech. This controversy parallels that found in the teaching of deaf children. The evidence from deaf children so far suggests that the teaching of manual communication does not inhibit speech and lip-reading from developing. Rather, it seems to have a facilitatory effect by providing a language base (Freeman, 1976). With autistic children, particularly the severely retarded, such alternatives merit further investigation (Carr, 1976).

Less success has been achieved in modifying some of the other skill deficits, particularly those of play and social development, although provision has often been made for these activities in the school timetable (Elgar, 1976; Bartak & Pickering, 1976). Most children improve somewhat in these areas (Rutter & Bartak, 1973) but limited social awareness and social skills are frequently still conspicuous handicaps of autistic adolescents and young adults (Rutter et al., 1967). These are skills which we should be studying more actively and trying to teach. There are now several studies indicating that social interaction and cooperative play may be successfully taught (using similar techniques to those found to work with other skills) to young behaviorally disturbed and psychotic children (e.g., Cooke & Apolloni, 1976; Romanczyk et al., 1975; and Strain & Wiegerink, 1975).

Two other important aspects of social development which are deficient in many autistic children are the ability to exercise control over their own behavior and the closely related ability to be part of a group. The possibilities of teaching other handicapped children to control their own behavior are being explored (e.g., Kurtz, 1976; Meichenbaum & Goodman, 1971). Though these techniques rely on the child having useful language, their potential value to older autistic children who do have language should not be overlooked.

Most of the research intervention with autistic children has relied heavily on a one-to-one teacher-child ratio (Lovaas *et al.*, 1976), and it is a common cry of teachers that they wish they could spend more time individually with each child. Bartak found that on all three units in his study children spent a large amount of time unoccupied. More research attention should be paid to the possibility of teaching the children appropriate group behavior. There is evidence that this is feasible, at least with some children (Halpern, 1970; Koegel & Rincover, 1974).

TEACHING METHODS

A structured approach to teaching is frequently advised for autistic children (e.g., Rutter, 1970; Wing, 1976; Bartak & Pickering, 1976; see also Chapters 28 and 29), and it is important to be clear about what is meant. At the molar level, the term "structure" is used to describe the organization of the school day so that there is a planned, well-organized timetable with time allocated to different activities (Rutter, 1970; Bartak & Pickering, 1976; Elgar, 1976). Within this overall planned day, flexibility is needed in drawing up separate and specific programs to suit the needs of each child (e.g., Elgar, 1976).

At a more molecular level, we talk of structuring any particular learning task. The stimuli impinging on the child and the responses he makes are controlled or structured by the teacher in such a way as to facilitate correct behavior (Bartak & Pickering, 1976). This is equivalent to using some but not all aspects of a behavioral or learning approach to teaching, but phrased in ordinary English rather than jargon.

The question of how far to structure the treatment or learning situation is central to the controversy over whether a psychotherapeutic or a learning approach is best for the autistic child. While Bartak is rightly cautious about claiming causal links, his study goes a long way toward demonstrating an association between the scholastic attainments of children and the degree of structure and task orientation of the school.

Structure and behavioral (mainly operant) approaches to teaching are often discussed as if they were very different. The reasons for this and the relationships between the two approaches warrant consideration. Several authors (e.g., Wing, 1976; Schopler, 1976) are disillusioned with and reject operant approaches for several reasons. First, there has been the failure of treatment effects to generalize or be maintained when the children move to a different environment (Lovaas *et al.*, 1973). Second, learning approaches are sometimes regarded as rigid, mechanistic, and inhuman. Third, operant approaches are often erroneously equated only with punishment. Finally, behavioral techniques are often seen as simply new ways of expressing old "commonsense" ideas. While many early laboratory studies can rightly be criticized on some of these grounds, they are not the essential intrinsic features of a behavioral approach. With the move toward treatment in the natural environment has come a general growing concern to find socially acceptable ways of implementing behavioral principles (e.g., Howlin *et al.*, 1973; Porterfield *et al.*, 1976). Problems of generalization and maintenance are not confined to behavioral interventions. The key features of a behavioral approach are: carrying out a functional analysis of the behavior in terms of the preceding stimuli and consequences; using this analysis as a basis for intervention or teaching; and monitoring progress and, if necessary, altering the training approach to insure that the child is learning (Gelfand and Hartmann, 1975). The case illustration of task analysis which Lansing and Schopler (Chapter 29) describe is an excellent example of sensitively applied behavioral principles. After deciding on a relevant task to teach, the child's behavior was observed and an appropriate level to begin teaching was selected. The task was broken down into small steps, each was taught, giving appropriate help or prompts (physical help, instructions, or demonstration), and the child was reinforced for each successful attempt.

Clearly, behavioral methods are not the only effective ways of going about teaching, but equally certainly it would be a pity if they were dismissed completely. This would be throwing the baby out with the bathwater. Behavioral principles have an important place in the classroom (e.g., Sherman & Bushell, 1975). In particular, they provide the most systematic basis available for answering questions on how to teach and motivate children. Of course, they do not specify the content or goals of training. As already discussed, a curriculum which draws from a knowledge of normal development and which is based on our social values of what is important seems best at present.

The main difference between structured and operant approaches is that only the latter explicitly considers the issue of motivating children. It is important but particularly difficult to provide optimal conditions of learning for the less-able children and for those who lack spontaneity and an intrinsic interest in learning. The explicit and systematic use of suitable extrinsic reinforcement can provide a means of motivating such children where all else fails. Hopefully, they will come to develop interest in new skills, or learn to appreciate more naturally occurring reinforcement such as praise and approval. Further research is needed on such issues.

The last reason for emphasizing the need to be explicit about the use of learning principles in education is that communication and understanding between workers in different settings would be improved if we all used the same terminology to describe the same phenomena. Psychology and education could both benefit from a common language.

SPECIAL EDUCATION

A central issue in special education is whether the needs of handicapped children are best met in special schools or within ordinary schools (Anderson, 1976). Bartak (Chapter 28) studied children in special schools, two of which were exclusively for autistic children. He found no advantages to autistic children in Unit B which catered to other behaviorally disturbed children as well. On the other hand, no disadvantages are reported either. The TEACCH project (Chapter 29) provides education for a wider range of children with communication problems in classrooms which are geographically part of public schools. It will be important to know what the problems and advantages of these arrangements are.

It is clear that autistic children have needs that cannot usually be met in the ordinary school system, though some children do cope there. What is less clear is whether autistic children need exclusive schools or whether there would be advantages in being part of a more heterogeneous school population. It would seem that this is likely, particularly if their social development is to receive more focused attention. Teachers would need to foster such interaction actively. The need for mixing with talking children has also been stressed (Rutter, 1970). The teaching skills needed for working with autistic children, the provision of other therapists and psychological services, are required for other developmentally handicapped children too.

Wherever autistic children are educated, it is clear that some special provision will usually be needed in the form of teacher expertise, extra classroom aides, specialist advice. In their recent discussion of the service needs, Wing and Wing (1976) consider that most autistic children, particularly those with difficult behavior, will need specialized units. "The ideal system seems to be to have both specialized units and the possibility of admitting autistic children to other kinds of schools so that the best place can be found for each child. He may have different needs at different stages of his school career so transfer should be easy and informal" (p. 303). At the present state of our knowledge, this seems good advice, with perhaps a question about whether the specialized units need to be exclusively for autistic children.

So far, attention has been directed to the education of children of school age only. It is clearly urgent to extend appropriate educational opportunities both to those under five and to the older autistic youngsters (e.g., Rutter, 1970; Wing & Wing, 1976). The aims and curriculum content will, of course, be different. For young adults, the main needs are for suitable vocational training and help with the social and behavioral handicaps that frequently prevent them from coping with employment.

Finally, it is essential to continue paying attention to the evaluation of educational practices and innovations. Only in this way are controversial theoretical issues likely to be replaced by a surer empirical foundation. This is necessary if we wish to achieve our aim of providing autistic children with the opportunities of developing their full potential.

REFERENCES

Anderson, E. M. Special schools or special schooling for the handicapped child? The debate in perspective. *Journal of Child Psychology and Psychiatry,* 1976, *17,* 151–156.

Bartak, L., & Pickering, G. Aims and methods of teaching. In M. P. Everard (Ed.), *An approach to teaching autistic children.* Oxford: Pergamon Press, 1976. Pp. 79–98.

Carr, J. The severely retarded autistic child. In L. Wing (Ed.), *Early childhood autism: Clinical, educational and social aspects* (2nd Ed.). Oxford: Pergamon Press, 1976. Pp. 247–270.

Cooke, T. P., & Apolloni, T. Developing positive social-emotional behaviors: A study of training and generalization effects. *Journal of Applied Behavior Analysis,* 1976, *9,* 65–78.

Elgar, S. Organization of a school for autistic children. In M. P. Everard (Ed.), *An approach to teaching autistic children.* Oxford: Pergamon Press, 1976. Pp. 121–132.

Freeman, R. The deaf child: Controversy over teaching methods. *Journal of Child Psychology and Psychiatry,* 1976, *17,* 229–232.

Gallagher, J. J., & Wiegerink, R. Educational strategies for the autistic child. *Journal of Autism and Childhood Schizophrenia,* 1976, *6,* 15–26.

Gelfand, D. M., & Hartmann, D. P. *Child behavior: Analysis and therapy.* Oxford: Pergamon Press, 1975.

Halpern, W. J. The schooling of autistic children: Preliminary findings. *American Journal of Orthopsychiatry, 1970, 40,* 665–676.

Hemsley, R., & Howlin, P. Managing behavior problems. In M. P. Everard (Ed.), *An approach to teaching autistic children.* Oxford: Pergamon Press, 1976.

Howlin, P., Marchant, R., Rutter, M., Berger, M., Hersov, L. & Yule, W. A home based approach to the treatment of autistic children. *Journal of Autism and Childhood Schizophrenia, 1973, 3,* 308–336.

Koegel, R. L., & Rincover, A. Treatment of psychotic children in the classroom environment. 1. Learning in a large group. *Journal of Applied Behavior Analysis,* 1974, *7,* 45–59.

Kurtz, P. D., & Neisworth, J. T. Self control possibilities for exceptional children. *Exceptional Children, 1976, 42,* 213–218.

Lockyer, L., & Rutter, M. A five to fifteen year follow-up study of infantile psychosis. III. Psychological aspects. *British Journal of Psychiatry, 1969, 115,* 865–882.

Lovaas, O. I., Koegel, R. L., Simmons, J. Q., & Long, J. Some generalization and follow-up measures on autistic children in behavior therapy. *Journal of Applied Behavior Analysis, 1973, 6,* 131–166.

Lovaas, O. I., Schreibman, L., & Koegel, R. L. A behavior modification approach to the treatment of autistic children. In E. Schopler & R. J. Reichler (Eds.), *Psychopathology and child development: Research and treatment.* New York: Plenum Press, 1976.

Meichenbaum, D. H., & Goodman, J. Training impulsive children to talk to themselves: A means of developing self-control. *Journal of Abnormal Psychology,* 1971, *77,* 115–126.

Porterfield, J. K., Herbert-Jackson, E., & Risley, T. R. Contingent observation: An effective and acceptable procedure for reducing disruptive behavior of young children in a group setting. *Journal of Applied Behavior Analysis,* 1976, *9,* 55–64.

Reichler, R. J., & Schopler, E. Developmental therapy: A program model for providing individual services in the community. In E. Schopler and R. J. Reichler (Eds.), *Psychopathology and child development: Research and treatment.* New York: Plenum Press, 1976.

Romanczyk, R. G., Diament, C., Goren, E. R., Trunell, G., & Harris, S. L. Increasing isolate and social play in severely disturbed children: Intervention and post-intervention effectiveness. *Journal of Autism and Childhood Schizophrenia,* 1975, *5,* 57–70.

Ruder, K. F., & Smith, M. D. Issues of language training. In R. L. Schiefelbusch & L. L. Lloyd (Eds.), *Language perspectives: Acquisition, retardation and intervention.* Baltimore, Maryland: University Park Press, 1974.

Rutter, M. Autism: Educational issues. *Special Education,* 1970, *59,* 6–10.

Rutter, M., & Bartak, L. Special educational treatment of autistic children: A comparative study—II. Follow-up findings and implications for services. *Journal of Child Psychology and Psychiatry,* 1973, *14,* 241–270.

Rutter, M., Greenfeld, D., & Lockyer, L. A five to fifteen year follow-up study of infantile psychosis—II. Social and behavioural outcome. *British Journal of Psychiatry,* 1967, *113,* 1183–1199.

Schopler, E. Towards reducing behavior problems in autistic children. In L. Wing (Ed.), *Early childhood autism: Clinical, educational and social aspects.* Oxford: Pergamon Press, 1976.

Sherman, J. A., & Bushell, D., Jr. Behavior modification as an educational technique. In F. D. Horowitz (Ed.), *Review of child development research.* Chicago: The University of Chicago Press, 1975.

Strain, P. S., & Wiegerink, R. The social play of two behaviorally disordered preschool children during four activities: A multiple baseline study. *Journal of Abnormal Child Psychology,* 1975, *3,* 61–69.

Wing, J. K., & Wing, L. Provision of Services. In L. Wing (Ed.), *Early childhood autism: Clinical, educational and social aspects.* Oxford: Pergamon Press, 1976.

Wing, L. The principles of remedial education for autistic children. In L. Wing (Ed.), *Early childhood autism: Clinical, educational and social aspects.* Oxford: Pergamon Press, 1976.

31

Long-Term Follow-Up of 100 "Atypical" Children of Normal Intelligence

JANET L. BROWN

This chapter presents selected material on 100 children who showed profound disturbances in interpersonal relationships during their preschool years, but who differed from most autistic children in that they developed average or better cognitive functioning.

The Putnam Children's Center, established in 1943, is a private, nonprofit, child guidance clinic. Until recently, only preschool children, age one to five, were accepted for diagnostic evaluation and treatment. From the start, the staff was actively interested in children with severe interpersonal disturbances, and collected comprehensive material on these children and their families, often seeing them for many years on a once or twice weekly outpatient basis. To date, about 400 preschool children, roughly 10% of all those evaluated, have been diagnosed at the Putnam Center as showing "atypical development." The essential criterion for this diagnosis is a profound disturbance in interpersonal relationships, as evidenced by an almost complete lack of mutuality in child-adult and child-child interaction. All the children so diagnosed showed major developmental distortions before age two. These children with atypical development show most or all of Creak's signs (Creak, 1964), and many show the characteristics of Kanner's early infantile autism (Kanner, 1943). Some are more similar to childhood schizophrenics, as described

JANET L. BROWN · James Jackson Putnam Children's Center, Boston, Massachusetts.

by Bender (1955). Children with aphasia or mental retardation who attempt appropriate interpersonal communication, verbal or nonverbal, are *not* included in this diagnosis. However, children with aphasia, mental retardation, whether functional or not, brain damage, and sensory handicaps such as blindness and deafness, *are* included if they show the cardinal symptom of profoundly distorted object relationships.

For the last six years, we have been summarizing the data on the total sample. Case records have been rated on over 450 variables, and parents and other appropriate sources contacted for follow-up information. Our sample is thus not only very large, but somewhat more inclusive than that studied by other investigators. As the Center is located in the inner city, and charges on a sliding scale or through third-party payment, it is convenient to lower-class clients, as well as attracting middle- and upper-class clients from the Boston suburbs. Also, because it is an outpatient clinic, we saw a number of children who were able to enter regular public school, and who probably would not have been seen in an inpatient setting.

SAMPLE SELECTION

As there has been relatively little attention in the literature to the substantial group of atypical children with intact cognitive functioning (see Brown, 1969, 1974; Kanner *et al.*, 1972), I selected the 100 children from the total sample of 400 who had average or better cognitive functioning and for whom we had follow-up material to midadolescence or later. Average or better cognitive functioning was defined as obtaining an IQ of 90 or better on a standard intelligence test such as the Stanford Binet or Wechsler Intelligence Scale for Children, and/or by graduation from regular high school at an appropriate age, and/or by employment at a regular job without special supervision or modified expectations.

The sample consists of 76 boys and 24 girls. Their modal age at the time of our diagnostic evaluation was three and one-half to four years, with 83% of the children evaluated before four and one-half. Two-thirds were from the upper two social classes (on a five-point scale). Sixteen were only children, 36 were first children, 34 were youngest children, and only 14 were middle children. All the children had some language at the time of their diagnostic evaluation except one who was less than two years old, but 25 of the children had only occasional single words. Only 19 of the children had more than occasional higher level construction or fantasy play, and 40 did not show any at all at the time of their diagnostic evaluation. Forty-two had a history of rocking or headbanging, but only 28 had additional motor mannerisms; 32 had never shown motor mannerisms. Of the 57 children who had neurological examinations, only 15 (26%) were at all questionable. Of the 61 who had EEGs, 43 (70%) were read as normal, 16 (26%) showed relatively minor abnormalities, and two had seizure activity. One child had had seizures during the first six months of life, one had a question of "abdominal epilepsy" in middle childhood, and one had a grand mal seizure at age 21. Thus this group of 100, selected for their intact cognitive functioning, was definitely superior to the remainder of our total sample of 400 in language and symbol formation, fewer mannerisms, and less organic involvement as documented by abnormal neurological or EEG examinations, or seizure history.

FOLLOW-UP RESULTS

Overall Outcome

The mean age of the subjects at the time of latest follow-up was 22.3 years; the oldest was 34. Sixty-one were over 20 years old, and were rated for status during both adolescence and young adulthood. Outcome status was rated on an 11-point scale which had been developed to assess the full range of outcomes for the larger sample, and which combined cognitive and personality functioning. (See Table 1). The results of this global rating for this sample with intact cognitive functioning are given in Table 2. It can be seen that subjects in this group clustered entirely in the best 5 outcome categories (as follows from the definition of the sample). The two subjects who were rated "5" in adolescence because their IQs fell in the dull normal range, were moved upward in young adulthood because they were able to engage in regular employment. The significant decrease in the percentage of adult subjects in category 3, "eruptive-schizoid," suggests that as the subjects became older, they tended to show either increased integration and high-level functioning, or became more like chronic schizophrenics with a decrease in cognitive flexibility and interpersonal skills. In contrast with the possibility of improvement for subjects rated "eruptive-schizoid" in adolescence, none of the subjects rated "schizophrenic" in adolescence moved out of this category as young adults.

The status of the 61 subjects, age 20 or older on follow-up, is given in Table 3. The regular jobs held by these young people were as follows: eight in clerical and office work, three in unskilled factory work, two in nursing, and one each as copywriter, bank teller, draftsman, computer programmer, postal clerk, library cataloguer, personnel manager, mechanic, house painter, mental health coordinator, jewelry maker, ad-

Table 1. Criteria for Outcome Ratings

Category	IQ	Learning	Social relationships	Surface pathology
1. Normal-neurotic	Average or better	Up to age	Accepted by peers	Neurotic or none
2. Brittle-schizoid	Average or better	Up to age	Friendships based on interests; considered "odd"	Schizoid
3. Eruptive-schizoid	Dull normal (80–89) or better	Somewhat retarded or better	Social drives, including sex, but too narcissistic for friendships	Primary process breakthroughs, "colorful"
4. Schizophrenic	Dull normal or better at some time (may lose capacity)	Somewhat retarded or better at some time	Usually talks "at" people; basically pregenital orientation	Pervasive thinking disorder
5. Passive-retarded (mild)	Dull (70–79)	Somewhat retarded; grade 2 at least	Parallel activity	Little
6. Eruptive-retarded (mild)	Dull (70–79)	Somewhat retarded; grade 2 at least	Some behavior and management problems	Breakthrough of aggressive or sexual behavior; somewhat "crazy"
7. Passive-retarded (moderate)	Educable (55–69) or less or untestable	Some formal learning; can print name, count	Definitely relates to specific people	Little
8. Eruptive-retarded (moderate)	Educable (55–69) or less or untestable	Some formal learning; can print name, count	Definitely relates to specific people	Definite mannerisms (hand movements, talking to self, etc.)
9. Passive-retarded (severe)	Trainable (25–54) or less or untestable	No formal learning but toilet trained, feeds self	Some evidence recognizes specific people	Little
10. Eruptive-retarded (severe)	Trainable (25–54) or less or untestable	No formal learning but toilet trained, feeds self	Some evidence recognizes specific people	Management problem; definitely "crazy"
11. Regressed/arrested	Less than 25 or untestable	No formal learning; no socialized behavior	No object relations	May be bizarre or not

ministrator in a federal agency, dietetic aide, Air Force, tennis teacher, and optician. The level of employment was usually below educational attainment, and interpersonal idiosyncracies seemed to interfere with employment more than they had with school performance. Areas of specialization in college or graduate school included four in art, three in mathematics, two in meteorology, two in astronomy, two in engineering, two in business, and one each in horticulture, chemistry, music, psychology, theology, broadcasting, law, zoology, and labor relations. The field of academic study was often based on a long-standing obsessional interest. Eight of the adult subjects were married (13%) and three were parents (two fathers and one mother of both a natural and an adopted child). Five others were engaged or involved in long-term dating relationships.

Kanner and Rimland Ratings

Case material for each subject was rated on the E2 Rimland scale (Rimland, 1964) and on a "Kanner scale." For the Kanner scale, each of Kanner's five basic criteria ("profound withdrawal of contact from people, obsessive desire for the preservation of sameness, skillful relation to objects, retention of intelligent and pensive physiognomy, and mutism or the kind of language that does not seem intended for the purpose of

Table 2. Adolescent and Adult Outcome

Category	Adolescent rating (N =%)	Adult rating (N)	(%)
1. Normal-neurotic	38	23	38
2. Brittle-schizoid	30	21	34
3. Eruptive-schizoid	21	7	11
4. Schizophrenic	9	10	17
5. Passive-retarded (mild)	2	0	0
	100	61	100

Table 3. Young Adult Status

	Living at home (N)	Living in community (N)	Residential placement (N)	Total (N)	(%)
Regular job	6	23	0	29	47
Student	7	15	0	22	36
Sheltered job	4	0	2	6	10
Day hospital	0	1	0	1	2
No program	3	0	0	3	5
N	20	39	2	61	
%	33	64	3		100

communication'') (Kanner, 1949, p. 416), was rated as ''highly characteristic'' (3 points), ''somewhat characteristic'' (1 point), or ''not characteristic'' (-2 points), and the five ratings were then summed for a
total ''Kanner score.''

In this sample, there was a high correlation between the Kanner and
Rimland scores, ($r = .82$), with children who scored higher on the Kanner
scale tending toward more ''autistic'' Rimland ratings. However, most of
the children scored quite high on the Kanner scale, but yet fell toward the
schizophrenic side of the Rimland scale. It was quite possible for a child
to have a high Kanner rating and still be classified ''schizophrenic'' by the
Rimland scale.

High Kanner ratings were significantly correlated with social classes
1, 2, and 4 (professional, semiprofessional, and skilled blue collar). They
were significantly negatively correlated with class 3, nonprofessional
white collar. Fathers of higher Kanner scoring children had significantly
more abilities in mathematics, mechanics, and music than did fathers of
lower Kanner scoring children. Mothers of higher Kanner scoring children had significantly more abilities in mathematics, music, and spatial
relations than did mothers of lower Kanner scoring children. These results are suggestive of a possible genetic influence with an hereditary
predisposition towards right brain skills. The higher Kanner scoring children had a significantly greater number of preoccupations, especially with
mechanical objects, spinning objects, and round objects. There was no
particular relationship between overall outcome rating and Kanner and
Rimland scores except for a consistently significant trend for ''eruptive
schizoid'' outcomes to be associated with low Kanner scores and high
''schizophrenic'' Rimland scores.

There were 13 children who did not show obsessional preoccupations
at the time of their diagnostic evaluation. Eight of these were included in
the group of 13 subjects with the best social outcomes (married or dating
steadily), while the other five became schizophrenic. From this finding,
one could speculate that the development of obsessional preoccupations
may be an adaptive as well as a defensive response to massive anxiety, a
''normal'' and perhaps ''necessary'' stage which the schizophrenic children never achieved, but which the most normal children in the group had
already passed through by the time we saw them. (Some support for this
point is the fact that three of the five children who had preoccupations but
developed good social outcomes were under age three when seen for their
diagnostic evaluations.)

A somewhat parallel finding held for speech. The best outcome in
adolescence or adulthood, a normal-neurotic rating, was significantly correlated with a rating of ''somewhat excessive'' speech at the time of the
diagnostic evaluation. However, if excessive talking was ''very charac-

teristic" at the time of preschool evaluation, indicating considerable un-
bound anxiety, the subjects showed significantly more schizoid functioning in adolescence. (There were not enough adult subjects in this category
to compute significance, but the trend is in the same direction.)

Family Characteristics

The high percentage of parents with special abilities has already been
mentioned (51% of the fathers and 45% of the mothers). Both parents
were also often rated as having obsessive-neurotic personalities (36% of
the fathers and 31% of the mothers). Fathers were next most likely to be
rated "normal" (28%) or narcissistic (14%) while mothers were more
likely to be rated chronically depressed (22%) or borderline psychotic
(22%). Fathers' mental health was significantly correlated with outcome.
Fathers rated psychotic, borderline psychotic, or narcissistic (22%) had
significantly more children rated eruptive schizoid or schizophrenic in
adolescence and adulthood than the remainder of the fathers who were
rated depressed, character-disordered, obsessive, hysterical, or normal.
The relationship was in the same direction, but not statistically significant, for the 46 mothers rated psychotic, borderline psychotic, or
depressed.

There were 35 families in which the siblings showed no disturbance.
In these families, the atypical child was significantly more likely to be the
oldest child. There were 48 families in which there were siblings older
than the atypical child. In 73% of these families, one or more of the older
siblings had some emotional problems; in 50% of the families, the emotional problems of the older sib or siblings were rated as major. In contrast, there were 50 families in which the atypical child had younger
siblings. In only 32% of these families was one or more of the younger
siblings found to have emotional problems, and only 14% of these were
considered major. This highly significant difference in the incidence of
sibling disturbance suggests that, in many of these families, there were
difficulties in appropriate child care before the arrival of the atypical
child. One suggestive finding is that there was a significant difference in
the percentage of mothers of older disturbed siblings who themselves had
suffered a loss of an important person (66%) compared with the percentage of mothers of younger disturbed siblings who had suffered a significant loss (33%).

From a theoretical viewpoint, *disruptive changes during the first year
of life* should influence subsequent development. Half the children had
had one or more of these changes such as moves, parental depression, or
family crises, but neither type nor number was related to relative outcome
in this better outcome sample.

Intelligence Test Patterns

Most of the children were given serial intelligence tests, and results were available from latency or adolescence when they were fully cooperative. As a group, they obtained IQs from 78 to 158 with a mean of 108. There were WISC scores for 31 of the subjects, and of these, 23 (74%), showed a higher Verbal than Performance IQ with a mean difference of 19.5 points. For 15 of the 23 children, the Verbal IQ was substantially higher than the Performance IQ by 12 or more points. Only eight subjects (26%) had WISC Performance IQs higher than Verbal IQs, and in only three of these was the difference greater than 12 or more points. Of the 63 children with Stanford Binet tests, 34 showed evidence for poor visual-motor functioning, while 29 did not. (There was no relevant information on visual-motor performance in the records of 6 children.) Thus *poor visual-motor functioning was found in almost two-thirds of the total group*.

A review of the literature on verbal/performance differences in better functioning autistic and schizophrenic children shows similar findings. Kanner (1949) reported a breakdown of IQ scores for only three cases. These were Sally (Susan) V IQ 119, P IQ 98; Fred, V IQ 126, P IQ 104; and Edward, "high on verbal, mediocre in performance." Bender (1973) found verbal scores "usually higher" in 37 subjects who made a relatively good social adjustment as adults. Wechsler and Jaros (1965), in a study of WISC scores of 100 schizophrenic boys aged 8 to 12 with IQs from 80 to 120, found Verbal IQs at least 16 points higher than Performance IQs in 15% of the group, but only in 2% of a control group matched for age and IQ. This large a difference between Verbal and Performance IQs was especially likely to occur in 10- and 11-year-olds (16% and 36%). Halpern (1964) reported on a ten-year study of one schizophrenic boy whose WISC scores at 14 were V IQ 114 and P IQ 74. Asperger (in Van Krevelen and Kuipers, 1962) spoke of a lack of "good spatial orientation" in his latency aged subjects with "autistic psychopathy." Lockyer and Rutter (1970) found that there was no difference between Verbal and Performance IQs in their mildly retarded best outcome group with normal speech. Rutter *et al.*, 1967, state "It was particularly striking that some of the children who had been markedly delayed in their language development in early childhood became verbally very fluent, and in adolescence their intellectual skills were largely verbal and mathematical. The pattern was confined to a small subgroup of intelligent, clumsy, psychotic children" (p. 1189). Rutter (1970) also mentions that five of the Maudsley children who acquired a reasonable level of language comprehension had a mean V IQ of 81 and a mean P IQ of 74 on follow-up. The only case I was able to locate in the literature in which the child had average or better intelligence and a higher P IQ than V IQ was one of Fish's (Fish *et al.*, 1966). Frank had a V IQ of 115 and a P IQ of 121, but it should be noted that his poorest score was on Coding, which is the purest visual-motor task on the WISC.

This review of the cognitive functioning of the average or better IQ children suggests that as these children with better potential overcome their language handicap, their verbal skills improve significantly while their performance skills show less change. Although this could be partly due to the much greater emphasis that parents, teachers, and therapists place on verbal skills, it is also possible that many of these children have underlying difficulties in visual-motor function, and by extension, in general body integration. This is certainly not a new finding, but one that has been overlooked or underemphasized because of the more obvious nature of the language handicap in the early development of the children, and its continuing importance in the more severely impaired autistic children.

Visual-motor and body integration disturbances may have organic or emotional etiologies, or a combination of these. The 18 children in this sample who showed the highest verbal/performance discrepancy had significantly more EEG abnormalities than the remainder of the children. In the preschool period they were significantly more likely to be rated as talking excessively, and on adolescent follow-up were more likely to be rated eruptive schizoid.

These findings appear consistent with the considerable body of literature that posits perceptual inconstancy, and more specifically, vestibular disturbance, as an etiological agent, and with other clinical findings that could be interpreted in this light. Fish (1976) reports on several infants with both visual-motor and vestibular dysfunction who later became schizophrenic. Colbert and Koegler (1961) describe the quality of "vortical movement" in the drawings of childhood schizophrenics. More speculatively, the contents of the fantasies of many atypical children (for example those described by Ekstein, 1966; and Tustin, 1973) suggest preoccupation with the relation of the child's body to basic spatial and gravitational features of the environment.

The theoretical and experimental basis for vestibular disturbance has been documented in detail by Ornitz and Ritvo (Ornitz, 1970, 1971, 1974; Ornitz & Ritvo, 1968a, b), building on the earlier work of Schilder (1964), Bender (1947, 1956), Colbert *et al.* (1959), Rachman and Berger (1963), and Rosenblut *et al.* (1960). Another supporting piece of evidence may lie in pharmacological work. Campbell (1975) reported that diphenhydramine, a drug with anti-motion sickness properties, has been found particularly effective with schizophrenic children of high intelligence. A number of therapists have begun to report success with techniques that involve active body stimulation. These include DesLauriers and Carlson (1969) and Miller and Miller (1973). Techniques for direct vestibular stimulation have been developed with particular precision by Ayres (1972, 1974; Ayres & Heskett, 1972) who is training physical therapists and developmental occupational therapists in sensory-motor integration. In our own study, some of the best outcome subjects seemed to be aware of their need for better body integration, and invested much effort in self-set athletic goals, although these activities were usually more difficult for them than for their normal peers. Children with marked visual-motor problems who avoided physical activities appeared more schizoid on follow-up than those who engaged in sports.

A clinical vignette from our study may highlight some of these issues. This girl was brought to the Putnam Center by her mother at age two with the complaint that the child was afraid to ride on streetcars. By history she had screamed in her carriage from age two months and was afraid to be picked up. The mother was reassured, but returned when the child was four with the complaint that she could not copy straight lines properly. At this time, the girl appeared very anxious, talked excessively "at" people, and clearly showed a thought disorder. Her disturbance was considered related to her mother's anxiety and she was treated for many years by verbal psychotherapy at our Center and elsewhere, with only minimal improvement. Although at age 11 her WISC V IQ was 126 while her P IQ was 89, there was no specific attention to her visual-motor problems. On follow-up at age 33, she was having difficulty with the visual-motor aspects of her clerical job and was afraid to drive a car "because I might kill somebody." In this case and many others where a similar discrepancy between verbal and performance functioning was found, it was explained as due to intrapsychic conflict over the expression of aggressive impulses, and vestibular and other perceptual dysfunction was not considered as a possible etiological agent.

SUMMARY

This chapter explores some characteristics of a group of 100 children with early autistic behavior who developed average or better cognitive functioning. Many scored high on Kanner criteria and came from middle-class families in which the parents had significant skills in music, mathematics, mechanics, and spatial relations. Most of the children developed obsessional defenses which appeared adaptive in binding anxiety, but limited their interpersonal relationships as adults. The loosening of these obsessional defenses is a major challenge for the psychoanalytically oriented therapist. A high degree of emotional disturbance was found in the older siblings of these children, suggesting that parents may have had difficulties with child rearing in general. Early losses of important relationships suffered by the parents appear to be one major factor contributing to parenting difficulties. In this particular sample of atypical children, visual-motor and other types of performance skills often lagged behind verbal skills after the early language handicap was overcome. It is suggested that these lags may be due to underlying difficulties in body integration and that the findings support other research on the importance of vestibular functioning. Specific remediation of these difficulties through sensory-motor integration techniques appears to be a promising avenue for furthering optimal outcome.

ACKNOWLEDGMENT

This work was supported by Grant MH-18438 from the National Institute of Mental Health.

REFERENCES

Ayres, A. J. *Sensory integration and learning disorders.* Los Angeles: Western Psychological Services, 1972.

Ayres, A. J. *The development of sensory integrative theory and practice.* Dubuque, Iowa: Kendall/Hunt, 1974.

Ayres, A. J., & Heskett, W. M. Sensory integrative dysfunction in a young schizophrenic girl. *Journal of Autism and Childhood Schizophrenia*, 1972, 2, 174–181.

Bender, L. Childhood schizophrenia: Clinical study of one hundred schizophrenic children. *American Journal of Orthopsychiatry*, 1947, 17, 40–56.

Bender, L. Twenty years of clinical research on schizophrenic children, with special reference to those under six years of age. In G. Caplan (Ed.), *Emotional problems of early childhood.* New York: Basic Books, 1955.

Bender, L. Schizophrenia in childhood—its recognition, description and treatment. *American Journal of Orthopsychiatry*, 1956, 26, 499–506.

Bender, L. The life course of children with schizophrenia. *American Journal of Psychiatry*, 1973, 130, 783–786.

Brown, J. L. Adolescent development of children with infantile psychosis. *Seminars in Psychiatry*, 1969, 1, 79–89.

Brown, J. L. *Contributions of the James Jackson Putnam Children's Center.* Paper presented at the American Orthopsychiatric Association Institute: Infantile autism, then and now—thirty years retrospective and reconsideration, San Francisco, April 1974.

Campbell, M. Psychopharmacology in childhood psychosis. *International Journal of Mental Health*, 1975, 4, 238–254.

Colbert, E. G., Koegler, R. R., & Markham, C. H. Vestibular dysfunction in childhood schizophrenia. *Archives of General Psychiatry*, 1959, 1, 600–617.

Colbert, E. G., & Koegler, R. R. The childhood schizophrenic in adolescence. *Psychiatric Quarterly*, 1961, 35, 693–701.

Creak, M. Schizophrenic syndrome in children. Further progress of a working party. *Developmental Medicine and Child Neurology*, 1964, 6, 530–535.

DesLauriers, A. M., & Carlson, C. F. *Your child is asleep.* Homewood, Ill.: The Dorsey Press, 1969.

Ekstein, R. *Children of time and space, of action and impulse.* New York: Appleton-Century-Crofts, 1966.

Fish, B. Biological disorders in infants at risk for schizophrenia. In E. Ritvo (Ed.), *Autism: Diagnosis, current research and management.* New York: Spectrum, 1976.

Fish, B., Wile, R., Shapiro, T., & Halpern, F. The prediction of schizophrenia in infancy: II. A ten-year follow-up report of predictions made at one month of age. In P. Hoch & J. Zubin (Eds.), *Psychopathology of schizophrenia.* New York: Grune and Stratton, 1966.

Halpern, E. *Conceptual development in a schizophrenic boy: The use of traditional intelligence scales and Piaget tasks.* Paper presented at Canadian Psychological Association, Halifax, June, 1964.

Kanner, L. Autistic disturbances of affective contact. *Nervous Child*, 1943, *2*, 217–250.

Kanner, L. Problems of nosology and psychodynamics of early infantile autism. *American Journal of Orthopsychiatry*, 1949, *19*, 416–426.

Kanner, L., Rodriguez, A., & Ashenden, B. How far can autistic children go in matters of social adaptation? *Journal of Autism and Childhood Schizophrenia*, 1972, *2*, 9–33.

Lockyer, L., & Rutter, M. A five- to fifteen-year follow-up study of infantile psychosis: IV. Patterns of cognitive ability. *British Journal of Social and Clinical Psychology*, 1970, *9*, 152–163.

Miller, A., & Miller, E. Cognitive developmental training with elevated boards and sign language. *Journal of Autism and Childhood Schizophrenia*, 1973, *3*, 65–85.

Ornitz, E. M. Vestibular dysfunction in schizophrenia and childhood autism. *Comprehensive Psychiatry*, 1970, *11*, 159–173.

Ornitz, E. M. Childhood autism: A disorder of sensori-motor integration. In M. Rutter (Ed.), *Infantile autism: Concepts, characteristics and treatment.* London: Churchill-Livingstone, 1971.

Ornitz, E. M. The modulation of sensory input and motor output in autistic children. *Journal of Autism and Childhood Schizophrenia*, 1974, *4*, 197–215.

Ornitz, E. M., & Ritvo, E. R. Neurophysiological mechanisms underlying perceptual inconstancy in autistic and schizophrenic children. *Archives of General Psychiatry*, 1968, *19*, 22–27. (a)

Ornitz, E. M., & Ritvo E. R. Perceptual inconstancy in early infantile autism. *Archives of General Psychiatry*, 1968, *18*, 77–98. (b)

Rachman, S., & Berger, M. Whirling and postural control in schizophrenic children. *Journal of Child Psychology and Psychiatry*, 1963, *4*, 137–155.

Rimland, B. *Infantile autism.* New York: Appleton-Century-Crofts, 1964.

Rosenblüt, B., Goldstein, R., & Landau, W. Vestibular response of some deaf and aphasic children. *Annals of Otology, Rhinology and Laryngology*, 1960, *69*, 747–755.

Rutter, M. Autistic children: Infancy to adulthood. *Seminars in Psychiatry*, 1970, *2*, 435–450.

Rutter, M., Greenfeld, D., & Lockyer, L. A five to fifteen year follow up study of infantile psychosis: II. Social and behavioural outcome. *British Journal of Psychiatry*, 1967, *113*, 1183–1199.

Schilder, P. *The image and appearance of the human body.* New York: Science Editions, John Wiley, 1964. Pp. 114–118, 167.

Tustin, F. *Autism and childhood psychosis.* New York: Jason Aronson, 1973.

Van Krevelen, D. A., & Kuipers, C. The psychopathology of autistic psychopathy. *Acta Paedopsychiatrica*, 1962, *29*, 22–31.

Wechsler, D., & Jaros, E. Schizophrenic patterns on the WISC. *Journal of Clinical Psychology*, 1965, *21*, 288–291.

32

Follow-Up Studies

V. LOTTER

Follow-up studies of autistic children have two general aims: (a) to describe what such individuals are like in later life (i.e., the course and outcome of the autistic condition), and (b) to identify factors which are associated with differences in course and outcome. An important secondary consequence of such investigations is the contribution they may make to classification. Since definitions of autism vary somewhat and the numbers of afflicted children are few, comparability among studies is at the same time both uncertain and especially necessary. This is particularly the case as a wide variety of terms have been applied to psychoses beginning in infancy or early childhood. Thus, in some centers the terms "childhood schizophrenia," "child psychosis," or "atypical child" have been used for disorders fulfilling the usually accepted diagnostic criteria "autism." On the other hand, these same terms have also been used for conditions which clearly are not autistic. Accordingly, an initial survey was made of all follow-up studies which referred to autistic, schizophrenic, or psychotic disorders beginning in early childhood in order to determine which could be taken to refer to autistic children.

FOLLOW-UP STUDIES OF PSYCHOTIC CHILDREN

Description of Studies

Altogether reports were found for 25 studies which described some aspects of the later progress of children classed as psychotic in their early years. Four reports were excluded from further consideration: three be-

---

V. LOTTER · University of Guelph, Guelph, Ontario, Canada.

cause they used criteria which were difficult to evaluate or because they reported results unsystematically (Lourie *et al.*, 1942–1943; Clardy, 1951; Bennet & Klein, 1966) and one because it mainly reported case descriptions. A further set of observations only considered cases with successful outcomes (Kanner *et al.*, 1972). These are not included in the main tabulation but, because they describe some cases included in a more complete report (Eisenberg, 1956), these will be drawn on in a later part of this chapter. No longitudinal studies of single cases are included.

The remaining 21 reports are listed in Appendix I. The studies are arranged chronologically by date of publication and are identified by the name of the originating center, by author, and by an identifying number to facilitate reference in the text. The 21 reports, from England, North America, and various parts of Europe, deal with some 1380 children of whom 1278 were followed up for various periods. Wherever possible the numbers both at intake and follow-up have been included to give some indication of case loss. Where this information was not available intake and follow-up numbers in the tables are the same, but do not necessarily mean 100% successful case finding at follow-up. Often it was not clear whether subclassification of cases preceded follow-up, and the summary may misrepresent some studies in this respect. Blank spaces in the tables mean no information was available.

The diversity among these investigations, so often commented upon, is immediately evident. Six aspects are worth noting.

Size of Samples

Four studies had samples of 100 or more cases at follow-up (2, 3, 4, 18); six had 50–99 cases (1, 7, 13, 16, 19, 21); four had 30–49 cases (6, 10, 11, 17); and seven had fewer than 30 cases (5, 8, 9, 12, 14, 15, 20).

Sources of Samples

Samples were drawn between 1955 and 1968 from accumulated records except in one instance (8) where cases were identified in a population survey. Four sources of records could plausibly be distinguished: hospitals, mainly outpatients (1, 2, 3, 5, 7, 15, 20); hospital inpatient units (6, 14, 16, 18, 19); children's clinics (4, 9, 12, 13, 21); residential schools (10, 11, 17).

Case Definition

Of necessity in most studies case identification relied on records of children diagnosed in some appropriate way. In most instances information was given to indicate what criteria were used in original diagnoses, but only in rare instances were the behaviors of the children described. Attempts to classify types of diagnoses resulted in four somewhat overlapping categories: first, there was some version of the Creak (1961) criteria (3, 8, 9, 17); second, there was Kanner's infantile autism, or

Bender's pseudodefective type (1, 2, 5, 7, 9, 12, 13, 16, 18, 19); third, in some cases the diagnosis "psychosis" or "schizophrenia" was asserted without elaboration (6, 10, 14); and fourth there was a variety of mostly single instances of: "atypical child"; Bender's pseudoneurotic type; Ingram's Chronic Brain Syndrome plus psychosis; Cameron's criteria; "the classical criteria used in Europe"; and a more or less undefinable category of psychosis called "other."

The 14 studies that used the Creak (1961), Kanner (1943), or Bender (1956) (pseudodefective) criteria for at least some of their cases included 786 (61%) of the children followed up. In some of the studies these criteria defined small subgroups for which follow-up information was not separately given (e.g., 2).

Reporting

In six studies results were reported informally with little or no detail (3, 11, 14, 15, 16, 20); in a further four some detailed information was given but with no attempt at systematic analysis (1, 9, 10, 13); and in eleven, results were subjected to more or less detailed systematic analysis (2, 4, 5, 6, 7, 8, 12, 17, 18, 19, 21).

Subclassification

In twelve studies some form of subclassification was attempted. In five, subclassification was by syndromes (2, 5, 6, 9, 19); in two according to an "organic/nonorganic" distinction (15, 17); in two, according to severity (13, 18); in one each by language competence (12), amount of rated behavior (8), and by mixed criteria of brain damage and outcome (16). In most studies the subcategories were formed at intake, and maintained at follow-up. In one study (15) intake and follow-up subclassifications were different.

Age of Onset

In six studies an attempt was made to specify age of onset of the disorders (1, 2, 5, 7, 8, 15). In a further eleven, onset could be inferred from diagnostic category (e.g., infantile autism), or age of admission (e.g., "preschool" or "before five years"). In altogether twelve studies, all the children included seemed to be abnormal before the age of five years. The remaining nine studies (2, 5, 6, 9, 10, 14, 15, 17, 19) included a wider age range or the issue of onset could not be determined from the information given.

A similar diversity is evident in the reports of outcome.

Age at Follow-Up and Duration of Follow-Up

In six studies at least some of the children had reached 20 years of age (1, 2, 3, 4, 7, 16); in twelve, at least some were in their teens (5, 6, 8, 10, 13, 14, 15, 17, 18, 19, 20, 21), and in two all the children were preadolescent (9, 12). Insufficient information as to age at follow-up was given in one study (11).

Minimum follow-up duration varied from one to twelve years. In eight studies duration was at least five years (2, 3, 7, 8, 9, 15, 16, 21); in two, average duration only was reported (17, 18); and in five, clear information about duration was not given (4, 12, 14, 19, 20).

Outcome Criteria

Most of the studies reported the numbers or proportions of cases in custodial care, in various kinds of schools, and whether any were employed. Few reported deaths. Psychological measurement of any kind of outcome was very rarely reported. Several studies attempted general judgments of social or educational adaptations. Comparability of follow-up placement is made uncertain by the ambiguity of unqualified terms like "school," or "at home," or unspecified judgments of "improved."

Outcome Findings

The studies included here all refer to "psychotic" children, more or less narrowly defined. The question is whether outcome status is related to case definition. Comparison of diverse findings may help to identify reasons for the diversity. Three criteria for outcome were chosen to explore this question, and the studies were arranged in three categories according to whether they reported a relatively high, medium, or low proportion of cases with a good outcome (Table 1).

Seven studies reported much better outcomes than all the others. The probable reasons are various but instructive to explore.

For example, in study 2, only 51 of the 115 cases has an onset before the age of five, cases with a "doubtful" diagnosis were included, and in the entire sample, only three cases were diagnosed "infantile autism." In study 4, cases were selected according to their judged likelihood of responding to treatment, while in study 21, the 59 cases included 20 with "later" onset, apparently corresponding to Bender's definition of "pseudoneurotic" type. In study 13, seventeen severely affected children were excluded from the analysis of outcome, so that only 20% of the "mildly" or "moderately" affected children were found to be in institutions. In study 6, the criterion for "school" was wider than in the other studies. Studies 11 and 15 provide too little information to allow any inference as to reasons for the relatively better outcome.

Two studies (8, 19) found very few cases "employed or in normal school." Since in both studies the oldest children were not yet 20 years old (thus minimizing the employment category), and the proportions in institutions were around 50%, it seems likely that the numbers in "normal school" may have been influenced by local educational practices.

From all of this, several conclusions are possible. First, in every study, however cases were defined, the range of outcomes was very wide.

Table 1. Proportions of Cases with Various Outcomes in 21 Follow-Up Studies of "Psychotic" Children

Proportion of good outcomes	Employed or in normal school	Institutional care	General adjustment	
			Good	Poor/very poor
High	31–52%[a]	20–21%[d]	42%[g]	20–33%[j]
Medium	10–17%[b]	35–48%[e]	14–17%[h] ⎫	62–74%[k]
Low	1.7–3.4%[c]	50–66%[f]	5–10%[i] ⎭	

[a]Study 2, 4a, 6, 15, 20, 21. [b]3, 7, 10, 12, 16. [c]8, 19. [d]13, 15. [e]3, 5, 8, 10, 18. [f]1, 6, 7, 9, 14, 16, 17, 19. [g]11. [h]3, 7, 8, 20. [i]1, 18. [j]11, 13. [k]1, 3, 7, 8, 17, 18. NOTE: Not all studies provided information for all three kinds of outcome.

Second, children with a relatively later onset had a generally better outcome. Third, children whose disorder was initially relatively less severe had a generally better outcome. Fourth, allowing for the imprecision of outcome criteria, and excluding only gross identifiable selection factors, similar proportions in most studies were found in institutions at follow-up (50-60%), or had a poor or very poor outcome (62-74%). Fifth, since outcome did differ in the different studies according to identifiable selection factors, systematic exploration of those factors should provide some basis for prediction of outcome.

FOLLOW-UP STUDIES OF AUTISTIC CHILDREN

Because the groups are so heterogeneous and because so many of the studies are unsystematic very little can be concluded about prognosis from any consideration of "psychotic" children as a whole. To provide greater precision we tried to distinguish those studies which appeared most similar in their diagnosis of autism and which were roughly comparable in terms of duration of follow-up and collection/reporting of data.

Thirteen studies were excluded from further analysis for reasons listed in Table 2.

The eight studies selected, together with the basis for selection are given in Table 3. Further details are provided in Appendix II. We have called these subjects "autistic" because case definition was related to the Kanner or Creak criteria. However, it should be appreciated that, even so, there is variation between the studies in how they applied the criteria.

Outcome Status

Between 5% and 17% of the autistic children had a good outcome as assessed from judgments of overall social adjustment.* Conversely,

---

Good outcome: normal or near normal social life and satisfactory functioning at school or work. *Fair outcome:* some social and educational progress in spite of significant, even marked abnormalities in behavior or interpersonal relationships. *Poor outcome:* severe handicap, no independent social progress. *Very poor outcome:* unable to lead any kind of independent existence.

*Table 2. Studies Excluded from Further Analyses on Grounds of
Diagnostic Heterogeneity*

Study	N at follow-up	Reasons for exclusion
2	115	Only three cases infantile autism. No separate analysis possible.
4	241	Wide criteria (atypical child); cases selected for response to treatment.
5	23	Only three cases infantile autism. Other cases varied in onset age.
9	29	Treatment evaluation. Most information concerns chronic brain syndrome group.
10	43	Uncertain diagnostic criteria and onset age.
11	40	Uncertain diagnostic criteria and onset age. Incomplete information.
12	25	Follow-up duration too short.
13	71	Diagnosis includes many pseudoneurotic; analysis excludes severe cases.
14	27	Uncertain criteria and onset age.
15	24	Uncertain criteria and onset age. Subclassifications confounded with outcome.
16	50	Subcategories confounded with outcome.
19	57	Follow-up duration uncertain. "Discharge" criterion of outcome not useful.
21	59	Diagnosis includes many "pseudoneurotic." Separate analyses not possible.

61-74% had a generally poor outcome (Table 4). Few were employed, although with increasing age, the prospects for this seem to improve. Around half were placed in institutions; this proportion also tended to rise with age.

Obviously placement depends on age and also on local patterns of available services. In all the studies, very few subjects were reported to be employed. The best evidence of the proportion likely to become independent in this sense comes from the second follow-up of study 7, in which all the subjects were over 15 years old. In that group eight (13%) had jobs, mostly unskilled.

At the other extreme, the proportions of subjects receiving institutional care varied from 39% to 74%. In most of the studies, the proportion was about half, ranging from 43% to 54%. The effect of age on institutional placement is evident in study 7: At the first follow-up, 44% were so placed, while 6 years later, the proportion had risen to 54%.

Table 3. *Studies Included as Comparable*

Study	N at follow-up	Case definition	Follow-up (years)		Reporting	Factors explored	Comparison group
			Age	Duration			
1	63	Infantile autism	12–25	Mean 9 4–20	Semiformal	Yes	No
3	100	"Like Kanner's cases"[a]	up to 28	5–15	Informal	No	No
6	26	Assume Creak criteria	mean 15.1	most 4	Detailed	Yes	Yes
7	63	Most cases infantile autism[b]	(a) mean[c] 16.5 (b) 15–29	5–15 12–20	Detailed	Yes	Yes
8	29	Version of Creak criteria	16–18	8	Detailed	Yes	Yes
17	46	Creak criteria	?	Mean 7	Detailed	Yes	No
18	120	Infantile autism	Mean 10.7–12.5	Mean 5.5–8	Detailed	Yes	Yes
20	27	Infantile autism	Mean 7.8	?	Informal	Yes	No

[a] Creak (1951), p. 550.
[b] Rutter (1970), p. 437.
[c] (a) First follow-up; (b) second follow-up.

Table 4.   Outcome in Autistic Children

			Outcome		
Study	N at follow-up	Good	Poor or very poor	Employed	Institution
1	63	5%	73%	—	—
3	100	17%	73%	5%	39%
6	26	—	—	0	74%
7(a)	63	14%	61%	3%	44%
(b)	63	17%	64%	13%	54%
8	29	14%	62%	4%	48%
18	27	14%	—	—	—

Comparison Groups

Four of the selected studies compared the outcome of their autistic cases with groups of other children variously defined (Table 5). More of the comparison children were judged to have a generally better outcome (33%-44% good) than the autistic groups, and fewer were in institutions at follow-up. The one exception to this generalization, in study 6, can plausibly be attributed to selection of the comparison cases. The most convincing evidence that autistic children have a worse prognosis than other handicapped children comes from study 7, in which the comparison cases were most carefully matched with the index cases, and all but two were successfully followed up. In that group, 33% of the comparison subjects over 15 years old were employed, and 39% were in long-stay hospitals.

Prediction of Outcome

Many variables have been explored in an attempt to predict which autistic children will do well, and to find reasons why outcomes are so different. Three measures were found to be related to outcome in most of the selected studies: some measure of IQ, some measure of speech use, and some measure of severity (Table 6).

It is evident that some degree of prediction is possible using a number of different indices. For practical purposes which index is most useful depends on how readily, and how early, the predictor measures can be made, as well as the practical relevance of the outcome criterion.

In study 7, the IQ score was the best predictor of outcome. The predictive usefulness of the presence of useful speech at five years of age depended in part, it was argued, on intellectual level, because the more severely retarded child is less likely to develop speech. Estimates of IQ in young autistic children derive very largely from scores of nonverbal tests. A low score on such tests (probably indicating the child's best level of functioning) is thus associated with a lesser likelihood of developing useful speech, and a poor outcome. A high nonverbal score, with no subsequent language development is of no predictive value, whereas if language subsequently does develop, the nonverbal score is a useful guide to later general IQ scores (Rutter, 1970). It seems that some combination of speech and IQ may be a more useful predictor than either separately.

The main limitations of speech and IQ as predictors seem to be: (1) early assessment of autistic children requires very expert psychometric testing which may not always be practicable; (2) IQ alone predicts best only those with poor outcome; (3) "useful speech" cannot by definition of the variable be assessed before the age of 5 years [although Rutter (1970) has argued that comprehension of speech in a nonspeaking child is a useful guide]; (4) the combination of speech and IQ requires subtle judgments; (5) high initial (nonverbal) IQ scores correlate with later estimates of intelligence which may however not be closely related to social competence which is the main practical aim of prediction.

In the light of these reservations, the predictor found most efficient in study 18 is of considerable interest. Instead of using an indirect index, these workers chose to predict directly the educational and employment prospects of their subjects. At intake each child was rated on a 5-point scale (Normal, Borderline, Educable, Trainable, Subtrainable) using simple observational criteria of common school-related skills, including, for example, whether the child could be managed in class. Similar ratings were made at follow-up, with the addition of items involving postschool training and employment.

This "Work-School Status" scale was found to be the best predictor of how well the child would function educationally at follow-up. For all their subjects combined (i.e., autistic and comparison group children) prediction on the 5-point scale was exact for over half, and within one step in 89% of cases.

This scale appears to have a number of advantages over other attempts to find a way to predict outcomes in autistic children; it is simple, requiring no sophisticated measurement devices or testing procedures; its results are based on the most complete information; it predicts characteristics of practical relevance; it seems to be useful for preschool chil-

Table 5. Outcome in Variously Defined Comparison Groups

Study	N and definition of comparison group	Outcome			
		Good	Poor or very poor	Employed	Institution
6	21 Borderline psychotic	—	—	10%	52%
	25 Subnormal	—	—	4%	72%
7	61 Matched age, sex, IQ	33%	36%	22%	35%
8	21 Some autistic features	38%	43%	17%	14%
18	26 Subnormal	44%	28%	—	10%

Table 6. *The Most Efficient Predictors of Outcome*

Study	Predictor variable and discrimination		
	IQ	Speech	Severity
1	—	No useful speech at 5. Good/Fair vs. Poor/Very Poor ($p < .001$)	—
6	Test scores at intake and follow-up, r .74 ($p < .05$)	—	—
7	Mean IQ at intake. Good/Fair vs. Poor/Very Poor at follow-up ($p < .001$)	No response to sounds at intake. Good vs. Fair at follow-up ($p < .02$) No useful speech at 5. Good vs. Fair at follow-up ($p < .02$)	No. of symptoms at intake. Good vs. Fair ($p < .01$)
8	High, middle, low scorers at intake. Good, Fair, Poor, or Very Poor at follow-up, r .77 ($p < .001$)	Speech ratings at intake. Four outcome categories r .86 ($p < .001$)	Ratings based on IQ and developm. retardation. Four outcome categories r .85 ($p < .001$)
17	Unscorable at intake. Categories of ego status at follow-up (χ^2: $p < .09$)	No useful speech at intake. Categories of ego status at follow-up (χ^2: $p < .03$) No useful speech at intake. Institutional placement at follow-up (χ^2: $p < .02$)	Ego status at intake. Ego status at follow-up ($p < .02$) Ego status at intake. Institutional placement at follow-up ($p < .02$)
18	Mean IQ High, Middle, Low groups at intake, significantly different from mean IQ at follow-up.	Speech ratings at intake significantly related to work-school status at follow-up	Diagnostic categories High, Middle, Low groups significantly related to work-school status at follow-up

dren; and it requires little interpretive correction for changes in age.

The predictor variables described above were most effective in distinguishing between broad groups of relatively better or worse outcome. An attempt was made in study 18, by a close scrutiny of the 20 individual cases with the best outcome, to find reasons for their relatively better progress. Generally, there was found only a repetition of the results of the whole group. Examination of the two children in that study who were considered "normal" at follow-up likewise revealed no consistent differences between them and the other 18 with the best outcome.

Four further variables have been reported to be significantly associated with outcome: amount of time spent in school (studies 7 and 8); ratings of social behavior (study 18); ratings of social maturity (study 8); and developmental milestones (study 8). Most of these variables were correlated less strongly with outcome than the variables already described. The strong relation between schooling and outcome, on the other hand, is ambiguous: The most likely explanation seems to be that the children thought likely to benefit most from schooling were retained in schools longest.

For a number of variables, there were conflicting findings, i.e., for the same variable a relation with outcome was found in one study and not in another. These were: sex; some index of brain dysfunction or damage; and the category "untestable" child. Apart from the variable "sex," problems of measurement could plausibly explain these differences. Although sex was statistically related to outcome only in study 8 (and not related in studies 7 and 17), it has been pointed out that among the children with the best outcome, there are very few girls.

It is worth noting that the three main predictor variables were also highly correlated with outcome in "nonautistic" comparison children in study 8.

Variables Not Associated with Outcome

No relation could be found between outcome and: home background, type of onset, late development of fits (study 7); and birth weight, perinatal complications, age of onset, social class, family mental illness (study 8).

Course

The general picture arising from the statistical results described above is of a disorder with a relatively stable course. Several additional observations are of interest.

A sizable proportion of autistic children develop fits for the first time in adolescence or early adult life. The best evidence comes from study 7 in which by the time of the second follow-up, when the subjects were 15-29 years old, 28% of those not epileptic in early childhood had developed fits; and in none of these children had brain damage been thought probable at intake (Rutter, 1970).

The importance of other changes during adolescence has been noted by Kanner and Rutter. Kanner (Kanner *et al.*, 1972) paid special attention to the life histories of eleven autistic children seen by him who were most successful in later life. He observed that during adolescence their progress seemed to begin to emerge for the first time as different from those who had a poorer outcome. As he puts it, "It was not until the early to middle teens when a remarkable change took place. Unlike most other autistic children, they became uneasily aware of their peculiarities and began to make a conscious effort to do something about them. This effort increased as they grew older" (Kanner *et al.*, 1972, p. 29).

In study 7, a small number (7 out of 64) of autistic children showed a progressive deterioration in adolescence, characterized by general intellectual decline. In 3, the deterioration was accompanied by epileptic fits (Rutter, 1970).

Study 17 was the only one to attempt a fairly full analysis at two different follow-up ages. At the first, conducted when the children were discharged from the center, proportionately more "nonorganic" than "organic" children had improved in ego-status to a "mild" level. At final follow-up, this "relationship between neurological integrity of the child and his adjustment level was no longer evident. This shift in relationship reflected the new condition of "nonorganic" children who declined in follow-up, and on the other hand, the number of "organic" children who improved" (Goldfarb, 1970, p. 58).

Apart from corroborating observations of shifts in the course of the disorders, these results, with those relating to the late onset of fits, point to some of the difficulties in attempts to relate measures of brain damage to outcome.

Substantial correlations between intake and outcome measures of IQ suggest a stable course: children with low scores in early childhood tend to have low scores at follow-up. In study 18, verbal and nonverbal scores were separated; it was found that mean Verbal IQs increased during follow-up in all diagnostic groups, while mean Performance IQs decreased in the Middle and Low autistic groups only. "The overall General IQ tended to remain the same. Thus, the stability of IQ over time depends on which IQ was measured and which group of children tested" (DeMyer *et al.*, 1973, p. 228).

Grossly impaired social relationships—described by such terms as "autism" or "extreme aloneness"—play an important part in the definition of these children. As they grow older, there is usually some improvement in social abilities, while intellectually, there is no improvement (study 7, 18). The belief that the basic problem is social withdrawal, and that a general improvement should result from amelioration of this symptom, is not consistent with these findings.

CONCLUSIONS

Only superficial and uncertain generalizations could be made from a review of all follow-up investigations of "psychotic" children, because the studies were too variable in case definition, intentions, methods, and data presentation. Instead, attention was focused on a subgroup of studies which were more nearly comparable in methodology and in dealing with children most of whom appeared to meet the main diagnostic criteria for infantile autism.

One important result of the follow-up work is the systematic demonstration that the range of outcomes is very wide; a small proportion of autistic children "emerge" sufficiently to be socially and vocationally independent, but for most the prospects are poor.

Moreover, the broad outlines of probable outcome can be predicted statistically from observations made at a quite early age. Outcome of the 50% or so severely affected children is most consistently predictable, most of them remaining severely handicapped in all respects up to the age of about 30 years, which is as far as the studies go. There seems to be little expectation of dramatic later improvement in their condition since nearly all now live in long-term institutions. Outcome in the remaining children was more variable; among these there were a few who achieved social independence, although attempts to find differences in their early histories which would anticipate this have not been successful. At the same time, evidence of changes in course as late as adolescence underline the importance of follow-up investigations continuing at least to early adulthood.

Probably the most fruitful result of the follow-up work is the impetus it gives to renewed attempts at subclassification for purposes of research into causes. For example, evidence of some kind of brain dysfunction or damage is much stronger in severely retarded autistic children than it is in those who at follow-up do relatively better. However, any systematic study of subgroups will mean dealing with even smaller numbers of children; which, in turn, makes even more urgent the need for the accumulation of findings among different researchers. The follow-up results reviewed here make it increasingly difficult to talk meaningfully about "autistic children" without qualification. Even general estimates of likely outcome are very different for statistically definable subgroups. A welcome move in the direction of uniformity has been made by the WHO study group on child psychiatry (Rutter *et al.*, 1969). It seems likely that the development of a more highly differentiated reporting scheme would make a greater contribution to case comparability than further attempts to refine behavioral criteria alone can do.

Appendix I.   Follow-Up Studies of Psychotic Children: Description of Intake

No.	Name of study	Syndrome	Number of cases
1(a)	Johns Hopkins Hospital, Baltimore (Eisenberg, 1956)	Infantile autism[a]	80
(b)	Johns Hopkins Hospital, Baltimore (Kanner, 1971)	Infantile autism[f]	11
2	Dept. of Child Psychiatry, Uppsala (Annell, 1963)	Child psychosis	62 schiz. syndr. 53 other psych.
3	Great Ormond St. Hospital, London (Creak, 1963a, b)	Schizophrenic syndrome of childhood	100
4(a)	Putnam Children's Center, Boston (Brown, 1963)	Atypical dev. (inf. psychosis)	135
(b)	Putnam Children's Center, Boston (Brown, 1969)	Atypical dev. (inf. psychosis)	241[b]
5	Lepinlahti Hosp., Helsinki (Alanen, 1964)	Child psychosis	3 inf. aut. 9 ch. schiz. 5 other psych. 6 reactive
6	Smith Hospital, Henley, England (Mittler, *et al.*, 1966)	Child psychosis	27
		Borderline psychosis	21
		Subnormal	25
7	Maudsley Hospital, London (Rutter, *et al.*, 1967) (Rutter, 1970)	Child psychosis	63 psychotic
			63 subnormal
8	Middlesex, Surrey, England (Lotter, 1974a, b)	Autistic	15 nuclear group 17 mixed group
			22 nonautistic
9	M.R. Clinic, Nebraska Psychiatric Institute (Menolascino & Eaton, 1967)	Child psychosis	22 CBS[d] plus psychoses 5 ch. schiz. 2 inf. aut.

[a]Kanner criteria.   [b]Includes many of the cases in 4(a).   [c]Age at survey.

Criteria and subclassification	Number followed up	Sex M	Sex F	Age admitted or diagnosed	Age onset
Known at least 4 yr, and at least 9 yr old; 17 c/s untraceable.	63	50	13		Infancy
Kanner's original 11 cases.	9	8	3	2–8 yr	Infancy
All admissions 1948–57 with psychotic syndrome. Include doubtful cases. Various criteria.	62￼ 53	94	21}	18 <7 yr } 97 7–11 yr}	11<2 yr 40 2–5 yr 64 5–11 yr
Admissions since 1945—not clear whether these are *all* admissions, or selected.	100	80	20		
All children admitted to the Center, and diagnosed inf. psychoses. Admission according to likely response to treatment.	129	105	31	Preschool	
All children diagnosed as "atypical" dev.; excl. those <9 yr old. Criteria incl. Kanner, Rank, Mahler, and Bender.	241			Preschool	
Clinic admissions 1952–60. All cases diagnosed psychotic <12 yr of age. Excl. neurological disease or cerebral trauma.	3 9 5 6	0 4 2 5	3 5 3 1	Inf. aut.—2–6 yr Schiz.—14 yr	<2 yr 3–6 yr <6 yr 2–12 yr
All children "firmly" diagnosed psychotic on admission, and subsequently discharged.	26			Mean 7.4 yr	
Seriously disturbed, with some psychotic features on admission; and subsequently discharged.	21			Mean 8.4 yr	
Subnormals (IQ <80), no psychotic features on admission, and subsequently discharged.	25			Mean 7.1 yr	
Diagnoses of child psychosis; schiz. syndrome; autism. Criteria of Cameron, Potter, and Anthony, 1950–58. Includes all admissions.	63	51	12	24 <5 yr 22 5–7 yr 17 7–11 yr	45 <2 yr 13 2–2½ yr 5 2½+
Comparison group of nonpsychotic referrals matched on age, sex, IQ.	61	51	12		
Children with "most" rated autistic behavior, selected from 54 children identified in an epidemiological survey as showing autistic behavior, 1963–64.	}29	11 12	4 5	8–10 yr[e]	10 <30 mo 19 2½–3 yr 3 3–4½ yr
Comparison group: survey children with some signs of the syndrome who failed to reach the behavioral criterion.	21	13	9	8–10 yr	
Selected from 616 admissions; criteria from Ingram, Creak, and Kanner.	22￼ 5 2 }	10 (2 c/s later excl.)	22}	Mean 4⁸⁄₁₂ All <8 yr	

[d]Chronic Brain Syndrome.   [e]Brain damage.   [f]Included in 1(a).

Appendix I.   Follow-Up Studies of Psychotic Children: Description of Intake (Cont.)

No.	Name of study	Syndrome	Number of cases
10	Children's Mental Health Center, New York (Gittelman & Birch, 1967)	Child schiz.	97
11	Orthogenic School, Chicago (Bettelheim, 1967)	Inf. autism	46
12	Bellevue Hospital, New York (Fish, *et al.,* 1968)	Child schiz.	32
13	West End Creche, Toronto (Havelkova, 1968)	Child psychosis	25 mild 29 moderate 17 severe
14	Bradley Hospital, Rhode Island (Davids *et al.,* 1968)	Child schiz.	10 ch. schiz. 17 personality disorder
15	Medical Academy, Gdansk, Poland (Sulestrowska, 1969)	Child schiz.	11 no B/D[c] 10 suspect B/D 9 certain B/D
16	Bellevue Hospital, New York (Bender, 1970)	Child schiz. with autism	50
17	Ittelson Center (Goldfarb, 1970)	Child schiz.	29 organic 19 nonorganic
18	Carter Memorial Hospital, Indiana (DeMyer *et al.,* 1973)	Autism	22 high aut. 53 mid aut. 51 low aut. 36 subnormal
19	Winnebago Hospital, Wisconsin (Treffert *et al.,* 1973)	Inf. autism	33
		Psychosis and autistic features	9
		Psychosis and organic complication	15
20	St. Pierre Hospital, Brussels (Beeckmans-Balle, 1973)	Inf. autism	27
21	West End Creche, Toronto (Rees & Taylor, 1975)	Child psychosis	59

[a] Kanner criteria.   [b] Includes many of the cases in 4(a).   [c] Age at survey.

Criteria and subclassification	Number followed up	Sex M	Sex F	Age admitted or diagnosed	Age onset
All admissions to League school over 9-year period. Study concerned only 44 former pupils.	43	79	18	Age at study 4½–20 yr	
All autistic children the school had worked with. Excl. 6 as too young or otherwise unsuitable or unavailable.	40	30	16		
All clinic admissions 1961–66. Correspond to Bender's "pseudodefective" category. Excl. children with focal CNS abns., or fits. Excl. 4 youngest children.	25	Ratio 3:1		<5 yr Mean 3.4 yr	
Admissions to treatment center with 7 additional cases from Children's Hospital. Criteria from Kanner, Bender, and Creak: autism or pseudoneurotic.	25 29 17			Preschool	
Discharges from hosp. unit. Excl. 5 girls; divide total group on basis of psychiat. diagnosis.	10} 17}	27	0	Mean 7.2 yr Mean 10.3 yr	
Cases with "verified" diagnosis of child schiz., admitted 1948–68. Excl. 62 c/s for various reasons.	24	23	7	<10 yr	87% preschool
Selected from 200 ch. schiz. cases as meeting criteria of infantile autism, 1935–50.	50	43	7	<12 yr	
All schiz. children discharged from school over 10 yr. Referral diagnosis plus concurring psychiatric diagnosis of psychosis. Excl. sensory and motor dysfunction, epilepsy, and mental retardation.	29 17	37	11	5–9 yr	
All referrals 1954–69 diagnosed as inf. autism. Details given only for children successfully followed-up.	21 52 47	male 71.4% male 67.3% male 74.5%		Mean 71.4 mo Mean 67.3 mo Mean 74.5 mo	
Comparison group, consecutive referrals 1962–69.	26	male 61.5%		Mean 61.5 mo.	
All 57 children admitted to the hospital before 12 yr of age, 1961–71. Inf. aut. diagnosis involved early onset and excl. organicity. The second group had later onset.	33	29	4	Mdn 6.6 yr (4–11)	"Early"
	9	9	0	Mdn 11 yr (7–12)	"Later"
	15	11	4	Mdn 9.5 yr (4–12 yr)	
Examined at hospital; diagnosed infantile autism after Kanner. Criteria explicitly stated.	27	21	6		
All admissions 1956–62 who were available for follow-up. No details of unavailable cases.	59	41	18	18 mo–5 yr mean 3 9/12	

d Chronic Brain Syndrome. *e* Brain damage. *f* Included in 1(a).

Appendix II. Follow-Up Studies of Autistic Children: Description of Outcome

Study	N successful follow-up and subcategories	Age at follow-up	Duration of follow-up (yr)	Sources of information at follow-up	Deaths (and age)	Work or university
1(a)	63 inf. aut.	13 <12 50 12–25	4–20 Mean 9	30 observed; records and letters		
(b)	9 inf. aut.	29–39	26–32	Records, reports, informal contact	1 (29yr)	1 bank teller 1 machinist 1 ward orderly
3	100 schiz. syndrome	Oldest 28	5–15	Records and informal contact	1 (10yr) 1 (25yr) 1 (28yr)	3 unskilled work, special consideration 2 normal work 1 college student
6	26 ch. psychosis	Mean 15.2	In most cases over 4 yr since discharge	Observation and records		0
	21 borderline	Mean 16.8				2
	25 subnormal	Mean 16.1				1
7(a)	63 psychotic	Mean 15.6	5–15 Mean 9.7	Observation, IQ tests, parent interview, behav. ratings	None	2 working
	61 comparison group	Mean 16.5	5–15 Mean 10.2		None	14 working
(b)[a]	63 psychotic	15–29 Mean 24.7	12–20	Postal questionnaire to parents and hospital records	2	1 computer oper. 1 clerk 6 unskilled work
8	29 autistic	16–18	8	27 observed plus records and parent interview in most cases	1 (10yr)	1 unskilled
	21 comparison	16–18	8	14 observed plus records and parent interview in most cases		5 unskilled
17	29 organic		Mean 7	36 observed; 10 records and family interview		
	17 nonorganic					
18	21 high aut.	Mean 149.6 mos	Mean 67.6 mos	94 observed; 52 telephone interviews of parent and reports	None	
	52 mid. aut.	Mean 149.8 mos	Mean 95 mos			
	47 low aut.	Mean 131.4 mos	Mean 70.7 mos			
	26 subnormal comparison group	Mean 118.8 mos	Mean 60.6 mos			
20	27 inf. aut.	Mean 7–8 (1 adult, the balance 1–14)		All (?) reexamined in clinic at follow-up		

[a] Second follow-up, same group.

Normal schooling	Special school/ training	At home unoccupied	Insti-tution	N outcome status	Late onset fits	Factors related to outcome
29 at home with parents; 34 in full-time residential settings				3 good 14 fair 46 poor		Useful speech at 5 yr $p < .001$
		0	5		2	
5 normally 1 pvt tuition 5 slow	41		39	17 recovered/ significantly improved 10 some improvement 73 poor outcome	7	
8 (30%) attended some kind of school for at least 12 months			20 (74%)	30% improved (job or school)		IQ intake/outcome $r = 0.74$ ($p < .05$)
12 (57%) attended some kind of school for at least 12 months			11 (52%)	57% improved (job or school)		IQ intake/outcome $r = 0.34$ (NS)
5 (20%) attended some kind of school for at least 12 months			18 (72%)	20% improved (job or school)		IQ intake/outcome $r = 0.49$ ($p < .05$)
3	21	9	28	9 good 16 fair 8 poor 30 very poor	10	IQ scores $p < .001$, Speech variables $p < .02$, Severity (symptom score) $p < .01$, Schooling at least 2 yr.
7	16	2	22	20 good 19 fair 7 poor 15 very poor		
3 at school or further education	12	3	34	17% good one-sixth fair 64% poor/ very poor	further 8	
	12	2	14	14% good 24% fair 14% poor 48% very poor	3	Significant relation to outcome: speech, IQ, SQ, severity, milestones, sex, neurol. impairment, years of schooling ($p < .05$ to $p < .001$)
1	8	3	4	38% good 19% fair 24% poor 19% very poor	2	All above relations significant except milestones, sex, and neurol. impairment.
14 at home			15	10 normal–mild		Main predictor was ego status on admission.
10 at home			9	5 normal–mild		
85.7% at home or seeking placement			14.3%	10% good 16% fair 24% poor 50% very poor		Best predictor of educ. outcome was performance on work/school items at initial examination. Also significantly related to o/c ratings were: IQ; severity; social and speech ratings; brain dysfunction.
46.0% at home or seeking placement			54.0%			
58.3% at home or seeking placement			41.7%			
90.3% at home or seeking placement			9.7%	44% good 28% fair 24% poor 4% very poor		
				4 good 11 mixed 6 poor	of 21 cases 5 yr or older	No detailed examination of factors. Suggests level of IQ is most important factor related to severity.

REFERENCES

Alanen, Y. O., Arajarvi, T., & Viitamaki, R. O. Psychoses in childhood. *Acta Psychiatrica Scandinavica*, 1964, *40* (Suppl. 174).

Annell, A. -L. The prognosis of psychotic syndromes in children. *Acta Psychiatrica Scandinavica*, 1963, *39*, 235–297.

Beeckmans-Balle, M. Le syndrome d'autisme infantile. *Acta Psychiatrica Belgica*, 1973, *73*, 537–555.

Bender, L. Schizophrenia in children—its recognition, description and treatment. *American Journal of Orthopsychiatry*, 1956, *56*, 499–506.

Bender, L. The life course of children with autism and mental retardation. In F. J. Menolascino (Ed.), *Psychiatric approaches to mental retardation*. New York: Basic Books, 1970.

Bennett, S., and Klein, H. R. Childhood schizophrenia: 30 years later. *American Journal of Psychiatry*, 1966, *122*, 1121–1124.

Bettelheim, B. *The empty fortress: Infantile autism and the birth of the self.* New York: Free Press, 1967.

Brown, J. L. Follow-up of children with atypical development (infantile psychosis). *American Journal of Orthopsychiatry*, 1963, *33*, 855–861.

Brown, J. L. Adolescent development of children with infantile psychosis. *Seminar in Psychiatry*, 1969, *1*, 79–89.

Clardy, E. R. A study of the development and course of schizophrenia in children. *Psychiatric Quarterly*, 1951, *25*, 81–90.

Colbert, E. G., & Koegler, R. R. The childhood schizophrenic in adolescence. *Psychiatric Quarterly*, 1961, *35*, 693–701.

Creak, M. Psychoses in childhood. *Journal of Mental Science*, 1951, *97*, 545–553.

Creak, M. Schizophrenic syndrome in childhood. *Developmental Medicine and Child Neurology*, 1961, *3*, 501–504.

Creak, M. Childhood psychosis: A review of 100 cases. *British Journal of Psychiatry*, 1963, *109*, 84–89. (a)

Creak, M. Follow-up cases (100) of schizophrenic syndrome of childhood. In Intervento alla Tavola Rotonda sul tema: *Catamnesi delle Psicosi Infantili* al II Congresso Europeo di Pedopsichiatria, Roma, 1963. Vol. 1. Assisi: Tipografia Polzuincola S. Maria Degli Angeli, 1963. (b)

Davids, A., Ryan, R., & Salvatore, P. D. Effectiveness of residential treatment for psychotic and other disturbed children. *American Journal of Orthopsychiatry*, 1968, *38*, 469–475.

DeMyer, M. K., Barton, S., DeMyer, W. E., Norton, J. A., Allen, J., & Steele, R. Prognosis in autism: A follow-up study. *Journal of Autism and Childhood Schizophrenia*, 1973, *3*, 199–246.

Eisenberg, L. The autistic child in adolescence. *American Journal of Psychiatry*, 1956, *112*, 607–612.

Fish, B., Shapiro, T., Campbell, M., & Wile, R. A. A classification of schizophrenic children under 5 years. *American Journal of Psychiatry*, 1968, *124*, 1415–1423.

Gittelman, M., & Birch, H. G. Childhood schizophrenia: Intellect, neurologic status, perinatal risk, prognosis and family pathology. *Archives of General Psychiatry*, 1967, *17*, 16–25.

Goldfarb, W. A follow-up investigation of schizophrenic children treated in residence. *Psychosocial Process*, 1970, *1*, 9–64.

Havelkova, M. Follow-up study of 71 children diagnosed as psychotic in preschool age. *American Journal of Orthopsychiatry*, 1968, *38*, 846–857.

Kanner, L. Follow-up study of 11 autistic children originally reported in 1943. *Journal of Autism and Childhood Schizophrenia,* 1971, *1,* 119–145.

Kanner, L. Autistic disturbances of affective contact. *Nervous Child,* 1943, *2,* 217–250.

Kanner, L., Rodriguez, A., & Ashenden, B. How far can autistic children go in matters of social adaptation? *Journal of Autism and Childhood Schizophrenia,* 1972, *2,* 9–33.

Lotter, V. Social adjustment and placement of autistic children in Middlesex: A follow-up study. *Journal of Autism and Childhood Schizophrenia,* 1974, *4,* 11–32. (a)

Lotter, V. Factors related to outcome in autistic children. *Journal of Autism and Childhood Schizophrenia,* 1974, *4,* 263–277. (b)

Lourie, R. S., Pacella, B. L., & Piotrowski, Z. A. Studies on the prognosis in schizophrenic-like psychoses in children. *American Journal of Psychiatry,* 1942/3, *99,* 542–552.

Menolascino, F. J., & Eaton, L. Psychoses of childhood: A five year follow-up study of experiences in a mental retardation clinic. *American Journal of Mental Deficiency,* 1967, *72,* 370–380.

Mittler, P., Gillies, S., & Jukes, E. Prognosis in psychotic children: Report of a follow-up study. *Journal of Mental Deficiency Research,* 1966, *10,* 73–83.

Rees, S. C., & Taylor, A. Prognostic antecedents and outcome in a follow-up study of children with a diagnosis of childhood psychosis. *Journal of Autism and Childhood Schizophrenia,* 1975, *5,* 309–322.

Rutter, M. Autistic children: Infancy to adulthood. *Seminars in Psychiatry,* 1970, *2,* 435–450.

Rutter, M., Greenfield, D., & Lockyer, L. A five to fifteen year follow-up study of infantile psychosis. II. Social and behavioral outcome. *British Journal of Psychiatry,* 1967, *113,* 1183–1199.

Rutter, M., Lebovici, S., Eisenberg, L., Sneznevskij, A. V., Sadoun, R., Brooke, E., & Lin, T-Y. A tri-axial classification of mental disorders in childhood. *Journal of Child Psychology and Psychiatry,* 1969, *10,* 41-61.

Sulestrowska, H. Studies on the causes of diagnostic and prognostic difficulties in childhood schizophrenia. *Polish Medical Journal,* 1969, *8,* 1505–1515.

Treffert, D. A., McAndrew, J. B., & Dreifuerst, P. An inpatient treatment program and outcome for 57 autistic and schizophrenic children. *Journal of Autism and Childhood Schizophrenia,* 1973, *3,* 138–153.

33

Developmental Issues and Prognosis

MICHAEL RUTTER

In science the best test of the validity of any research finding is *not* whether the results reach statistical significance but rather whether they can be replicated by independent investigators using similar methods with comparable new populations of subjects. On this criterion our knowledge on the prognosis of autism rests on firm foundations. The results of prospective and follow-up studies of well-diagnosed groups of autistic children are highly consistent. Lotter (Chapter 32) has admirably summarized the key findings and outlined many of the outstanding issues.

As he has shown, from as early as age 4 or 5 years, it is now possible to provide a reasonably good actuarial prediction of the likely outcome in broad social terms. Moreover, it has been clearly shown that the most accurate predictions are based on *cognitive* variables—especially IQ, language skills, and school performance. Autistic children vary extremely widely on these measures and their social outcome is comparatively varied. At one extreme are those individuals whose autism is accompanied by severe global mental retardation (as shown by a nonverbal IQ below 50 or an inability to score even after repeated testing under good conditions). Almost all of these autistic persons remain grossly handicapped, the great majority never gain communicative language, they are unable to hold jobs, and most end up in some form of long-stay institution (Rutter, 1970). Furthermore, many develop epileptic fits during adolescence. A gloomy outlook indeed.

However, the picture at the other extreme with autistic children of

MICHAEL RUTTER · Department of Child and Adolescent Psychiatry, Institute of Psychiatry, De Crespigny Park, Denmark Hill, London SE5 8AF, England.

normal nonverbal intelligence is quite different. We found that most made worthwhile educational progress, most become reasonably proficient in their use of spoken language, and about half went on to productive paid employment in the open labor market (Bartak & Rutter, 1976). Only a few developed seizures. Brown's findings (see Chapter 31) on a more broadly defined group of "atypical" children are equally encouraging.

While IQ scores provide the best single means of prediction, assessments of prognosis can be usefully improved by taking into account measures of language, play and school performance as well as nonverbal intelligence (see Chapter 29). Taken together, these findings mean that it is possible to give parents a rough and ready estimate of the range of developmental progress which is likely to be possible for their autistic child. Even so, there are considerable limitations to our prognostic assessments. First, the predictions remain quite crude with much variation yet to be explained. The outcome for mildly retarded or normally intelligent autistic children is less easy to predict than the outcome for the more severely handicapped. Some do much better than expected and some do much worse. Why? Second, while we can provide statistical forecasts, this is not enough. We need also to know *how* development takes place and *why* things occur in the way that they do. Third, actuarial estimates based on *past* experience are not necessarily good guides to the improvement likely to be possible for today's children with the help of currently available methods of treatment. The purpose of this chapter is to consider some of the practical and theoretical questions which arise from these issues.

PROGNOSIS FOR THE MILDLY HANDICAPPED

Many uncertainties remain with regard to the outcome of autistic children of normal nonverbal intelligence or at most mild mental retardation. These include: why speech sometimes fails to develop; why lack of speech is associated with such a poor social prognosis; and the effect of visuo-spatial defects.

Failure to Develop Spoken Language

In general, an autistic child's score on nonverbal tests of intelligence provides a fair guide to his chances of developing communicative skills in spoken language. Very few severely retarded autistic children gain a useful level of spoken language whereas the great majority of those with a nonverbal IQ in the normal range do so (Rutter, 1970). Nevertheless, some of the intellectually able group never learn to speak. It is important to consider what differentiates this subgroup and how the child who will not gain language can be identified at an early age.

Systematic research data are lacking and are much needed. However, clinical experience provides some leads on variables to be studied. In the children we have seen, the IQ as such is of little predictive value within the normal range. Some of those who have failed completely to gain spoken language have had very high nonverbal scores. On the other hand, the child's qualities of play do seem to provide a useful guide. The prognosis is less good if, in spite of a nonverbal IQ in the normal range the child does *not* show spontaneous constructive activities (e.g., making recognizable objects with "leggo" or building blocks); does *not* show a varied, appropriate, and meaningful use of toys (e.g., making toy cars race each other, bathing dolls, etc.); does *not* actively explore his environment (finding out how toys work, how to tune the radio, etc.); and does *not* imitate the actions of others (e.g., copying his parents when they use the vacuum cleaner or mow the lawn). The relative importance of these different aspects of play is not known but all have been thought to be involved in the early development of language (Sheridan, 1969; Rutter, 1972).

The child's level of language comprehension may also be used as an indicator of the child's chances of gaining useful skills in spoken language. Clinical assessment is difficult because the intelligent child often picks up cues from gesture, social context, and what is usually expected of him. It is essential when assessing language comprehension to check that the situation is not one which offers nonverbal clues as to what is being asked. More systematic quantified measurement of the child's understanding of language is available through the Reynell Developmental Language Scales (1969).

Lastly, but probably to a lesser extent, some guide to language is provided by the child's use of babble. If the nonspeaking autistic child is using a rich and varied range of babble with speech cadences and communicative intent, then the development of unspoken language is probably not far off. If there is not much babble that does not necessarily mean a bad prognosis but it is another indicator that the child is probably still some way off learning to talk.

Of course, the value of all these items in predicting language development has not yet been properly evaluated. Accordingly, the suggestions given must be treated as provisional and tentative. On the other hand, it seems highly probable that the predictors will be found somewhere in the areas of play, imitation, language comprehension, and babble.

Effects of Lack of Speech

Autistic children who fail to gain useful spoken language tend to have a poor social prognosis. In my experience, very few have made a good social adjustment and none has obtained a regular paid job. It is necessary to consider why this is so. It might be thought that a lack of speech is such a handicap in itself that it provides sufficient explanation. But it does not. A follow-up study of normally intelligent *aphasic* children who failed to gain useful speech in spite of attending a good special school showed that most made a fairly satisfactory social adjustment and most were able to hold a steady job (Moor House School, 1969). Similarly, deaf mutes may become well-functioning adults in spite of the fact that they cannot speak. A lack of speech as such need not prevent a good social outcome. The important differences between aphasia and deafness is that the nonspeaking autistic adult lacks far more than speech. Usually there is a global problem in communication (the nonspeaking aphasic adult often has considerable skills in gesture and written language whereas the autistic does not) and in thinking. As a result the language impairment is often associated with a variety of social and behavioral problems in addition to the lack of speech.

It should be added that even in those who go on to acquire spoken language, the extent of the language impairment serves as a prognostic indicator. Our own studies (Rutter, 1970) and those of others (Kanner *et al.*, 1972) have shown that the best outcomes are largely restricted to children who never showed a profound lack of response to sounds, who gained useful speech by age 5 years, and whose phase of echolalia proved transient.

Visuo-Spatial Deficits

Most of the prognostic indicators in autism work in the expected direction. That is, the better the child's IQ, language, or play, the better the prognosis. However, there is one which seems to work the opposite way with a *deficit* associated with a *good* outcome. Brown (Chapter 31) has emphasized that poor visual-motor functioning was characteristic of two thirds of her group of better functioning atypical children and suggested that sensorimotor integration techniques may therefore be important in treatment. There are difficulties in interpreting this finding in view of the diagnostic heterogeneity of the sample, the uncertainty as to whether the visuo-motor deficit applied to autistic children (the one case history given does not sound like Kanner's syndrome), and the lack of evidence as to whether or not visuo-spatial deficits were associated with a good outcome *within* the sample studied. Nevertheless, as Brown (Chapter 31) points out, both our own follow-up study (Rutter *et al.*, 1967) and other investigations have noted that clumsy autistic children with visuo-spatial problems often overcome their early language handicaps more completely than other children and have a good social outcome.

The meaning of this observation, still to be confirmed, remains obscure. One possibility is that the visuo-motor difficulties merely mean that the language disorder is more nearly "dysphasic" than autistic in character. Our comparison of autistic and "dysphasic" children showed that the "dysphasic" children were more clumsy (Bartak *et al.*, 1975) and more likely to have articulation problems (Boucher, 1977) in spite of having a less severe and less extensive language handicap. In other words, it may be that the clumsy autistic children do well, not because of their visuo-spatial defect, but because the defect happens to be associated with a *lesser* and different language problem. Certainly, in our own group the clumsy children tended *not* to show the more gross manifestations of an autistic-type language disability. Whether or not this is the whole story remains to be determined. However, in the meanwhile, it should be noted that Brown's observations may carry the *opposite* implications to the one she gives. In short, the visuo-spatial problems may not predispose to autism but rather, within an autistic group, they may predispose to *recovery* from autism. Whether treatment aimed to improve visuo-motor functioning would help or hinder social development is an arguable point.

TREATMENT AND THE MODIFIABILITY OF DEVELOPMENT

In his review of follow-up studies, Lotter (Chapter 32) paid very little attention to the effects of treatment. Indeed, almost the only mention is his suggestion that the correlation between schooling and outcome is most likely to be an artefact of educational selection procedures. On the other hand, there are numerous chapters in this book (20 to 30) which describe apparently successful therapeutic interventions. Do they make any difference to the children's ultimate level of development and of social adjustment? The evidence suggests that some treatments do—within limits.

Limiting Factors in the Child's Handicap

The importance of limiting factors which derive from the severity of the child's basic handicap is shown by numerous studies. There are four key findings in this connection. First, the child's measured IQ in early childhood has been found to be associated with major differences in *biological*, as well as social, outcome (Bartak & Rutter, 1976). Thus, in the Maudsley Hospital follow-up 36% of autistic children with an initial IQ

below 70 developed epileptic seizures whereas of those with an IQ of 70 +
none did so. This suggests that cognitive performance is a function, in
part, of differences in the degree of biological impairment. Second, IQ
measures not only predict outcome in groups of autistic children given
little active treatment but also they do so in those given skilled help over
many years (Rutter & Bartak, 1973). Individual differences in the extent
of handicap appear to correlate with outcome regardless of how well or
how long the child is treated. Third, no form of treatment has been found
which can enable the most handicapped autistic children (those without
speech and with an IQ below 50 even after repeated skilled testing) to
achieve social independence. Fourth, as discussed below, treatment
seems to make little difference to the children's overall cognitive development. Clearly, we must conclude that individual differences in the
degree of basic handicap play a major role in determining how later development will proceed.

Benefits of Treatment

However, this definitely does *not* mean that treatment makes no
difference. There are several well-controlled studies which show clinically worthwhile (as well as statistically significant) benefits from
therapeutic interventions. Thus, our own evaluation of contrasting types
of special education demonstrated marked differences in outcome (Rutter
& Bartak, 1973—see also Chapter 28). We found that scholastic progess
was best in an autistic unit with large amounts of specific teaching in a
well-controlled classroom situation. Systematic teaching in a well-ordered
environment with an appropriate task orientation was associated not
only with greater scholastic attainment but also with more play with
other children. This finding was *not* an artefact of selective intake as the
differences between units applied to children with initially similar
characteristics.

Similarly, our evaluation of a home-based approach to treatment
showed it to be superior to other approaches in both the short-term and
long-term (see Chapter 26). The autistic children whose parents had been
helped to utilize behavioral techniques in a developmental context
showed considerably less social and behavioral handicaps than children in
the matched control groups treated in other ways. Appropriate treatment
can and does make a difference to prognosis.

Nevertheless, it is important to be clear just what is altered by treatment. In both studies, we found that treatment made *no difference* to the
child's intellectual level. Cognitive development in autistic children may
well be held back by environmental disadvantage or privation but, given
an ordinary upbringing at home, treatment seems to make little additional

difference to intellectual growth. Furthermore, it appears that treatment makes only a modest impact on long-term language development. In the educational study (Rutter & Bartak, 1973), although there were differences between the units studied in the children's acquisition of spoken language, the differences fell well short of statistical significance. Similarly, in the home-based treatment project (Chapter 26), the long-term effects on language were much less marked than the improvements in social and behavioral problems. Behavioral treatments do have an important short-term impact on speech acquisition, but in the longer term the main effect is on the social usage of language skills rather than on basic language competence (although this, too, may be improved to some extent). In short, treatment has least effect on those cognitive deficits which appear most fundamental to autism (see Chapter 1) and which are most strongly associated with outcome.

On the other hand, treatment does make a real difference with respect to socialization, communication, behavior, and scholastic attainment. Gains in these areas may be crucial in determining overall social independence and adjustment. Two points need to be made in this connection. First, we do not yet know how *much* difference treatment can make to prognosis. In particular, knowledge is lacking on whether really early therapeutic intervention could prevent many of the later social deficits and behavioral abnormalities. Certainly it seems that many of the best results are obtained with children treated from the very early years of life and some of the worst results occur with those in whom active treatment does not start until severe abnormalities of behavior are well established. On the other hand, comparisons of well-matched groups which differ only in the age when treatment starts are lacking.

Secondly, in many of the autistic children who have the best eventual outcome improvement continues not only through late adolescence but also well into early adult life. Kanner *et al.* (1972) commented on the important changes which took place during the teens and our educational study (Rutter & Bartak, 1973) noted the scholastic progress which took place during adolescence. Help at this time is particularly necessary. Also it may need to take a somewhat different form to that employed during the early years. Many of the more mildly handicapped autistic people develop an awareness of their own limitations and handicaps during adolescence. Often they are deeply distressed by the gap between their newly found desire to make close friendships and their continuing incompetence in social relationships. Emerging sexuality and the frequent impossibility of expressing this in normal heterosexual relationships adds an additional dimension to their unhappiness. We have found that a psychotherapeutic approach to help them understand and come to terms with their feelings and to find ways of coping with their disabilities is sometimes helpful. In addition, we have used systematic social skills training through role-playing and videotape feedback to help some autistic adolescents and adults learn what is required to make social contacts and hold a conversation. This has involved a particular focus on showing the autistic individuals how to *observe* social and conversational cues—a skill that they have strikingly lacked. Marked excitability, anxiety, and tension have sometimes impeded social interactions and in these cases a regularly taken small dose of one of the phenothiazines has often been helpful. None of these therapeutic approaches in early adult life has been systematically evaluated but they provide leads as to what might be beneficial.

Occasionally, autistic children regress during adolescence, becoming inert and less communicative, and sometimes this is associated with a general intellectual decline (Rutter *et al.*, 1971). We have investigated a number of such cases and so far have found no explanation of their deterioration in either physical abnormality or psychosocial stress. The reasons for the change in course remain obscure. In the great majority of cases the deterioration is *not* progressive. Rather, a plateau is reached, sometimes followed by improvement, but only, occasionally, a return to the level prior to the regression.

The transition from school to work is usually a difficult one even for the autistic person who has made most progress. Our own experience, like that of Brown (Chapter 31), is that behavioral problems, rather than work skills as such, made employment difficult. In part it is a question of learning to adhere to work routines; in part it is a matter of learning to adapt to changes; and in part it is the need to avoid socially embarrassing behaviors. Several autistic young men have lost jobs largely because their workmates were driven to distraction by talk about bus routes, timetables, or other obsessional interests. In almost all cases jobs have had to be simpler than the autistic person's IQ and educational attainments would suggest, because although they could do the work tasks they lacked the flexibility and initiative essential for higher status employment. The recipe for successful jobs has usually included a predictable routine, day-to-day contact with only a limited number of people, a supervisor who can exercise suitable personal guidance and control, and experience of previous work. We have found that often there are many job failures before a steady job is obtained and held. There are several autistic men who now have regular work but who did not hold down regular employment until their mid-20s.

CONCLUSION

Our increasing knowledge on the "natural history" of infantile autism has given us a better understanding of some of the factors in the child which influence prognosis. However, also it has raised new questions about the nature of autism. In particular, it emphasizes our ignorance about the nature of the social difficulties experienced by the intelligent autistic children with the best outcome. By the time they reach adult life most of them have good language skills, they have a normal level of intelligence, there is no thought disorder or psychotic disturbance, they *want* social relationships, and yet still they have marked and persistent social difficulties. Why? What is the nature of their disability and why does it persist beyond most of their other handicaps? Is there a critical period for the development of social relationships? Could good early treatment prevent these later social problems? Perhaps the most important result of all "natural history" studies should be the call to therapists to *alter* the course of the disorder. How far is the development of autistic children an inevitable consequence of their basic biological handicaps and how far is it a result of our inability to intervene effectively? To answer that challenge remains an important goal for the future.

REFERENCES

Bartak, L., & Rutter, M. Differences between mentally retarded and normally intelligent autistic children. *Journal of Autism and Childhood Schizophrenia*, 1976, *6*, 109–120.

Bartak, L., Rutter, M., & Cox, A. A comparative study of infantile autism and specific developmental receptive language disorder. I. The children. *British Journal of Psychiatry*, 1975, *126*, 127–145.

Boucher, J. Articulation in early childhood autism. *Journal of Autism and Childhood Schizophrenia*, 1977, *6*, 297–302.

Kanner, L., Rodriguez, A., & Ashenden, B. How far can autistic children go in matters of social adaptation? *Journal of Autism and Childhood Schizophrenia*, 1972, *2*, 9–33.

Moor House School. *A Report on a Follow-up in 1969 of Ten Receptive Aphasic Ex-Pupils of Moor House School.* Moor House School, 1969.

Reynell, J. *Reynell Developmental Language Scales.* Slough: N. F. E. R. Publishing, 1969.

Rutter, M. Autistic children: Infancy to adulthood. *Seminars in Psychiatry*, 1970, *2*, 435–450.

Rutter, M. Clinical assessment of language disorders in the young child. In M. Rutter & J. A. M. Martin (Eds.), *The child with delayed speech.* Clinics in Developmental Medicine No. 43. London: SIMP/Heinemann, 1972.

Rutter, M., & Bartak, L. Special educational treatment of autistic children: A comparative study. II. Follow-up findings and implications for services. *Journal of Child Psychology and Psychiatry*, 1973, *14*, 241–270.

Rutter, M., Bartak, L., & Newman, S. Autism—A central disorder of cognition and language. In M. Rutter (Ed.), *Infantile autism: Concepts, characteristics and treatment.* London: Churchill-Livingstone, 1971.

Rutter, M., Greenfeld, D., & Lockyer, L. A five-to-fifteen-year follow-up study of infantile psychosis. II. Social and behavioural outcome. *British Journal of Psychiatry*, 1967, *113*, 1183–1199.

Sheridan, M. Playthings in the development of language. *Health Trends*, 1969, *1*, 7–10.

34

Subgroups Vary with Selection Purpose

ERIC SCHOPLER and MICHAEL RUTTER

The work on autism, brought together in this volume, is not complete. The search for causal factors continues. New ground has been broken in the field of genetics. The focus on physiological, neurological, and biochemical factors is being sharpened. Blind alleys followed in the search for social and family contributions to the disorder are gradually being replaced by data and rational inferences. The efficacies of different treatments are still being tested, and optimum educational strategies are evolving. However, while some clear steps toward enlightenment have been taken, this devastating disorder is still shrouded in too much darkness.

With these concluding comments it is fitting that we review the problems of classification. After all, the definition and shorthand designation of autism can mean no more and no less than what is known about the condition. For all practical purposes our scientific search had its beginning with Kanner's (1943) lucid description and observation of a small group of children. This starting point is confirmed in that virtually every investigator in this volume refers to it and cites it.

In spite of Kanner's clinical clarity, the confusion in the field about the severely disturbed children has been undeniable. It is attested to by a proliferation of labels, including childhood schizophrenia (Bender, 1956),

ERIC SCHOPLER · Department of Psychiatry, Division of Health Affairs, University of North Carolina School of Medicine, Chapel Hill, North Carolina.   MICHAEL RUTTER · Department of Child and Adolescent Psychiatry, Institute of Psychiatry, De Crespigny Park, Denmark Hill, London SE5 8AF, England.

symbiotic psychosis (Mahler, 1952), borderline psychosis (Ekstein & Wallerstein, 1954), and atypical child (Rank, 1949). Each label had its roots in a particular view of the nature and causation of autism. No professional consensus developed on how to differentiate severely disturbed children with these labels. The first step out of this morass was taken by Creak's working party (1961, 1964). They attempted to specify the disorders shared by the children designated by these different labels. This resulted in the well-known "nine points." They sought to clarify the behavioral and clinical characteristics which transcended the labels, and drew attention to the need to study comparable groups of children. However, their approach was unsatisfactory in practice because several of the points were loosely described and were open to widely differing interpretations. Moreover, the points could be combined in an immense variety of ways. Some investigators tried to resolve this problem by describing their sample in terms of a minimum of four points. But since the four points were not the same for all the children in the sample, the resulting clarification was more apparent than real.

A second step toward clarifying the diagnostic confusion was taken by Rutter (1968) who combined a survey of the literature with a keen critical analysis of existing empirical evidence. This is updated in the first chapter of this volume. When this knowledge is summarized, four essential characteristics of autism stand out.These are (1) impairment in human relatedness, (2) impaired language, ranging from absence of speech to bizarre and deviant language patterns, (3) peculiar motor behavior ranging from stereotypies to more complex behavioral repetition, and (4) early onset, before 30 months of age. These are the general characteristics on which there is the greatest consensus and which can readily be applied to large groups in different centers. However, as described by Kanner, Creak, and others, there are many additional specific characteristics such as gaze avoidance, mental retardation with islets of normal functioning, unusual sensitivities, and pronoun reversal. These specific features may well reflect causal mechanisms underlying the disorder and they are no less important than the four central defining characteristics. The reason for not including all possible autistic behaviors in the diagnostic criteria is simply that they are unevenly distributed. Each symptom appears in some autistic children and not in others, or it is evident at only some phases of development. Any attempt to define autism by certain selected specific characteristics results in too much variability from one sample to another. Under these conditions each center develops its own definition of autism. Communication between workers is seriously impaired and definitions come to be accepted on ideological rather than rational grounds. As this is still the case to some extent, the question arises: Is this not still the same confusion that existed under the period of multilabels, only with different behaviors replacing the labels?

In fact real progress has been made. The different groups of autistic children referred to in the chapters of this volume can be described in ways which are replicable and understandable by all. Confusion has been greatly reduced by two measures. First, as already noted, there has been the broad agreement on overall defining criteria. Second, there has been the use of the multiaxial system of classification developed under the World Health Organization (Rutter *et al.*, 1975, 1976). With this approach diagnostic disagreements among psychiatrists were significantly reduced without oversimplifying diagnostic decisions.

MULTIAXIAL SYSTEMS

The chief value of a multiaxial system is that it makes clear distinctions between the various different elements (or axes) involved in diagnosis. Thus, in the case of autism it is very important to know whether the child is also mentally retarded. It is necessary to determine intellectual level because it is an important predictor of later development and outcome (see Chapters 1 and 33). In the past, classification systems seemed to demand a choice between autism *or* mental retardation. This was absurd because the child might have one, both, or neither problem. With the multiaxial approach this difficulty is avoided by making a *separate* decision about the clinical psychiatric syndrome (i.e., whether the child has autism, schizophrenia, manic-depressive psychosis, or whatever) and secondly about intellectual level (i.e., whether, regardless of the syndrome, intelligence is normal, mildly retarded, severely retarded, etc.).

The same issue arises with causal influences. In the past, diagnoses involved statements about etiology. This works well when there is a one-to-one association between cause and syndrome—as with measles, Down's syndrome, and the like. On the other hand, it does not work well either when causation is multifactorial (as is usual in psychiatry) or when the etiology is unknown or disputed. Thus, if a child has congenital rubella but shows the clinical picture of autism his disorder might be grouped with the organic brain syndrome rather than the functional psychoses. But then what do you do about the autistic child with no known medical condition who develops epileptic fits in adolescence? Does he change diagnosis as he enters the teenage period? Clearly that would be silly but equally it is crucial for both research and clinical purposes to know if autism is due to some diagnosable disease or organic disorder. Exactly the same issues apply to knowledge about associated or etiological psychosocial influences. The solution in the multiaxial system is to record each separately under independent axes. In short, there is a diagnostic coding which indicates that the child shows the behavioral syndrome of autism, another coding which indicates the intellectual level, a third coding which deals with associated or causal medical conditions (such as phenylketonuria, congenital rubella, etc.) and a fourth which describes abnormal psychosocial situations when they occur. The approach reduces ambiguity and provides greater information by making clear the different elements or axes in diagnosis. The multiaxial concept has applications which are broader than psychiatry (Sartorius, 1976).

PURPOSES OF CLASSIFICATION

Just as classification involves different axes, so also it is needed for different purposes. This is illustrated by the chapters in this volume which focus on cognitive mechanisms, biological and physiological functions, family factors, and various treatment and educational goals. These different purposes require emphases on different characteristics of autistic children. It would not be appropriate to utilize precisely the same groupings for all purposes. To the contrary, such a requirement is both unrealistic and unnecessarily restrictive. On the other hand, it is essential that both the selection purpose and the defining characteristics are made explicit so that the rationale for sampling is clear and replicable. The multipurposes and consequent sampling characteristics illustrated in this volume include several different organizational principles.

DEFINITION BY SYNDROME

For most purposes, there is an initial grouping by overall descriptive characteristics or syndrome. This is the first of the five axes developed for the World Health Organization—the one that refers to "clinical psychiatric syndrome." Its purpose is to provide generally agreed upon overall defining characteristics of the syndrome under consideration. In the case of autism, the four necessary conditions for diagnosis are outlined by Rutter in the first chapter of this volume. Such conditions are required in order to be sure that when people talk about "autism" they are referring to the same thing. The criteria delineate the ball park from which autistic subjects are drawn. It is not surprising, therefore, that these four conditions are included, more often than any others, in the chapters in this volume. Nevertheless, these provide only a beginning rather than an ending. For most purposes it is necessary to concentrate on narrower subsamples. It is necessary and equally crucial that the process by which the narrowing takes place is made explicit according to replicable criteria. It is also essential that attention is paid to the necessary restrictions and qualifications which follow from the focus on a specific subgroup of autistic children.

On the other hand, it is not helpful when groups are defined in an arbitrary and idiosyncratic way. This confuses the diagnostic process and to that extent invalidates or at least reduces the value of the research. Whenever otherwise well-controlled studies give rise to apparently contradictory findings it is essential to check whether they refer to the same groups of children.

Thus, the admirable review of biochemical and hematologic studies in Chapter 11 notes several investigations with inconsistent findings on the efflux of serotonin in the blood platelets of autistic children. Boullin *et al.* (1971) found significant differences between autistic and other children whereas Yuwiler *et al.* (1975) found no differences in serotonin efflux. But the Boullin group selected their autistic sample from the Rimland E-2 checklist, whereas Yuwiler *et al.* utilized Ornitz and Ritvo's (1968) diagnostic criteria based on the concept of perceptual inconstancy. The disagreement in findings could be due to differences in assay techniques or to differences in subject selection. The only satisfactory way to resolve this question is to utilize *both* techniques with a group of autistic children which includes subgroups meeting *both* the Rimland and the Ornitz and Ritvo criteria. One of the many useful outcomes of the St. Gallen meeting is a joint project in which both groups of investigators are now collaborating to do just that.

Another example of a similar kind is provided by McAdoo and DeMyer (Chapter 17). They were concerned to replicate the Goldfarb (1961) study finding that parents of "organic" children are less pathological in their personality characteristics than parents of "nonorganic" autistic children. McAdoo and DeMyer do not consider the distinction between organic and functional as an important one and do not use it in their own work. Nevertheless, in order to make the replication they used Goldfarb's criteria for organicity to distinguish two samples comparable to his.

DEFINITION BY SUBGROUP

Classification or selection priorities may be influenced or defined by the research issue under investigation. Several examples are included in the chapters of this book. The value of this sharper focus is that it may provide a more clear-cut answer to research questions. However, necessarily this also means that there must be caution in generalizing the findings outside the specific subgroup studied.

Frequently, the subgroups have been narrowed in terms of intellectual level. Thus Hemsley *et al.* (Chapter 26) restricted their study of home-based treatment to autistic children with a nonverbal IQ of 60 or greater. This was necessary because it had already been shown that the prognosis was very much worse for severely retarded children. Accordingly (because the treated group had to be fairly small by statistical standards), it was desirable to reduce individual differences in order to obtain a better estimate of the efficacy of treatment. The cost of doing this was that the investigators were not able to determine whether the same treatment principles applied to retarded autistic children. Lansing and Schopler (Chapter 29) took the opposite strategy with opposite strengths and weaknesses. Thus, they were able to show that their treatment methods *were* applicable to children with all levels of handicaps, but equally the price paid was a less stringent evaluation of the efficacy of specific treatment methods within their therapeutic program. Neither research strategy is sufficient on its own. Rather, both are needed and complement one another.

Cantwell and his colleagues (Chapter 18) also confined their attention to autistic children without severe mental retardation (in their case to children with a nonverbal IQ of at least 70). They were interested in exploring the idea that family influences may play a part in the environmental causation of autism. If psychogenic factors were important in the etiology of autism they were most likely to be so in the autistic children of normal intelligence. This is so because other studies had clearly shown that many severely retarded autistic children had overt organic pathology. In fact, Cantwell's findings were generally negative in spite of his sample selection which suggests that autism is rarely, if ever, primarily due to abnormal family influences of a psychogenic kind.

Arising from his hypotheses about the neurological basis of autism, DeLong (Chapter 14) anticipates two separate autistic syndromes. One is the left medial temporal syndrome associated with higher IQ function and the other is the bilateral syndrome associated with profound retardation in all cortical functions. Clearly future autopsy studies based on this hypothesis must place priority on the differentiation between autistic individuals with high and low intellectual functioning.

In Chapter 12, Coleman notes that antibodies to the herpes virus were found unexpectedly often in her group of autistic children. As the herpes virus is said to have a special predilection for the temporal lobe area of the brain, viral studies should perhaps focus particularly on autistic children whose histories or characteristics suggest temporal lobe pathology.

DEFINITION BY SPECIFIC BEHAVIORS

Sometimes it may be desirable to focus on groups of children defined in terms of specific behaviors rather than syndromes. Thus, behavior modification programs have often been concerned with the alteration of specific items of behavior rather than with the treatment of the condition as a whole. Lovaas' successful modification of self-destructive behavior in autistic children (Lovaas *et al.*, 1965) had to be demonstrated with children who showed this problem. The results probably have more relevance to the elimination of this symptom in any severely disturbed child than to the treatment of autistic children (only a few of whom are seriously self-destructive).

The therapeutic use of drugs has been frequently abused (see Chapter 23). Often multiple drugs are administered without adequate monitoring, making it virtually impossible to observe either positive or negative drug effects. At present it appears that there is no drug which is effective in treating autism. On the other hand there may be drugs which are effective in diminishing certain symptoms shown by some autistic children. Such symptoms may not be central to autism, in which case subsequent drug research might then be directed at responders who are appropriately selected for their nonautistic characteristics (these might include overactivity, anxiety, or agitation).

Biochemical studies may also need to focus on specific symptoms. Coleman (Chapter 12) observed that hyperactive children showed an abnormal platelet serotonin concentration which returned to normal when the child became more manageable and quiet. Rodnight (Chapter 13) suggests that platelet function may be modified as a result of some unknown peripheral consequence of hyperactivity. The mechanisms involved in this hypothesis will be best explored with subjects who show different levels of hyperactivity, a characteristic relevant but not central to autism.

DEFINITION BY HYPOTHESIZED VARIABLE

When there is a well-developed hypothesis about some underlying abnormality or causal mechanism, it may be advantageous to start with a sample selected on the basis of that abnormality rather than to investigate the syndrome as such. Thus, Hermelin (Chapter 9) argues that instead of adopting a distinct disease concept it may be preferable to define certain specific underlying mental operations and then to go on to find out how these are linked with a variety of different pathologies. In her own work, this strategy was of value in showing both similarities and differences between autistic children on the one hand and blind and deaf children on the other.

Rutter and his colleagues (see Chapter 6) used a similar approach in the case of language abnormalities. They did not start with a study of autistic children. Rather they took *all* children (regardless of diagnosis) who presented a severe impairment in the development of language comprehension. In this way (by comparisons within the overall group) it was possible to determine which aspects of language functioning were associated with autism and which ones were not. For the same reasons, in their genetic study Folstein and Rutter (Chapter 15) did not restrict their attention to the extent to which twin pairs showed concordance on autism. Instead they asked the question—Which characteristics were genetically linked with autism? Their findings suggested that although autism was probably not mainly inherited as a specific disease, it was inherited to some extent as part of a broader cognitive deficit.

In his research, Ornitz has concentrated on disturbances in the interaction between the visual and vestibular sensory systems. It appears that the severity of this abnormality is correlated with the severity of motility disorder. As a consequence the Ornitz studies show a selective priority for autistic children with severe motility dysfunction (hand flapping, whirling, rocking, toe walking, etc.). The question arises whether the vestibular abnormality underlies the motility dysfunction or underlies the autistic syndrome as a whole. One way of examining this matter would be to study both autistic and nonautistic children with motility disorders in order to determine whether the vestibular dysfunction is associated with autism regardless of motility problems or rather whether it is associated with motility disturbances irrespective of whether the children are autistic.

GROUPINGS FOR THERAPEUTIC AND ADMINISTRATIVE PURPOSES

So far in this chapter we have been primarily concerned with the use of classification for research purposes. We have emphasized the need for a generally agreed upon system of diagnosis and classification. Only in this way is it possible for clinicians in different centers to communicate with one another. A common approach to classification is also needed for research. Progress is slow and misunderstandings rampant if one worker's investigation of "autism" concerns a group which differs in general diagnostic characteristics from another worker's group of "autistic" children. Equally, however, we have tried to show how research requires a progressive narrowing down of issues and questions. In order to do this successfully it is necessary to focus on subgroups or to define samples in terms of a specific behavior or abnormality rather than in terms of any syndrome or disease.

Advances have also followed from the realization that a single diagnosis or term provides a quite inadequate means of description. Autism is a complex disorder with several different dimensions or axes of description. For all purposes—clinical, research, or administrative— it is not enough to know that a given child or group of children suffer from autism. It is also essential to know their intellectual level, psychosocial circumstances, and whether they have a diagnosable medical disorder. These axes are generally applicable and hence are included as part of the regular routine coding of disorders in the multiaxial scheme developed for the World Health Organization.

However, they are not the only important axes. For example, in planning services it may also be important to know about the severity of social impairment (an additional axis currently being studied by the WHO). In adults the dimension of personality functioning takes on an importance which it does not have in childhood at a stage when personality is still forming and developing. In societies where health and education services are determined in part by people's ability to pay for them, rather than by individual need, administrators will routinely need information on the family's financial resources and the cost of treatment. If the central question had to do with insurance then the classification would have to include risk factors for incapacity.

It is also crucial to appreciate that a diagnostic classification provides the lowest common denominator applicable to all. In clinical practice a diagnostic formulation involves much more than that (Rutter, 1975). It is necessary to consider each child's idiosyncratic features and the ways in which he is different as well as the ways in which he is similar to others. Thus, in Chapter 29, Lansing and Schopler point out the importance of individual characteristics in selecting children to fit particular curricula. The therapists must have answers to questions such as: Which of the child's behaviors will disrupt other children? At what developmental level are his language and cognitive skills? How much individual attention will he require to function in the classroom? Is the current classroom situation sufficiently settled and stable for it to contain an additional severely disturbed child?

CONCLUSION

In this final chapter we have considered the history of confusion in the diagnosis and classification of autism. A limiting factor has been our ignorance about the condition. The advances in knowledge and understanding over the last two decades have led to more rational approaches to classification which in turn facilitate future research. There is a better appreciation of the cognitive mechanisms which are impaired in autism and progress is being made in identifying the affected biological structures and behavioral systems, together with the interaction between the two.

This knowledge has been associated with real advances in treatment so that effective methods are now available to help autistic children and their families. Of course, autism remains a serious handicapping disorder and our ability to influence autistic children's development is severely limited. Nevertheless, a start has been made. From the broader definition of autism in Chapter 1, it has become increasingly clear that many specific mechanisms underlie the disorder. These mechanisms can be studied more effectively and reported with more validity when samples are selected for appropriate characteristics and when the selection purpose is clearly specified. This procedure is used in many studies in this volume. It appears to be the direction for defining subclusters of children whose disorders and treatment will be understood increasingly in the future. The task is far from complete but the research reported in this volume not only indicates the vigor and liveliness of current work but also points out some of the directions to be followed in clinical practice and research over the next decade.

REFERENCES

Bender, L. Schizophrenia in childhood—its recognition, description, and treatment. *American Journal of Orthopsychiatry*, 1956, *26*, 499–506.

Boullin, D., Coleman, M., O'Brien, R., & Rimland, B. Laboratory prediction of infantile autism based on 5-hydroxytryptamine efflux from blood platelets and their correlation with the Rimland E-2 score. *Journal of Autism and Childhood Schizophrenia*, 1971, *1*, 163–171.

Creak, M. Schizophrenic syndrome in childhood. Progress report of a working party. *British Medical Journal*, 1961, *2*, 889–890.

Creak, M. Schizophrenic syndrome in childhood. Further progress report of a working party. *Developmental Medicine and Child Neurology*, 1964, *6*, 530–535.

Ekstein, R., & Wallerstein, J. Observations on the psychology of borderline and psychotic children. *Psychoanalytic study of the child* (Vol. 9). New York: International Universities Press, 1954.

Goldfarb, W. *Childhood schizophrenia.* Cambridge, Mass.: Harvard University Press, 1961.

Kanner, L. Autistic disturbances of affective contact. *Nervous Child*, 1943, *2*, 217–250.

Lovaas, O. I., Freitag, G., Gold, V. J., & Kassorla, I. C. Experimental studies in childhood schizophrenia—Analysis of self destructive behavior. *Journal of Experimental Child Psychology*, 1965, *2*, 67–84.

Mahler, M. On child psychosis and schizophrenia. Autistic and symbiotic psychoses. *Psychoanalytic study of the child* (Vol. 7). New York: International Universities Press, 1952.

Ornitz, E., & Ritvo, E. Perceptual inconstancy in early infantile autism. *Archives of General Psychiatry*, 1968, *18*, 79–98.

Rank, B. Adaptation of the psychoanalytic technique for the treatment of young children with atypical development. *American Journal of Orthopsychiatry*, 1949, *19*, 130–139.

Rutter, M. Concepts of autism: A review of research. *Journal of Child Psychology and Psychiatry*, 1968, *9*, 1–25.

Rutter, M. *Helping troubled children.* Harmondsworth: Penguin. New York: Plenum Press, 1975.

Rutter, M., Shaffer, D., & Shepherd, M. *A multi-axial classification of child psychiatric disorders.* Geneva: WHO, 1975.

Rutter, M., Shaffer, D., & Sturge, C. *A guide to a multiaxial classification scheme for psychiatric disorders in childhood and adolescence.* London: Institute of Psychiatry, 1976.

Sartorius, N. Classification. An international perspective. *Psychiatric Annals,* 1976, *6,* 24–35.

Yuwiler, A., Ritvo, E., Geller, E., Glossman, R., Schneiderman, G., & Matsuno, D. Uptake and efflux of serotonin from platelets of autistic and non-autistic children. *Journal of Autism and Childhood Schizophrenia,* 1975, *5,* 83–98.

Author Index

Aarkrog, T., 16, 20, 21, 212, 217
Abramson, H., 354
Ackerman, A., 378
Adey, W. R., 139
Ainsworth, M. D., 28, 40, 42
Alanen, Y. O., 274, 279, 291, 488, 494
Alexander, D., 378
Allen, E., 411, 438
Allen, H., 304, 311
Allen, J., 21, 43, 68, 102, 266, 279, 291, 292, 494
Alpern, G. D., 6, 20, 21, 244, 249, 366
Alpert, M., 425, 435
Anderson, E. M., 459, 460
Anderson, L. T., 350, 351
Annell, A. L., 488, 494
Anthony, E. J., 243, 244, 249, 253, 266, 273, 275, 289, 291
Apolloni, T., 456, 460
Arajarvi, T., 291, 494
Argyle, M., 67, 68
Arnold, C. R., 435
Aschan, G., 132, 136
Ashenden, B., 22, 353, 474, 495
Asperger, H., 7, 20
Attneave, F., 150, 154
Axelrod, R., 182, 354
Axline, M., 317, 325
Ayd, F. J., 347, 350
Ayres, A. J., 442, 449, 450, 472, 473
Ayres, F. J., 345, 350
Azrin, N. H., 419, 421

Baer, D. M., 156, 161
Baker, B. L., 375, 377
Baker, J., 410
Baker, L., 102, 269, 282, 291, 292
Bakker, D. J., 157, 161
Bakwin, H., 2, 20, 220, 237
Bald, D., 183
Baltaxe, C., 434, 437
Baroff, G. S., 379, 410, 421
Barrera, S. E., 238
Bartak, L., 5, 9-11, 18, 20, 21, 24, 28, 42, 61, 65, 68, 71, 82, 83, 89, 91, 92, 93, 95, 101, 102, 103, 220, 235, 237, 238, 240, 244, 250, 266, 283, 288, 291, 292, 386, 408, 410, 423, 426, 434, 435, 437, 453–457, 460, 461, 498, 501, 502, 503, 505
Barton, S., 21, 43, 68, 102, 266, 292, 366, 494
Basamania, B., 291
Bateman, P. P. G., 334
Bayley, H. C., 292
Bayley, N., 307, 311
Beaulieu, M. R., 294
Becker, W. C., 375, 377, 428, 435
Beeckmans-Balle, M., 490, 494
Begab, M., 25
Behrens, M. C., 280, 281, 284, 291
Bell, R., 287, 291
Bellugi, U., 72, 83

Belmont, L., 158, 161
Benbrook, F., 295
Bender, L., 2, 14, 21, 124, 136, 169, 178, 181, 244, 249, 272, 274, 276, 279, 287, 288, 291, 338, 344, 350, 356, 360, 361, 366, 464, 470, 471, 473, 478, 490, 494, 507, 516
Bene, E. A., 272, 275, 291
Benjamin, F., 354
Bennett, S., 476, 494
Bennette, E. L., 312
Benor, D., 354
Benson, L., 150, 154
Berberich, J. P., 367, 409
Berendes, H., 230, 238
Berger, M., 157, 160, 162, 293, 379, 380, 405, 408, 409, 410, 421, 424, 438, 460, 471, 474
Bergman, P., 117, 123, 136
Bergstedt, M., 132, 136
Berkhout, J. I., 139
Berkowitz, B. P., 375, 378, 408, 480
Berlin, I., 271, 296, 303, 304, 307, 308, 311, 312, 327, 329, 330, 334, 335
Bernal, M. E., 122, 136, 137, 375, 378
Bettelheim, B., 271, 280, 291, 314, 325, 361, 363, 366, 424, 435, 490, 494
Bhagavan, H. N., 189
Birch, H. G., 5, 22, 158, 161, 289, 293, 311, 443,

519

Birch, H. G., *(cont'd.)*
451, 490, 494
Black, M., 429, 435
Blanton, R. L., 103
Bleuler, B., 2, 21
Block, J., 252, 266, 275,
291
Bloom, W., 354
Boatman, M. J., 270, 291
Boesen, L., 16, 21
Boesen, V., 212, 217
Bomberg, D., 14, 21
Bonvillian, J. D., 380, 408
Böök, J. A., 220, 238
Boomer, D., 312
Born, B., 72, 84
Borrow, N. A., 249
Bosch, G., 2, 21, 476
Boucher, J., 94, 100, 102,
213, 218, 501, 505
Boullin, D. J., 17, 21, 166,
169, 170, 171, 181, 189,
190, 198, 347, 350, 510,
516
Boulton, D., 23
Bouthilet, L., 312
Bowen, M., 181, 274, 280,
291
Bower, T., 113, 116, 331,
334
Bowers, M., 198, 351
Bowlby, J., 28, 42, 305,
306, 311, 328, 331, 333,
334
Bradford-Hill, A., 341, 351
Brambilla, F., 346, 350
Brandt, T., 125, 136
Brannigan, C., 67, 68
Brask, B. H., 220, 338
Brazelton, T. B., 53, 60,
307, 311
Brehm, S., 437, 452
Breuer, H., 351
Breuer, H., Jr., 351
Brierley, L. M., 45, 241
Brinkmann, L., 211, 218
Brodsky, G., 380, 409
Broer, H., 199
Bronner, A. F., 303, 312
Brooke, E., 24, 25, 452, 495
Brown, G., 276, 283, 292,
295, 387, 408

Brown, J. H., 125, 137
Brown, J. L., 220, 238, 463,
464, 473, 488, 494
Brown, M. B., 23, 25, 138,
139, 250, 335
Brown, N. A., 199
Brown, R., 72, 83, 376, 378,
379, 408
Brown, R. I., 435, 436
Browning, E. R., 424, 434,
435
Browning, R. M., 380, 408
Bruch, H., 220, 238
Bruner, J., 53, 58, 60, 112,
116, 316, 323, 325, 326
Bryant, P., 157, 161
Bryson, C. Q., 6, 9, 21, 22,
64, 68, 72, 83, 182, 366,
441, 450
Buchness, R., 139
Bucy, P., 218
Burdock, E. I., 181, 340,
350
Burland, B. J., 292
Burns, P., 312, 335
Bushell, D., Jr., 458, 461
Buss, C. L., 295
Byassee, J., 280, 292

Calcagno, P. L., 191
Callias, M., 294, 453
Cameron, K., 181
Campbell, E. J. M., 3, 21
Campbell, M., 166, 167,
168, 181, 189, 198, 337,
338, 339, 341, 342, 343,
344, 345, 346, 347, 348,
350, 351, 352, 353, 471,
473, 494
Cantwell, D., 93, 102,
269, 276, 283, 284, 286,
288, 290, 291, 292, 293,
295, 297, 299, 301, 332,
354, 385, 408
Caparulo, B., 102, 181
Carlson, C. F., 100, 102,
315, 316, 325, 471, 473
Carr, E. G., 424, 434, 435
Carr, E. M., 138
Carr, J., 456, 460
Case, Q., 353
Cashdan, A., 12, 21

Castell, R., 64, 65, 67, 68
Castells, S., 355
Cates, N., 199
Chan, T. L., 347, 350
Chapman, A. H., 220, 238
Chassan, J. B., 340, 351
Chess, S., 9, 15, 18, 21, 25,
38, 42, 86, 91, 102, 104,
187, 198, 221, 236, 237,
238, 243, 249, 339, 351,
359, 366
Chiarandini, I., 367
Chomsky, N., 363, 366
Christensen, A., 424, 436
Churchill, D. W., 5, 9, 21,
22, 28, 42, 50, 60, 64, 66,
68, 71, 72, 73, 75, 83, 84,
86, 92, 100, 102, 182,
366, 408, 434, 436
Churchill, J. A., 231, 241
Clancy, H., 54, 60
Clardy, E. R., 476, 494
Clark, F. C., 249
Clark, P. D., 5, 21, 65, 68
Clarke, A. D. B., 40, 42,
331, 334
Clarke, A. M., 330, 334
Cleeman, J., 176, 182
Clements, M., 295
Cline, D. W., 221, 238
Closs, K., 204, 205
Cobrinik, L., 71, 83, 350
Cohen, D., 164, 174, 179,
181, 190, 198
Cohen, D. J., 92, 101, 102,
347, 351
Cohen, H. J., 15, 25, 38, 44,
221, 236, 240, 243, 250
Cohen, I. L., 350, 351
Cohen, J., 87, 91, 102
Colbert, E. G., 124, 125,
136, 471, 473, 494
Colby, K. M., 365, 366
Cole, P. G., 434, 436
Cole, J. O., 354
Coleman, M., 21, 173, 178,
181, 185, 187, 189, 190,
193, 195, 196, 198, 204,
205, 350, 351, 359, 366,
516
Collins, P. J., 181, 198, 350,
351

Collins, W. E., 125, 132, 136
Conners, C. K., 341, 352
Connolly, F. J., 255, 266
Cook, M. G., 67, 68
Cooke, T. P., 456, 460
Cooper, B., 274, 278, 292
Cooper, E., 292
Cooper, J. E., 272, 292, 408
Corbett, J., 7, 21, 33, 42
Cordwell, A., 425, 436
Coss, R. G., 50, 60, 64, 66, 68, 69
Costello, J. M., 379, 410
Coulter, S. K., 408
Covert, A., 66, 69, 100, 102
Cowan, P. A., 5, 21
Cowen, M. A., 137, 182, 249
Cowie, K., 181
Cowie, V., 233, 244
Cox, A., 21, 28, 41, 42, 83, 102, 237, 238, 272, 276, 279, 282, 286, 288, 289, 291, 292, 408, 435
Cravioto, J., 305, 311
Creak, M., 2, 7, 8, 21, 38, 43, 120, 136, 167, 181, 209, 218, 220, 238, 243, 249, 274, 278, 286, 288, 292, 329, 334, 338, 351, 440, 450, 463, 473, 476, 477, 481, 488, 494, 508, 516
Creedon, M., 442, 449, 450
Critchley, M., 38, 43
Cross, L. A., 16, 23
Cummings, S. J., 287, 292
Cunningham, M. A., 9, 21, 86, 102, 434, 436
Curnow, R. N., 220, 234, 236, 238
Currie, K., 67, 68

Dahlstrom, L. E., 263, 266
Dahlstrom, W. G., 266
Dalby, M., 212, 218
Dalgleish, B., 434, 436, 441, 450
Daniel, D., 354
da Prada, M., 205
Darr, G. C., 2, 21
David, R., 181, 350, 351

Davids, A., 279, 292, 424, 436, 490, 494
Davies, P., 137
Davis, D. R., 12, 21
Davis, K., 331, 334
Dawley, A., 311, 304
Day, R., 230, 238
De La Cruz, F., 25
de la Pena, A., 138
Delay, J., 337, 351
De Licardie, E. R., 311
Delle Ore, G., 212, 218
DeLong, G. R., 22, 218, 249
DeMyer, M. K., 5, 8, 9, 21, 27, 28, 29, 41, 43, 65, 66, 68, 72, 74, 79, 83, 89, 102, 107, 116, 137, 139, 159, 161, 166, 178, 182, 251, 253, 266, 272, 286, 291, 292, 297, 298, 365, 366, 408, 441, 451, 486, 490, 494
DeMyer, W. E., 21, 43, 68, 102, 366, 494
Deniker, P., 337, 351
De Sanctis, S., 338, 351
DeSausure, F., 363, 366
DesLauriers, A. M., 100, 102, 313, 315, 316, 325, 471, 473
Despert, J. L., 1, 22, 270, 292
DeVilliers, J. G., 72, 84
DeVito, E., 181, 198, 351
Dewey, M. A., 40, 43, 65, 68
Diamant, C., 437, 461
Dichgans, J., 125, 136, 139
Dicks, D., 213, 218
Dietrich, L., 23, 138, 335
DiMascio, A., 347, 351, 352, 354
Dix, M. R., 125, 136
Dixon, C., 86, 102
Doll, E. A., 223, 238
Domino, E., 182
Donnelly, E. M., 292, 280
Dooher, L., 352
Douglas, J. W., 273, 283, 292, 387, 408
Douglas, V. I., 17, 22
Dowling, J., 352
Drage, J. S., 230, 238

Dratman, M. L., 410, 438
Dreifuerst, P., 495
Dubenoff, B., 442, 451
Dundas, M. H., 425, 436
Dunn, L. M., 442, 449, 451
Dunn, P. M., 230, 233, 238
Dutton, G., 348, 351
Dysinger, R., 291

Earl, C. J. C., 38, 43
Eaton, L., 488, 495
Ebstein, R. P., 198
Eimas, P., 113, 116
Eisenberg, L., 2, 8, 22, 24, 27, 40, 41, 43, 44, 219, 239, 252, 266, 272, 273, 292, 338, 341, 347, 352, 357, 366, 452, 476, 488, 494, 495
Ekstein, R., 471, 474, 507, 516
Elgar, S., 425, 435, 436, 450, 451, 453, 456, 457, 460
Embrey, M. G., 294
Endicott, J., 293, 334
Engelhardt, D. M., 342, 343, 348, 352, 353, 355
Engelman, E., 442, 449, 451
Erikson, E., 416, 421
Erikson, E. H., 323, 325
Erikson, M. T., 264, 266
Escalona, S. K., 117, 123, 136
Esman, A. H., 252, 266
Estes, W. K., 213, 218
Etemad, J. G., 21, 329, 334
Euper, J. A., 449, 451
Evans, M. A., 410
Everard, M. P., 40, 43, 65, 68
Eviatar, A., 137

Falstein, E., 312
Faretra, G., 124, 136, 338, 343, 350, 352
Farley, A. J., 138, 249
Farmer, C., 355
Feldman, W., 175, 183
Fenichel, C., 360, 366

Fernandez, C., 132, 137
Fernandez, P. B., 21, 102, 198, 238
Ferris, S., 351
Ferster, C. B., 28, 43, 72, 84, 280, 292, 421
Fieldsteel, N., 291
Firestone, P., 378
Fischer, I., 424, 436
Fish, B., 104, 124, 136, 168, 182, 275, 292, 338, 339, 340, 341, 342, 343, 344, 346, 350, 351, 352, 353, 359, 360, 361, 362, 366, 367, 470, 474, 490, 494
Flach, F. F., 354
Florsheim, J., 279, 293
Floyd, A., 351, 352
Folstein, S., 99, 101, 102, 219, 233, 238, 386, 408
Forrest, S. J., 43, 137
Forsythe, A. B., 138
Fouts, R. S., 365, 367, 380, 408
Fowle, A., 176, 182
Fox, R., 59, 60
Fraknoi, J., 280, 293
Frank, T., 294
Fraser, C., 287, 293
Freedman, A. M., 124, 136
Freedman, D. G., 66, 68, 136, 164, 167, 183
Freedman, D. X., 199, 347, 354
Freeman, B. J., 163, 182, 364, 367, 435
Freeman, R., 456, 460
Freibrun, R., 437
Freitag, G., 409, 421, 516
Friedman, E., 181, 187, 198, 351
Frith, U., 71, 72, 84, 122, 137, 441, 451
Frostig, M., 360, 367, 442, 451
Fujimori, M., 182
Fulwiler, R. L., 365, 367, 380, 408
Furer, M., 361, 367, 425, 437
Furneaux, B., 426, 436

Gair, D. S., 2, 23
Gajzago, C., 23, 338, 353
Gallagher, J. J., 453, 460
Gallant, D., 354
Garcia, B., 244, 250, 293, 294
Gardner, E. A., 354
Gardner, G. E., 304, 312
Gardner, J., 66, 68, 282, 287, 293
Garside, R. F., 249, 293, 294
Gedda, L., 223, 238
Geertz, C., 48, 60
Gelfand, D. M., 458, 460
Geller, B., 351
Geller, E., 25, 163, 182, 183, 199, 354, 517
Gelman, R., 109, 116
Geneiser, N., 181, 351
Genova, P., 326
Gerner, E. W., 353
Gerson, S., 351
Gerver, J. M., 230, 238
Geschwind, N., 213, 218
Gift, T., 355
Gilkey, K., 83, 182
Giller, D. R., 352
Gillies, S. M., 13, 22, 23, 348, 355, 495
Gilson, R. D., 139
Ginsberg, G. L., 104
Gittelman, M., 5, 22, 176, 182, 289, 293, 445, 451, 490, 494
Gittelsohn, A. M., 233, 238
Glahn, T. J., 378
Glanville, B., 424, 436
Gleser, G. C., 341, 353
Glossman, R., 25, 183, 199, 517
Gold, E., 183
Gold, V. J., 409, 421, 516
Goldberg, J. M., 132, 137, 244, 249
Goldfarb, L., 354
Goldfarb, N., 274, 291, 293, 332, 436
Goldfarb, W., 8, 18, 22, 63, 68, 117, 137, 160, 161, 192, 198, 243, 250, 251, 252, 260, 265, 266, 270, 274, 275, 277, 278, 280,

Goldfarb, W., *(cont'd.)* 281, 282, 286, 288, 289, 291, 293, 294, 311, 334, 335, 338, 353, 361, 367, 424, 436, 486, 490, 494, 511, 516
Goldstein, M., 189, 198
Goldstein, R., 474
Gonzales, S., 273, 276, 286, 289, 293
Goodman, J., 456, 461
Goodman, L., 360, 367, 442, 451
Goodwin, M., 182
Goodwin, M. S., 120, 137, 243, 249, 365, 367
Goodwin, T. C., 137, 177, 182, 243, 249, 365, 367
Goren, E. R., 437, 461
Gottesman, I. I., 219, 220, 227, 238
Gould, J., 32, 33, 43, 44, 241
Graf, K., 219, 240
Graham, P., 24, 335
Gram, L. F., 344, 353
Gray, B., 442, 449, 451
Graziano, A. M., 375, 378, 379, 380, 408, 424, 436
Green, W., 181, 351
Green, W. H., 351
Greenbaum, G., 295, 345, 353
Greenberg, E., 238
Greenfeld, D., 24, 104, 240, 250, 354, 437, 461, 494, 495
Greenman, G., 66, 68
Greenspan, L., 181, 198, 351
Group for the Advancement of Psychiatry, 353
Grugett, A. E., 272, 275, 277, 279, 287, 288, 291
Guedry, F. E. Jr., 139
Guerrini, A., 350
Gullion, M. E., 375, 378, 383, 409
Guthrie, D., 138, 249

Haber, K., 239
Hagamen, M. B., 355

Hailey, A., 44
Halpern, W. I., 71, 84, 379, 408, 424, 436, 457, 460, 474
Hammerlynch, L., 378
Hanley, L., 378
Hanson, D. R., 219, 220, 238
Hardesty, A. S., 340, 350, 351
Hargrave, E., 365, 367
Harkins, J. P., 287, 296
Harper, J. A., 271, 296
Harris, F., 410, 411, 438
Harris, M., 48, 60
Harris, R., 21, 42
Harris, S. L., 437, 461
Harthman, D., 138, 460
Hartmann, D., 458, 460
Hartung, R. O., 441, 451
Hauser, S. L., 16, 22, 207, 218, 246, 247, 249
Havelkova, M., 220, 238, 424, 436, 490, 494
Hawiger, A., 205
Hawiger, J., 203, 205
Hawkins, D. R., 314, 325
Healy, W., 303, 312
Heeley, A., 173, 174, 182
Heifetz, L. J., 375, 377
Heims, L., 266, 294, 436
Heinrich, A., 211, 218
Held, R., 150, 154
Heller, T., 7, 22
Helme, W. H., 124, 136
Hemsley, D. R., 156, 162
Hemsley, R. L., 284, 293, 386, 392, 408, 456, 460
Henn, V. S., 125, 139
Henry, J., 416, 421
Herbert–Jackson, E., 461
Hermelin, B., 9, 13, 18, 22, 28, 43, 64, 68, 71, 72, 84, 88, 90, 91, 100, 102, 103, 110, 116, 120, 122, 137, 141, 145, 147, 148, 150, 153, 154, 157, 161, 218, 315, 325, 329, 330, 333, 334, 434, 436, 441, 451
Herrick, J., 266, 294
Hersh, S. P., 240, 355
Hersov, L., 293, 379, 408,

Hersov, L., (*cont'd.*) 409, 421, 460
Herzberg, B., 194, 278, 293
Heskett, W. M., 360, 366, 472, 473
Hewett, F. M., 364, 367, 440, 451
Hills, M., 96, 102
Himwich, H. E., 172, 182, 347, 353
Hinde, R. A., 330, 334
Hingtgen, J. N., 5, 6, 21, 22, 72, 84, 366, 378, 408, 414, 421
Hinton, G. G., 99, 233, 238
Hirsch, S. E., 281, 290, 293, 298, 301
Hirschberg, J. C., 312
Hockett, C. F., 100, 102
Hoddinott, B. A., 21, 220, 240
Hodges, J., 330, 335
Hoffman, S. P., 352, 353, 355
Holbrook, D., 378
Hollander, C. S., 348, 351
Hollister, L. E., 348, 353
Holroyd, J., 272, 293
Holt, S., 228, 238
Honrubia, V., 132, 137
Hood, J. D., 125, 136, 137
Horn, G., 332, 334
Horne, D., 442, 451
Hornykiewicz, O., 205
Howard, C., 182, 354
Howlin, P. A., 23, 63, 282, 284, 288, 293, 379, 380, 381, 383, 387, 390, 391, 395, 408, 409, 413, 421, 456, 458
Hughson, E. A., 435, 436
Humphrey, M., 249
Hunter, J. A., 230, 238
Hurwie, M. J., 355
Hutt, C., 9, 22, 29, 43, 49, 60, 64, 65, 66, 68, 102, 120, 123, 137, 329, 334
Hutt, S. J., 9, 22, 29, 43, 64, 66, 67, 68, 100, 102, 137, 329, 334

Ingleby, J. D., 283, 294, 386, 409
Inhelder, B., 143, 154
Ini, S., 181, 220, 238, 274, 278, 286, 292
Irwin, S., 338, 345, 353
Isaev, D. N., 7, 22, 23
Itard, J. G., 440, 451

Jackson, R., 435, 436
Jaffe, J., 66, 69
Jaros, E., 470, 474
Jenkins, R., 182
Jervis, G. A., 38, 43
Johnson, A., 312
Johnson, C. A., 378
Johnson, E., 195
Johnson, S. M., 375, 376, 378, 424, 436
Johnson, W. T., 181, 198, 351
Johnston, M., 410, 411, 438
Jones, G. M., 132, 133, 137
Jones, N. B., 39, 43
Jorgensen, S., 174, 182
Jukes, E., 23, 495
Julia, H., 312, 335
Jusseaume, P., 137

Kagan, V. E., 7, 22
Kales, A., 182, 354
Kallman, F. J., 187, 198, 220, 238
Kalmijn, M., 351
Kamp, L. N. J., 220, 239
Kanner, L., 1, 2, 3, 4, 5, 7, 8, 9, 11, 15, 22, 27, 28, 29, 40, 41, 43, 44, 47, 54, 60, 63, 69, 85, 86, 102, 155, 161, 219, 222, 238, 244, 249, 252, 266, 269, 273, 278, 294, 297, 301, 303, 311, 313, 318, 324, 325, 338, 339, 347, 353, 439, 451, 463, 464, 467, 470, 474, 476, 477, 486, 488, 495, 503, 505, 507, 516
Kaplan, B. S., 363, 368

Karon, B. P., 275, 287, 288, 294
Kasanin, J., 274, 295
Kassorla, I. C., 409, 421, 516
Katz, R. C., 375, 378
Kaufman, I., 252, 266, 274, 278, 294, 425, 436
Kean, J. M., 220, 239
Keating, W. J., 239
Keeler, W. R., 38, 44, 220, 239, 243, 249
Kelly, P., 295
Kendall, R. E.. 292
Kendon, A., 56, 60
Kent, L. R., 442, 449, 451
Kepecs, J. C., 295
Kidd, J. S. H., 249
Kim, S., 354
Kimberlin, C. C., 6, 20, 21, 366
King, P. D., 281, 294
Kinsbourne, M., 437, 452
Kipper, D., 183
Kirkland, J., 434, 437
Klebanoff, L. B., 253, 264, 266, 273, 286, 287, 294
Klein, D., 282, 294
Klein, H. R., 476, 494
Kling, A., 213, 218
Kluver, H., 212, 218
Knight, E., 294
Knobloch, H., 233, 238, 239, 354
Koegel, R. L., 66, 69, 82, 84, 100, 102, 367, 376, 378, 409, 437, 451, 457, 460
Koegler, R. R., 124, 136, 471, 473, 494
Koenigsknecht, R. A., 437
Koh, C., 350
Kohn, M., 266
Kolodny, H. D., 354
Koluchova, J., 40, 44, 331, 334
Kolvin, I., 7, 13, 15, 22, 38, 44, 120, 125, 137, 243, 249, 276, 279, 286, 287, 293, 294, 329, 335, 338, 353
Koper, A., 183

Korein, J., 341, 350, 351, 353
Korn, S. J., 21, 102, 198, 238
Kotsopoulos, S., 220, 239
Kraemer, H. C., 365, 366
Kram, E., 139
Krauser, R., 182
Krech, D., 305, 312
Krieger, H. P., 125, 138
Krinsky, L., 354
Krug, O., 303, 312
Kuchar, E., 410
Kuipers, C., 7, 25, 224, 240, 270, 474
Kurtz, G. R., 230, 239
Kurtz, P. D., 434, 437, 456, 460

Lacherbruch, P. A., 96, 102, 130, 137
Laffont, F., 122, 137
LaFranchi, S., 138, 139
Landau, W., 474
Landgrebe, D., 189, 195, 196
Langer, S. K., 324, 326
Lansing, M. D., 424, 438, 439
Lasagna, L., 354
Laufer, M. W., 2, 23
Laukhuf, C., 103
Lawson, A., 283, 292, 294, 386, 408, 409
Leberfeld, D. T., 437
Lebovici, L., 24, 452, 495
Lee, D., 22, 43, 102, 137, 334
Lee, H. Y., 139
Lee, J., 189
Lee, J. C. M., 138
Lee, L. L., 434, 437
Lee, Y. H., 138
Leff, J. P., 281, 290, 293, 298, 301
Lehman, E., 220, 239
Leiken, S. J., 355
Leitch, I. M., 293
Lelord, G., 122, 137
Lennard, H. L., 281, 294
Lenneberg, E., 84, 92, 324, 326, 331, 335, 363, 367

Lennox, C., 277, 286, 294
Lesser, S. R., 239, 244, 249
LeVann, L. J., 343, 353
Levidow, L., 353
Levy, D., 293
Lichstein, K. L., 364, 367
Lin, T–Y, 24, 25, 452, 495
Lintzenich, J., 182
Lipman, R. S., 337, 340, 353
Lipsett, L. P., 305, 312
Lockyer, L., 4, 5, 6, 13, 15, 23, 24, 65, 69, 86, 87, 89, 90, 102, 104, 160, 161, 240, 244, 249, 250, 272, 279, 288, 294, 295, 338, 354, 423, 437, 453, 460, 461, 474, 495
Loen, L., 84
Loftin, J., 276, 286, 288, 296, 361, 367
Loftus, J. B., 239
Lombrosco, C. T., 230, 239
Long, J. S., 378, 409, 437, 460
Lordi, W., 295
Lotter, V., 2, 8, 15, 23, 32, 35, 44, 65, 69, 89, 102, 220, 233, 239, 278, 279, 294, 315, 326, 475, 488, 495
Lourie, R. S., 312, 476, 495
Lovaas, O. I., 67, 69, 220, 239, 315, 323, 326, 355, 364, 367, 368, 369, 378, 379, 380, 404, 409, 413, 414, 421, 424, 435, 437, 443, 451, 454, 456, 457, 460, 513, 516
Lovitt, T. C., 437
Lowe, M., 33, 35, 44, 272, 279, 294
Lowery, V. E., 355
Lucas, A., 182
Lucas, J., 169, 178, 183

McAdoo, W. G., 251, 255, 266, 276, 297, 298, 332
McAndrew, J. B., 348, 353, 495
McArthur, D., 272, 293

McBride, G., 54, 60
McCabe, K. D., 240, 355
McConnell, O. L., 67, 69, 379, 409
McCulloch, M. J., 28, 44, 100, 103, 122, 137
MacKeith, R., 181
Macmillan, M. B., 17, 23
McNay, A., 249
McNeill, D., 72, 84
McPherson, H., 410
McQuaid, P. E., 220, 239
Madsen, C. H., 435
Mahanad, D., 189, 191
Mahler, M., 304, 312, 363, 367, 425, 437, 507, 516
Maier, K., 379, 409
Makita, K., 7, 13, 23
Malamud, N., 7, 23
Malcolm, R., 132, 133, 137
Mancall, E. L., 218
Mann, L., 367, 442, 451
Marchant, R., 12, 23, 293, 381, 389, 408, 409, 421, 460
Margolis, R. A., 355
Markham, C. H., 136, 139, 473
Markley, J., 437
Marlowe, W. B., 213, 218
Marshall, G. R., 379, 409
Marshall, J. E., 125, 137
Martin, J. A. M., 14, 24
Martin, M., 410
Mash, E., 278, 375
Mason, A., 138, 139
Mathis, M. I., 379, 409
Matsuno, D., 25, 183, 517
Maxwell, A. E., 87, 91, 103
Mayer-Gross, W., 187, 198
Meecham, M. J., 223, 239, 409
Mees, H., 378, 411
Meichenbaum, D. H., 456, 461
Meier, M., 350
Melchior, J. C., 211, 218
Mellerup, E., 182
Meltzer, D., 324, 326
Menolascino, F. J., 243, 249, 488, 495
Menyuk, P., 82, 84, 105, 108, 109, 113, 116

Merlis, S., 355
Metz, J. R., 379, 409
Metzer, H., 238
Meyer, R. E., 218
Meyer, R. G., 274, 275, 294
Meyers, D. I., 252, 265, 266, 276, 278, 281, 286, 287, 288, 289, 291, 293, 294, 297, 301, 332, 335
Mickey, M. R., 96, 102
Mielke, D. H., 354
Milham, S., 238, 253
Millar, S., 362, 367
Miller, A., 380, 409, 449, 451, 471, 474
Miller, E., 380, 409, 449, 451, 471, 474
Miller, R. T., 14, 15, 16, 17, 23, 367
Miller, W. H., 122, 136, 137
Milner, B., 214, 215, 218
Milstein, V., 139
Minde, K. K., 340, 353
Miners, W., 378
Mitchell, G., 181
Mittler, P. J., 5, 23, 233, 234, 239, 240, 488, 495
Mnukhin, S. A., 7, 23
Moerk, E., 287, 294
Money, J., 243, 249
Montenegro, H., 25
Montgomery, J., 435
Moore, D. J., 290, 295
Moore, M., 233, 239
Moor House School, 500, 505
Morehead, S. D., 139
Morrison, A. R., 125, 139
Mulhern, S., 437
Murphy, L. B., 323, 326
Murphy, R., 139
Murrell, S., 280, 292

Narasimachari, N., 172, 182
Nash, H., 340, 353
Nasrallah, H., 354
Naughton, J. M., 72, 84
Neisworth, J. T., 434, 437, 460
Nelson, K. E., 380, 408
Netley, C., 276, 277, 295

Newman, S., 24, 28, 42, 61, 84, 103, 240, 266, 292
Newsom, C. D., 160, 161, 162
Newson, E., 48, 56, 59, 60
Newson, J., 48, 56, 59, 60
Ney, P., 424, 437
Nichtern, S., 238
Nielsen, R., 211, 218
Niemenson, G., 378
Norton, J. A., 22, 43, 68, 102, 266, 291, 292, 494
Nyman, L., 266

O'Brien, R. A., 21, 170, 181, 189, 198, 350, 516
O'Connell, E. J., 287, 295
O'Connor, N., 9, 13, 18, 22, 28, 43, 64, 68, 72, 84, 88, 90, 91, 102, 103, 120, 122, 137, 142, 145, 147, 148, 149, 153, 154, 157, 161, 215, 218, 315, 326, 329, 330, 333, 334, 436, 441, 451
O'Dell, S., 380, 409
Odom, P. B., 101, 103
Offord, D. Q., 16, 23
Ogdon, D. D., 275, 288, 295
O'Gorman, G., 14, 23, 181, 221, 239, 270, 295, 481
Ohtsuka, Y., 125, 138
O'Leary, K. D., 434, 437
Oman, C. M., 132, 133, 139
O'Neal, P. L., 424, 437
Onheiber, P., 120, 137
Oppenheim, A. N., 253, 266, 272, 286, 295
Oppenheim, R., 441, 451
Orford, F., 181
Ornitz, E. M., 3, 9, 14, 23, 25, 28, 44, 86, 100, 103, 117, 119, 120, 121, 123, 124, 125, 126, 130, 131, 137, 138, 139, 158, 162, 165, 169, 182, 199, 219, 220, 239, 243, 245, 247, 248, 249, 250, 279, 295, 329, 335, 345, 347, 353, 471, 474, 511, 516
Osborn, J., 442, 449, 451

Osofsky, J. D., 287, 295
Ottinger, D., 72, 84
Ounsted, C., 9, 22, 43, 44,
  64, 65, 66, 102, 137, 249,
  294, 334, 335
Oxford, D. R., 187, 199

Pace, Z. S., 379, 408
Palbesky, A. E., 437
Palella, B. L., 495
Pampiglione, G., 38, 43,
  120, 136, 329, 334
Pangalila–Ratulangi, E. A.,
  343, 354
Panman, L. M., 138
Parmelle, T., 138
Partington, M., 190, 199
Pasamanick, B., 233, 239,
  338, 354
Patterson, G. R., 375, 378,
  380, 383, 409
Pauling, L. C., 314, 325
Peery, J., 69
Penati, G., 346, 350
Perel, J. M., 355
Perkins, M., 378
Perloff, B., 355, 367, 409
Perry, D., 23
Peterfreund, O., 279, 293
Petersen, O., 218
Peterson, M., 188
Petti, T., 351
Pfaltz, C. R., 125, 138
Pfeiffer, E., 425, 435
Phipps, E., 183
Piaget, J., 143, 154, 324,
  326, 362, 367
Pickering, G., 434, 435, 453,
  455, 456, 457, 460
Piffko, P., 125, 138
Piotrowski, Z. A., 495
Pitfield, M., 253, 266, 272,
  286, 295
Pletscher, A., 203, 205
Plotkin, S., 182, 183, 199,
  354
Polan, C. G., 220, 339
Polevoy, N., 351
Polizos, P., 348, 352, 353,
  355

Pollack, M., 125, 138, 282,
  294
Pompeiano, O., 125, 139
Pontius, N., 21, 83, 182,
  266, 291, 292, 366
Poole, D., 343, 354
Popper, K., 3, 23
Porterfield, J. K., 458, 461
Pribram, K., 441, 451
Prior, M., 12, 23, 338, 353
Pronovost, W., 86, 92, 103
Pulashi, M. A., 317, 326
Purchatzke, G., 355
Putnam, M., 274, 280, 295,
  312, 364
Putnam, N. H., 138

Quay, H., 280, 295
Quinn, P. O., 192, 198
Quinton, D., 273, 292, 295

Rabin, K., 163
Rabinovitch, R., 183
Rachman, S., 160, 161, 471,
  474
Rafaelson, O. J., 182, 344,
  353
Rank, B., 175, 182, 272,
  274, 280, 295, 304, 312,
  508, 516
Rapin, I., 124, 139
Rapoport, J. L., 192, 198
Rashkins, S., 182, 354
Ratusnick, C. M., 449, 451
Ratusnick, D. L., 449, 451
Raven, J. C., 91, 103
Ray, R. S., 378
Read, J., 173, 183
Reed, G. F., 28, 44
Rees, J., 9, 25, 28, 44, 67,
  69, 273, 296, 330, 335
Rees, S. C., 89, 103, 423,
  437, 490, 495
Reichler, R. J., 333, 335,
  361, 367, 380, 409, 410,
  413, 414, 416, 419, 421,
  437, 440, 443, 444, 445,
  451, 452, 453, 461
Reiser, D., 271, 295
Rendle–Short, J., 3, 23

Rexford, E. N., 321
Reynell, J., 33, 44, 224,
  239, 408, 499, 505
Ricciardi, F., 350
Rice, C., 275, 287, 295
Richards, B., 50, 61, 64, 69,
  406, 410
Richardson, L. M., 249, 294
Richer, J. M., 29, 43, 44, 47,
  49, 50, 51, 56, 60, 61, 64,
  69, 137, 406, 410
Richman, N., 235, 240
Ricks, D. M., 3, 10, 23, 25,
  29, 40, 41, 44, 47, 61, 65,
  69, 100, 103, 106, 116,
  333, 335, 406, 410
Rie, H. E., 292
Riesen, A., 331, 335
Riggi, F., 350
Rimland, B., 15, 17, 21, 24,
  181, 182, 195, 198, 239,
  243, 250, 290, 295, 314,
  326, 345, 354, 414, 421,
  440, 451, 467, 474, 516
Rincover, A., 457, 460
Risley, T., 72, 84, 161, 378,
  379, 410, 411, 438, 461
Ritvo, E. R., 3, 23, 25, 126,
  131, 133, 139, 163, 165,
  166, 167, 168, 169, 176,
  182, 183, 187, 189, 198,
  199, 239, 247, 249, 250,
  279, 295, 335, 344, 347,
  353, 354, 367, 471, 474,
  511, 516, 517
Ritz, M., 295
Rivera–Calimlin, L., 347,
  354
Roberts, A., 104, 287, 293
Roberts, G., 173, 174, 182
Robertson, J., 328, 335
Robins, L. N., 14, 24, 424,
  437
Robinson, J. F., 42, 312
Robson, K., 66, 69
Rodnight, R., 201
Rodriguez, A., 22, 353, 474,
  495
Roeske, N., 182
Roffman, L. S., 198
Rogers, W., 181
Rolo, A., 344, 354

Romanczyk, R. G., 424, 434, 437, 456, 461
Roney, M., 410
Roniger, J. J., 354
Rose, A. L., 230, 239
Rose, S. P. R., 334
Rosenberg, S., 287, 296
Rosenblatt, J. S., 358, 366
Rosenblum, E., 436
Rosenblut, B., 471, 474
Rosenzweig, M. R., 312, 331, 332, 335
Rosman, N. P., 22, 218, 249
Ross, I. S., 7, 24
Roth, M., 44, 137, 249, 335
Rovner, L., 139
Rowland, V., 139
Ruder, K. F., 455, 461
Rusir, D., 266, 294
Russell, S., 423, 424, 435, 437
Russo, D. C., 377, 378
Rutt, C. N., 187, 199
Ruttenberg, B. A., 54, 61, 271, 280, 293, 295, 362, 368, 404, 407, 410, 438
Rutter, M., 1, 2, 4-6, 8, 10, 11, 13-25, 27, 28, 40, 42, 44, 47, 59, 61, 65, 67-69, 71, 82-92, 99, 101-104, 116, 126, 139, 157, 161, 169, 182, 183, 219-222, 225, 233-235, 237-239, 244, 245, 249, 250, 253, 262, 266, 269, 271-273, 276-279, 283, 288, 291-295, 315, 326-331, 333, 335, 338, 347, 354, 357, 367, 378, 380-382, 384, 386, 387, 404, 408-410, 414, 421, 423, 426, 427, 435, 437, 440, 452-457, 459-461, 470, 474, 481, 483, 486-488, 495, 497-505, 507-509, 516, 517
Ryan, B., 442, 449, 451
Ryan, R., 436, 494

Sachar, E. J., 350
Sadoun, R., 24, 452, 495
Saeger, K., 182, 199

Sage, P., 294
Sakalis, G., 351
Saladino, C., 176, 183
Salvatore, P. D., 436, 494
Samit, C., 350
Sander, L. W., 305, 312, 332, 335
Sanders, B. J., 408
Sanders, F. A., 17, 22
Sankar, B., 183, 199
Sankar, D. V. S., 350
Sartorius, N., 25, 510, 517
Sarvis, M. A., 243, 250, 271, 295
Satterfield, B. T., 354
Satterfield, J. H., 338, 354
Sauvage, D., 137
Scanlon, J. P., 435, 437
Schaeffer, B., 69, 239, 355, 367, 409
Schain, R. J., 3, 25, 164, 167, 183, 199, 243, 250, 347, 354
Schico, A., 182, 354
Schiele, B. C., 348, 354
Schilder, P., 471, 474
Schmidt, J., 112, 116
Schneiderman, G., 25, 183, 517
Schneirla, T. C., 358, 367
Scholl, M., 293
Schopler, E., 13, 25, 72, 84, 276, 286, 288, 296, 297, 333, 335, 361, 367, 380, 409, 410, 413, 414, 416, 419, 421, 434, 437, 439, 440, 441, 443, 444, 445, 451, 452, 453, 457, 461, 507
Schover, L. R., 160, 161, 162
Schreibman, L., 162, 364, 367, 435, 460
Schweier, H., 182
Segal, J., 305, 312
Seidal, U. P., 219, 239
Seigle, G. M., 287, 296
Semenek, G., 189
Senden, M. von, 147, 154
Shader, R. I., 347, 351, 352, 354
Shaffer, D., 25, 517
Shapiro, A., 181
Shapiro, T., 86, 104, 341,

Shapiro, T., *(cont'd.)* 346, 350, 351, 352, 353, 357, 366, 474, 494
Sharpe, L., 292
Shatz, M., 109, 116
Shaw, C., 173, 174, 183
Shaw, D. A., 378
Shaywitz, B., 102, 181, 198, 351
Sheik, D. A., 290, 295
Shepherd, M., 25, 517
Sheridan, M., 33, 35, 44, 499, 505
Sherman, A. C., 348, 354
Sherman, J. A., 458, 461
Sherwin, A., 220, 240, 254, 345
Sherwood, U., 323, 325
Shields, J., 227, 238
Simmons, J. Q. III, 69, 199, 239, 344, 354, 355, 378, 409, 434, 437, 451, 460
Simon, G. B., 348, 355
Simon, J. B., 240
Simon, R., 292
Simpson, G., 354
Sinclair-de-Zwart, H., 364, 367
Singer, M., 272, 275, 287, 296
Siva-Sankar, D., 168, 176, 177, 183, 189, 199
Sivertsen, B., 138
Skinner, B. F., 414, 421
Slagle, S., 295
Slater, E., 233, 239
Sloane, H. N., 379, 410
Sloman, L., 410
Small, A., 181, 351, 352
Small, A. M., 350
Small, J. G., 120, 121, 122, 139
Smith, C., 220, 238, 240
Smith, J. O., 437, 442, 449, 451
Smith, M. D., 455, 461
Sneznevsky, A. V., 24, 495
Snow, C. E., 287, 296
Snyder, L. K., 434, 437
Solow, E., 182
Soltys, J. J., 352
Soper, H. H., 244, 249
Sorosky, A. D., 9, 23, 25,

Sorosky. A. D., *(cont'd.)*
131, 138, 139, 243, 250,
335
Spector, S., 189
Speers, R. W., 424, 438
Spence, M. A., 187, 199
Spencer, B. L., 220, 239
Spitzer, R. L., 293, 334
Spradlin, J. E., 287, 296
Sprague, R. L., 341, 355
Stechler, G., 312, 335
Steele, R., 21, 43, 68, 102,
266, 494
Stephant, J. L., 137
Stern, O., 54, 61, 69, 362,
368
Stevens, J., 451
Stevenson, J., 235. 240
Stockwell, C. W., 132, 139
Stokes, P. E., 354
Stone, F., 181
Strain, P. S., 456, 461
Strauss, J., 354
Strellioff, D., 137
Stroh, G., 181
Strotzka, H., 25
Sturge, C., 517
Stutsman, R., 31, 44, 224,
240, 385, 410
Stutte, H., 220, 240
Sulestrowska, H., 490, 495
Sulzbacher, S. I., 379, 410
Sussenwein, F., 333, 335,
380, 381, 382, 410
Sutton, E., 173, 183
Sverd, J., 355
Sweeney, N., 84
Swisher, L., 365, 367
Sylva, K., 323, 326
Symmes, J., 194
Szurek, S. A., 21, 270, 271,
291, 296, 304, 305, 307,
312, 327, 329, 330, 334,
335

Taft, L. T., 15, 25, 38, 44,
221, 236, 240, 243, 250
Tanguay, P. E., 138
Tarjan, G., 19, 25, 347, 355
Tate, B. G., 379, 410, 421
Taveras, J. M., 211, 218

Taylor, A., 89, 103, 437,
490, 495
Taylor, J. E., 423, 426, 438,
441, 449, 452
Terzian, H., 212, 218
Thomas, D. R., 435
Thomas, E. R., 295
Thomas, J. J., 218
Thygesen, P., 218
Tinbergen, E. A., 3, 14, 25,
29, 44, 64, 67, 69, 270,
280, 296, 406, 410
Tinbergen, N., 3, 14, 25, 29,
44, 64, 67, 69, 270, 280,
296, 406, 410
Tizard, B., 9, 25, 28, 44, 64,
67, 69, 273, 296, 330, 335
Tizard, J., 25, 424, 438
Tjernström, Ö., 126, 139
Torrey, E. F., 188, 230, 240,
338, 355
Tranzer, J. P., 205
Treffert, D., 188, 199, 233,
240, 279, 296, 353, 490,
495
Trimble, M., 21
Trost, F. C., 378, 408
Trunnell, G., 461
Tu, J., 190, 199
Tubbs, U. K., 13, 25, 86,
104
Turner, D. A., 340, 355
Tustin, F., 271, 296, 471,
474

Uhlenhuth, B. H., 355

Vaillant, G. E., 2, 25, 220,
240
Vaizey, M. J., 2, 22, 49, 60,
64, 67, 68
van Krevelen, D. A., 2, 7, 25,
224, 237, 240, 470, 474
Verhees, B., 219, 240
Viitamaki, R., 291, 494
Vine, I., 66, 69
Vrono, M., 2, 7, 13, 27

Waelder, R., 362, 368
Waizer, J., 343, 352, 353,
354

Wakstein, D. J., 103
Wakstein, M. P., 103
Walker, H. A., 192, 193
Wallerstein, J., 507, 516
Walter, D. O., 122, 139
Walter, R. D., 138, 239
Ward, A. J., 280, 296
Ward, P. H., 137
Ward, S., 182
Ward, T. F., 220, 240
Warrington, E. K., 100, 102,
213, 218
Wasman, M., 120, 139
Watson, J. B., 414, 421
Weber, D., 220, 240
Webster, C. D., 380, 410
Wechsler, D., 224, 240, 385,
410, 470, 474
Wechter, V., 138
Weiss, G. C., 340, 353
Weiss, H., 72, 84
Welsh, G. S., 266
Wenar, C., 362, 368, 404,
407, 410, 424, 438
Wendt, G. R., 125, 132, 139
Werner, H., 363, 368
Werry, J. S., 262, 266, 280,
295, 341, 355
Wetzel, R. J., 379, 380, 410
White, B., 323, 326
White, P. T., 137
Whitmore, K., 25
Whittam, H., 233, 240
Widelitz, M., 175, 183
Wiegerink, R., 453, 456,
460, 461
Wiener, J. M., 349
Wikler, L., 199
Wile, R., 353, 474, 494
Wilhelm, H., 82, 84
Willer, L., 266, 294, 436
Willerman, L., 231, 241
Williams, C., 100, 103, 137
Williams, L., 28, 44
Williams, S., 270, 296, 426,
438
Wilson, C., 138
Winett, R. A., 434, 438
Wing, J. K., 2, 25, 31. 44,
271, 296, 338, 355, 459,
461

Wing, L., 3, 9, 10, 23, 25,
27, 28, 29, 31, 32, 33, 40,
41, 44, 45, 47, 49, 59, 61,
63, 65, 67, 69. 71, 84, 86,
92, 100, 103, 106, 116,
153, 154, 220, 233, 241,
243, 250, 316, 326, 333,
335, 406, 410, 426, 438,
440, 441, 445, 452, 453,
455, 457, 459, 461
Winkler, R. C., 434, 438
Winsberg, B. G., 348, 355
Witmer, H. L., 303, 312
Wolf, E. G., 379, 410, 424,
438
Wolf, M. M., 72, 84, 161,
375, 378, 379, 410, 411,
438

Wolff, H., 211, 218
Wolff, S., 9, 25, 86, 104
Wolkind, S. N., 28, 45, 273,
296
Wolpert, A., 342, 343, 355
Wood, E. M., 211, 218
Worden, F. G., 2, 21
World Health Organization
(WHO), 338, 355
Wright, B. A., 21
Wright, S. W., 355
Wynne, L., 272, 275, 287,
296
Wypych, M., 191

Yahalom, I., 295
Yang, E., 21, 291

Yannet, H., 3, 25, 164, 165,
183, 199, 243, 250
Yeates, S. R., 45, 241
Yepes, L., 355
Youdim, M. B. H., 189
Young, D., 378
Young, G., 164, 181
Young, L. R., 125, 132, 139
Yudkin, S., 181
Yudkovitz, E., 293
Yule, W., 23, 24, 156, 157,
162, 292, 293, 330, 335,
378, 380, 405, 408, 409,
411, 421, 424, 438, 460
Yuwiler, A., 17, 25, 163,
165, 166, 171, 182, 183,
189, 190, 199, 354, 511,
517

Subject Index

Abstraction, 88, 90
Addison's disease, 243
Adenosine triphosphate (ATP), 170
Adolescence
  changes during, 486, 503
  sexuality, 503
  See also Follow-up studies
Adrenoleukodystrophy, 211
Adventitious movements, 193-194
Affective contact, lack of, 34
Affective processing, 316
Age of onset, 2, 6-7
  and cognitive defect, 154
  and follow-up studies, 477
Aggression toward autistic children, 49
Allergic responses, 178
American Sign Language, 365
Amphetamines, dextro- and levo-, 345
Amygdala, 217
Anthropology, 59-60
Approach behavior of child, 49, 64
  compared with normal children, 52
Arithmetic skills. *See* Educational attainment
Arousal
  and EEG, 329
  of midbrain limbic system, 316
  See also Overarousal
Ascending reticular formation, 316, 328-329
Asperger's Syndrome. *See* Autistic psychopathy
Assessment for education, 445-446
Attachment behavior, 9, 306
Attentional behavior, operant control of, 387
"Atypical" children, 463-474
Autistic psychopathy, 7, 224, 470
Automated teaching, 365

Autonomic responses
  galvanic skin response, 122
  heart-rate variability, 123
  studies, 119, 122-123
Aversive techniques, 364-365
Avoidance motivation, 50, 55. *See also* Social avoidance

Bannister-Fransella Grid Test, 277
Behavioral techniques, 156. *See also* Behavior modification; operant techniques
Behavior analysis, applied, 156
Behavior interactions, 64
Behavior modification, 314, 364-365
  and degree of impairment, 415
  and developmental appropriateness, 381, 405, 419
  and family stress, 390
  follow-up
    6-month, 391-398, 502
    18-month, 399-403, 502
  generalization of treatment, 369, 380, 403
  history of, 414
  long term benefits, 382
  North Carolina program, 416-417
  parental commitment, 405, 414-417
  with parents as therapists, 369-378, 380-411, 413-421
  practical procedures with parents, 371-374, 382-383, 405
  techniques of training parents, 375
  See also Operant techniques
Behavior Rating Instrument for Autistic Children (BRIAC), 362
"Bell and Pad," 381, 390
Biochemical studies, 163-183, 185, 201-205

  behavioral rating scales, use of, 180
  diagnostic specificity, 179, 511, 513

Biochemical studies (*cont'd*)
  dietary manipulations, 180
  difficulties in experimental design, 163,
    180, 202
  research strategies, 204
Biological hazards, 230
Blind children, 66
Blood cortisol level and decreased eo-
  sinophil count, 178
Blood lead level, 179
Brain pathology, 7
Brain stem function, 100
Broken homes, 272
Bufotenin, 171-172
Butyrophenones, 343
  haloperidol, 338, 343, 348

Catecholaminergic system, 164, 174-176,
  189
Catechol-O-methyl transferase, 191
Cattell 16PF, 276
Celiac disease, 178, 196-197, 243
Cerebral lateralization, 215-217
Cerebral lipoidosis, 243
Cerebrospinal fluid, 163, 175, 190, 348
Changes with age, 40. *See also* Follow-up
  studies; Prognosis
Characteristics of autistic children, 28, 508.
  See also Classification; Definition;
  Diagnostic criteria
Childhood schizophrenia, compared with
  autism, 13, 15
Child rearing patterns, 253, 271
Children's Behavior Inventory, 340
Children's Psychiatric Rating Scale, 340
CIBA foundation, 1970 symposium, 157
Classification of, 507-517
  by hypothesized variable, 513-514
  multiaxial, 19, 509-515
  purposes of, 510
  by specific behaviors, 512-513
  by subgroup, 511-512
  by syndrome, 510-511
  for therapeutic and administrative
    purposes, 514-515
Clinical variability and neurophysiologic
  studies, 128-130
Clumsy children, 471, 501
Codes
  and education, 441
  double storage, 146

Codes (*cont'd*)
  image, 142
  spatial, 143-154
  temporal, 143-154
Cognitive defect
  as central problem, 88, 98
  and language, 47, 85-92
  and treatment, 315, 360-361
  related to social abnormalities, 100
Cognitive processes
  and language, 152
  codes, 142-154
  qualitative analysis, 144
  rules, use of, 153
  symbolic operation, 143
Columbia Mental Maturity Scale, 97
Communication, linguistic vs. nonlinguistic
  organization, 113. *See also* Language
Communication interactions, 56, 114-115
Concept formation, 28
  and education, 441
Concordance in twins, 226
Conditionability, 81, 108-109
Conditioning, 72
  of language, 72, 78
  See also Behavior modification; Operant
    techniques
Congenital syphilis, 15
Conners Parent-Teacher Questionnaire, 340
Control groups, 141
  blind children, 142, 144, 147
  choice of, 156-160
  deaf children, 142, 144-146
  in family studies, 287
  mental age, 159-160
  normal children, 82, 142, 145
Convulsive therapy, 361
Coping skills of parents, 398
Copper serum level, 191
Cornell Medical Index, 276
Counseling. *See* Psychotherapy
Cranial circumference, 188, 192
Cross modal integration and education, 441
Culture
  conceptions of, 48
  partial noncommunication of, 47-60
  and symbols, 48, 60
Curriculum
  content of, 455-457
  individualized, 439-452
    developmental therapy model, 444
    materials, 449

Curriculum (*cont'd*)
  individualized (*cont'd*)
    and new educational approaches, 449-
      450
    problems involved, 443
    school and home, 446
    task analysis, 447-448, 458
    therapist qualities, 444-445
Cytomegalic inclusion virus, 188

Definition, 3
  and education, 440-441
Dermatoglyphic studies, 192-193
Desensitization and phobias, 390, 405
DesLauriers—Carlson hypothesis, 315-316,
      329
Developmental language disorder compared
      with autism, 14, 92-100, 106-108
Diagnostic criteria, 2-5, 17, 19, 119, 178,
      186, 209, 224
  Creak, 209, 508
  DeMyer classification, 74
  differences in, 29
  and follow-up studies, 475
  Rimland classification. *See* E2 scale
Dihydroindolones, 343
  molindone, 343, 347
N,N-Dimethyltryptamine, 172
Diphenylbutylpiperidines, 344
  pimozide, 344
Discriminant function analysis, 95
Discriminative stimuli
  ability to use, 81
  use of, 76
Dopamine-beta-hydroxylase, 191
Down's syndrome, 15, 31, 38, 42, 67, 194
Drug studies, methodological issues, 339-
      340, 358-360. *See also* Pharmaco-
      therapy
Dysmaturation, 359-360

E2 scale, 17, 166, 169, 467
Early environment and development, 305,
      350
  critical periods, 331
  of brain, 331
Early intervention, 306
Echolalia, 11, 79, 82, 86, 93, 96
  and behavior modification, 391-392
  and normal imitative linguistic behavior, 109
  and paired associate learning, 112
  immediate vs. delayed, 106

Edinburgh Articulation Test, 94
Education
  behavioral principles, 457-458
  and Europe, 315
  goals of, 443, 453-461
  group and individual therapy, 424
  operant methods, 424, 443, 457-458
  preliminary assessment, 445-446
  programmed teaching, 424
  See also Special education
Educational attainment
  and child characteristics, 431-433, 454
  as function of type of school, 431-432
  and IQ, 445, 453
  measures of, 428, 448-449
Educational milieu therapies, 360-361
EEG (Electroencephalogram)
  and high IQ, 471
  studies, 119, 120, 134, 179, 210-212
Ego development, 274, 307
Elective mutism, 14, 86
Electroconvulsive therapy, 361
Empathy, 10
Employment, 465-467, 480, 504
"Empty clinging," 50
English Picture Vocabulary Test, 428
Enuresis, use of 'Bell and Pad,' 381, 390
Environmental trauma, 271
Epidemiological studies, 27-42, 65, 188
Epilepsy, 8, 18, 89, 190, 485-486, 498
Error patterns, 75
  in language training, 75-80
Ethological theories, 29, 64
  human ethology, 49
Etiology, 28, 101, 251, 507, 509
  and treatment, 327-335
  hierarchy of, 246
Experimental research vs. treatment in
      behavior modification, 417-420
Eye contact, 33, 41, 63, 65. *See also* Gaze
      avoidance
Eysenck Personality Inventory, 276

Familial autism, 195
Family adequacy, 281
Family counseling, 383
Family factors, 269-296
  control groups in the study of, 287
  diagnosis in the study of, 287
  influence, modes of, 289-290
  methodological issues, 288-289, 297-301
  overview, 285-296

Family factors (*cont'd*)
  and psychoanalytic studies, 300-301
  and schizophrenia, 299
  stress, 332
Family history, 194
  neuropsychopathology, 209
Family interaction, 252
Family measures, 386-387
Fels Parent Behavior Rating Scales, 280
Ferreira and Winter Unrevealed Differences
    Task, 280
Fish's Scale, 340
Follow-up studies, 463-495
  age
    at follow-up, 477
    on onset, 477
  of children with high IQ, 463-474
  diagnosis, 475-476
  institutional placement, 480
  methodology, 475-479
  outcome
    criteria, 478-479
    prediction of, 482
  sample size, 476
  social relationships, 486
  summary of 20 studies, 488-493
Functional analysis, 382
Fundamental disorder. *See* Primary
    handicap

Gaze avoidance, 3, 9. *See also* Eye contact
Genetic influences, 219-241, 245
  and environmental interaction, 236-237
  family studies, 219-220
  twin studies, 231-236, 514
Gesture, 11, 93, 380
Global Clinical Impressions, 340
Goldstein-Scheerer Object Sorting Test, 276
Grammar, 11, 94

Hallucinations, 14
Haloperidol and behavior therapy, 338
Hematologic studies, 176-179
  and diagnostic specificity, 511
  maturational lag, 176, 177
  serum magnesium level, 176
  serum potassium level, 176
Heredity, 230-236, 514. *See also* Genetic
    influences
Herpes simplex virus, 188
Histamine Wheal Test, 177
Hollingshead—Redlich index, 179

Homovanillic acid, 349
Huntington's Chorea, 202
Hydrocephalus, 211
5-Hydroxyindoleacetic acid, 173, 174-175,
    190, 349
Hyperuricosuria, 195
Hypothyroidism, 187
"Hypsarrhythmia," 15

Image, use of term, 144. *See also* Codes
Imitation
  and education, 441
  verbal, nonverbal, and behavior modifica-
    tion, 373-374
Individual differences, and language, 402,
    407
Individual therapy, 361
Indoleacetic acid, 173
Indoleamines, 164-172
Infantile spasm syndrome, 190, 243
Infectious agents, 188
Inner language, 324
  and "inner pictures," 143
Inorganic phosphate, 177
"Insistence on sameness," 11-12
Intelligence, 5, 18
  high and low, 91, 463-474
  patterning of, 470-472
  and prognosis, 482-484, 501
  and scholastic achievement, 89
  stability of, 486
Intelligent children, 463-474
  autism, degree of, 467-468
  employment of, 465-467
  family characteristics, 469
  follow-up, 465
  intelligence test patterns, 470
Interhemispheric transfer, 215
Internal representation of objects and ac-
    tions, 153
Interviews,
  of nurses, 32
  of parents, 32, 223, 253, 283, 386, 445
  of teachers, 32
IQ. *See* Intelligence
Ittelson Center Scales, 284

Kanner's syndrome
  causation, 3
  characteristics, 1, 3, 27
  criteria, 34
  and other infantile psychoses, 17

Kernicterus syndrome, 190
Kinesthetic feedback, 122, 161
  and movement distance, 151
  optokinetic stimulation, 125
  and position, 150
Klinefelter's syndrome, 167

Language, 10-11, 105
  acquisition and use, 105
  analysis of, 386
  cerebral lateralization, 215-216
  individual differences, 402
  and motivation, 88
  nine-word language (9WL), 71-83
  normal development, 108-110
  prepositions, use of, 76-78, 82
  pronouns, use of, 79, 82
  scholastic achievement, 89
  spontaneous, 380
  stimulus-response chains, 106, 363
  twin studies, 99
  vs. speech, 363
Language abnormalities, 2, 47, 99
  and social abnormalities, 99
  and social avoidance, 40
Language, age
  and eye contact, 36
  and sociability, 35
  and stereotypies, 37
  and symbolic play, 37-38
Language comprehension
  behavior modification, 380
  and prognosis in the mildly handicapped,
    498-499
  and stereotypies, 39
  and symbolic play, 39
Language deficit, 28
  and deviance, 99
  and IQ, 87
  and prognosis in the mildly handicappted,
    498-499
Language development
  and conditioning, 72, 78
  of normal children, 83, 108-110
  and operant techniques, 364, 373-374,
    388, 391-393, 400-402, 407, 503
  and prognosis, 357
  and "theraplay," 317
Language disorder
  beyond conditioning, 71
  boundaries of, 90-92, 115
  as central problem, 71, 86, 98, 110

Language disorder (*cont'd*)
  and cognition, 152
  extent of, 99, 108
  family history, 99
  neurological substrates, 108
  type of, 86-88
  underlying cognitive disorder, 100
Lesch–Nyhan Disease, 202
Levoamphetamine, 168
Levodopa, 344
Leyton Obsessional Index, 276
Limbic system, arousal of, 316
Lithium carbonate, 344
Longitudinal studies, 107. *See also* Follow-
  up studies
LSD, 344

Macronutrients, 178-179
  free fatty acids, 179
  plasma glucose level, 179
Malabsorption, 177-178, 197
Marriage, involving autistic adults, 467
Maryland Parental Attitude Scale, 273
Mecham Language Scale, 223, 385
Megavitamin therapy, 17, 314, 345
Memory
  and cognitive deficit, 112-114
  deficit, 100
  and education, 441
  immediate, 145, 153
  meaning in, use of, 88
  and temporal lobe, 215
  See also Recall
Mental age, 6, 20, 34, 63
  control groups, 159-160
  nonverbal
    and eye contact, 36
    and sociability, 35
    and stereotypies, 37
    and symbolic play, 37-38
Mental retardation, 244
  compared with autism, 13, 17, 67
  and diagnosis, 509, 511-512
Merrill–Palmer Scale, 33, 224, 385, 428
Metaphorical language, 93
  and behavior modification, 391-392
Methodology, 151-161
  of biochemical studies, 163, 180, 202, 204
  controls for mental age, 159-160
  of drug studies, 339-340, 358-360
  of family studies, 288-289, 297-301
  of follow-up studies, 475-479

Methodology (*cont'd*)
  of parent–child interaction studies, 282,
    288-289
  single case design, 156, 340
  of twin studies, 232-234
5-Methoxy-N-dimethyltryptamine, 172
Methysergide, 344
Minimal brain dysfunction, 8
Misclassification, 98
MMPI (Minnesota Multiphasic Personality
    Inventory), 254-265, 275, 276
Modeling procedures and parent training,
    375, 418
Monoamineoxidase, 168
Mother
  emotional detachment of, 270, 280
  gesture, use of, 395
  speech, 395
Mother–infant interaction
  compared with blind children, 66
  compared with normal children, 53-54
  individuation-separation, 363
Mothers' speech and behavior modification,
    395
Motivational factors, 5
Movements, adventitious, 193-194
Multiaxial classification, 19, 509-515
Multiple discriminations, 80, 82

Neale Analysis of Reading Ability, 428
Negativism, 5
Neologisms, 86, 96
Neurodevelopmentally directed therapies,
    360
Neurofibromatosis, 212
Neuroleptics, 341-344, 345-347
Neurological disorder, 8, 16, 18-19
Neurophysiological studies, 207-218
Neurotransmitters, 164, 202
Nonverbal communication, 39. *See also*
    Gestures
Norepinephrine, 169, 189
Normal controls, lack of, 82
Novelty, 41
Numerical taxonomy, 17
Nystagmus
  caloric, 126
  electronystagmography, 125
  postrotational, 125-131
  primary vs. secondary, 131-134
  vestibular, 126

Objects, use of, functional vs. symbolic,
    108, 110, 113
Object attachment, 12, 389
Observation and Handicap Register, 30
Observation techniques, 386, 392
Obsessive questioning, 34, 90
Onset, age of. *See* Age of onset
Operant techniques
  and education, 424, 443
  and language, 364, 373-374, 388, 391-
    393, 400-402, 407, 419-420
  and specific behaviors, positive and
    negative, 379, 413
  simple vs. complex behavior, 379, 413,
    418
  See also Behavior modification
Organic brain dysfunction, 85
Organic conditions and autism
  encephalitis, 38, 41
  infantile spasms, 38, 41
  kernicterus, 38
  meningitis, 38, 41
  PKU, 38
  retrolental fibroplasia, 38
  rubella, 38, 41
  tuberose sclerosis, 38, 41
Organicity, Goldfarb's criteria, 254
Organic vs. nonorganic varieties of autism,
    254, 290, 298, 304, 511
  and outcome, 486
Orienting behavior, 157
Outcome criteria
  in follow-up studies, 478
  and "Work-School Status," 483
Overarousal, 29
Overselective attention, 160

Paired associate learning, 112
Parental Attitude Research Instrument
    (PARI), 273
Parental blame, 307-308, 361
Parental characteristics, 41
  'communicative impact,' 275
  and developmental receptive aphasia,
    272, 276
  deviance and disorder, 273-278, 298-301
  and Down's Syndrome, 272
  hypothyroidism, 187
  of intelligent children, 469
  IQ, 270, 278-279
  and parent training, 375

Parental characteristics (*cont'd*)
  personality, 251-267, 273-278
  questions, use of, 281
  schizophrenia, 275, 277-278
  social class, 270, 278-279
  social introversion, 257, 263
  stress, 270-273
  thought disorder, 276-277
Parental counselling, 333
Parental depression, 271, 469
"Parental perplexity," 274, 280. *See also*
    Parental blame
Parental rejection, 272, 280
Parental stimulation, 280
Parental warmth, 276, 280
Parent-child interaction, 270, 279-285
  attention, amount of, 283
  and behavior modification, 394-398
  and developmental receptive aphasia,
    283-285
  "double-bind," 281
  familiarity, 282
  and language, 282, 284-285
  measures of, 387, 394-395, 445
  methodological issues, 282, 288-289
  See also Family interaction
Parents
  psychotherapeutic work with, 303-312,
    329, 333
  as therapists, 369-378, 380-411, 413-421,
    444
Parent training procedures, evaluation of,
    376, 382, 393-403, 404, 416
Parkinson's disease, 202
Peabody Picture Vocabulary Test, 94, 97
Perceptual abnormalities, 3, 91, 361
  and education, 443
Perceptual motor skills and education, 443
Perceptual training, 442
Perinatal influences, 230-231, 237, 243
Personality of parents, 251-267, 273-278
Pharmacotherapy, 337-355
  and the biological organism, 358-360
  and other therapies, 337-338
  untoward effects, 347-349
  See also Drug studies
Phenothiazines, 178, 341-343, 346-347,
    504
  chlorpromazine, 341, 346
  diphenhydramine, 342
  fluphenazine, 342

Phenothiazines (*cont'd*)
  thioridazine, 341
  trifluoperazine, 341
Phenylketonuria, 15, 187, 202
Physical stigmata, 192
Piagetian framework, 144
"Pink spot," 175-176
Platelet uptake, 168-171
Play, 10, 313-316
  imaginative, 144
  and prognosis, 498-499
  as a reinforcer, 323
  in school, 429, 456
  stereotyped symbolic, 33, 34, 37-38
  symbolic and representational, 30, 33,
    63, 91, 107, 362-363
Play therapy, 317, 361
Pneumoencephalography, 207-218, 246-
    247
Postural whirling, 160
Preconception history, 187. *See also* Parental
    characteristics
Pregnancy complications, 177, 230-231
  prenatal insult, 209
Prelanguage skills, 10-11, 58
  babble, 10
  and prognosis, 499
  and treatment, 333
Preoccupations, 12
Primary handicap, 63
  and education, 441
Prognosis
  for 'bright' children, 436-474, 498
  and IQ, 89, 482, 497
  lack of speech, effect of, 500
  and language, 89, 498
  and mental retardation, 18
  for the mildly handicapped, 498-501
  nonpredictive variables, 485
  and play, 498-499
  and school performance, 498
  and treatment, 501-504
  See also Follow-up studies
Projective tests with parents, 275
Pronoun reversal, 11, 86, 93, 96
Pseudoneurosis, 478
Pseudoretardation, 244
Psychoanalysis, 314, 361
Psychoeducational Profile, 445
Psychogenic factors, 262, 269
Psychosocial disorder, 16, 100, 110

Psychosocial influences, 290
  and twin studies, 231
Psychotherapy
  case study, 309-310
  with children, 304-305
  with parents, 303-312, 329
Psychotic vs. nonpsychotic children, 32-42
"Pure autism," 90
Purine metabolism, 198

Questionnaires
  and diagnosis, 166
  and retrospective data, 187
Questions, repetitive, 34, 90

Raphé system, 164
Rapid eye movements, 123, 125
Ratings of behavior, 33
Raven's Coloured Progressive Matrices, 91
R.B.C. cholinesterase activity, 178
Reading accuracy. *See* Educational attainment
Recall, temporal vs. spatial, 145. *See also* Memory
Reinforcement, 365
Reinforcers
  normal vs. "artificial," 370
  social, 381
Repetitive questions, 34, 90
Reticular formation, 316, 328-329
Retrolental fibroplasia, 243
Reynell Language Scales, 33, 97, 224, 385
Rimland scale. *See* E2 Scale
Rituals
  and behavior modification, 393
  and mothers' behavior, 397
  See also Stereotypies
Role playing and social skills training, 504
Rubella, 15, 19, 190, 243, 359
  virus, 188-189

Schilder's disease, 190
Schizophrenia in later life, 465-466
Scholastic achievement
  and IQ, 89
  and language, 89
  and treatment, 503
Selective attention, 82
Self-destructive behavior and behavioral techniques, 364
Sensitivity to sounds, 92, 96, 107, 113

Sensory evoked responses
  auditory, 121
  contingent, 121
  contingent negative variation, 122
  noncontigent, 120
  studies, 119
Sensory input, modulation of, 117-119
Sensory processes and experience, 142
Sequencing, 88, 90-91
  spatial, 111
  temporal, 110
Serotonin
  circadian rhythmicity of, 166, 189
  level in blood, 164, 189, 346
  and developmental age, 165, 169
  N-dimethylation of, 171-172
  and Down's Syndrome, 170
  efflux of, 168-171, 180, 190
  and IQ, 167
  and L-Dopa, 167-168
  and platelet level, 165, 180, 203
Sexuality, 503
"Shared understandings," 59
Sign Language, American, 365
Single case design, 156
  in drug studies, 340
Sociability score, 35, 63
  and language age, 35, 59
  and mental age, 35
  and nonverbal age, 35
  and organic condition, 38
Social attention, 54
Social avoidance, 48-60
  and communication skills, 56, 59
  compared with normal children, 52
  immediate causal factors, 50
  immediate effects of, 51-53
  reflexive mechanism, 52-54
  threshold of, 49
  See also Social withdrawal
Social imitation, 10
Social introversion of parents, 257, 263
Social relationships, 9, 28, 65
  and communication, 48
  See also Social avoidance; Social withdrawal; Social responsiveness
Social responsiveness, complexity of, 64, 67
Social skill training, use of videotape, 504
Social withdrawal, 2, 41, 85, 280
  active vs. passive, 2, 34, 65
  aloofness, 34

Social withdrawal (*cont'd*)
  and cognitive defect, 65
  factors related to, 66
  and language, 65
  as primary handicap, 486-487
  See also Social avoidance
Spatial rotation and orientation, 148-150
Special education
  and behavior at home, 430, 433, 446,
      454, 457
  and behavior in test situation, 430
  degree of structure, 427, 434, 457-458
  evaluation of, 423
  history of, 439-440
  language programs, 442, 456
  in ordinary schools, 459
  regressive techniques, 427
  social development, 456
  solitary activities, 429
  and speech, 430
  teacher-child ratio, 427, 457
  techniques, 441-443, 457-459
  for under-five-year-olds, 459-460
  unoccupied time, 429, 454
  See also Education; Special schools
Special Schools
  for autistic children exclusively, 459
  comparison of 3 schools, 426-435, 453,
      502
  and follow-up studies, 423
  formal teaching, amount of, 428
  variation, areas of, 425-426
  See also Education; Special education
Specific hyperaminoaciduria, 174
Speech delivery, 90, 107
Spitzer-Endicott interview schedule, 278
Stability of disorder, 485
"Standard Day Interview," 387
Stanford-Binet, 464
Stereotyped movements, 31, 33
  arm flapping, 33, 37
  elaborate, 33, 34, 37
  finger flicking, 33, 37
  severity, 34
Stereotyped utterances, 93
Stimulus encoding, 28. *See also* Coding
Stress, severe early parental, 270-273
Structure in educational setting, 427,
      434
Symbols
  and culture, 60

Symbols (*cont'd*)
  use of, 47-49, 91, 313-316, 323-325,
      362-363
  use of objects, 108, 110, 113
Symptomatology
  details, 8-12
  specificity, 4-8
  See also Autism, characteristics, clas-
      sification, *and* definition; Diagnostic
      criteria
Symptoms. *See* Autism, characteristics,
      classification, *and* definition; Diag-
      nostic criteria
Syndrome
  homogeneity, 63
  subclassification, 16, 487
Syntax, 71, 72. *See also* Grammar

Tactual discrimination, 148-150
Tantrums and 'time out' procedures, 383
Tardive dyskinesia, 348
Task difficulty, 66
Teacher behavior, 428
  tone of voice, 429
Teacher-child interaction, measures of, 428
"Teaching homes," 376-377
Temporal horn, 210-211
Temporal lobe, 246-247. *See also* Temporal
      lobe lesions
Temporal lobe lesions
  effects of, 212-214, 244
  and Kluver-Bucy syndrome, 212, 214,
      217
  and Korsakoff's psychosis, 213-214, 217
  and verbal learning, 214
Testability, 6
"Theraplay," 316-325, 329, 363
  case studies, 317-325
Thioxanthenes, 343
  thiothixene, 343, 348
"Time out" procedures and tantrums, 383
Toxoplasmosis gondii, 188-189
Transcephalic Direct Current Potential,
      121, 177
Treatment
  benefits of, 502
  efficacy, 329
  evaluation of, 357
  home-based, 380-411, 502
  and IQ, 502
  and language, 503

Treatment (*cont'd*)
  limiting factors of child's handicap, 501-
    502
  and prognosis, 501-504
  and socialization, 503
  strategies of, 332-333
  vs. experimental approach in behavior
    modification, 417-420
Treatment and Education of Autistic and
    Related Communications Handicapped
    Children (TEACCH), 361, 439, 443,
    454-455, 459
Tricyclic antidepressants, 345
Tricyclic dibenzoxazepines, 343-344
  loxapine succinate, 343
Triiodothyronine, 345
Tryptophan
  and amino acid excretion, 173
  and phenylketonuria, 173
  side effects, 173
  metabolism, 173-174, 189
Tuberose sclerosis, 8, 211
Twin studies, 219-241
  and early stress, 272
  hereditary influences, 234-236
  psychosocial influences, 231
  sampling, 232-234
Type of disorder, 15-16

Uric acid metabolism, 359

Validity of autistic syndrome, 12-13, 15
Ventricular system, 210-211
Vestibular responses, studies of, 119, 514
Vestibular system, 123-128
  adaptation of, 132-133
  central versus peripheral, 132
  coriolis stimulation, 124
  labyrinthine stimulation, 124
  relevant clinical observations, 123-125
  visuo-vestibular studies, 125, 130, 135, 471
Videotape
  used with parent training, 375, 418
  used in social skill training, 504
Vineland Social Maturity Scale, 97, 223
Visuo-spatial deficits and prognosis, 471,
    500-501

Wechsler Intelligence Scale for Children
    (WISC), 94, 97, 224, 428, 464
"Wild boy of Aveyron," 440
"Work-School Status" as predictor of
    educational outcome, 483
World Health Organization (WHO), 7, 487,
    509-511

Zinc, serum level, 191, 204
Zygosity, 222-241